Third Edition

Principles of
Electrical Machine Design

With C++ (Windows & Linux)

S.K. SEN

B.E, Ph.D. (Lond), DIC, FIE, FNAE, D.Sc., D. Eng.
Fellow, Imperial College, London
Formerly, Prof. & Head, Electrical Engineering,
Bengal Engineering College, Howrah.
Formerly, Vice-Chancellor, Jadavpur University, Kolkata

OXFORD & IBH PUBLISHING CO. PVT. LTD.

New Delhi

Oxford & IBH Publishing Company Pvt. Ltd.
113-B Shahpur Jat, *Asian Games Village Side*
New Delhi 110 049, India

Fax: (011) 4151 7559
Email: oxford@oxford-ibh.in
Website: www.oxford-ibh.in

Printed at Mudrak, New Delhi.

ISBN 978-81-204-1770-0

1-N3-05

TO

MY WIFE

Preface to the Third Edition

When the book was written in early 1980s, FORTRAN-80 computing language was in vogue, and C and C++ were in a nascent stage. Since then, C and C++ languages under Windows operating system has become popular, mainframe computers have been replaced by desktop computers.

Since the DOS-based FORTRAN has been abandoned in most of the colleges and universities all over the world in favour of LINUX open source software, we thought of adding some programs in a separate chapter using this programming language will be useful.

In addition, we have added a chapter on the design of single phase induction motor, as per the wish of several readers, in which all the programs have been written in C++ within the LINUX OS. Since single phase motors cover a wide range and may require the space of a full book, we chose to restrict with only split-phase and capacitor-start motors and with double-revolving field theory.

I take this opportunity in acknowledging the help of Sm. Anasua Chakraborty of Sri Aurobindo institute of Culture, Tollygunj, Kolkata and Sm. Bula Paul of the Future Foundation School, Kolkata.

October, 2013   **S.K. Sen**

Preface to the First Edition

The curriculum development in Electrical Engineering (Power) has in recent years been around the inclusion of subjects such as, digital computer, numerical methods and computer programming, microprocessors. Such inclusion obviously necessitated reduction in course-details in certain areas, and in the area of electrical machines design, the subject is now covered mostly in one semester with 3 hours of lectures, 1 tutorial and 3 sessionals per week. Again the recent policy of the Ministry of Human Resources (Department of Education), Government of India of providing funds for a mini-computer in almost all engineering colleges in the country has enabled institutions to impart programming as a compulsory course. Teachers in Electrical Machines design, as such, can take this opportunity of updating the curricula by introduction of some basic elements of computerised design and analysis in the undergraduate course. Thus, the one-semester course on electrical machines design can have a coverage of basic principles supported by materials giving an insight to the computerised design and analysis.

This book has been planned on the above thought. Computer programs written in FORTRAN-80 language have been added, in an effort to illustrate how the logic, can be established on any problem. It is hoped that this will help in training the students in logical thinking so that complicated problems on design analysis and synthesis can be tackled by them. These programs are in no way optimal as the main purpose has been to carry the students through a program with each step of logical thinking.

Contents

List of Quantity Symbols

Symbol	Meaning	Unit
A_b	Brush contact surface	m^2
A_e	Gross core area (transformer)	m^2
A_{con}	½ × (total conductor section per window) For 3-phase transformer	m^2
A_{cond}	½ × (total conductor section) for single-phase transformer	m^2
A_{cu}	Copper area in a slot	m^2
A_i	Net core area (transformer)	m^2
A_l	Area of leakage flux-path	m^2
A_{sl}	Slot area	m^2
A_t	Cooling surface area of tank	m^2
A_w	Window area (transformer)	m^2
B	Flux-density	T
B_{av}	Average flux density	T
B_p, B_m	Peak flux-density	T
B_t	True tooth-density	T
B'_t	Apparent tooth density	T
B_{30}	Flux-density at 30° elec. from direct-axis	T
C	Capacitance	Farad
C	Number of commutator segments	
D	Armature diameter	m
D_{am}	Diameter of circle through the centre of balls of ball bearing	m

Symbol	Meaning	Unit
D_{dm}	Diameter of bearing for roller bearing	m
D_m	Mean diameter (transformer coils)	m
E_t	Terminal voltage	V
E_T	Voltage per turn (transformer)	V
E_o	Induced voltage	V
E_{ol}	Induced voltage in transformer primary	V
E_l	Transformer primary voltage per phase	V
F	Magnetomotive force	Amp. turn
F_A	Mean axial force	Newton
F_e	Coefficient of emissivity	
	= 1 for perfect black body	
	< 1 for other bodies	
F_t	Field m.m.f.	Amp. turn.
P_g	Gap m.m.f.	Amp. turn
F_p	Pole m.m.f.	Amp. turn
F_R	Mean radial force	Newton
G_C	Specific weight of conducting material	kg. m^{-3}
H	Presure-drop	kg. m^{-2}
H	Magnetising force	Amp. turn. m^{-1}
H_{sic}	Height of insulated conductor	m
I	Current	A
I_a	Armature current	A
I_c	Current per path	A
I_f	Field current	A
I_i	Leakage current through insulation	A
I_s	Current through a conductor in a slot	A
I_w	Loss component of current	A
I_1	Transformer primary current per phase	
I_2	Transformer secondary current per phase	A
J	Current density	A. m^2
J_1	Primary current-density (transformer)	A. m^2
J_2	Secondary current-density (transformer)	A. m^2
K_a	Thermal conductance	W. mm^2. °C^{-1}
K_{AQ}	Output coefficient (transformer)	
K_{EQ}	Voltage coefficient (transformer)	
K_{ed}	Proportionally constant connected with specific eddy-current loss	
K_h	Proportionality constant connected with specific hysteresis loss	

Symbol	Meaning	Unit
K_R	Rogowski coefficient	
K_w	Winding factor	Henry
L	Inductance	Henry
L_a	Armature length	m
L_A	Overall rotor length	m
L_a	Modified armature length due to duct	m
L_b	Length of bearing	m
L_i	Net length of armature core	m
P	Power	W
P	Load on bearing	Newton
P_b	Brush pressure	$N. m^{-2}$
P_L	Loss power	kW
Q	kVa per phase	
Q	Volume of air flowing through a machine	m^3
R	Resistance	ohm
R	Radius	m
R_{th}	Thermal resistance	$°C. W^{-1}$
R_f	Field resistance	ohm
S	Surface area	m^2
S	Number of slots	
S_f	Cooling surface for field coil	
S_1	Number of stator slots	
S_2	Number of rotor slots	
T	Temperature	°C
T_a	Armature turns	
T_a	Effective armature turns	
T_e	Cooling time-constant	sec.
T_f	Fluid temperature	°C
T_f	Field turns	
T_H	Heating time constant	sec.
T_w	Wall temperature	°C
T_1	Primary turns	
T_2	Secondary turns	
U	Potential	
V	Air velocity	$m. Sec^{-1}$
V_c	Peripheral speed of commutator	$m. sec^{-1}$
V_i	Potential difference across insulation	Volt
V_r	Rotor peripheral speed	$m. sec^{-1}$
V_s	Peripheral speed of slip-ring	$m. sec^{-1}$

Symbol	Meaning	Unit
W_{bf}	Brush-friction loss	W
W_n	Dielectric loss	W
W_{ed}	Specific eddy-current loss	$W.\,kg^{-1}$
W_H	Specific hysteresis loss	$W.\,kg^{-1}$
W_L	Total loss	W
W_{if}	I^2R-loss field winding	W
W_r	Specific $i^2 R$-loss	$W.\,kg^{-1}$
W_w	Windage loss	W
X_j	Transformer leakage reactance-total in l.v. winding terms	ohm
Y	Pole-pitch	m
Z_a	Number of armature conductors	
Z_s	Number of conductors per slot	
a	A constant	
a	Froelich constant	
a	Cross-sectional area	m^2
a	Number of parallel paths	
a_e	Conductor section	m^2
a_c	Specific electric loading	$A.\,m^{-1}$
b	A constant	
b	Froelich constant	
$b_F($	Width of insulation	m
b_s	Width of salient pole	m
b_s	Width of conductor in slot	m
C_{cl}	Width of clearance between core circumscribing circle and l.v. winding	m
C_f	Heat transfer coefficient-stationary field	$W.\,m^2.$
C_{hh}	Width of clearance between two consecutive h.v. h.v. windings of two phases	m
C_o	Width of duct between l.v. and h.v. windings	m
c_1	Width of l.v. winding	m
c_2	Width of h.v. winding	m
d	Diameter of core-circumscribing circle	m
d_c	Depth of field coil-salient pole	m
d_e	Width of pole-body insulation	m
d_s	Number of disc sections	
d_{ico}	Internal diameter of coil	m
e	Instantaneous voltage	V
f	Frequency	Hz

Symbol	Meaning	Unit
f_a	Field form (or distribution) factor	
f_r	Rotor frequency	Hz
f_s	Supply frequency	Hz
h	Convection heat-transfer coefficient	
h	Height	m
h_c	Height of conductor in slot	m
hs	Width of coil-end insulation-salient field pole	m
h_s	Height of slot	m
h_w	Window height (transformer)	m
h_{yh}	Clearance between yoke and h.v. winding	m
h_{yl}	Clearance between yoke and l.v. winding	m
k	Thermal conductivity	W. m^{-1}
k_b	Breadth factor	
k_c	Eddy-loss factor	
k_c	Carter's coefficient	
k_a	Carter's coefficient for duct	
k_g	Factor by which the gap-density is increased due to the presence of slots	
k_i	Stacking factor	
k_s	Factor determining the current at maximum efficiency	
k_w	Window space-factor (for transformer)	
l	Length of flux-path	m.
h	Gap length	m.
h	Length of leakage flux-path	in.
l_P	Length of salient pole	m.
	Length of heat flow path	m.
m	Number of layers	
m	Number of phases	
m_c	Mass of copper	m^3
m_i	Mass of iron	m^3
n	An exponent	
n_c	Number of coil sections	
n_a	Number of ducts	
n_r	Rotor speed	r.p.s.
P	Number of poles	
q_c	Heat transfer rate (conduction)	W
q_{conv}	Amount of heat dissiuated (convection)	W
q_{rad}	Amount of beat dissipated (radiation)	W

Symbol	Meaning	Unit
r	Radius	m
R	Magnetic reluctance	
s	Mean length of Turn	m
s_1	Mean length of turn l.v. winding	m.
s_2	Mean length of turn h.v. winding	m.
t	Thickness	m.
t	Time	sec.
t_{in}	Inlet temperature	°C
u_f	Field voltage per coil	V
v_r	Peripheral speed	m. rad
u_r	Width of duct	m.
w_o	Width at opening of a slot	m.
w_s	Width of slot	m.
w_t	Width of tooth	m.
x_m	Magnetising or mutual reactance	ohm
x_o	Overhang leakage reactance	ohm
x_z	Zigzag leakage reactance	ohm
x_1	Leakage reactance of l.v. winding	ohm
x_2	Leakage reactance of h.v. winding	ohm
y_{eg}	Pitch for equalising rings	
y_f	Front pitch	
y_{ph}	Phase pitch	
y_R	Winding pitch	
y_s	Slot pitch	m.
y_w	Coil span	
y_x	Coil span / pole-pitch	
x_f	Conductor section field winding	mm^2
ϕ	Flux	Wb
ϕ	Maximum flux	Wb.
ϕ_s	Flux in slot	Wb.
ϕ_T	Flux in tooth	Wb.
ϕ_t	Total flux per pole	Wb.
ϕ_y	Flux per slot pitch	Wb.
δ	Dielectric loss-angle	
ε_x	Per unit leakage reactance	
γ	Slot pitch pertaining emf polygon	°electrical
Ψ	Flux linkage	Wb-turn.
π	Perneance oi leakage fluk-path	
λ	Permeance coefficient	

Symbol	Meaning	Unit
μ	Coefficient of friction	
μ_o	Permeability of free space $= 4 \pi\, 10^{-7}$	
μ_r	Relative permeability	
η	Efficiency	
η_f	Fan efficiency	
θ	Temperature	$^\circ$C
θ_m	Maximum temperature-rise	$^\circ$C
ω_s	Angular frequency of supply	rad. sec^{-1}
ρ	Resistivity	ohm. in.
ρ_{th}	Thermal resistivity	m. $^\circ$C. W-1
σ	Stephan-Boltzmann constant $= 5.7 \times 10^8$ W. m^{-2} $^\circ$K^4	
τ	Pole-pitch	m
τ	Temperature-rise	$^\circ$C
τ_a	Temperature-rise of air	$^\circ$C
τ_o	Initial temperature-rise	$^\circ$C
τ_s	Steady-state temperature-rise	$^\circ$C

Subscripts *al* and *cu* stand respectively for aluminium and copper.

CHAPTER **1**

Basic Considerations

1.1 Design of Electrical apparatus consists mainly of furnishing data for the manufacture of the apparatus to suit a given specification using available materials economically. A designer should strive to obtain,

(i) lower cost, (ii) smaller size, (iii) wider temperature limits of operatibility and (iv) lower weight,
by judicious use of materials at his disposal.

In the design, a designer is required to have certain basic data of primary importance on which he will have to develop his design. Based on the given specification he chooses basic materials–magnetic, conducting and insulating, depending on their characteristics, availability and cost.

The design procedure starts with the assumption of certain basic quantities such as flux-density and ampere-conductors per metre or current-density (Figure 1.1). From the designed data the parameters of the apparatus are determined so that its performance and temperature-rise are calculated which are then compared with the given specification. If no satisfactory result is obtained the basic assumed quantities are suitably modified till the result is upto satisfaction. The design process is thus iterative whether done by hand or by computer.

1.2 COMPUTERS IN DESIGN

There are several benefits to be gained through the use of computer in the design of machines. A knowledge of such benefits is essential so as to assess the cost-effectiveness of such use.

(a) Once a program has been developed and fully implemented on a computer all future designs are nothing but routine computations largely independent of designer's skill. Highly trained designers are thus relieved of routine tasks and may be utilised for developmental work.

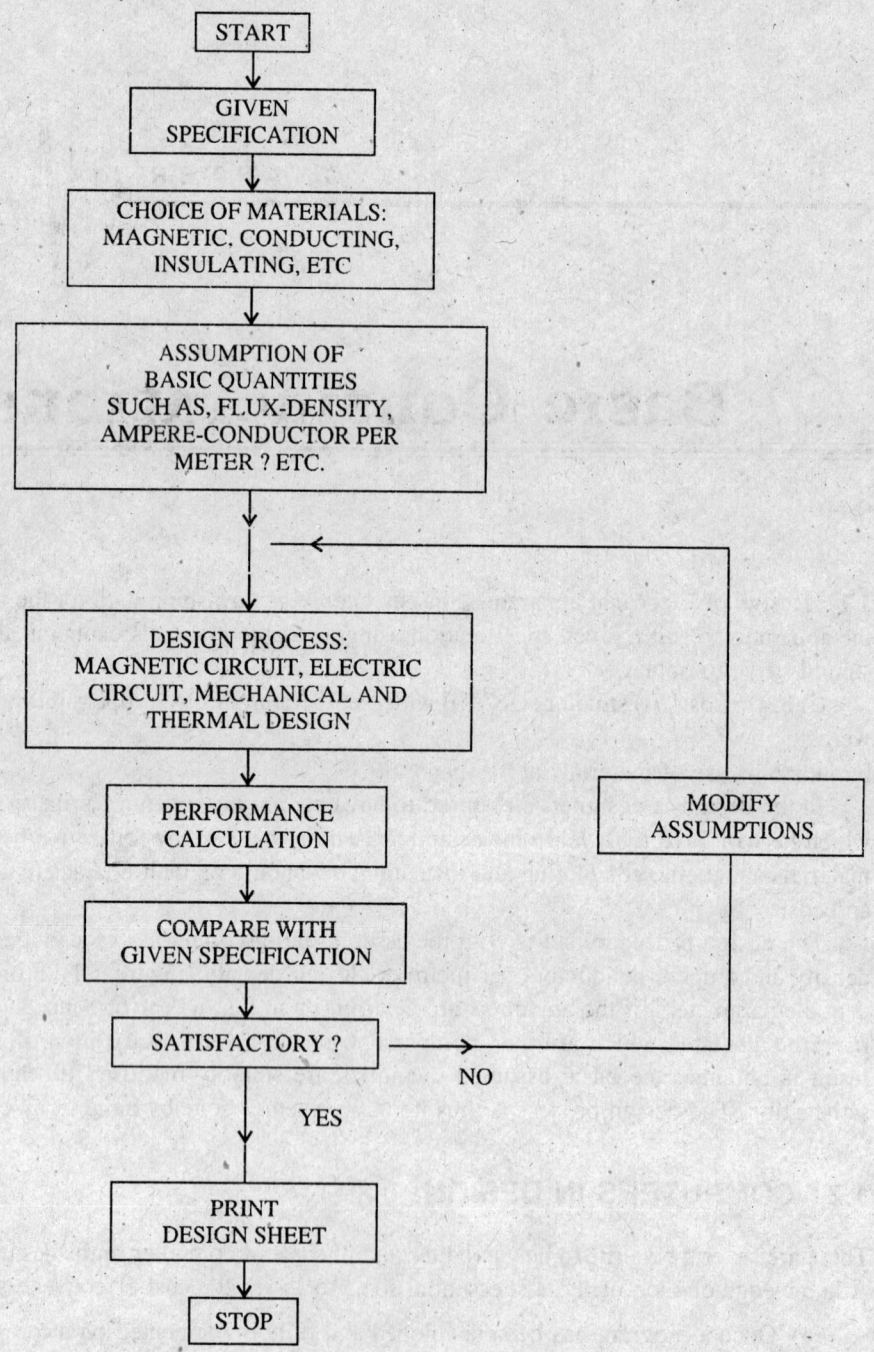

Figure 1.1   Design flow-chart.

(b) A computer can only work on exact information. Though it can reduce empiricism and handle non-linearities, it has neither the 'feel' nor the intuition of a designer. Feeding exact information to the computer means formulation of mathematical relationships between various functional variables, keeping in mind their relative importance in the design. Thus, for successful use of computer in design it is essential that the design principles are thoroughly understood. Again, computer can be effectively used as a means of this understanding, so that the mathematical relationships between the variables could be more correct.

(c) In many applications, 'total computerisation' of design is not very effective because of the difficulty in formulating accurate functional relationship's between the constructional variables. Instead, 'partial computerisation' gives better result, in which a designer breaks down the complete design process into several parts, assigns routine tasks to the computer to obtain a series of intermediate results. He uses his skill and judgement for the decisions which are dependent upon experience and human ability to detect trends, based on which he feeds further information to the computer to arrive at a final solution. Such interactive method generally gives a better design since it combines skill and judgement of the designer with the fast computing power of the computer.

(d) Computer's ability to furnish optimum design by sorting through a large number of different combinations is a welcoming feature. However, optimisation through total computerisation is often difficult specially with electrical machines due to a large number of available frame sizes, magnetic materials and wire gauges. Such approach involves so much logic that formulation is difficult and often requires storing of relatively large amount of data. Moreover, it is very difficult to get a number of designers to agree with one formulation of logic as the best one.

On the contrary, optimisation through continuous interaction between the designer and the computer has been more effective specially for electrical machines, though it is costlier in terms of operation time.

The process of design of electrical machine can be broadly divided into a few major problems :-

  (i) Design of magnetic circuit,
 (ii) Design of electric circuit,
(iii) Mechanical design,
(iv) Thermal design,

followed by performance analysis.

For computer-aided design, the above problems are often treated separately, even broken down into simple elements and considered as individual problem. The results are then combined.

A simple generalised design procedure is outlined in the flow-chart of Figure 1.1

(a) *Given specification consists* of performance requirements as defined by customer's need and Indian Standard Specifications.

(b) Based on given specification, the designer chooses materials—magnetic, conducting and insulating, for electrical design and other materials for frame, bearing, etc. For this, the designer must be conversant with the characteristics, availability and cost of materials needed as so to feed the computer with relevant informations.

(c) *Assumption of basic design parameters* such as, flux-density (Specific magnetic loading), ampere-conductor per meter (Specific electric loading), space-factor, stacking factor, etc. is then made and fed to the computer.

(d) *Design process* consists of analysis calculations to determine the various dimensions of magnetic and electric circuits, thermal and mechanical designs.

(e) *Predetermination of performance* of the machine is then made based on the calculated dimensions. This means calculation of machine parameters from mechanical dimensions obtained through the design process followed by calculation of performance under no-load and load conditions, determination of temperature-rise, cost etc.

(f) The next procedure is the *comparison between the calculated performance and customer's requirement.* If not satisfactory (which is generally the case at the first instance), the designer has to modify the basic assumptions so as to bring the final design closer to the objective. Such modification is not generally a simple task for there are many input parameters that can be changed, and needs skill and intuition of a designer.

1.3 STANDARDISATION AND STANDARDS

Standardisation and Standard Specifications play an important part in the choice, *design*, manufacture, and operation of any apparatus. Standardisation of apparatus presents definite advantages over that made to order.

To the manufacturer, it means reduction in cost as economy results when a number of objects are built at the same time. A planning, a production line can thus be established.

To the user, standardisation means interchangibility of equipment and spares.

To the designer, it means rigidity. The customer can not be given the whole benefit of technical possibilities and up-to-date experience till the Standards are modified.

Published Standard Specifications

In an effort to standardisation, all countries have established national rules which are revised as and when required. Such work is conducted by organisations which include representatives of manufacturers and users : in India by the Indian Standards Institution; in Great Bretain by the British Standards Institution ; in the U.S.A. by the American Standards Association. Again International Electrotechnical Commission (IEC) created jointly by various countries publishes internationally-accepted recommendations so that a product can be sold in other countries.

A few important Indian and IEC Standards related to design of transformers and electrical machines are given below :–

Indian Standards These are issued by Indian Standards Institution, Manak Bhavan 9, Bahadur Shah Zafar Road New Delhi 110001.

1. IS 1271-1958 : Classification of insulating materials for electrical machinery and apparatus in relation to their thermal stability in service.
2. IS 1885 (Part XXVIII) - 1973 : Electrotechnical vocabulary—Transformers.
3. IS 2026-1962 : Power Transformers.
4. IS 1180-1972 : Three-phase distribution transformers upto and including 100-kVA 11-kV outdoor type.

5. IS 3639-1966 : Fittings and accessories for Power transformer.

6. IS 4722-1968 : Rotating electrical machines.

7. IS 1885 (Part XXXV)-1973 : Electrotechnical vocabulary–Rotating machines.

8. IS 325-1970 : Three-phase induction motors.

9. IS 4691-1968 : Degrees of protection provided by enclosures for rotating electrical machinery.

10. IS 6362-1971 : Designation of methods of cooling for rotating electrical machines.

11. IS 5422-1969 : Turbine-type generators.

12. IS 1231-1974 :Dimensions of three–ph. foot-mounted induction motors

13. IS 3003 (Part I)-1966 and 3003 (Part II)-1969 : Carbon brushes for electrical machine.

14. IS 5571-1970 : Guide for selection of electrical equipment for hazardous areas.

15. IS 3682-1966 : Flame-proof alternating current motors for use in mines.

16. IS 4800-1968 : Specification for enamelled round wires.

17. IS 6160-1971 : Specification for rectangular conductors for electrical machines.

IEC Standards : These are issued by the International Electrotechnical Commission, 1 Rue de Varembe', Geneva, Switzerland.

1. IEC 27-1971 : Letter symbols to be used in electrical technology, Part I – General.

2. IEC 38-1975 : IEC standard voltages.

3. IEC 50-1973 onwards : International electrotechnical vocabulary.

4. IEC 85-1957 : Recommendations for the classification of materials for the insulation of electrical machinery and apparatus in relation to their thermal stability in service.

5. IEC 76-1976 ; Power transformers.

1.4 SPECIFICATION

(A) Transformer*: Important specifications are : -

Volt-ampere (kVA or MVA); Voltage ratio (primary volt/secondary volt on no-load); Currents (primary & secondary); Number of phases; Frequency; Class (power/distribution); Percentage impedance; Load loss at 75-C; Connections–h.v. and l.v. windings; Maximum temperature-rise ; Per-cent-tappings; vector group reference (for three-phase transformers).

(B) Rotating Machines†: Important specifications are : –

(i) Direct current machine; Generator or Motor; Type of field excitation; Rated output power; Rated voltage ; Rated current :-Speed; Field exciting voltage and current; Type of enclosure; Type of duty (for motor–continuous/ intermittent/short-time).

(ii) Alternating current machine : Generator or motor; Type (Induction/Synchronous/ Commutator); Frequency; Number of phases; Connection (Star/Delta); Rated output power (kW/MW); Rated voltage; Rated current; Speed; Type of rotor winding (for induction motor-squirrel-cage/slip-ring); Method of starting (for motor); Exciter rating (for synchronous machine); Type of duty; Type of enclosure; Type of cooling.

---

* IS 2026 – 1962 & 1180 – 1964
† IS 4722 – 1968

Additional informations furnished may be :–Short-circuit ratio and under-excited kVar/ MVar (for synchronous generator); Overspeed on throwing off full-load with governor not in operation (for large power system generator); Breakway torque (for motor).

1.5   DESIGN AND CONSTRUCTIONAL ELEMENTS.

(A) Transformer: Important design and constructional elements are :–

   (a) magnetic circuit, consisting of limb and yoke;
   (b) electric circuit, consisting of low- and high-voltage windings;
   (c) coil formers, on which the windings art wound; insulations–conductor, interwinding, and between windings and limb/yoke; spacers in core and winding to provide ducts for coolant;
   (d) tank and auxiliaries, such as, heat exchanger, conservator, bushing, etc.

Limb and yoke consist of assembly of flat laminated steel sheets of suitable thickness. Laminations are assembled tight by compression to minimise vibration under magnetic repulsion forces, and noise. The laminations are separated from each other with thin layer of insulating film applied during the manufacturing process to reduce eddy-current losses. For large transformers, additional coating of thin insulation varnish or phosphate is normally required. The sheets are generally interleaved (art. 1.6.2) to avoid continuous break in the magnetic circuit.

The assembly of limb and yoke are so made that the top yoke is removable to admit the windings. Figure 1.2 shows some typical transformer frames.

The limb and yoke sections may be rectangular, square, or stepped (Figure 1.3) the number of steps in the yoke being generally less than that in the limb.

[*Note* : For large 3-phase transformers the five-limb pattern is commonly adopted since this reduces yoke depth and thus overall height of the transformer. This is necessary since height is usually restricted by transport considerations. Additional advantage of using this pattern is improved cooling of the yoke.]

Windings form the electric circuit and consist of conducting material–copper or aluminium, and insulation. In large transformers, capacitive protection rings and electrostatic shielding are provided for protection against voltage surges.

Figure 1.4   shows typical arrangement of l.v. and h.v.windings on the transformer limb. The l.v. winding is usually placed adjacent to the limb so that the insulation requirement between the winding and limb is less.

Figure 1.5 shows tank, tank-profile, conservator, heat exchanger and other auxiliaries. Tanks are normally of rectangular shape constructed from mild steel. Cylindrical shapes are often used for smaller sizes. For very large transformers, tanks are often made from aluminium with stiffeners to increase rigidity. The tank contains the core and winding assembly immersed in oil which serves as both dielectric and cooling medium. The cold oil enters at the bottom of the tank, extracts heat from winding and core, and its temperature rises. Hot oil moves upwards, flows into the heat exchanger and loses temperature. Conservator is an expansion chamber mounted either on the tank or on the heat exchanger so as to accomodate (i) increased volume of oil due to thermal expansion, and (ii) pressure–and gas-activated (Buchholtz) relay, an effective means of transformer protection.

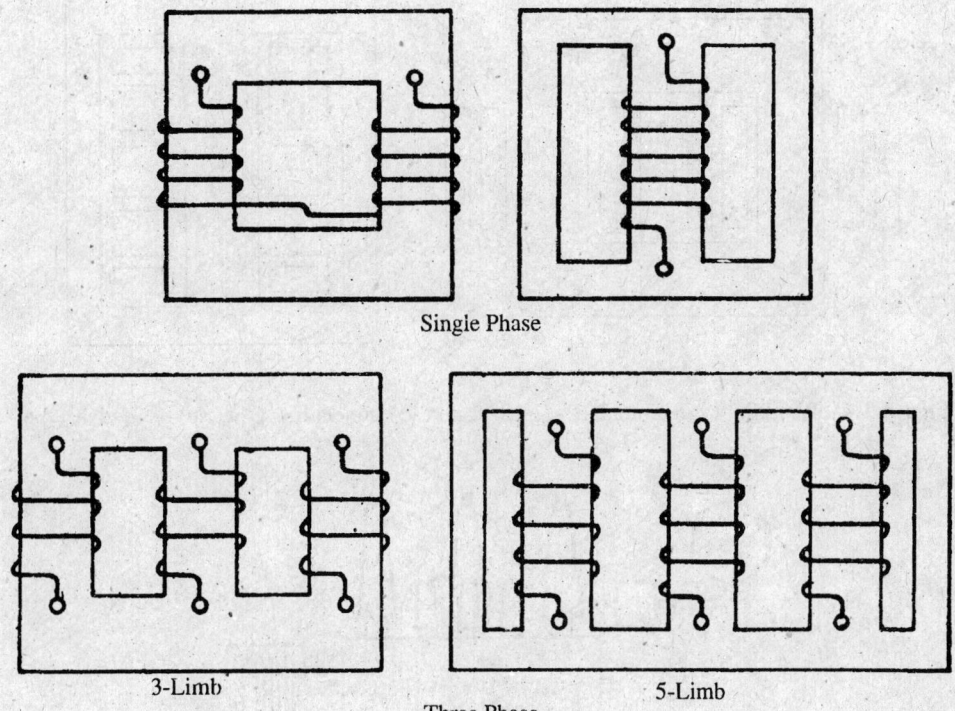

Figure 1.2 Transformer Frames.

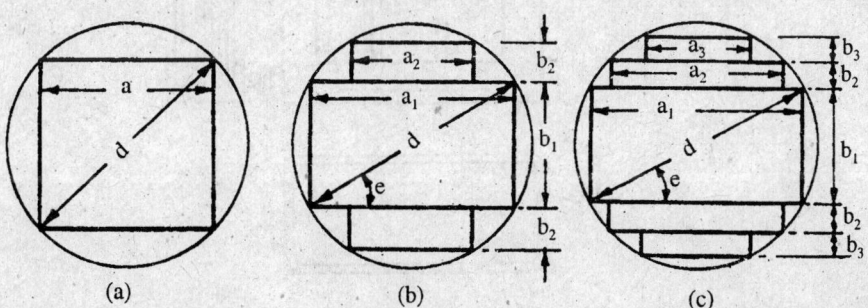

Figure 1.3 Tranfsormer core-sections, (a) square, (b) two-stepped, (c) three-stepped.

(B) *Rotating Machines*

Important constructional elements of a rotating machine (Fig 1.6) are :–

(a) stator and rotor cores separated by gap;

(b) stator and rotor windings with insulations;

(c) frame;

(d) shaft and bearings;

(e) commutator, slip-rings, brushes, brush-holder, brushgear, as and when required;

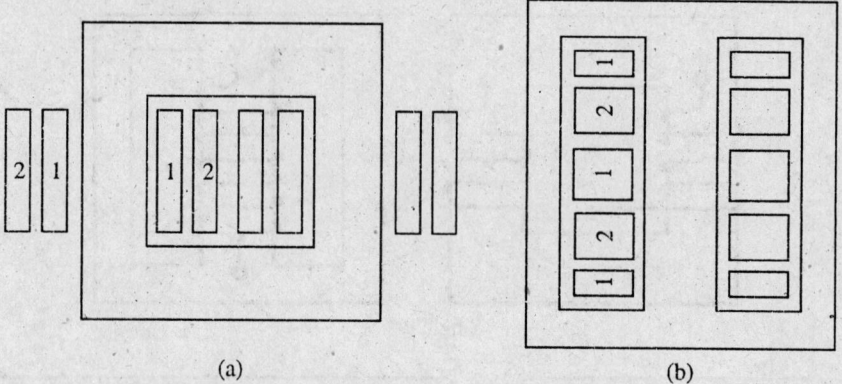

Figure 1.4 Winding arrangements, 1-1. v,   2-h.v. (a) concentric type, (b) sandwich type.

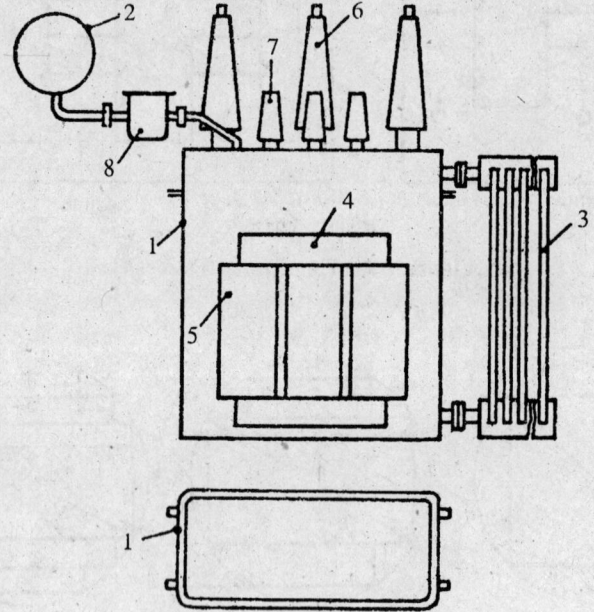

Figure 1.5 Transformer tank and auxiliaries, 1–tank, 2–conservator, 3–heat exchanger, 4–core, 5–winding., 6–h. v. bushing, 7–1. v. bushing, 8–Buchholz relay.

    (f) cooling arrangement, such as, fan mounted on the shaft, provision of ventilating ducts, etc.

    In d. c. machine, the armature is on the rotor consisting of an assembly of laminations 0.4 to 0.45 mm. thick insulated from each other by thin coating of varnish or paper. There are slots uniformly around the armature periphery to house the armature winding. The field poles, generally unlaminated and made of forged steel are bolted to the stator frame. In almost all d.c. machines the magnetic circuit completes through the frame made of cast steel or rolled steel

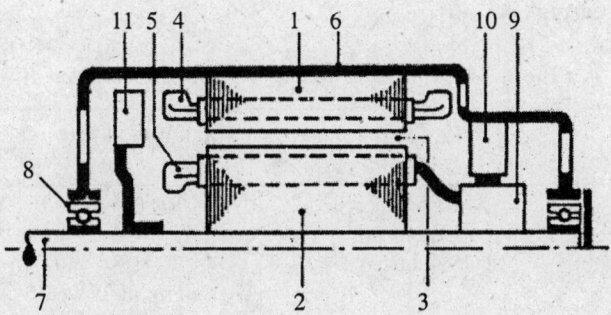

Figure 1.6 Rotating machines-constructional elements, 1–stator core, 2–rotor core, 3–gap, 4–stator winding, 5–rotor winding, 6–frame, 7–shaft, 8–bearing, 9–commutator, 10–brush and brush–gear, 11–fan.

(Figure 1.7). The pole-shoes are laminated and fixed to the pole-core by counter-shunk screws. In very small machines, the entire stator is usually laminated-yoke, poles aud shoes forming part, of the same lamination (Figure 1.8), 1 mm. to 1.5 mm. thick.

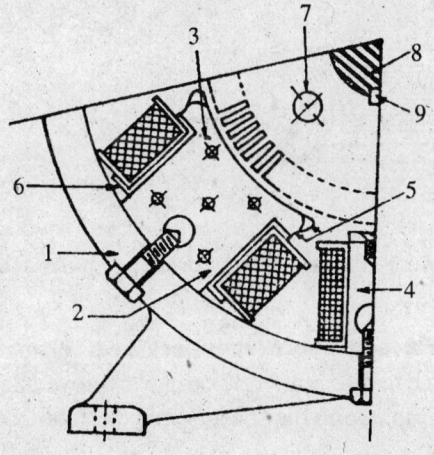

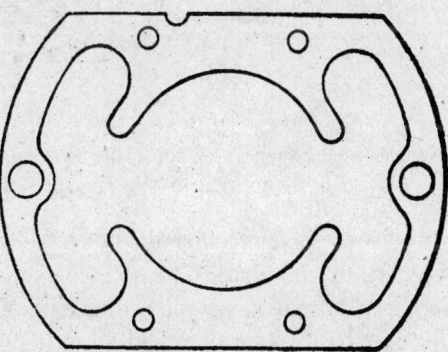

Figure 1.7 D.C. machine, 1–yoke, 2–pole core, 3–pole shoe, 4–interpole, 5–interleaved assembly, 6–insulated shield, 7–axial duct, 8–shaft.

Figure 1.8 Stator lamination–small d.c. machine.

In synchronous and induction machines, the armature core generally forms the stator. The core is laminated with slots placed uniformly around the inner periphery to house the armature, winding.

The rotor of a synchronous machine can be either salient-pole or non-salient pole type (Table 1.1). Except the water-wheel generator, all other classes are horizontal-shaft type, whereas waterwheel generators are generally vertical shaft type. Thus, waterwheel generators require thrust- and guide-bearings.

Table 1.1 Classification of synchronous Machines

Class	Type	Maximum rating to – date	Speed range
Turbo-generator	Non-salient pole	1000 MW	upto 3600 rpm.
Water-wheel generator	Salient-pole	750 MW	90 to 1000 rpm.
Engine-driven generator	Salient-pole	20 MW (Diesel engine driven)	upto 1500 rpm.
		50 MW (Gas turbine driven)	upto 5000 rpm.
Motor	Salient/non-salient pole		
Compensator	Salient /non-salient	100 MVAr	upto 1000 rpm.

In turbo-generators, large centrifugal force due to high peripheral speed necessitates the use of rotor made of massive steel forging. The field winding is distributed in slots which are thus deep and are milled from the rotor cylinder. (Figure 1.9)

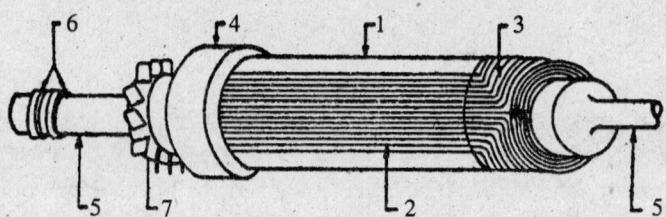

Figure 1.9   Turbogenerator rotor, 1–rotor body, 2–winding in slots, 3–rotor overhang, 4–retaining ring, 5–journal, 6–slip-rings, 7–blower.

In a salient-pole machine the poles are formed by an assembly of thin steel sheets clamped between heavy end-plates. The poles are suitably bolted to the rotor body or may be dovetailed. The rotor body may be forged or made up of discs shrunk on to the shaft or may be fabricated from a cast steel spider mounted oti to the shaft.

For an induction motor, the rotor body is fully laminated with slots uniformly punched around the rotor periphery. For smaller sizes the rotor lamination assembly is keyed to the shaft and suitably clamped. For larger sizes, the core is held on a fabricated shaft.

In magnetic circuit calculation, gap forms the most important element. Flux distribution in the gap is important and in salient-pole machines, the gap length at the pole tip is increased to avoid excessive flux distortion.

The electric circuit comprises the stator and rotor windings. The armature windings in rotating machines and the rotor winding of slip-ring induction motors are usually of copper placed in slots with suitable insulations on the conductor, between turns and between the coils and the armature core. The armature windings are generally of double layers. The field exciting windings on salient poles are concentrated type, generally wound on edge (Figure 5.2) on coil formers mounted suitably on the pole body. The squirrel-cage winding of an induction motor are formed by copper or aluminium bars placed in rotor slots without any insulation. The bars are shortcircuited at each end by endrings welded or brazed to the bars. (Figure 1.10) For small motors, rotor bars and endrings are of aluminium formed by integral die-casting.

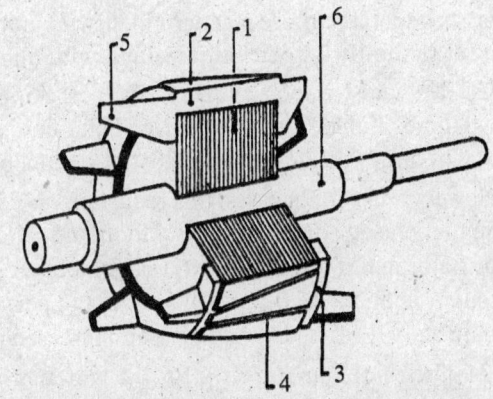

Figure 1.10 Squirrel-cage rotor. 1–lamination, 2–bar section, 3–end ring, 4–skewed bar, 5–buikt-in fan 6–shaft.

The vast majority of rotating machines are air-cooled. The cooling system is closely associated with ventilation schemes and enclosures of the machines, and the most common system for low and medium-size machines is by using fan mounted on the shaft and ventilating ducts–radial and axial. In large turbogenerators, hydrogen cooling and water cooling are being used with considerable advantage (chapter 2.).

1.6 MATERIALS

1.6.1 CONDUCTING MATERIALS

Copper is the most important conducting material used in electrical machines. However, because existing copper deposits are fast exhausting and the price of copper fluctuates widely, aluminium is progressively replacing copper in many applications. Important points of comparison between copper and aluminium are given in Table 1.2.

Table 1.2

SI. No.	Item	Copper	Aluminium (times that of copper)
1.	Resistivity* (ohm. mm^2. m^{-1})	0.0214	1.64
2.	Specific weight (kg. mm^{-2}, m^{-1})	8.89	0.33
3.	Thermal conductivity (W. m^{-1}. °C^{-1})	350	0.57
4.	Specific heat (J. Kg.$^{-1}$ °C^{-1})	400	2.3
5.	Coefficient of linear expansion at 20°C per degree celsius	$17\times10^{=16}$	1.35
6.	Melting point (°C)	1083	0.6

* at 75°C temperature.

[*Note* : (1) Specific resistance of aluminium is 1.64 times that of copper Thus, for the same i^2r- loss a general purpose standard industrial motor or generator will have roughly a reduction of power rating by 22 percent if wound with aluminium*.

(2) The most important advantage of using aluminium is that it is approximately 3.3 times lighter than copper ; and aluminium being cheaper (in rupees per tonne) results in large reduction in the cost of conductors.

However, in certain cases, this reduction may be offset by the increased volume of windings needing increased amount for insulation and labour. For example, for the same temperature-rise of stator and rotor, and for the same number and width of stator slots, the slot depth with aluminium is about 1.20 times that with copper. This means increased eddy-current loss (by approximately 1.62 times) and also an increased leakage reactance.

(3) Use of aluminium strip for l.v. winding, and in some cases, of aluminium foil for the h.v. windings of distribution transformers are popular in USA and certain European countries. Problems concerning manufacturing methods and jointing have been solved.*

One important advantage of using aluminium for transformers is its immunity to attack by transformer oil, and oil in aluminium-wound transformers deteriorate more slowly than in copper-wound ones.]

1.6.1.1 Superconducting Material

Superconductivity, a phenomenon discovered by Dutch Scientist Onnee, in 1911 exhibits the property of certain elements and compounds that their resistivity sharply decreases to practically zero value when, (i) temperature is brought down below 'transition'temperature; (ii) the magnetic field external to the super conductor or created by the current flowing in it is below a certain value.

For example, Niobium-tin compound (Nb_3Sn) has a transition temperature of 18.1°K and its critical flux-density (above which the super conductivity ceases to exist) is about 20 Tesla. Some important superconducting material and their transition temperatures are given in Table 1.3. Again, in a ring of superconducting material, the current once induced will continue to flow unchanged without additional power being supplied.

That is, with a superconductor under cryogenic conditions, a very strong magnetic field can be obtained with hardly any power consumed.

Superconductor in transformer

The use of superconductors in transformers and rotating machines depends on the comparative gain in reduction of I^2R- losses against the cost for provision of cryogenic conditions. A study on the use of superconducting transformer winding* revealed that in a 570 MVA transformer,

---

* $R_{al}/R_{cu} = 1.64$; $I^2_{al}\ R_{al} = I^2_{cu}\ R_{cu}$; $I_{al}/I_{cu} = (R_{cu}/R_{al})^{1/2}$
     $P_{al}/P_{cu} = (E_t.I_{al})/(E_t.I_{cu}) = (R_{cu}/R_{al})^{1/2} = (1/1.64)^{1/2} = 0.78$

* T. Pelican : "Modern Distribution Transformers with aluminium foil winding", Brown Bovari Review, vol. 54 (7), 1967, pp. 376-381.

  M. McCormick : "Aluminium foil transformer coils", Electrical Review, vol. 183, 1968; pp. 452-454.

* K. J. H. Wilkinson, "Superconductive windings in Power Transformers", Proceedings, IEE, vol. 110, 1963, p. 2271.

Table 1.3 Superconducting Element and Compound

Element	Transition temperature (°K)
Titanium	0.49
Zinc	0.82
Aluminium	1.20
Tin	3.73
Mercury	4.16
Vanadium	5.1
Lead	7.22
Niobium	8.0
Compound	
Nb_2Zr	10.8
V_3Si	17.1
Nb_3Al	18.0
Nb_3Sn	18.1

(i) a superconducting winding will have to operate at a leakage flux-density of 3.5 T;

(ii) I^2R- loss of 3 MW can be a eliminated;

(iii) core loss can be reduced by 50 KW by decreasing tho core mass from 100 to 65 Mg.

(iv) refrigeration power requirement will be about 0.5 MW.

(v) cost of development and of refrigeration plant might balance out the saving obtained in the reduction of losses.

That is, the use of superconducting transformer winding is not yet economically viable.

Superconductor in large turbogenerator

A study with respect to the feasibility of using superconducting materials for the rotor winding of a 660 MW turbogenerator has been made by Lorch†. Since the stator current is alternating, superconductors cannot be used for the stator winding. For the rotor, the study is based on the following recommendations :–

(i) The rotor body is a hollow cylinder of glass reinforced epoxy resin with steel ends;

[Note : A normal forged steel rotor will be of no advantage at the high flux-density developed by superconductivity. Moreover, such a rotor will conduct heat from the bearing which is a disadvantage since cryogenic condition is to be set up.]

(ii) Slots are provided on the rotor cylinder in which superconducting winding wrapped in glass-epoxy resin is placed.

(iii) The coolant for the coils is helium entering and leaving the rotor axially by which the temperature of the coils can be brought down to 5°K.

(iv) A thin copper cylinder is fitted to the rotor and is thermally insulated from it by a pealed stainless steel vacuum enclosure maintained at a temperature between 80 and 135°K. The copper cylinder is required to act as a screen to protect the rotor from harmful a.c. magnetic fields.

---

† Lorch, H. O., "Feasibility of Turbogenerator with superconducting rotor and conventional stator", Proceedings IEE, vol. 120; 1973, p. 221.

(v) A stationary system of cylinders within the stator bore with liquid nitrogen flowing through it forms a cold screen at about 77°K.

(vi) The gap between the stationary and rotating members is filled with hydrogen at 0.1 mm mercury pressure to reduce the windage loss.

The above recommendations if implemented, will result in,

(a) a very low excitation requirement, The exciter rating would be of the order of 2.5 MW for a few seconds only as the initial excitation to establish the required magnetic field energy.

(b) a peak density as high as 5 Tesla, at which the stator core must be nonmagnetic.

However, as in the case of transformer, the use of superconductors in large power system generators depend on cost effectiveness, and it has teen envisaged that such a step will be economically viable for generators above 1000 MW rating. A designer's conception of synchronous generator with superconducting field winding is shown in Figure 1.11.

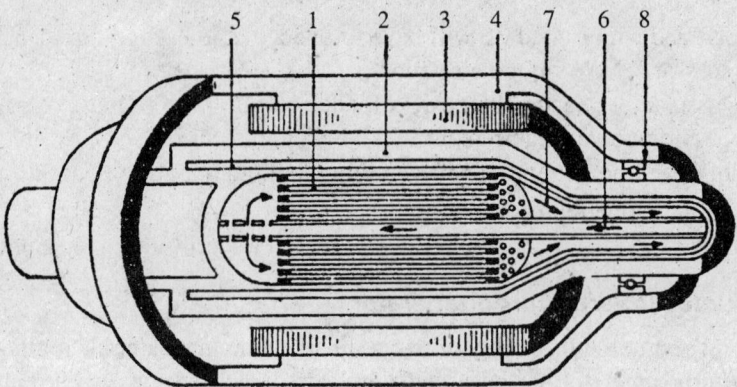

Figure 1.11 Synchronous generator with superconducting winding, 1–Superconducting field winding, 2–rotor body, 3–stator core, 4–iron shielding, 5–vacuum insulation, 6–helium inlet, 7– helium outlet, 8–bearing.

1.6.1.2 For direct-cooled synchronous machines, stator conductors are made hollow to allow the coolant to flow through the conductors and extract heat directly from the source. Hard-drawn copper tubes or electrolytic copper with about 4 per cent silver is used.

1.6.2 Magnetic Material

The electric sheet steel is the most, important core material for electrical machines. Addition of silicon, 1.8 to 3 per cent to the iron

(i) increases the resistivity in steel almost in direct proportion to the silicon content (Figure 1.12);

(ii) reduces hysteresis and eddy-current losses;

(iii) increases magnetic permeability of steel in weaker magnetic fields, but reduces it in stronger fields;

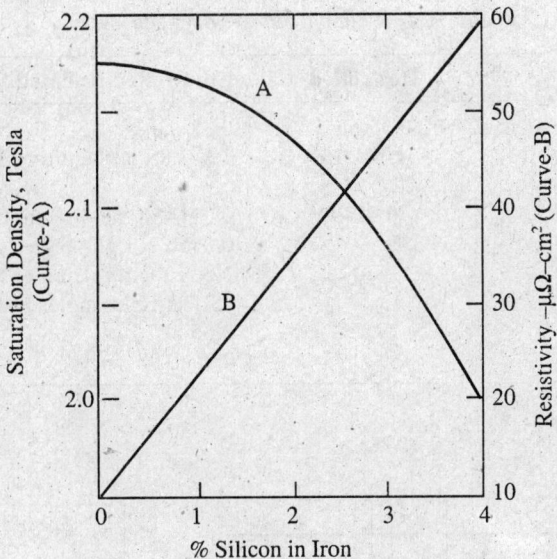

Figure 1.12 Effect of addition of silicon to iron.

(iv) abates 'aging'* of steel;

(v) impairs certain mechanical properties of steel, specially its machinability.

Electric sheet steel is mainly of two types :–

(1) Hot rolled (HRS);

(2) Cold-rolled Oriented (CROS).

In contrast to hot-rolled steel, CROS sheets have improved magnetic properties (lower specific loss and magnetic field intensity at working flux-density), when magnetic flux is directed along the direction of rolling. Due to rolling under cold conditions and subsequent annealing in an atmosphere of hydrogen, iron crystals are made to prevail in the direction of rolling thereby imparting to the steel a sharply defined anisotropy. This property has led to its successful use in transformers and lately, in certain rotating machines. Considerable reduction in core dimension and weight are achieved by use of CROS, i.e. by 20 to 25 per cent in power transformers and about 40 percent in radio transformers.

Table 1.4 and Figure 1.13 give a comparison between the two types. HRS and CROS sheets are available in many grades.

Hot-rolled sheets suffer from an important imperfection in comparison with cold-rolled sheets relating to the variation of thickness. For example, a typical sample of HRS sheet –1.0 m × 0.30 m. having a nominal thickness of 0.33 mm. has a thickness variation from 0,03 to 0.0387 mm. Such imperfection results in imperfect assembly of core with alternatively tight and slack zones.

In contrast to above, cold-rolled sheets of nominal thickness 0.33 mm. show a variation of about 0.0123 mm. i.e. less than one-third of that of hot-rolled sheets.

---

* 'Aging' is expressed by increased iron losses with time.

Table 1.4 Comparison of Magnetic Sheets.

	Hot-rolled	Cold-rolled oriented
Sheet thickness	0.35 mm.	0.33 mm.
Maximum working flux-density, Tesla	1.1–1.15	1.6–1.85 in the direction of rolling.
Specific loss,* W. cm^{-3}	0.0249	0.0114 in the direction of roll; 0.0256 at right angles to the direction of roll.
Ampere-turn per cm*	20	1.3 in the direction of roll. 18.1 at right angles to the direction of roll.
Density gm. cm^{-3}	7.55	7.65
Stacking factor	0.9–0.95	0.95–0.98

* at 1.45 Tesla and 50 Hz.

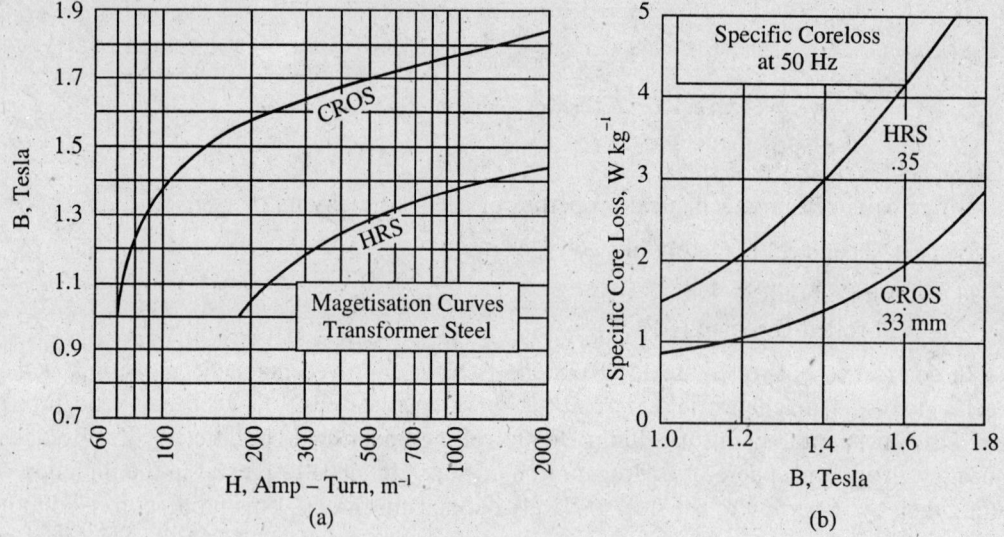

Figure 1.13 Characteristic curves for HRS and CROS.

[*Note* : The working of HRS sheets into proper shapes and sizes is done with the help of power guillotines and multiple presses. With CROS such process impairs the magnetic properties and this can be somewhat avoided if sheets are given stress-relief anneal at a temperature of about 800°C in an inert atmosphere so as to avoid oxidation and carbon contamination. For HRS, a thin coating of varnish or kaolin serves as surface insulation so that eddy-currents are checked. Such thin coating is normally not necessary for CROS as the phosphate-basic coating used during the annealing process serves as an insulation. However, in transformers above 10 MVA, varnish insulation is used.]

CROS sheet is costlier than the HRS, involves complicated process of construction and assembly of core, and improved method of machining the laminations. Care must be taken with respect to the direction of rolling during the assembly of laminations as both specific loss and ampere-turn per cm. increases rapidly with the direction of flux-path away from the direction of rolling (Figure 1.14).

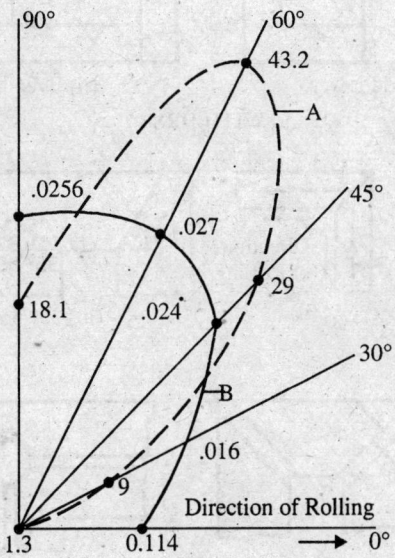

Figure 1.14   Effect of direction of rolling. A - amp - turn/cm, B - Sp. loss.

CROS in transformer   Figure 1.15 (a) shows the assembly and interleaving of hot-rolled laminations in three-phase core of a tranformer. Such method of assembly if used with CROS laminations will give rise to increased losses at the corners of the assembly since the flux bends at the corners (Figure 1.15 b-i) and loss increases as the direction of flux-path is away from the direction of rolling. A corner-cut as in Figure 1.15 (b-ii) will definitely improve the use of CROS but this leads to continuous break at the corners. Mitre joints (Figure 1.15 b-iii & viv) are used to overcome this problem. Further, with mitre, specific iron loss is lesser than that with square-cut corners since uneven flux-distribution at the joints, bolt-holes, etc. are less pronounced with mitre. For transformers in which tbe strip width required is larger than the available standard, the core is splitted up into two sections connected by bridges for effective maintenance of flux-path in the central limb (Figure 1.15 b-v).

CROS in rotating machines   In rotating machines such as induction and synchronous, the direction of flux-path in teeth and core below teeth is as shown in Figure 1.16. Efficient use of CROS in armature core of synchronous machine, and stator and rotor core of induction motor is difficult to achieve because,

(i) The flux-path is radial along the teeth and circumferential in the core below teeth. Thus, if the direction of orientation is made somewhat radial, specific loss in teeth will be reduced by about (0.0114/0.0249) = 46 percent, whereas there will be an increase in specific loss in the core below teeth by about (0.0256/ 0.0249) = 103 percent.

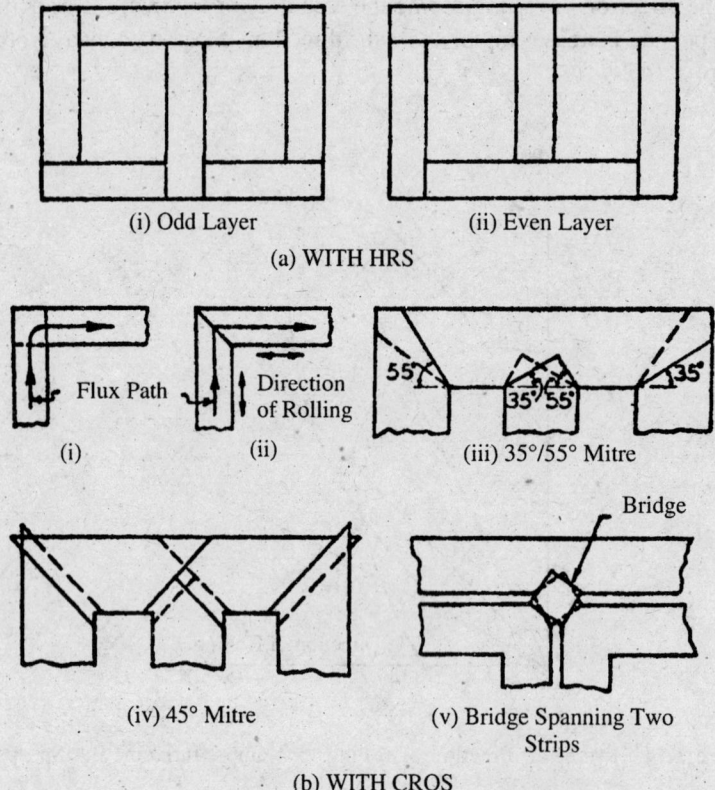

Figure 1.15 Assembly and interleaving of transformer laminations.

    (ii)   Axis of one tooth makes angle with that of the next, and if the direction of rolling is made to coincide with the axis of any one tooth, it will be at some angles with the other. That is, the reduction in specific loss will vary from one tooth to another.

  (iii) In small induction and synchronous machines, the complete armature core is stamped out of a single sheet (Figure 1.16) and use of CROS sheets of a particular directional rolling will result in reduction of losses in certain regions and increase in other, and net result will be of disadvantage.

From the above discussion, it can be concluded that the use of CROS in rotating machines depends on its judicious use in consideration of direction of rolling. For example, for machines with large diameter and large number of poles (as in water-wheel generator) variation of angle between the axis of one tooth and that of the next is not pronounced and use of CROS with segmental core laminations and radial direction of grain oiientation (Figure 1.17a) has proved to be of definite advantage*. On the other hand, in large induction motors (2, 4 and 6 poles) and in 2-pole turbogenerators, the bulk of the core loss is in the core below teeth and segmental

*   J. H. Walker : Large A. C. Machines. Published by Bharat Heavy Electrical Limited.

core with rolling direction tangential to the bore-circle (Figure 1.17b) has often given rise to an overall reduction in core losses.

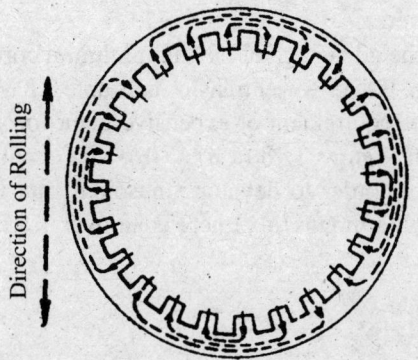

Figure 1.16   A complete stator lamination.

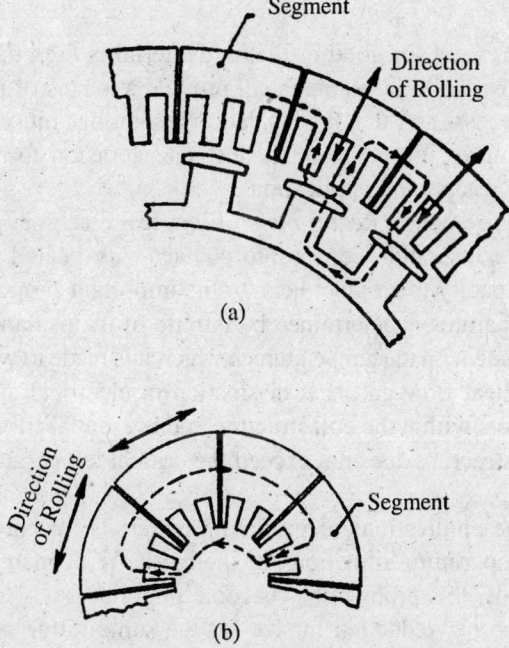

Figure 1.17   Segmental-core lamination-assembly. (a) water-wheel generator, (b) turbogenerator and induction motor.

1.6.2.1   *Strip-wound transformer core using CROS*

It has been shown above that if the flux is made to 'flow' in the direction of orientation of grains in CROS sheets high permeability and low hysteresis can be achieved. Use of strip-wound cores manufactured by winding the material in the form of continuous strip on suitable mandrels, allows the flux to travel in the direction of orientation. Such winding operation imposes mechanical stress on the material which can be removed by annealing the formed cores in a controlled atmosphere.

Strip-wound cores have the advantage of having minimum core loss and is widely used in small distribution transformers. Use in power transformers upto 3.3 MVA has also been reported[†] Important disadvantage is the requirement of expensive plant for annealing the entire formed core. Further, as evident from Figures 1.18 (a-ii) & (b), flux cannot readily transfer from one loop to another ; and since, in order to develop sinusoidal flux in each winding individual loop-fluxes develop large 3rd harmonic flux, there is an overall increase in core losses which can be as high as 33 percent.

1.6.3. Insulating Materials*

There are quite a large variety of insulating materials that are used for the insulation of current-carrying parts in electrical machines. Some common insulating materials are :–

paper, pressphan, card board, empire cloth, mica, asbestos, micanite, transformer oil, varnish, etc.

The fundamental requirement for good insulating material is *high dielectric strength*.

Again, it is necessary that solid insulation shall provide a means of physically separating the conducting part. This necessitates that the insulation assemblies must have the mechanical properties to withstand compression, tension, flexing, and abrasion to the extent depending upon each type of machine design and application.

The third important requirement is *heat transfer*. Electrical machines generate heat in their normal operation, and since insulation comes into contact with heated parts its response to high temperature and its capacity to conduct heat are its important properties.

Life of an electrical apparatus is determined by the life of its insulation which apart from mechanical damage, is dependent on the temperature at which it is made to work. The complicated nature of heat generation, heat flow and heat dissipation in electrical apparatus leads to the existence of thermal gradients within the coil structure and is esential for a designer to see that the hottest spot in the coil-structure does not exceed the critical temperature for the insulation used.

Thus in considering the applications of insulating materials, recent progress has placed major emphasis on the temperature-limitations of their use. With increased applications of synthetic insulating materials, this problem has become important.

Again, efficiency in use i.e. reduction in size for the same rating and reduction in cost demand operation of a machine at higher temperature and higher voltage, thereby demanding the use of insulating materials to the practical limits of their temperature and voltage.

Table 1.5 gives the thermal classification of insulating materials as adopted by the International Electro-technical Commission (IEC 85–1957).

† Gordy, T. D. and Sotnervillc, G. : Single-phase Power Transformer Formed cores, Transaction AIEE, Vol. 69, 1950, p. 1384.
* F. M. Clark, Insulating Materials for Design and Engineering Practice. John Wiley.

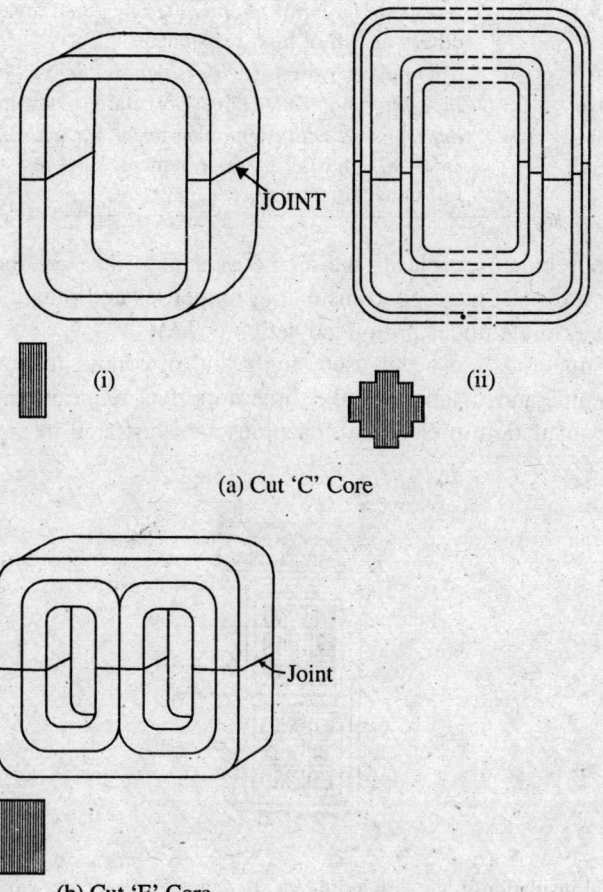

(a) Cut 'C' Core

(b) Cut 'E' Core

Figure 1.18 Strip-wound cores.

Table 1.5 Thermal Classification of insulating materials.

Class	Limiting working temperature °C	Typical materials
Y	90	*Organic, fibre materials on cellulose base*, such as, paper, pressboard, cotton etc.; natural silk (not impregnated), ther mo-plastics.
A	105	*Fibre materials of class Y impregnated* with lacquers (varnishes) or compounds, or dipped in liquid dielectrics ; enamel of wires coated with varnishes and phenolic resins.
E	120	*Enamelled wire insulations on base of* polyvylformal, poly-urethane and epoky resins, moulding powder plastics on phenolic formaldehyde, etc.
B	130	*Inorganic material* (such as, mica, glass, asbestos) *impregnated* or glued with organic binder possessing ordinary heat resistance (on bases of drying oils, bitumen, shellac, bakelite etc.)

F	155	*Inorganic material impregnated* or glued together with epoxy or other varnish of high resistance.
H	180	Mica, glass, asbestos *with silicon binder, silicon resin.*
C	Above 180	*Inorganic materials* (mica, porcelain, glass, mycalex, etc.) *not impregnated with organic material but with inorganic binding material,* such as, glass or cement. Heat resistant micanite, Teflon fall under this category.

[*Note* : (1) Impregnation replaces the air between the fibres of the insulating material. Impregnating material must have good insulating properties and must neither flow at normal temperature nor deteriorate under prolonged action of heat.

(2) Insulating materials are never used singly, but in combination which promotes their usability at high temperatures satisfying the three important requirements mentioned above.]

Typical use of insulation in electrical machines are illustrated in Figure 1.19.

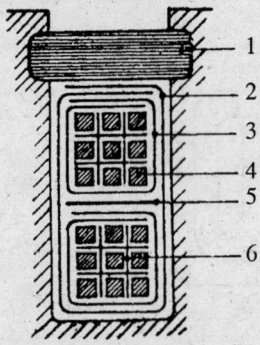

Figure 1.19 Typical insulation in d.c. armature slot. 1–wedge, 2–paper slot-lining, 3–paper – varnished cambric, 4–d.c.c. conductor, 5,6–paper separator.

1.7 LOSSES IN ELECTRICAL MACHINES

Study of various losses occuring in electrical machines is important as the losses determine the heating, temperature-rise and efficiency of the machine. The localised losses in a machine chiefly determine the temperature-rise whereas the total losses determine the efficiency.

1.7.1 Variation of Losses with Load

The total losses W_L can often be expressed as a function of the power-rating P by an equation of the form,

$$W_L = a + b\,P^2 \tag{1.1}$$

a and b being constants (Figure 1.20)

Equation (1.1) is practically valid for machines with somewhat constant flux (transformer) or, machines with somewhat constant flux and speed (synchronous machines and induction

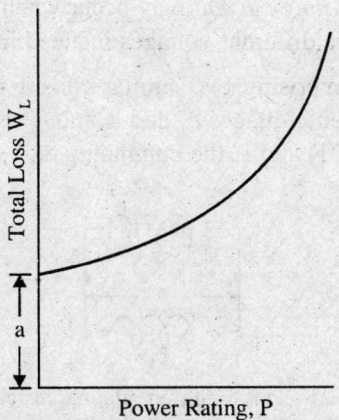

Figure 1.20 Variation of total loss.

motor), 'a' represents the constant losses (iron, friction and windage) and the second term, the variable losses which are proportional to the square of the load.

1.7.2 Electrical Losses

(i) I^2R- losses in a coil due to the flow of uniform electric current in it. Such losses depend on the temperatures of conductors which affect resistivity. It is customary to use 75°C as the normal operating temperature of a machine and state the losses corresponding to this reference value.

For a coil of T-turns and mean length of turn s metre wound with conductor of section area a metre^2, and resistivity ρ ohm. m, the coil resistance is given by,

$$R = \rho \, T \, s / a \text{ ohm.} \qquad (1.2)$$

Specific i^2r-loss, defined as the I^2R-loss watt per kg. of conductor material, is given by,

$$W_r = J^2\rho/G_c \text{ watt. kg}^{-1} \qquad (1.3)$$

where, J amp.m^{-2} is the current-density, and G_c kg.m^{-3}, the specific weight of conductor material.

From Table 1.2, at 75°C, for copper, $\rho = 0.0214$ ohm.mm^2.m^{-1} and $G_c = 8.89$ g.mm^{-2} .m^{-1}

$$W_r = 2.41 \times 10^{-12}J^2 \text{ W.kg}^{-1} \qquad (1.3a)$$

and similarly, for aluminium,

$$W_r = 11.84 \times 10^{-12}J^2 \text{ W.kg}^{-1} \qquad (1.3b)$$

[*Note :* The resistivity of aluminium being about 1.6 times and the density being about 1/3rd that of copper, for equal length of a coil and losses, aluminium conductor-section will be about 1.6 times that of copper. This means increased volume of aluminium required. But the weight of aluminium will be only 1/2 that of copper. For a coil on the rotor of a machine, this means lesser centrifugal stress.]

(ii) I^2R- *loss due to circulating current* may occur when two or more coils in a machine are in parallel and have different voltages induced in them.

Consider a d.c. circuit comprising two similar coils 1 and 2 in parallel each having a resistance R but slightly different voltages e_1 and e_2 induced in them. With currents i_1 and i_2 flowing in the coils (Figure 1.21) and e, the common p.d.,

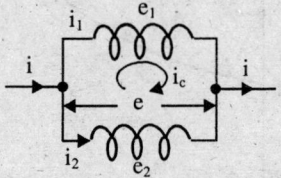

Figure 1.21   Pertaining circulating current in coils.

$$i = i_1 + i_2$$
$$e = e_1 - Ri_1$$
$$= e_2 - Ri_2$$

Defining, average e.m.f. $e_{av} = \frac{1}{2}(e_1 + e_2)$.

$$e = e_{av} - \frac{1}{2}Ri$$

$$i_1 = \frac{1}{R}(e_1 - e_{av}) + \frac{1}{2}i$$

and the circulating current,

$$i_c = \frac{1}{R}(e_1 - e_{av}) \tag{1.4}$$

Equation (1.4) shows that a circulating current is superposed on the normal curent can be very large if R is very small. It can cause losses which at first sight seem to be inexplicable since they correspond to a current which does not appear in the main circuit.

(iii) I^2R-*loss due to a current varying* according to a given law.

For any current $i = f(t)$ in a coil of resistance R in time T sec, the losses are expressed in terms of effective current I_{rms} which would produce the same heating in time T as the varying current. That is

$$I_{rms}^2 = \frac{1}{T}\int_0^T i^2 dt \tag{1.5}$$

The following two cases may be considered :-

Case I : i varying linearly from I_1 to I_2 in time T sec. (Figure 1.22a).

$$i = \frac{1}{T}(I_2 - I_1)t + I_1, \qquad 0 \le i \le T$$

giving,
$$I_{rms} = \left[\frac{1}{3} (I^2_1 + I^2_2 + I_1.I_2) \right]^{1/2}$$
(1.7)

Case II : i varying as in Figure (1.22b).

$$i = I_1, 0 \le t \le T'$$
(1.7)

giving,
$$I_{rms} = I_1 \cdot (T'/T)$$

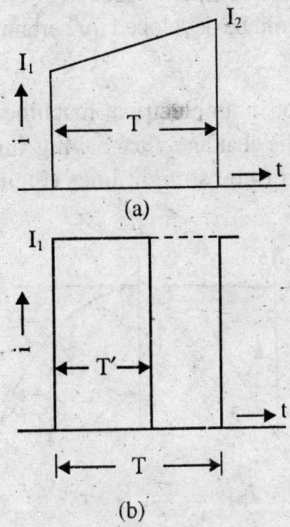

(a)

(b)

Figure 1.22 Pertaining I^2R-loss due to varying current.

(iv) *I^2R–loss due to superimposed currents.*

Case I : A coil carrying a direct current I_1 and a sinusoidal alternating current I_2 rms.,

$$i = I_1 + \sqrt{2} I_2 \sin \omega_s t$$

whence
$$I_{rms} = (I^2_1 + I^2_2)^{1/2}$$
(1.8)

The total I^2R- loss is thus the sum of the losses due to the direct current and the alternating current flowing independently.

Case II : A coil carrying a non-sinusoidal alternating current. Resolving the non-sinusoidal current into its fundamental and harmonic components,

$$i = \sqrt{2} I_1 \sin \omega_s t + \sqrt{2} I_2 \sin(2\omega_s t + \phi_1) + \sqrt{2} I_3 \sin(2\omega_s t + \phi_2) + \cdots$$

whence,
$$I_{rms} = \left(I^2_1 + I^2_2 + I^2_3 + ... \right)^{1/2}$$
(1.9)

That is, the total I^2R- loss is the sum of the losses due to each component current flowing independently.

(v) *Eddy-current loss in conductors*

Eddy-currents induced in conductors placed in an alternating magnetic field result in losses.

In an isolated conductor, parasitic eddy-currents are induced due to its own field (developed as a result of current flowing through the conductor). The phenomenon, known as *skin effect* results in conductor current flowing more readily in the outer layers of the conductor.

In machines and transformers however, the conductors are not isolated (e.g. in machines, the conductors are embedded in slots), and the proximity of ferromagnetic material intensifies the effects of alternating leakage fluxes arid causes increased eddy-current losses. Determination of eddy-current loss in such cases is complex because of complicated nature of leakage paths. However, simplified expressions could be developed for certain cases, making suitable assumptions, as indicate below.

Case I : Single conductor in slot in an electrical machine.

Assuming the iron path of the leakage flux having infinite permeability, the flux-paths within the slots can be considered to be straight lines (Figure 1.23a).

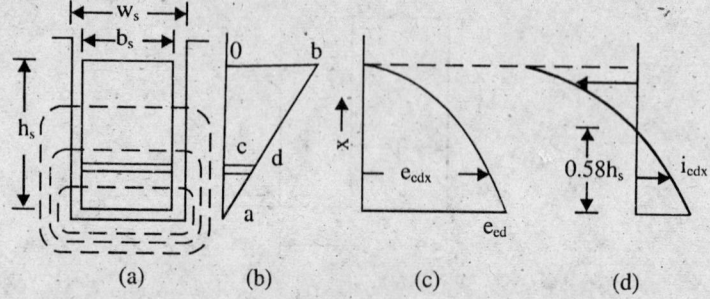

Figure 1.23 Eddy-current in single conductor in a slot.

For simplicity, consider a parallel-sided open slot with the conductor carrying a uniformly distributed current of maximum value i_m, alternating at a frequency f.

The mmf is zero at the bottom a of the conductor increasing uniformly to a value F at the top b (Figure 1.23b). At the top, the flux-density $B_b = \mu_o i_m/w_s$. so that the total flux per metre length of the conductor,

$$= \text{area } oab = \frac{1}{2} \mu_o i_m h_s / w_s.$$

Considering an elemental lamina dx and noting that the lamina is linked by the flux above it, which is equal to the area *obdc*,

$$= \frac{1}{2}(h_s - x) \cdot (ob + cd)$$

$$= \frac{1}{2}(h_s - x)\left(\mu_o \frac{i_m}{w_s} + \frac{x}{h_s} \mu_o \frac{i_m}{w_s}\right)$$

$$= \frac{1}{2} \mu_o \frac{i_m}{w_s}\left(h_s - \frac{x^2}{h_s}\right),$$

the maximum value of eddy-voltage induced in the lamina,

$$e_{edx} = 2\pi f \frac{1}{2} \mu_o \frac{i_m}{w_s} \left(h_s - \frac{x^2}{h_s} \right)$$

$$= \pi f \mu_o \frac{i_m}{w_s} \left(h_s - \frac{x^2}{h_s} \right) \qquad (1.10)$$

A plot of eddy-voltage eedx against x as obtained from equation (1.10) is shown in Figure 1.23(c). The average value of the eddy-voltage,

$$e_{edav} = \frac{1}{h_s} \int_o^{hs} \pi f \mu_o \frac{i_m}{w_s} \left(h_s - \frac{x^2}{h_s} \right) dx$$

$$= \frac{2}{3} \pi f \mu_o \frac{i_m}{w_s} h_s \qquad (1.11)$$

which occurs at $\qquad x = x_1$ where, $h_s - \dfrac{x_1^2}{h_s} = \dfrac{2}{3} h_s$

That is, $\qquad\qquad\qquad x_1 = 0.58\, h_s \qquad\qquad\qquad (1.12)$

At $x = x_1$, there will be no circulating eddy-current. But between $x = 0$ and x_1 the eddy-current will flow in one direction and between $x_1 = x_1$ and h_s, in the opposite direction (Figure 1.23d). Such current flow will result in distortion of the originally assumed flux distribution.

Neglecting the distortion of flux and noting that eddy-current in the lamina is due to ($e_{edx} - e_{edav}$), we have, the maximum value of eddy-current in the elemental lamina dx,

$$i_{edx} = \frac{1}{r_x} (e_{edx} - e_{edav})$$

where, r_x is the resistance of the elemental lamina per unit length $= \dfrac{\rho}{b_s dx}$

That is, $\qquad\qquad i_{edx} = \dfrac{xf\mu_o b_s}{\rho w_s} i_m \left(\dfrac{h_s}{3} - \dfrac{x^2}{h_s} \right) dx \qquad\qquad (1.13)$

A plot of i_{edx} against x is shown in Figure 1.22(d).

The eddy loss due to the above current is proportional to i^2_{edx}.

It is of interest to determine the eddy-loss as a fraction of the $I^2 R$-loss due to the current $i_m \cdot \dfrac{dx}{h_s}$ in the element. That is, (eddy loss/I^2R-loss) in the elemental lamina dx,

$$= \left(\frac{\pi f \mu_o b_s}{\rho w_s} \right)^2 h_s^2 \left(\frac{h_s}{3} - \frac{x^2}{h_s} \right)^2$$

$$= \alpha^4 h_3^2 \left(\frac{h_s}{3} - \frac{x^2}{h_s} \right)^2$$

Integrating the ratio over the conductor height h_s, we have, average $\dfrac{eddy \text{ loss}}{I^2R\text{ - loss}}$ in the conductor

$$= \frac{\alpha^4 h_s^2}{h_s} \int_0^{h_s} \left(\frac{h_s}{3} - \frac{x^2}{h_s} \right) dx$$

$$= \frac{4}{4.5} a^4 h_s^4.$$

That is, the I^2R-loss due to the conductor current is augmented by a factor kc (generally called, eddy-loss factor or ratio), where,

$$k_c = \frac{I^2 R \text{ } loss + eddy \text{ loss}}{I^2 R \text{ loss}}$$

$$= 1 + \frac{4}{4.5} (\alpha \text{ } h_s)^4 \qquad \qquad (1.14)$$

At $f = 50$ Hz and at a temperature of 75°C,

for copper, $\alpha = 96 \text{ } (b_s/w_s)^{1/2}$ $\qquad\qquad$ (1.14a)

for aluminium, $\alpha = 75 \text{ } (b_s/w_s)^{1/2}$ $\qquad\qquad$ (114b)

Note :

(1) Equation (1.14) shows that the eddy-Joss is proportional to (conductor-depth)^4. Thus, the loss can be reduced by having reduced conductor depth.

(2) In cases where the design requirement is that the conductor-depth has to be large, sectionalised conductors (Figure 1.24a) are used with each section insulated from the other but connected in parallel. Eddy-current is confined to individual layer (Figure 1.24d) and production of excessive eddy-voltage is avoided (Figure 1.24c).

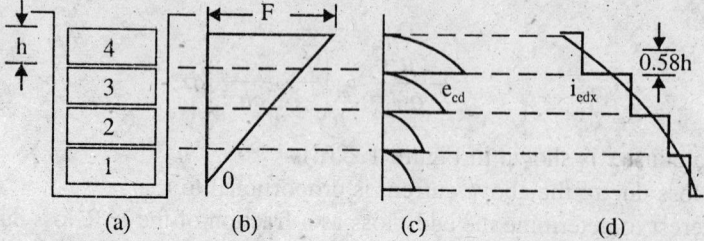

Figure 1.24   Eddy-currnet in sectionalised conductor in a slot.

Following the analysis as above, the eddy-loss ratio in the p-th layer is obtained approximately as,

$$k_{cp} = 1 + (ah_c)^4 \frac{p(p-1)}{3} \qquad\qquad (1.15)$$

The average loss ratio, considering all the m layers each of height h_c is,

$$k_{cav} = 1 + \frac{m^2}{9}(a\,h_c)^4 \qquad (1.16)$$

(3) Equation (1.16) is valid so long the conductors are not connected in parallel. However in practice, parallel connection is essential so as to distribute the total current. But, since each layer occupies a different position in the slot, it is necessary to ensure complete balance between individual subdivisions, to keep circulating current to a minimum (art 1.7.2 (ii)). This is achieved by twisting or transposition (art. 7.3.1).)

Case II : Transformer Coils.

For transformers of normal design, eddy-current loss in conductors can be approximately obtained using equations (1.13) and (1.14) suitably, replacing b_s by H_{wdg}, ω_s by h_w, and h_s by c_1 (Figure 1.25).

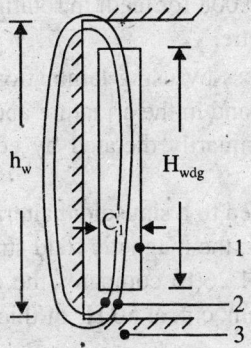

Figure 1.25 Eddy-current in transformer coils, 1–coil, 2–leakage flux, 3–yoke.

That is, $$k_{cav} = 1 + \frac{4}{45}(a\,c_1)^4 \qquad (1.17)$$

where, $$a = 96\,(H_{wdg}/h_w)^{1/2} \text{ for copper} \qquad (1.18a)$$
$$= 75\,(H_{wag}/h_w)^{1/2} \text{ for aluminium} \qquad (1.18b)$$

Note : (1) As shown above, the eddy-loss is proportional to c_1^4. Subdivision of conductors into insulated sections with suitable transposition may be used to reduce the loss.

(2) In general, with suitable design, eddy-loss may be kept to a small value, about 5 percent of I^2R-loss at 75°C for small transformers and about 15 percent for larger units.

1.7.3 Magnetic Losses

Iron losses occur in the portions of magnetic circuit which are subjected to varying or alternating magnetic fields. The magnetic losses consist of,

(i) *Hysteresis loss :* For sinusoidally varying flux-density of peak value B_m, the specific loss,

$$W_h = K_h\,f\,B_m^n \text{ watt per kg.} \qquad (1.19)$$

where, K_h is a proportionality constant which depends on the characteristics and volume of the magnetic material.

$$= 0.039 \text{ for medium and large transformers}$$
$$= 0.063 \text{ for rotating machines;}$$

and n, the exponent, ranging between 1.5 and 2.35.

[*Note* : Quite often, $n = 2$ is used for estimation of losses in electrical machinery.]

(ii) *Eddy-current loss* : For sinusoidally varying flux of peak density B_m, the specific eddy-current loss.

$$W_{ed} = K_{ed} B^2_m f^2 t^2 \text{ watt per kg.} \tag{1.20}$$

where, K_{ed} is another proportionality constant depending on the volume of the magnetic material and its resistivity,

$$= 0.0026 \text{ for medium and large transformers;}$$
$$= 0.005 \text{ for core and } 0.008 \text{ for teeth in rotating machines; and } t \text{ is the thickness of lamination in metre.}$$

The reduction of eddy-current loss by using laminations is a well-established method in transformers, a.c. rotating machines, and in the armature and pole-shoes of d.c. machines. The choice of lamination thickness is primarily dictated by eddy-current loss considerations as illustrated in Figures 1.26(a) and (b).

In a ferro-magnetic sheet subjected to a sinusoidal alternating flux, the permeability is not constant. For any particular thickness, the magnetic field strength reduces with the increase in permeability (Figure 1.26a). Figure 1.26(b) compares the degree of penetration of flux into HRS sheet with resistivity of 50 μ-ohm. cm. at 50 Hz of thicknesses 0.5 mm. and 2.0 mm., the

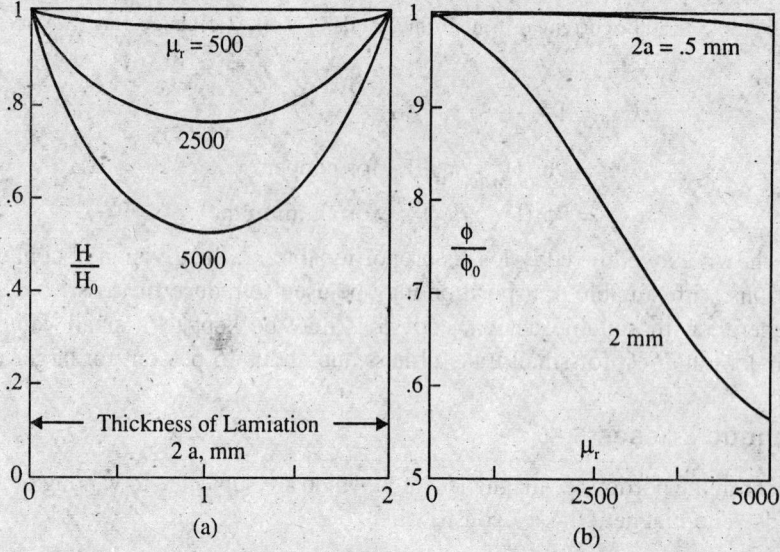

Figure 1.26 Choice of lamination thickness–(a) flux penetration for varying μ_r, (b) flux vs. μ_r. H_o–surface intensity, Φ_o–flux for uniform induction. H_1, Φ–respectively intensity and flux at any point inside.

ordinate representing the ratio of flux through the cross-section to the flux which would pass if the induction is uniform. It can be seen that for laminations of thickness upto 0.5 mm., the influence of permeability is negligible.

Note :

(1) The choice of lamination thickness on the lower side is guided by material handling consideration. Very thin laminations have low eddy-current loss but are difficult to handle, and in rotating machines, teeth would bend too easily.

(2) The factor K_{ed} is inversely proportional to resistivity of the material and thus to limit the eddy loss, high resistivity material may be used.

(3) Preferred lamination thickness, as per ISS, are :–
non-oriented sheets : 0.35, 0.40, 0.45, 0.50, 0.63, 1.0 mm;
oriented sheet : 0.33 mm.

(iii) Apart from above, there are minor magnetic losses, such as, (a) stray losses, due to e.m.f. induced in adjacent conductors and metallic parts, e.g. end plates, end connectors, slot wedges, ventilating duct fingers, binding wires, etc.

(b) losses in teeth due to the non-sinusoidal field-form in the gap and losses due to tooth pulsation because of slot-opening.

These losses may cause high and inadmissible local temperature.

1.7.4 Supplementary Losses

(i) An insulating material subjected to an alternating field is the seat of losses distributed throughout its volume. The phenomenon can be depicted by a parallel combination of a capacitance C and a resistance R (Figure 1.27a), the latter giving rise to losses. If I_i be the total current through the dielectric and

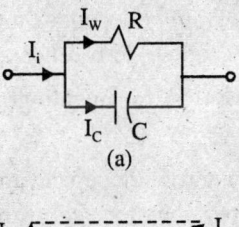

(a)

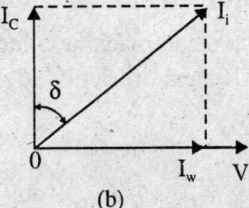

(b)

Figure 1.27 Pertaining dielectric loss.

V_i, the potential difference to which the insulation has been subjected to the dielectric loss

$$W_{di} = V_i \cdot I_w,$$

where, I_w is the loss component of current $I_i = \dfrac{V_i}{R}$

$$= I_c \tan \delta;$$

δ being the dielectric loss-angle $= \tan^{-1} \{I_w / I_c\}$.

That is, $$W_{di} = 2\,\pi f\, v_i^2\, C \tan \delta \qquad\qquad (1.21)$$

[*Note* : Dielectric losses in insulation distributed per unit, volume by conduction and periodic variation of the electric field are somewhat analogous to hysteresis loss in iron in a magnetic field. For most insulating materials such losses increase rapidly with temperature and voltage-gradient (voltage V_i / insulation thickness).]

(ii) *Losses at the brushes* (a) *Frictional* losses depend on contact surface area of the brushes with the commutator or slip-rings, the peripheral speed, and the brush-pressure. The following empirical equations may be used :–

For direct current machines

Brush-friction loss, $$W_{bf}' = \mu P_b\, V_c\, A_b \text{ watt} \qquad\qquad (1.22)$$

where, μ, the coefficient of friction depends somewhat on the nature of commutator surface, the peripheral speed V_c metre, sec^{-1} of commutator, and the material of the brush. Normally, μ ranges between 0.15 and 0.30.

P_b, the brush pressure, generally varies between 10,000 and 20,000 Newton, m^{-2}; A_b, the brush-contact surface, metre^2.

For machines with slip-rings

Brush-friction loss, $$W_{bf} = 3000\, V_s\, A_b \text{ watt} \qquad\qquad (1.23)$$

V_s being the peripheral speed of slip-ring, metre, sec.$^{-1}$.

(b) Losses *due to voltage-drop at contact* depends on the quality of brushes, and is equal to the product of the voltage-drop at the brushes and the current in the brushes.

The voltage-drop can be taken as : 0.55 for metallographite, 1.4 for electro-graphite, 2.0 for hard and 2.5 for soft brushes.

(iii) Mechanical losses due to bearing friction depend approximately on the type of bearing, load on the bearing, and the peripheral speed of the bearing.

Ball bearings are generally used in small machines, and roller bearings in medium-power machines. The latter has a smaller coefficient of friction (μ), reduced axial length, and requires less maintenance.

With ball and roller bearings, bearing friction loss,

$$W_b = \mu P V_b / D_{dm} \text{ watt} \qquad\qquad (1.24)$$

where, P is the load on the bearing, Newton;

V_b, the peripheral speed of the bearing, metre, sec–;

and D_{dm}, the diameter of the circle through the centre of the ball for ball bearing; and diameter of the bearing for roller bearing; in metre.

μ may be taken as 1.5×10^{-6} for ball and 2.55×10^{-6} for roller bearing. With sleeve bearing,

$$W_b = 3000 \, L_b \, V_b \, D_{dm} \text{ watt} \tag{1.25}$$

where, L_b, metre, is the length of the bearing.

(iv) *Windage losses* due to air friction and ventilation depend on the type of ventilation, and is roughly proportional to the overall rotor surface area ($= \pi D \, L_A$, L_A being the overall axial length of the rotor including overhang; D, the diameter of the armature) and the peripheral speed.

For machines having rotor peripheral speed $V_r < 50$ metre, sec.$^{-1}$,

the windage loss, $\qquad\qquad W_w = 0.08 \, V_r^{\,n} \, L_A \, D$ watt $\qquad\qquad$ (1.26)

exponent n being equal to 2.0 for machines without fan, 2.5–3.0 for machines with fan.

For closed circuit ventilation schemes with air as the primary coolant (chapter 2), the windage loss can be determined from the following equations :—

$$W_w = (QHn_f) \, 10^{-2} \text{ kw} \tag{1.27a}$$

where Q volume of air flowing through the machine required to give the prescribed temperature-rise

$$= 0.815 \frac{P_L}{\tau_a} \left(\frac{273 - t_{in}}{289} \right) \text{m}^3 \text{ sec}^{-1} \tag{1.27b}$$

P_L = losses to be dissipated, kW;
t_{in} = temperature of inlet air, °C;
τ_a = temperature-rise of air, °C.
H = pressure-drop across the machine and air cooler, kg. m-2;
η_f = fan efficiency.

1.8   TEMPERATURE-RISE

The losses in an electrical machine are converted into heat and cause the temperature of the machine to rise. Of the three materials viz. conducting, magnetic, and insulating used in electrical machines, the insulating material is the most vulnerable to temperature-rise, and it is often said that 'every 10°C rise in temperature in an electrical machine cuts the life of insulation by 50 percent'. Higher temperature gradually oxidise and carbonise the insulating materials and thereby restricts the output of the machine. Again, the limits of temperature-rise is closely linked with cooling, ventilation and enclosure of the machine.

1.8.1   Class of Duty

An analysis of various classes of duty as per IS will indicate that from the point of view of temperature-rise, the eight types of duty as classified in IS (Figure 1.28) can be broadly divided into two groups :—

(i) Continuous duty (Type S_1), in which the duration of operation of the machine under load is such that the machine reaches its thermal equilibrium (i.e. the maximum steady-state temperature-rise).

(ii) 'Non-continuous' duties (all types other than type S_1 i.e. type S_2 : Short-time duty; S_3 : Intermittent periodic duty; S_4 = Intermediate periodic duty with starting ; S_5 : Intermittent periodic duty with starting and braking; S_6 : Continuous duty with intermediate periodic loading; S_7 : Continuous duty with starting and braking; and type S_8 : Continuous duty with periodic speed changes) in which the duty cycle is shorter than the continuous duty, such that the machine never reaches its thermal equilibrium during the duty cycle.

From the user's point of view, choice of larger machine than what is required means operation at lower efficiency and unnecessary additional expenditure. Again, use of type motor (say, 10 kW) for duties other than continuous but of the same kW and class of insulation means lower maximum temperature-rise, and pessimistic use (though the life of the machine will be longer).

From the point of view of design, machines under the class of 'non-continuous' duty should be designed for higher curent-density (about 2 to 2.5 times for short-time duty upto 5 mins.) than that of continuous duty class (type S_1), since the machine can be allowed to attain the allowable maximum temperature-rise quicker than under continuous duty.

It is obvious that for maximum utilisation of material a designer should strive to attain the condition so that the maximum allowable temperature-rise is attained for the specific type of duty for which the machine is required.

1.8.2   Limits of Temperature-rise

It is a usual practice to indicate the allowable temperature-rise data on the name-plates of rotating machines. For example, the name-plate may show 50°C which means that the average temperature of the machine is 50°C above the ambient temperature.

[*Note* : (1) *Ambient temperature* means the temperature of the coolant at its inlet. This, for a large majority of rotating machines which are air-cooled, is the temperature of the surrounding air. For an oil-cooled transformer with cooling tubes or external radiator, the ambient temperature is the temperature of the oil at the bottom -end of the tubes or outlet end of the radiator respectively, through which cold oil enters the tank.

(2) With higher ambient temperature the capacity of the machine is reduced, since insulation used in the machine can withstand a maximum temperature depending on its class (Table 1.5), and thus with higher ambient temperature, the margin of safe temperature-rise to which the machine can be subjected to is reduced.

(3) The reference ambient temperatures as per IS are :-

(i) Transformer, with air as the coolant :
maximum ambient air temperature : 45°C;
maximum daily average ambient temperature :35°C;
maximum yearly weighted average ambient temperature :32°C;

(ii) Transformer, with water as the coolant : maximum ambient temperature : 30°C;

(iii) Rotating machines :

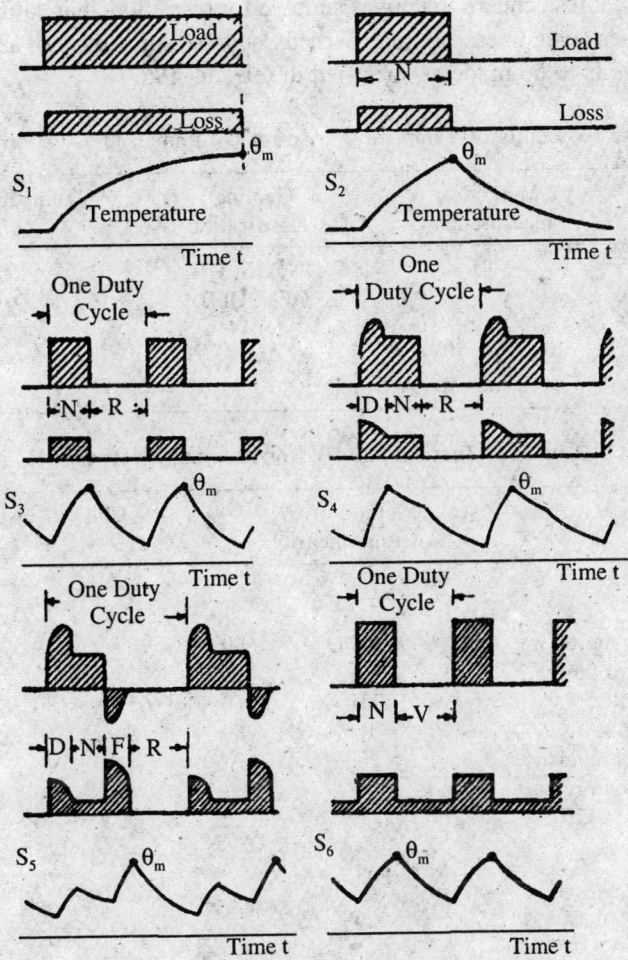

Figure 1.28 Classes of duty for rotating machines.
N–operation under rated load.
D–starting.
R–rest and de-energized,
F–electric braking,
V–operation on no-load.

For class of insulation A, E, B, F, H and altitude below 1000 metres : 40°C]

The limits of temperature-rise as adopted by the Indian Standard Institution are given in Table 1.6-for oil-immersed transformer, and Table 1.7–for rotating electrical machines. Three methods of determining the temperature-rise of windings and other parts are recognised :-

(1) Thermometer method (T H),
(2) Resistance method (R), and
(3) Embedded temperature-detector method (ETD).

[*Note :* (1) The tables concern machines suitable for operation at an altitude not exceeding 1000 metres above sea level. In case altitude exceeds, suitable adjustment of ambient temperature and temperature-rise is to be made as specified in relevant IS.

Table 1.6   Limits of Temperature-rise for Oil-immersed Transformers,

	Method of measurement	Cooling classification	Temperature-rise. °C
1. Winding	TH	ON, OB, OW	55
		OFN, OFB	60
		OFW	65
2. Oil	TH in top oil	All	45

Table 1.7   Limits of Temperature-rise for Rotating Electrical Machanies.

Part	Method of measurement	Class of insulation A	E	B	F	H
A. C. Winding						
1. Air-cooled						
(a) Output 5 MW or more, or, core-length 1 m or more	R, ETD	60	70	80	100	125
(b) Of machines smaller	TH	50	65	70	85	105
than in (a)	R	60	75	80	100	125
2. Indirect Hydrogen-cooled :						
Absolute hydrogen-pressure						
20.28×10^4 to 28.84×10^4				80	100	
29.74×10^4				75	95	
39.35×10^4	ETD			70	90	
49.36×10^4				65	85	
59.16×10^4				62	82	
Commutator Winding	TH	50	65	70	85	105
	R	60	75	80	100	125
D. C. Fied Winding						
1. Low resistance, single layer;						125
and compensating winding.	TH,R	60	75	80	100	
2. Single layer winding with						513
exposed bare surface	TH,R	65	80	90	110	
3. Of a.c. and d.c. machines other	TH	50	65	70	85	105
than (1) and (2) above	R	60	75	80	100	125
Other Parts						
1. Permanently short-circuited insulated winding ; Magnetic cores and other parts in contact with windings	TH	60	75	80	100	125
2. Commutator and slip rings	TH	60	70	80	90	100

2. For rotating electrical machines,
    (a) with class Y, the limits of permissible temperature-rise are 15°C lower than those with class A.
    (b) Suitable adjustment of the limits of temperature-rise of stator winding is to be made as per IS, for rated voltage in excess of 11 kv.
    (c) When the machine is designed to operate with a cooling medium temperature other than 40°C, suitable adjustment of the permissible temperature-rise shall be made as specified in relevant IS.]

SHORT QUESTIONS

1.1 Mention 4 goals which a designer should strive to reach.

1.2 Write a typical flow-chart for the design of electrical machines.

1.3 Compare 'total computerisation' of design with 'partial computerisation'.

1.4 Mention 2 points in favour of standardisation.

1.5 'To the designer, standardisation means rigidity'-explain.

1.6 Mention important specifications of (i) transformer, (ii) d.c. machine, (iii) induction motor.

1.7 What is the function of steel core in electrical machines ?

1.8 What are the important design and constructional elements of a transformer ?

1.9 Why is the 5-limb pattern commonly adopted for a large three phase transformer ?

1.10 What are the functions of a conservator ?

1.11 Mention 2 purposes which oil serves in an oil-immersed transformer.

1.12 State important constructional elements of a rotating electrical machine.

1.13 Why are turbo-generators of non-salient pole type, and water-wheel generators of salient-pole type ?

1.14 Compare copper and aluminium as conducting materials for electrical machines.

1.15 Show that, for the same i^2r-losses, a general purpose standard electrical machine wound with aluminium has a power rating of approximately 78% of that wound with copper.

1.16 What is Superconductivity ? Name 3 important superconducting compounds.

1.17 Indicate the feasibility of using superconductor in large (i) transformers, (ii) turbogenerators.

1.18 State 5 points in favour of adding silicon (about 3%) to iron in electric steel.

1.19 Compare HRS and CROS.

1.20 Write a note on the use of CROS in (i) transformers, (ii) rotating machines.

1.21 State 2 advantages and 2 disadvantage in using strip-wound transformer core with CROS.

1.22 What is Mitre ?

1.23 Mention 3 fundamental requirements for a good insulating material.

1.24 What is varnish impregnation ?

1.25 Enumerate the component losses in a (i) transformer, (ii) rotating machine.

1.26 Compute the i^2r-loss in a coil when it carries (i) a direct current and a superposed sinusoidal alternating current; (a) a non-sinusoidal alternating current.

1.27 What is the necessity of using sectionalised conductors in large machines ?

1.28 Why is transposition or twisting of conductors used ?

1.29 Mention 8 classes of duty for electrical machines.

1.30 Name 3 recognised methods of determining temperature rise of windings and other parts in electrical machines.

1.31 State 'true' or 'false' :-

   (i) In large transformers, capacitive protection rings and electro-static shieldings are provided for protection against voltage surges.

  (ii) Laminations must be assembled tight to reduce eddy-current losses.

 (iii) Tight lamination assembly reduces vibration and noise.

  (iv) In turbogenerators, large centrifugal forces necessitate the use of rotor cores assembled from thin steel plates.

   (v) Cold-rolled sheets as compared to hot-rolled has larger magnetic field intensity at a working flux-density.

  (vi) Studies have shown that superconductors are not economical for large transformers.

 (vii) Hot rolling results in lesser variation in thickness of sheets as compared with cold rolling.

(viii) Class E insulation has a higher limiting working temperature as compared with class B.

  (ix) Eddy-current loss in a conductor is reduced with increased conductor depth in the slot of a rotating machine.

   (x) The magnetic field strength in a ferro-magnetic sheet of a particular thickness subjected to sinusoidal flux increases with the increase in permeability.

  (xi) Machines with non-continuous duty should be designed with larger current density.

 (xii) With higher ambient temperature, the capacity of a machine for the same class of insulation, is increased.

## CHAPTER 2

Heating and Cooling

2.1 As pointed out in chapter 1, temperature-rise forms a basic factor in the design of electrical machines, and the heating is due to the losses occuring in the machine. As heat is developed in a machine its temperature rises, and if it is unable to dissipate heat to the surrounding environment, its temperature will increase to an extremely high value. However, the rate of heat transfer from the outer surface of the machine to the surrounding medium increases with the rise in temperature until after a certain time, the temperature ceases to rise. This occurs when the amount of heat dissipated into the surrounding atmosphere 'per unit time' equals the heat evolved in that time (heat balance equation).

2.2 MECHANISM OF HEAT TRANSFER

The mechanism of heat transfer in electrical machines depends on three well-known modes, viz. *conduction*, *convection*, and *radiation*. The heat generated in core and windings is transferred to the surrounding air by one or more of the above modes.

For example, in an oil-immersed transformer (Figure 2.1) the transfer of heat can be in the following five stages :-

1. Heat transfer by conduction from the interior of the winding or the magnetic core to the external surface in contact with oil.
2. Heat transfer front the surface of core and winding to the oil by convection.
3. Heat transfer from oil to the tank-wall and cooling tubes by convection.
4. Transmission of heat energy by conduction through the thickness of tank-wall and cooling tube-wall.

[*Note* : The temperature-difference between inner and outer walls is not more than 1°C and this stage is generally neglected.]

5. Heat transfer from the tank wall and tube wall to the surrounding air by natural convection and thermal radiation.

Typical temperature distribution in various parts iu a transformer is shown in Figure 2.1.

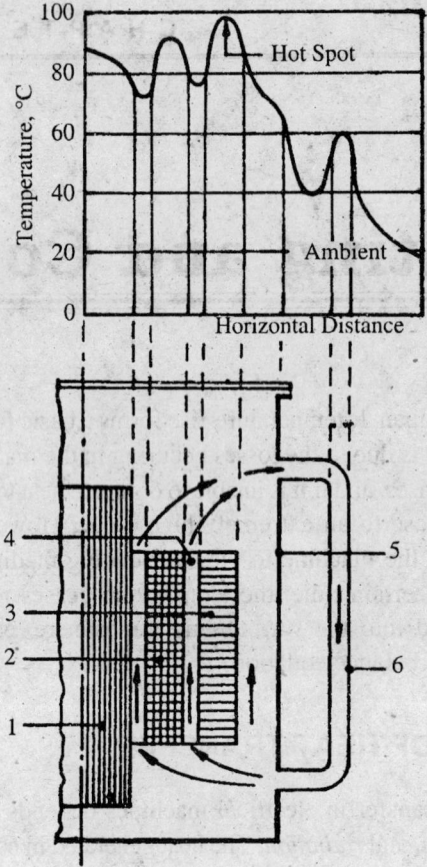

Figure 2.1   Oil-cooled transformer–heat flow and temperature-distribution. 1 – core, 2 – l.v. winding, 3 – h.v. winding, 4 – duct between l.v. and h.v. windings, 5 – tank, 6 – cooling tube.

2.2.1 Conduction

When a temperature-gradient exists in a body, there is a transfer of heat energy from the high temperature region to the low temperature region. The energy transfer is by the mechanism of conduction through the material of the body, and by Fourier Law the rate of heat transfer per unit area is proportional to the temperature-gradient. That is, the heat transfer-rate,

$$q_c = -kS \frac{\delta T}{\delta x} \text{ watts} \qquad (2.1)$$

where, $(\delta T/\delta x)$ is the temperature-gradient in the direction of heat flow in °C. m^{-1}; k is a constant of proportionality, known as thermal conductivity of the material of the body. in W.m^{-1}. °C^{-1}; S is the surface area of beat transfer in m^2; the minus sign indicates that heat flow is from high temperature to low temperature.

The values of thermal conductivity of various important materials used in electrical machines are given in Table 2.1.

Table 2.1   Thermal Conductivity k (W $\cdot$ m^{-1}·°C^{-1})

Material	k	Material	k
Conducting		*Magnetic*	
Coper	350 - 375	Steel : along lamination	50- 63
Aluminium	130 - 200	across lamination	10- 20
		Steel-varnished :	
		across lamination	1.20
Insulating			
Impregnated cotton	0.285		
Micanite	0.165	*Cooling*	
Asbestos	0.250	Air	0.03
Empire cloth	0.250	Hydrogen	0.197
Varnished paper	0.165	Transformer oil	0.143
Mica	0.340	Water	0.60

2.2.1.1   *Steady-state Equations*

Steady-state conduction heat-flow equations as can be found in any standard text book on heat transfer are :-

1. One dimensional flow in cylindrical co-ordinates (with heat generation inside) :

$$\frac{d^2T}{dr^2} - \frac{1}{r}\frac{dT}{dr} - \frac{q_e}{k} = 0 \qquad (2.2)$$

   where, q_c is the heat energy generated per unit volume.

2. One dimensional flow in rectangular co-ordinates (with heat generation inside) :

$$\frac{d^2T}{dx^2} - \frac{q_e}{k} = 0 \qquad (2.3)$$

Example 2.1

A current of 500 amperes is passed through a round conductor of diameter 9.45 mm and length 90 cm. The resistivity of the conductor material can be assumed constant at 2.80 × 10^{-8} ohm.m. If the outer surface is maintained at a constant temperature of 20°C, compute the hot-spot temperature. Assume k for the conductor material to be 130 W·m^{-1}. °C^{-1}.

Solution

The hot-spot temperature, which is the largest temperature within a conductor is at its centre. It can be obtained by solving equn. (2.2) with the following boundary conditions :-

(i) $T = T_w$ at radius $r = R$.

Thus, heat balance equation at $r = R$,

heat generated $= q_c \, \pi \, R^2 L =$ heat dissipated $= -k \, 2 \, \pi \, RL \, \left.\dfrac{\delta T}{\delta r}\right|_{r-R}$

or,
$$\left.\frac{\delta T}{\delta r}\right|_{r-R} = -\frac{1}{2} q_c \frac{R}{k} \tag{2.4}$$

(ii) T is continuous at the centre of the conductor,

Noting that,
$$r \frac{d^2T}{dr^2} - \frac{dT}{dr} = \frac{d}{dr}\left(r \frac{dT}{dr}\right),$$

we have, on integration,

$$T = -\frac{qc \, r^2}{4k} - C_1 \ln r - C_2 \tag{2.5}$$

C_1 and C_2 being constants.

or,
$$\frac{dT}{dr} = -\frac{q_c r}{2k} + \frac{C_1}{r} \tag{2.6}$$

Comparing equations (2.4) and (2.6) at $r = R$

$$-\frac{\dot{q}_c R}{2k} = \frac{\dot{q}_c R}{2k} - \frac{C_1}{R}$$

or,
$$C_1 = 0.$$

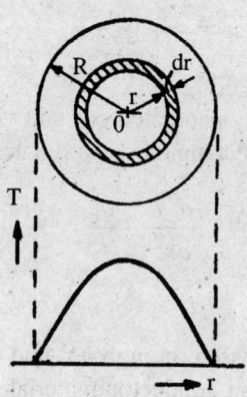

Figure 2.2 Temperature-gradient of radial heat-flow.

Again, at $r = R$ in equation (2.5),

$$T = T_w = \frac{\dot{q}_c R^2}{4\,k} + C_2$$

$$C_2 = T_w - \frac{\dot{q}_c \, R^2}{4 \, k}$$

That is,
$$T = \frac{1}{4} \frac{\dot{q}_c}{k} (R^2 - r^2) + Tw \quad °C \tag{2.7}$$

At the centre, with $r = 0$,

$$T = T_o = \frac{1}{4} \dot{q}_e \frac{R^2}{k} - T_w \tag{2.8}$$

Equations (2.7) and (2.8) indicate that the temperature-gradient of radial heat-flow follows parabola-law. This is illustrated in Figure 2.2. For the given data,

$$\text{power loss} = I^2 R = 500^2 \frac{2.8 \times 10^{-8} \times 0.9}{\frac{\pi}{4} 9.45^2 \times 10^{-6}}$$

$$= 89.822 \text{ W}$$

$$= \dot{q} c \times R^2 L$$

or,
$$\dot{q}_c R^2 = \frac{89.822}{\pi \times 0.9} = 31.768 \text{ W·m}^{-1}$$

$$T = \frac{31.768}{4 \times 130} + 20 = 20.061° C$$

[*Note :* Table 2.1 shows k for air much smaller than the solid insulating materials indicating that the presence of air pockets in an assembly of conducting and insulating materials are detrimental.]

Example 2.2

The conductor in example 2.1 is covered with class A insulation (k = 0.12 W.m^{-1}. °C^{-1}) of thickness 0.05 mm. Neglecting the effect of conducting material (since its k is much larger than that of the insulating material) compute the temperature-difference between the inner and outer surfaces of the insulation.

Solution

With the boundary conditions (Figure, 2.3) :
$$\text{at } r = r_1, \ T = T_1,$$
and
$$\text{at } r = r_2, \ T = T_2,$$
we have, from equn. (2.1),

$$q_c = \frac{2 \pi k L (T_1 - T_2)}{In (r_1 / r_2)} \text{ W} \tag{2.9}$$

Using the given data : $r_1 = 0.004725$ m, $r_2 = 0.005225$ m and heat loss (from example 2.1) = 89.822 W , we have, the required temp. diff. $(T_1 - T_2) = 13.314°C$

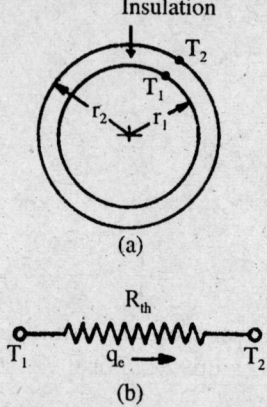

(a)

(b)

Figure 2.3 Pertaining Example 2.2.

[*Note* : (1) Equation (2.9) can be approximated putting $r_2 = r_1 + t$ t being the thickness of insulation, and letting $\log (1 + x) = x - \dfrac{1}{3} x^2 + \dfrac{1}{5} x^3 - \ldots, -1 < x \le 1$.

Neglecting higher powers of x,

$$\dot{q}_e = \frac{2 \pi k L (T_1 - T_2)}{t / r_1}$$

or,

$$T_1 - T_2 = \frac{\dot{q}_c t}{2 \pi r_1 k L} \tag{2,10}$$

With the given data, $T_1 - T_2 = 14.007 \ °C$

(2) *Electrical analog* of one-dimensional heat flow :
Expressing equn. (2.9) as,

$$\dot{q}_c = \frac{T_1 - T_2}{R_{th}} = \frac{\text{thermal potential difference}}{\text{thermal resistance}}$$

where, thermal resistance, $R_{th} = \rho_{th} (l_{th}/A)$ due to heat-flow across a cross-sectional area A and length of heat-flow path l_{th};

$\rho_{th} = 1 / k$, the thermal resistivity.

An electric circuit analogy based on equn. (2.11) is shown in Fig, 2.3(b)]

Example 2.3

The width of a transformer core-plate is 500 mm. The plate has a specilic core-loss of 4 W.kg^{-1}, a density of 7500 kg.m^{-3}, and a thermal resistivity of 0.01 m. °C.W^{-1} Assuming the core loss to be uniformly distributed and a stacking factor of 0.9, and the heat flow is along the laminae, compute the maximum core temperature when the two surfaces are maintained at a temperature of 45°C.

Solution

The boundary conditions (Figure 2.4) are :

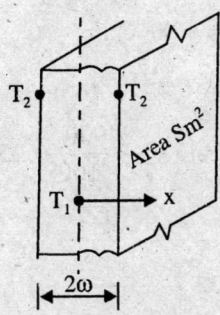

Figure 2.4 Pertaining Example 2.3

$$\text{at } x = 0, T = T_1;$$
$$\text{at } x = \pm \omega, T = T_2.$$

Integrating equation (2.3),

$$\frac{dT}{dx} = -\frac{\dot{q}_c}{k} x + C_1 \qquad (2.12a)$$

But, from the heat balance equation,

$$\dot{q}_c \, S.2w = -k.2S \left.\frac{dT}{dx}\right|_{x=\pm w}$$

or, $$\left.\frac{dT}{dx}\right|_{x=\pm w} = -\dot{q}_c \frac{w}{k} \qquad (2.12b)$$

Comparing equns. (2.12a) and (2.12b),

we have, at $x = + w, C_1 = 0.$

Hence, $$\frac{dT}{dx} = -\frac{\dot{q}_c}{k} x \qquad \qquad \dots (2.13)$$

Integrating equn. (2.13),

$$T = -\frac{\dot{q}_c}{k} \frac{x^2}{2} - C_2$$

Putting the first boundary condition.

we have, at $$x = 0, C_2 = T_1$$

That is, $$T = T_1 - \frac{\dot{q}_c}{2k} x^2 \qquad (2.14)$$

Using the given data,

heat developed $= 4 \times 7500 \times 0.9 \times 2 \text{ w S watts}$

and heat dissipated $=$ qe.2S. u watts

Equating, $q_c = 27{,}000 \text{ W.m}^{-3}$

Again, $k = \dfrac{1}{0.010} = 100 \text{ W.m}^{-1}.\,{}^{\circ}C^{-1}$; and $T = 45{}^{\circ}C$

Thus, at $x = 0.25$ m, from equn. (2.14),

$$T_1 = 45 + \frac{27000}{2 \times 100}\, 0.25^2$$

$$= 53{\cdot}44 \ {}^{\circ}C$$

2.2.2 Convection

If a heated body is exposed to ambient room air without'an external source of motion, there will be a movement of air due to the density gradients near the body. This is called natural or free convection in contrast to forced or artificial convection which is experienced in the case of a fan blowing air over a hot body The physical mechanism of convection is related to the heat conduction through thin layer of fluid adjacent to the heat transfer surface. Examples of heat transfer by natural convection is 'Air Natural (AN)' cooling under natural thermal head of oil-immersed transformer, and those by forced convection are : cooling of rotating machines by fan, cooling of turbo-alternator by hydrogen unner pressure, cooling of transformer by blasting air on the external radiator, etc.

The process of convection heat transfer is complex since it is dependent on large number of variables such as, the temperature-difference between the surface and coolant, height, orientation, configuration and condition of the heat dissipating surface ; the thermal conductivity, density, viscosity, coefficient of volume expansion of the fluid, etc.

The overall effect of natural convection can be expressed by an empirical equation,

$$q_{conv} = h.S.\,(T_f - T_w)^n \text{ watt} \qquad\qquad (2.15)$$

where, h is the convection heat-transfer coefficient;

T_f and T_w are respectively the fluid-and wall-temperature;
and exponent n depends on the dissipating surface, varying generally between 1.0 and 1.7.

For artificial convection by air blown over the surface, h in equation (2.15) is modified as,

$$h' = h\,(1 - C_a.\,V^{1/2}) \qquad\qquad (2.16)$$

where, C_a is an empirical coefficient ($= 1.3$ for uniform blast over the entire surface, and drops in value for non-uniform blast);

V is the velocity of air relative to the cooling surface, m. sec-1.

2.2.3 Radiation

In contrast to the mechanism of conduction and convection, where the transfer of heat energy through a medium is involved, the radiation heat-transfer involves the mechanism of propagation ot electromagnetic energy. For the study of such type of heat transfer, the concept of an *ideal radiator or black body* is introduced, which radiates energy at a rate proportional to the 4th power of the absolute temperature. When a very small spherical body, placed within another body which is a large black spherical cavity, exchanges heat by radiation, the net heat-exchange is given by,

$$q_{rad} = \sigma Fe\, S \left(\theta_1^4 - \theta_2^4 \right) \text{ watt} \tag{2.17}$$

where, σ is a proportionality constant known as *Stephen-Boltzmann constant*, of value 5.7×10^{-8} watt.m^{-2}. $^\circ$K^4 ;

F_e, the coefficient of emissivity (= 1 for perfect black body, and < 1 for other bodies);

θ_1 and θ_2 are the absolute temperatures of the inner and outer bodies respectively.

[*Note* : The concept of black body originates from the fact that black surfaces such as a piece of metal covered with carbon black radiates heat approximately to the θ^4 – law.]

For problems relating electrical machines, equn. (2.15) can be simplified as below*:

Expanding, $\quad q_{rad} = \sigma F_e\, S\, (\theta_1 - \theta_2)\, (\theta^3{}_1 + \theta^2{}_1 \cdot \theta_2 + \theta_1 . \theta^2{}_2 + \theta^3{}_2)$

$$= \sigma F_e\, S\, (273 + T_1 - 273 - T_2)\, 0_r$$

$$= \lambda\, S\, (T_1 - T_2) \text{ watt} \tag{2.18}$$

where, $\lambda = \sigma\, \theta_r\, F_e$, the modified constant of proportionality, is dependent on the temperature $T_1^0 C$ and $T_2^0 C$. It varies between. 5.0 and 7.5 for an ambient temperature of 25°C with $(T_1 - T_2)$ varying between 5 and 80.

2.3 TRANSIENT HEATING

For simplified study of the heating process in an electrical apparatus, the following assumptions are made :-

(1) The apparatus is assumed to be a homogenous body having the same temperature In all parts in which heat is evolved and also in the parts in contact with the coolant. That is, the heat conductivity of the apparatus is assumed to be infinity.

[*Note* : This assumption is not at all tenable for either transformers or rotating machines, but however, leads to simplified result. An electrical machine has various identifiable parts whose temperature-rise may be very uneven, specially under intermittent and shock loads.]

(2) The rate of heat transfer to the coolant is proportional to the temperature difference. This assumption amounts to the fact that the main factors governing the rate of heat transfer are heat conduction which is proportional to the temperature difference (Fourier Law) ;and convection and radiation heat-transfer are negligible.

Denoting, Q = Total heat evolved by the electrical machine per unit time, J.sec^{-1};

C = Heat capacity of the machine, i.e. the amount of heat required to raise the temperature by 1°C, J. °C^{-1};

A = Coefficient of heat transmission of the machine, i.e. the amount of heat dissipated to the surrounding medium per unit time per unit temperature difference, J. sec^{-1}. °C^{-1}.

τ = Temperature-rise of the machine above the ambient °C.

---

* Kostenko, M and Poitrovsy, L : Electrical Machines Part II, pp 138

[*Note* : (1) If G be the mass of body of the machine (kg) ; Cp, the specific heat (J. $kg^{-1}.°C^{-1}$),

$$C = G.\, C_P \qquad\qquad (2.19a)$$

(2) If λ be the heat eruissivity ($W.m^{-2} °C^{-1}$)

$$A = S.\, \lambda \qquad\qquad (2.19b)$$

Heat balance equation under constant load can be expressed as,

$$Q\, dt = A\, \tau\, dt + C.d\tau \text{ Joule} \qquad\qquad (2.20)$$

or,

$$dt = \frac{c.\, d\tau}{Q - A\tau}$$

Integrating,

$$t = -\frac{C}{A}\, \ln\,(Q - A\tau) - C_1$$

C_1 being the integrating constant, and can be determined from the known initial condition: at $t = 0$, $\tau = 0$.

Thus,

$$t = -\frac{C}{A}\, \ln \frac{Q - A\tau}{Q - A\tau_0}$$

or,

$$e^{(-A/C)t} = \frac{Q - A\tau}{Q - A\tau_0}$$

leading to,

$$\tau = \tau_s\,(1 - e^{-t/T}{}_H) + \tau_0\,e^{-t/T}{}_H \qquad\qquad (2.21)$$

where

$$T_H = \frac{C}{A}, \text{ the heating time-constant} \qquad\qquad (2.22a)$$

and

$$T_s = \frac{Q}{A}, \text{ the steady-state temperature-rise, °C} \qquad\qquad (2.22b)$$

If the temperature-rise at $t = 0$ is zero, i.e. the apparatus is initially at the ambient temperature, $\tau_0 = 0$,

$$\tau = \tau_s\,(1 - e^{-t/T}H)°C \qquad\qquad (2.23)$$

Cooling   The temperature-rise vs. time characteristic for cooling of electrical apparatus can be obtained from equn. (2.21) when it cools down from a steady-state temperature rise τ_{1s} to a final temperature-rise τ_{2s}, as below :-

$$\tau = \tau_2 s\,(1 - e^{-t/T}c\,) + \tau_{1s}\,e^{-t/T}c \qquad\qquad (2.24)$$

where, Tc is the cooling time-constant.

For the particular case, when the apparatus cools down to the ambient temperature, $\tau_{2s} = 0$, and

$$\tau = \tau_1 s\,e^{-t/T}c \qquad\qquad (2.25)$$

[*Note* : (1) The heating time-constant for normal oil-immersed power transformer is 2 to 4 hours ; that for a 15 kW well-ventilated motor may be 20 minutes. For larger motors or totally-enclosed motors, the heating time-constant may be a few hours.

(2) The cooling time-constant for motors with shaft-mounted fans is 2 to 3 times the heating time-constant, because the cooling conditions are poorer at lower speeds and at standstill.

(3) Experiments have shown that the theoretical temperature-rise/ time curve fails to coincide with the experimental curve. This discrepancy is mainly at the beginning of the heating process when the actual temperature-rise is faster than that predicted theoretically (Figure 2.5)]

Example 2.4

A 15 kW squirrel-cage induction motor having maximum efficiency of 90 percent on continuous full-load has a temperature-rise of 41.8°C after 30 minutes and 50°C after one hour under the above operating conditions.

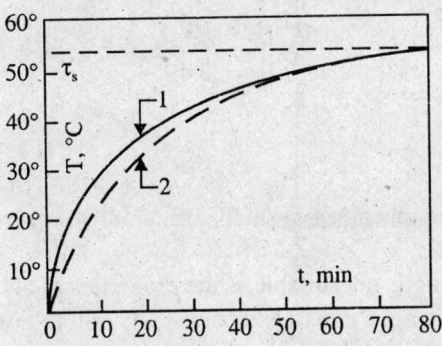

Figure 2.5 Temperature-time curves. 1–experimental, 2–theoretical.

(a) Compute its final steady-state temperature-rise on continuous load, and the heating time-constant.

(b) If the motor is operated on a short-time load for 6 mins., estimate the temperature - rise.

(c) Compute the maximum overload that can be applied on a short-time 6-minute rating so that the temperature-rise is the same as the final steady-state temperature-rise on continuous load.

Solution

(a)
$$41.8 = \tau_s (1 - e^{-1/2T_H})$$
$$50.0 = \tau_s (1 - e^{-1/T_H})$$

Dividing, $\quad\quad\quad 1.196 = 1 + e^{1/2T_H}$

or, $\quad\quad\quad\quad T_H = 0.307 \text{ hours.}$

Thus, $\quad\quad\quad\quad \tau_s = 51.99 \text{ °C}$

(b) On short-time load for 6 minutes,

temperature-rise $\quad\quad \tau = 51.99 (1 - e^{-0.1/0.307})$

$$= 14.45 \text{ °C.}$$

(c) Maximum steady-state temperature-rise on short-time rating so as to obtain 51.99 °C after 6 minutes,

$$51.99 = \tau_{ssh} (1 - e^{-0.1/0.307})$$

or, $\quad\quad\quad\quad \tau_{ssh} = 187 \text{ °C}$

Thus, $\quad\dfrac{\text{losses on short- time overload}}{\text{losses on continuous full-load}} = \dfrac{187.0}{51.99} = 3.6$

If the required overload be α times the full-load,

$$\frac{P_i + \alpha^2 . P_c}{P_i + P_c} = 3.6$$

P_i and P_c being respectively iron-losses and I^2 r-losses on full-load.
But, $\quad P_i = Pc$, since maximum efficiency occurs on continuous full-load.

That is, $\qquad\qquad\qquad\qquad \alpha^2 P_c = 6.2\, Pc$

or, $\qquad\qquad\qquad\qquad\qquad \alpha = 2.49$

2.4 COOLING

For transformers and rotating machines of small size, air at ambient temperature is the cooling medium.

For large units, air cooling is not suitable as the properties of air as a coolant are inefficient and the apparatus may be damaged due to excessive temperature-rise and extremely hugh 'hot-spot' temperature may be developed. For example, comparing two transformers A and B, A having K-times linear dimensions of the other, but otherwise similar, we have,

(i) Core-section of $A = K^2$. core-section of B ;-
       conductor section of A, and thus the window area of A = K^2. that of B.

(ii) For the same electric and magnetic loadings,

kVA-rating of $A = K^4$. that of B.

since, kVA-rating = a constant × core area × window area (Chapter 3).

(iii) Since loss = specific loss × volume,
        loss of $A = K^3$. that of B.

(iv) Surface area of heat dissipation of A

$= K^2 \times$ that of B.

Thus, by increasing the linear dimensions K-times, the rating is apparently increased by K^4– times, but the loss (and thus heat) to be dissipated per unit surface area has also increased by K-times. Thus, the method of natural air-cooling fails, and forced cooling method must be adopted.

2.4.1 Transformer

2.4.1.1 *Cooling Method* (i) Dry type : The methods of cooling for dry type transformers are- 'Air Natural' (abbreviation AN) and 'Air Blast' (AB). Dry type transformers include small transformers of rating upto 5 kVA, and special medium-size transformers which are used in cases where installation of an 'oil immersed' transformer may lead to fire hazard, e.g. coal mine transformer. 2.5 MVA, 3-phase, 16,500/575 Volt, 50 Hz dry-type AN-cooled transformer is in use in India, as a power-station auxiliary.

Air-cooling is most efficient when the transformer is without any enclosure. However, quite often, environmental factors may necessitate use of enclosure thereby reducing the efficiency in cooling. Also, the rating of the transformer for the same maximum temperature-rise is reduced.

Heat dissipation in an enclosed transformer is due to :

(a) by conduction through wall (the roof and floor areas are generally neglected);

(b) by convection in a ventilated enclosure.

The following approximate formula may be used for calculating the temperature-rise :

$$qeT = K_a \Delta \theta_w \qquad W \cdot m^{-2} \qquad (2.26)$$

where, qeT is the amount of heat dissipated per unit area of heat dissipating surface, K_a, the thermal conductance of wall, and $\Delta\theta_w$, the temperature difference through wall in °C.

For metallic enclosure, $Ka = 4.5 \ W.m^{-2}.°C^{-1}$,

Cooling is considerably improved with airblast. Propellor fans are generally preferred. Larger sweep with slower speeds are quieter. Usually the fan is placed at the inlet of ventilation so that maintenance is easier. In dusty environments, air filter can be used at the intake end.

Direct-cooled dry-type transformers have been designed in which cooling ducts are provided in the windings using hollow conductors, thereby enabling cooling air to extract heat directly from the conductors.

(ii) Most transformers are oil-immersed. Oil serves the dual purpose as liquid insulant and coolant. As an insulant, it should provide good insulation between various parts, and as a coolant, should effectively extract heat resulting from various losses. Typical characteristics of oil should be :-

(a) maximum viscosity : $40 \ mm^2 \ S^{-1}$ at 20°C,

(b) Pour point : –30°C,

(c) maximum acidity of new oil is that which can be neutralised by 0.03 mg KOG per gram,

(d) must not break down below 40 kV when a rising voltage at 2 kV/sec is applied between two spheres 13 mm in diameter and at 2.5 mm apart.

The cooling methods recommended[†] for these 'Wet-type' transformers are given in Table 2.2.

From Table 2.2, it can be seen that cooling of oil-immersed transformers can be basically classified under two heads–Natural circulation, and Forced circulation. The following cases may be considered -

Case I Natural circulation

(1) *Methods*: (a) Oil-immersed transformers, normally upto a rating of about 50 kVA may have only smooth tank walls without additional means of cooling (Figure 2.6a). Experiments have shown that, (i) plain tank surface is most efficient for heat dissipation;

(ii) heat transfer is by both convection and radiation;

---

† IS 2026 of 1962.

Table 2.2   Cooling of Oil-immersed Transformer

Oil Circulation	Cooling Method	Abbreviation	
		Mineral oil	Synthetic liquid
By Natural Thermal	Air Natural	ON	SN
Head only	Air Blast	OB	SB
	Water	OW	SW
Forced by Pump	Air Natural	OFN	SFN
	Air Blast	OFB	SFB
	Water	OFW	SFW

[*Note :* (1) In Table 2.2 'O' indicates use of mineral oil (refined crude petrolium), 'S' indicates use of synthetic liquid such as, Pyranol. Askarel, Sovotol, etc.

(2) Quite often, oil-immersed transformers are built with a combination of natural and forced cooling such as, ON/OFN, ON/OFB, SN/SFN, etc.

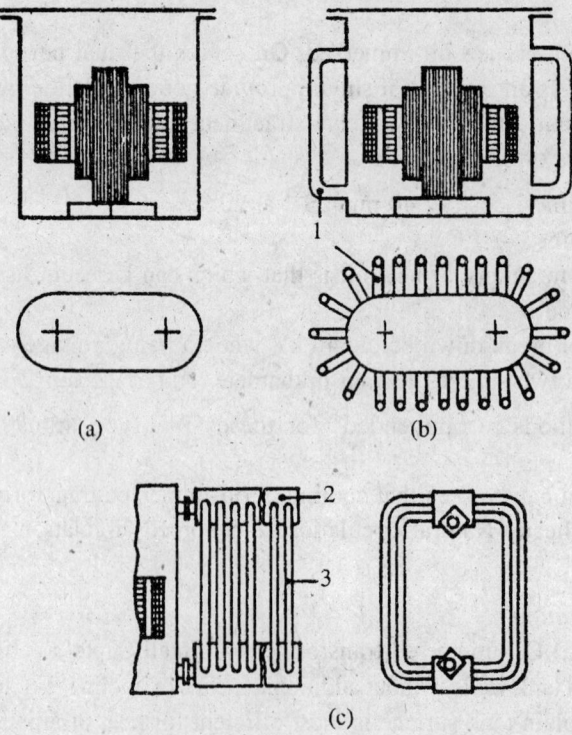

(a)   (b)

(c)

Figure 2.6   Transformer cooling methods-natural ciculation. 1–tubes welded to the tank, 2 – header, 3– heat exchanger tube.

(iii) plain sheet steel tanks have heat transfer properties which can be expressed by the following empirical equation[†] relating heat dissipation g_{eT} watt. metre^{-2} of heat dissipating surface and the mean top-oil temperature-rise $\Delta\theta_o$,

$$q_{eT} = K_T \cdot \Delta\theta_0^{1.25} \qquad (2.27a)$$

$$= \frac{P_i + P_c}{A_t} \qquad \text{.... (2.27b)}$$

where, K_T is an empirical coefficent dependent on the type of tank and associated cooling surfa.ee. For computation of effective cooling surface area A_t for plain tank, the usual practice is to take the vertical surface area of the tank ignoring the top and bottom surfaces. P_i and P_c are respectively the no-load and load losses of the transformer.

$$\text{For plain tank, KT} \approx 5.14 \qquad (2.28)$$

For maximum top-oil temperature-rise of 45°C as per IS, minimum value of $A_t \approx 1.70 \times 10^{-3}$ m^2. W^{-1} and $q_{eT} \approx 600$ W.m^{-2}

(b) For transformers, usually above 50 kVA, use of plain tank is not sufficient and the effective surface area of heat dissipation is required to be increased.

Between 50 and about 1600 kVA, external tubes of circular or elliptical section welded to the tank-wall in two or three rows (Figure 2.6b) are used, depending on the rating of the transformer. Alternatively, external radiators built from tubes set into headers are used.

For tank with tube, $K_T \approx 2.9$, minimum At per watt $\approx 3 \times 10^{-3}$ m^2.W^{-1} and $q_{eT} \approx 338$ W.m^{-2} for $\Delta\theta_o = 45°C$. $\qquad (2.29)$

(c) Use of tubular radiators-single or double rows (Figure 2.6c) all round the tank is somewhat essential for transformers above 1600 kVA, as a further increase in the number of welded tube-rows on the tank proves to be little effective. Generally, tubes of elliptical sections are preferred as larger number of tubes could be accomodated.

With radiators, heat transfer is mainly by natural convection, and K_T in equation (2.27a) is approximately 1.56. This gives $q_{eT} = 182$ W.m^{-2}, and minimum $At = 5.5 \times 10^{-3}$ m^2.w^{-1} for a maximum top-oil temperature-rise of 45°C $\qquad (2.30)$

[*Note* ; Transformers with above cooling methods fall under the class ON or SN.]

(d) OB-class of transformers have blast of air over the tank surface, and

(e) in the OW-class, secondary cooling is effected by running cold water through coils of pipe in direct contact with the oil. Oil flows outside the pipe- coils and cooling is by natural convection.

Example 2.5

A distribution transformer has the following tank-dimensions :-
Length = 1.15 m, Width = 0.6 m, Height = 1.4 m.
On continuous maximum rating, the guaranteed no-load and load losses are respectively 750 W and 4750 W.

---

† Feinberg, R : Modern Power Transformer Practice, The Macmillian Press, page 295.

The transformer is to be cooled by external tubes welded to the tank, the tubes being 0.05 m in diameter. Compute the number of tubes required for a mean top-oil temperature-rise of 45°C and indicate a possible setting for tubes, the distance between the tube centres being 0.08 m. For calculation of effective cooling surface for tubes, average length may be taken as 1.10 m.

Solution

From equn. 2.29, for mean top-oil temperature rise of 45°C.

$$\text{minimum } At = 3 \times 10^{-3} \text{ m}^2. \text{ W}^{-1}$$

$$= 3 \times 10^{-3} (750 + 4750) \text{ m}^2 = 16.5 \text{ m}^2$$

Effective surface area of the tank = $2(1.15 + 0.6) \times 1.4 = 4.9$ m^2

That is, minimum cooling surface to be provided for by the tubes

$$= 16.5 - 4.90 = 11.6 \text{ m}^2.$$

But, effective surface area of each tube = $\pi \times 0.05 \times 1.10$

$$= 0.173 \text{ m}^2,$$

Hence, minimum number of tubes required = 11.6/0.173 = 67.

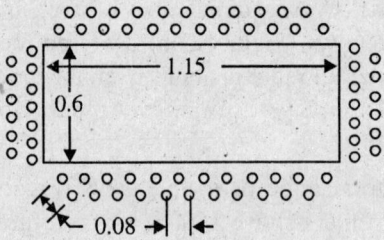

Figure 2.7 Pertaining Example 2.5

Figure 2.7   Indicates a possible setting for the tubes :-
Lengthwise, each side   : 1st row ... 12 nos of tube ;
                            2nd row ... 11 nos. of tube.
Widthwise, each side   : 1st row ... 7 nos;
                            2nd row ... 6 nos.
For computer program, see Figure 6.7.

Example 2.6

If the transformer of example 2.5 has a guaranteed total loss of 10900 W, it is to be cooled by external heat-exchangers. The available standard radiator with 4 tubes has an effective cooling surface of 0.49 m^2 per metre height of the tubes. Compute the number of radiators required.

Solution

From equn. 2.30, for mean top-oil temperature-rise of 45°C,

$$\text{minimum A per watt} = 5.5 \times 10^{-3} \text{ m}^2. \text{ W}^{-1}$$

$$= 5.5 \times 10^{-3} \times 10900$$

$$= 60 \text{ m}^2.$$

Effective surface area of tank (from example 2.5) is 4.4 m². Hence, minimum cooling surface to be provided for by radiators

$$= 60 - 4.4$$
$$= 55.6 \text{ m}^2.$$

Effective cooling surface per radiator assuming its height equal to (tank height – 0.1 m) = 0.49 × 1.3

$$= 0.637 \text{ ma.}$$

Hence, the number of radiators required $= \dfrac{55.6}{0.637}$

$$= 87.3$$

Choose 88.

(2) *Cooling System* :   The cooling system for this category of oil-immersed transformers is by closed-circuit circulation of oil by natural means. The mechanism of heat transfer for such case has been broadly discussed in art. 2.2 with reference to Figure 2.1. Under disconnected condition, oil in the transformer has the same temperature throughout and as such, there is no circulation.

On load, oil absorbs heat developed in core and windings. Its density falls causing flow of hot oil towards the top of the transformer. Hot oil finds path to the cooling tubes or radiators, loses heat, and the cold oil enters the tank at the bottom, thus completing a closed-circuit circulation. For effective extraction of heat from core and windings, the following provisions are generally made :-

(i) Provision of duct between core and windings, and ducts between windings (Figure 2.1);

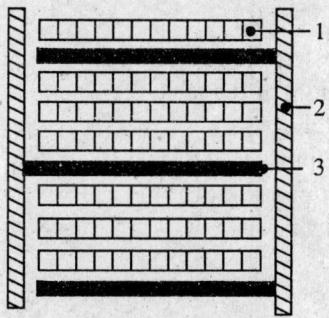

Figure 2.8 Guided oil-flow in transformer winding. 1–coil, 2–cylinder, 3–baffle.

(ii) For larger transformers, ducts are provided in the core ;
(iii) Provision of ducts in windings supported by suitable baffles (Figure 2.8) for guided circulation of oil.

The circulation of oil as described above can thus be attributed to a *'thermal head,* in the tank and a *'gravitational head'* in the cooling tubes. The mechanism of oil-flow, apparently simple, is actually quite complex mainly due to viscous drag at the walls of the tubes, ducts and tank resulting in partial heat barriers. Complicated velocity-and temperaure-profiles are developed and intricate thermodynamic problems arise.

However, oil circulation can be helped considerably if a *'density head'* is created. Such density head can be developed if the 'centre of cooling' (C_1C_2 in Figure 2.9) is maintained above the 'centre of losses' (C_3C_4). The centre of cooling is defined as the horizontal central-line of the radiator tubes and the centre of losses, that of transformer core and windings.

The effect of the density head can be studied with the help of h/θ–diagram (Figure 2.10) drawn with the following assumptions :-

(i)   The height-temperature (h/θ) relation in tank as well as in heat-exchanger is linear;
(ii) The rate of oil-flow is directly proportional to the head;
(iii) The rate of loss dissipation is proportional to the difference between mean winding temperature and mean oil temperature.
(iv) Heat losses between a and d, and b and c (Figure 2.9b) are negligible.

The straight line ab in Figure 2.10 indicates the h/θ-relation of oil in tank, temperature at 'a' being θ_i, and that at 'b', θ_o. With reference to the assumption (iv) above, temperatures at c and d are respectively θ_0 and θ_i, depicting the h/θ-relation in oil in radiators or tubes.

The area abcd may give an estimate of the effective head of oil circulation. Such an estimate though very approximate will give an insight as to how the rate of oil flow can be controlled by raising or lowering the heat exchanger.

Assuming that the temperature-difference Δ between oil and winding is constant for a particular position of the heat exchanger,

$$\Delta \propto \text{area abcd.} (\theta_o - \theta_i) \tag{2.31}$$

Example 2.7 illustrates a method of estimation of the effect of variation of height between the centres of cooling and losses on the mean top-oil temperature-rise.

Example 2.7

A 1500 – kVA transformer has the following data :-
Temperature-rise of oil at the inlet of the tank   ... 25 °C
Mean temperature-rise of windings   ... 65 °C
Height between outlet ard inlet pipes to the tank   ... 1.6 m.
Height between inlet and outlet pipes to the
    headers of the heat-exchanger   ... 2.0 m.

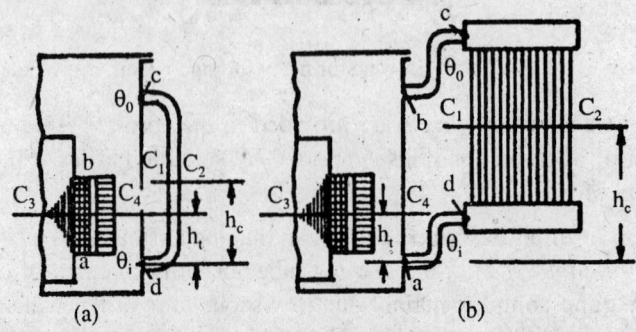

Figure 2.9 Production of density head. (a) with cooling tubes, (b) with heat exchanger.

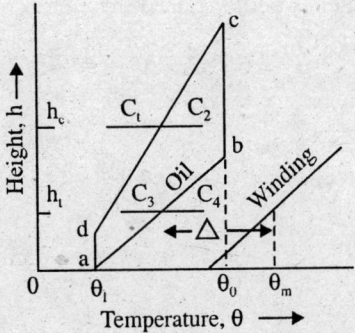

Figure 2.10 h/θ diagram.

When the heat-exchanger outlet and tank inlet pipes are at the same level, the mean top-oil temperature-rise is 55°C. Estimate the height through which the heat-exchanger is to be bodily raised so that the mean top-oil temperature-rise is brought down to 45°C. for the same mean temperature-rise of the windings.

Solution

(i) Figure 2.11(a) shows the h/θ diagrams for oil and windings when the heat-exchanger outlet and tank inlet pipes are at the same level.

Hence,

$$\theta_m = \Delta_1 + \frac{1}{2}(55 - 25) + 25 = 65°C$$

or,

$$\Delta_1 = 25°C$$

$$\text{Area } abc = \frac{1}{2}(2.0 - 1.6)(55 - 25) = 6\,\text{m.}°C$$

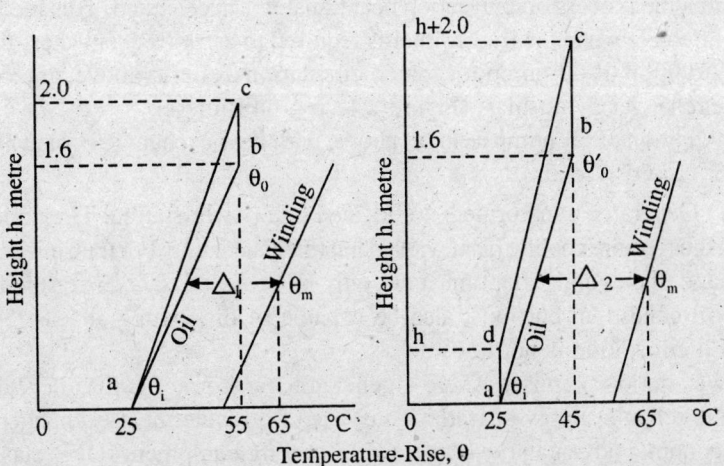

Figure 2.11 Pertaining Example 2.7

(ii) When the heat-exchanger is bodily raised through h metre (Figure 2.11b),

$$\theta_m = 65°C = \Delta_2 + \frac{1}{2}(45 - 25) + 25$$

or, $$\Delta_2 = 30°C$$

$$\text{Area abcd} = \frac{1}{2}(h + 0.4 + h) \cdot (45 - 25)$$

$$= \frac{1}{2}(4.0 + 20h)\ m \cdot °C$$

From equation (2.31).

$$\frac{\Delta_2}{\Delta_1} = \frac{\text{area abcd}}{\text{area abc}} \frac{\theta_0' - \theta_i}{\theta_0 - \theta_i}$$

or, $$\frac{30}{25} = \frac{4 + 20h}{6} \cdot \frac{45 - 25}{55 - 25}$$

or, $$h = 0.34\ m.$$

[*Note* : It is generally found that good circulation of oil is obtained if the ratio of the height of the centre of heating to the height of the centre of cooling above the bottom entry pipe (respectively h_t and h_c in Figure 2.0) is less than 0.8.]

Case II Forced circulation.

Cooling of transformers by 'natural circulation' becomes ineffective above a rating of about 2.5 MVA, and cooling must be through 'Forced circulation', Power system transformers, (generator as well as transmission) fall in this category

(1) *Methods*

(a) Addition of oil-pump between the heat-exchanger outlet pipe and the inlet to the tank (Figure 2.12a–OFN class of transformer) of an ON-cooled transformer so that the process of convection beat transfer is accelerated. This method is specially effective when the transformer is required to carry frequent short-time peak loads.

(b) Addition of oil-pump for forced circulation as in (a) above, and cooling of heat-exchanger by air blast (Figure 2.12 b – OFB class).

(c) Addition of oil-pump as in (a) above, with heat-exchanger cooled by water (Figure 2.12c-OFW class).

[*Note* : (1) Generator transformers are located at a power station. They operate primarily at full load and thus require cooling plant with an intermediate rating. For transmission transformers, on the other hand, the cooling plant must not only be designed to cater for the full load losses under forced-oil condition, but must also be capable of dissipating at least 50 of the losses when forced-oil circulation is not in use.

(2) In power station premises where a generator transformer is installed, there is usually an abundant and reliable supply of water. Thus, use of oil-water heat-exchangers (OFW class) may lead to economic advantage over the oil-air cooling equipment (OFB class).

(3) For transmission transformers, either OFB or OFW may be used depending on its MVA-rating. Most transformers upto 50 MVA may be ON-cooled with fans on detachable heat-exchanger.]

(2) *Cooling System*

(a) Forced oil circulation by addition of airblast on the oil-air heat-exchanger of an ON-class transformer can increase its rating by 20 to 30%. The increase in rating is dependent on (i) the rate of airflow, (ii) efficiency of cooling. Various choices such as, vertical cooling, lateral cooling (Figure 2.12c) may be tried to maximise the effective cooling.

(b) Addition of both oil-pump and fan may increase the ON-rating by 33 to 65%. Generally the pumps employed are totally oil-submerged impeller type driven by a squirrel-cage motor, and facilities are provided so as to permit the pump to be withdrawn from the oil-circuit without the need to remove oil either from the heat-exchanger or the transformer.

(c) With OFW-cooling, the velocity and quantity of oil required to maintain the maximum top-oil temperature-rise within the permissible limit are functions of, (i) the average temperature difference between winding and oil, (ii) *dynamic head* of the oil circuit as measured across the inlet and outlet of the oil pump.

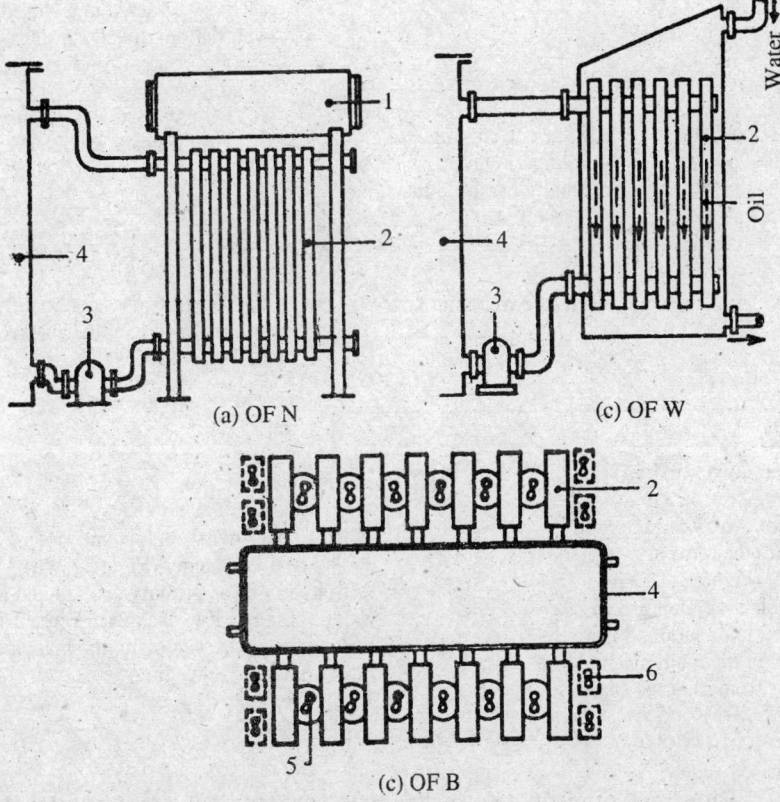

(a) OF N   (c) OF W

(c) OF B

Figure 2.12 Transformer cooling methods–forced circulation. 1 – header, 2 – tubes, 3 – pump, 4 – tank, 5 – fans for vertical cooling, 6 – fans for lateral cooling.

Similarly, the velocity and quantity of water required through a given heat-exchanger required to maintain a desired temperature at the oil-inlet of the transformer are functions of,

    (i) the temperature-difference between the oil-inlet and the outlet at the heat-exchangers,

    (ii) the temperature of the cooling water,

    (iii) dynamic head across the water-side of the heat-exchanger.

(d) The rate of oil-flow and the pump-capacity are determined by empirical equations based on studies made on the cooling method. Precautions must be taken to keep oil and water systems separate and as such oil pressure is maintained at a higher level than the water pressure.

2.4.2 Rotating Machines

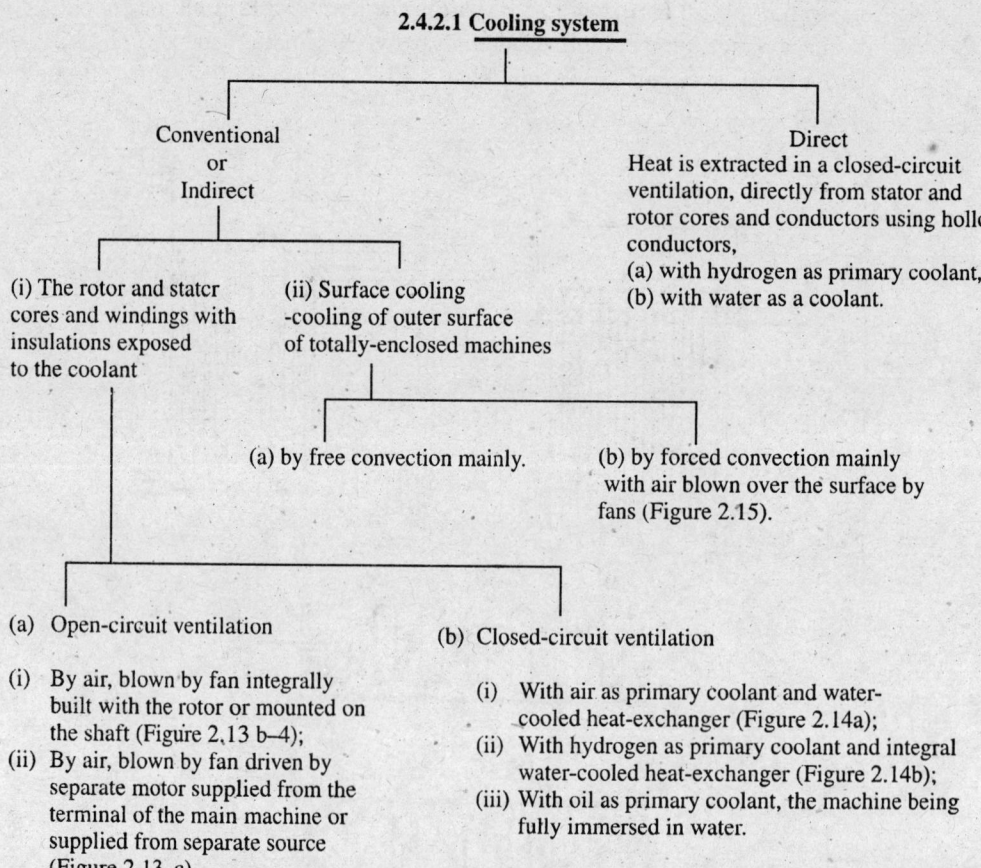

2.4.2.1 Cooling system

Conventional or Indirect

Direct
Heat is extracted in a closed-circuit ventilation, directly from stator and rotor cores and conductors using hollow conductors,
(a) with hydrogen as primary coolant,
(b) with water as a coolant.

(i) The rotor and stator cores and windings with insulations exposed to the coolant

(ii) Surface cooling -cooling of outer surface of totally-enclosed machines

(a) by free convection mainly.

(b) by forced convection mainly with air blown over the surface by fans (Figure 2.15).

(a) Open-circuit ventilation

(i) By air, blown by fan integrally built with the rotor or mounted on the shaft (Figure 2.13 b–4);

(ii) By air, blown by fan driven by separate motor supplied from the terminal of the main machine or supplied from separate source (Figure 2.13–c).

(b) Closed-circuit ventilation

(i) With air as primary coolant and water-cooled heat-exchanger (Figure 2.14a);

(ii) With hydrogen as primary coolant and integral water-cooled heat-exchanger (Figure 2.14b);

(iii) With oil as primary coolant, the machine being fully immersed in water.

[*Note* :   (1)   Direct cooling is more effective than conventional cooling and is used in large machines.

         (2)   Direct-cooled machines are often referred to as 'Super-charged', 'Inner-cooled' or 'Conductor-cooled' machines.]

2.4.2.2 *Coolant*

Important coolants are :-

Air,   hydrogen,   and   water.

(a) Air

  (i) The most common and universally used cooling medium. But above 60 MW air as coolant is not economic since large amount of coolant is required and fan-power requirement is also large. For example, a 60 MW turbo-generator with a total loss of 1 M W needs about 160 tonnes per hour of cooling air and a fan power of 0.1 MW.

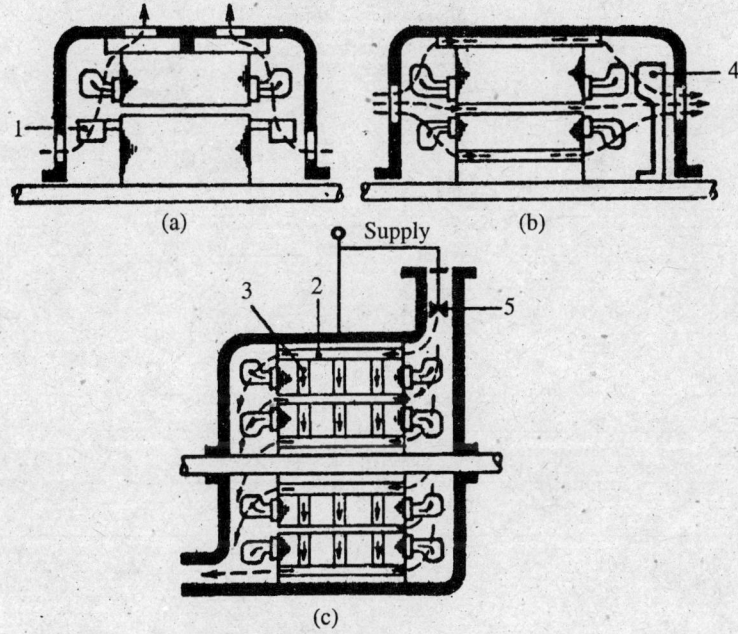

Figure 2.13 Open-circuit ventilation. 1 – built-in fan, 2 – axial duct, 3 – radial duct, 4 – fan on rotor shaft, 5 – separately-excited fan.

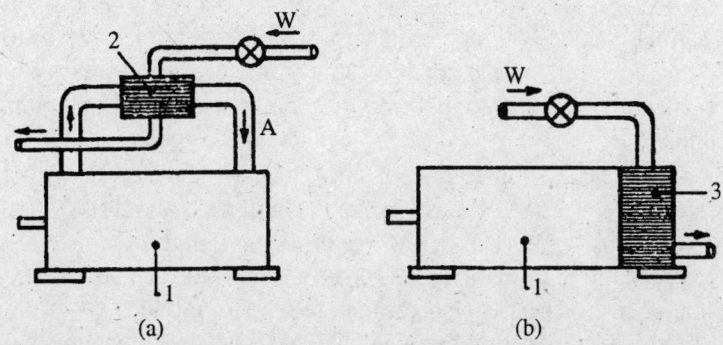

Figure 2.14 Closed-circuit ventilation, (a) air as primary coolant, (b) hydrogen as primary coolant. 1– generator, 2– external water-cooled heat-exchanger, 3– integral water cooled heat-exchanger. A : air, W : water.

(ii) For high voltage machines, air cooling may fail due to corona action.

(b) *Hydrogen Advantages :-*

(i) Density 7% that of air (Table 2.3); that is, the power requirement to circulate hydrogen is approximately 7% of that required by air.

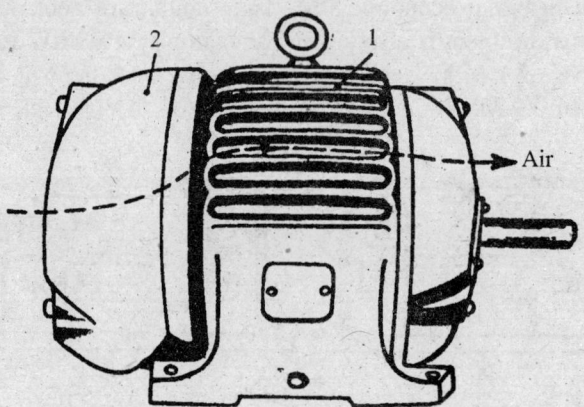

Figure 2.15 Surface ventilation–TEFC motor. 1-ribbed surface, 2–end-cover on the fan.

Table 2.3 Comparison of Coolants

Coolant	Thermal conductivity W. m⁻¹.C⁻¹	Specific heat J. kg⁻¹.C⁻¹	Density kg.m⁻³	Coefficient of volume expansion per °C
Air*	0.03	1000	1.3	3660
Hydrogen*	0.197	3400	0.09	3660
Water	0.60	4200	1000	–
Oil	0.143	1900	850	1000

* at atmospheric pressure.

(ii) Thermal conductivity approximately 6.5 times that of air at 1 atm. pressure. As compared to air, rating for the same frame size increases by about 15% at a pressure of 1 atm., by about 30% at 2 atm. and 40% at 3 atm. with indirect cooling.

(iii) Specific heat 3.4 times that of air. Thus, with hydrogen pressure just above atmosphere, a 120 MW turbo-generator will need about 0.03 m of hydrogen per MW per day, and 0.1 m per M W per day at 2 atm. pressure.

(iv) Increased life of insulation as air-pockets are avoided and hot-spots are eliminated.

(v) Smaller size of heat-exchanger than required with air.

(vi) Less noise, and friction and windage losses as the rotor moves in a medium of low density. There is an improvement in efficiency by 0.5–1% in a 120 M W generator.

(vii) Fire hazards are eliminated as hydrogen does not support burning so long as hydrogen : air mixture is more than 3 : 1.

Disadvantages :- Hydrogen mixed with air (with a ratio less than 3 : 1) forms an explosive mixture. Frames of hydrogen-cooled machines must be sufficiently strong to withstand internal explosion, and all joints must be leak-proof.

[*Note :* (1) Oil-film gas seals are used to ensure gas-tight stator frame. (2) Internal explosion is avoided by (i) using carbon dioxide to replace air before introduction of hydrogen into the casing while filling up. Similarly, carbon dioxide is used to replace hydrogen before emptying the casing, (ii) maintaining hydrogen pressure above atmospheric (generally 1 to 2 kg.cm^{-2} so that any leak will be outwards.]

(c) *Water*

   (i) Direct cooling using hydrogen in both stator and rotor needs large gas flow rate and pressure which is not feasible for a large machine. For example, for a 1000 MW turbogenerator, 15 m^3 sec^{-1} of hydrogen is required at a pressure of only 0.4 kg.cm^{-2}, and this absorbs only 5 MW loss. Again, 14 MW loss in a 1000 MW machine can be dissipated by direct hydrogen cooling, but fan power requirement is about 1.4 MW. On the other hand, with water, the same amount of loss can be dissipated by the use of a pump of about 100 kW rating.

   (ii) Water cooling permits lighter casing, eliminates fans and shaft seals, and allows the stator and rotor to be separated by an insulatirg cylinder in the gap.

   (iii) Water cooling has been used for field winding of large waterwheel generators. The mass of cooling water require is only 25% that of air at atmospheric pressure for the same cooling effect.

2.4.2.3 *Ducts*

Effective cooling of electrical machines, both direct and indirect, is based on the use of ducts (i) in stator and rotor cores, (ii) in slots, and in conductors.

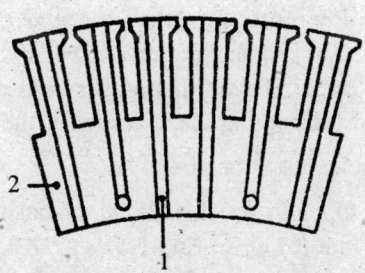

Figure 2.16 Ventilatin space, 1–spacer welded to lamination, 2–segmental lamination.

(i) The ducts in the stator and rotor of a machine could be *radial* and *axial* (Figure 2.13). For radial ventilation ducts, the entire lamiuation assembly in stator and rotor is divided into a number of packets (Figure 2.16) and between two consecutive packets, ventilation spacers are introduced. The spacers consist of small pieces of I-sections welded to the core plates, so that radial ducts are formed when assembled. For axial vent-ducts, axial holes are punched through the laminations.

(ii) Extraction of heat from cores and windings by coolants flowing over the windings and through ventilating ducts has been found to be insufficient for machines of larger ratings. In large machines, apart from the use of radial and axial vent ducts to extract core losses, it has been of utmost necessity to extract heat directly from the conductors which are the seat of I^2R-losses and consequent heating. Use of (a) ducts in slots, and (b) hollow conductors have been advantageous (Figure 2.17).

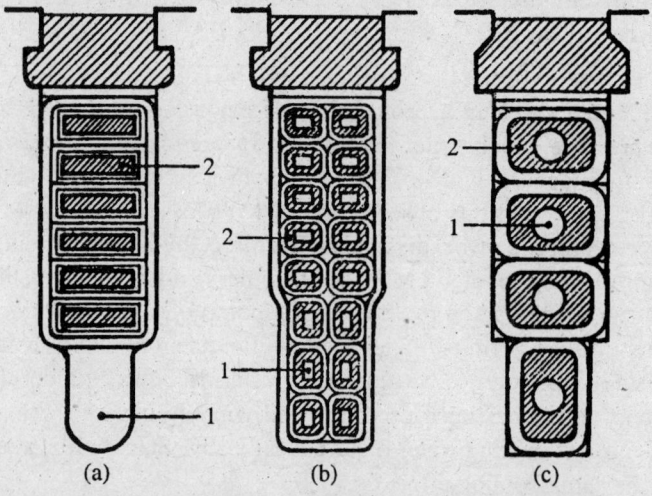

(a)   (b)   (c)

Figure 2.17 Direct cooling of conductors. 1–duct, 2–conductor.

2.4.2.4 *Open-circuit Ventilation*

Open-circuit ventilation schemes are generally employed for air-cooled machines of small and medium sizes, in which the coolant is made to enter the machine from one end, allowed to flow through the vent-ducts and over the winding overhangs. In the process of extraction of heat, its temperature rises, and the hot coolant is then allowed to be released into the atmosphere through the other end (Figure 2.13). The ventilation schemes as such, are closely associated with the type of enclosures and the cooling circuit used for the machines.

Enclosure The types of enclosure as per I.S. 4722 -1?68, are :-

(1) *Open type, such as,* (a) Open Pedestal (OP), which has the rotor on pedestals mounted independent of the machine frame (b) Open End-bracket (OEB), which has end-shields with bearings as integral part.

[Note : In both cases, the stator and rolor ends are in contact with ambient air, freely in the first case and through wide openings in the endshields in the second.

(2) *Protected type;* (a) Screen Protected (SP) - Figure 2.18(a), in which the end-shields have openings suitably guarded by screens such as wire mesh, expanded metal, etc.; (b) Drip Proof (DP)- Figure 2.18(b) & Splash Proof (SPLP).]

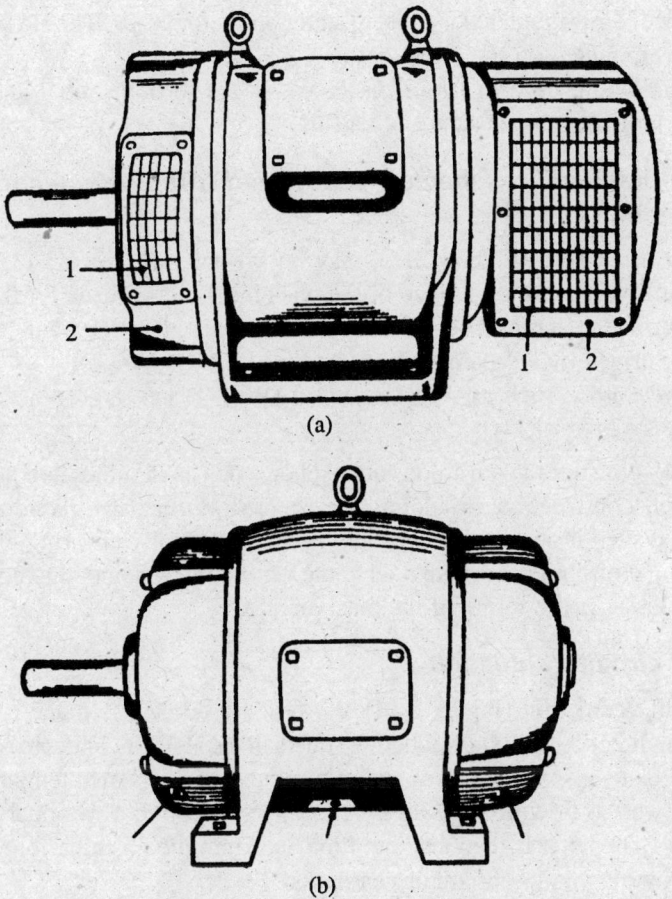

Figure 2.18 Enclosures : (a) screen-protected, (b) drip-proof. 1 – wire mesh, 2 – end cover, 3– air inlet, 4 - air outlet.

Hose Proof (HSP), in which the end shield openings are protected against falling water or dirt, vertically in the first case, at any angle between the vertical and 100° from it for the second, and jets of water in the third case.

(3) *Ventilated type* such as, (a) Pipe Ventilated (PV) or Duct ventilated (DVD), in which there is no opening in the end-shields except flanged apertures through which ventilating air may be let in and let out.

Cooling Circuit: Figure 2.13 shows typical cooling-air circuits such as, *Radial, Axial*, and *Mixed* (combined radial and axial) for open-circuit ventilation schemes.

(i) Radial ventilation, suitable for machines of low rating (about 20 kW) is due to natural centrifugal action of the rotating member. The process can be helped by fans mounted on the rotor, and suitable guides fixed to the stator frame

(ii) Axial ventilation, suitable for moderate size machines with higher speed ratings is due to the flow of air through axial ducts in stator and rotor cores. The flow is sustained by use of fans mounted on the rotor shaft.

(iii) Mixed ventilation, used in larger machines, is due to both radial and axial cooling ducts and judicious use of fans and baffles.

2.4.2.5 *Surface Ventilation schemes, associated with surface cooling system consists of,*

(1) *Totally-enclosed type* (Figure 2.15), may be (a) naturally cooled (TE), or by a fan blowing air over the outer surface of the machine which is usually ribbed so that the cooling surface is more. The fan may either be fixed to the shaft of the machine (TEFC) or driven by a separate drive motor (TESAC).

(2) *Special enclosures*, such as, Weather Proof (WP), Water Tight (WT), Submersible, and Flame Proof (FLP) types.

[*Note* : (1) The FLP–type used in hazardous atmosphere such as in mines and petroleum plants is a totally-enclosed construction. All joints are flanged so that any flame due to internal explosion is extinguished in its passage through the flange thereby ensuring external safety.

(2) The rating of a machine can be related to the ventilation scheme–the more complicated is the scheme, the lesser is the rating of the machine.]

2.4.2.6 *Closed-circuit Ventilation*

In the closed-circuit ventilation (Figure 2.14), the primary coolant is made to flow over the core and winding (indirect cooling) or through the ducts in the slots or conductors (Figure 2.17)–(direct cooling), so as to absorb the losses and in turn increasing its own temperature. The hot coolant is then allowed to flow through a heat-exchanger in which a secondary coolant cools the primary coolant. The latter is then made to flow back into the machine for fresh extraction of heat, thereby providing a closed-circuit ventilation.

Enclosures : Closed-circuit ventilation, schemes for both indirect and direct cooling systems are associated with the following types of enclosures :-

(1) *Totally-enclosed Closed-air Circuit type*, in which air as primary coolant is passed through heat-exchanger and is cooled by a secondary coolant such as, air (CACA), or water (CACW);

(2) Totally-enclosed Closed-gas Circuit type (CGCW), in which the primary coolant is any gas (such as hydrogen in a power system turbo-generator), which in turn is cooled in water-cooled gas coolers.

SHORT QUESTIONS

2.1 Mention 5 possible stages of heat transfer in an oil-immersed transformer.

2.2 Sketch a typical temperature distribution in various parts in an oil-immersed transformer.

2.3 State Fourier Law of conductive heat-transfer.

2.4 Define 'thermal resistance'.

2.5 What causes oil to circulate in the tank and heat-exchanger ?

2.6 Why do large transformers require elaborate cooling method ?

2.7 Explain the terms :- Direct cooling system; TEFC; Super-charged machine; Screen-protected enclosure; FP-enclosure.

2.8 Distinguish between : Open-circuit & closed-circuit ventilation; Radial & axial ducts; Direct-and indirect-cooling; Splash-proof & Drip-proof.

2.9 Compare Air, Hydrogen, and Water as cooling medium for rotating electrical machines.

2.10 State 'true' or 'false' :-

   (i) Hydrogen cooling leads to lesser friction & windage losses in a rotating machine.

  (ii) Hydrogen does not support burning with hydrogen to air mixture less than 3:1.

 (iii) Indirect air cooling is feasible in a 120-MW turbo-alternator.

 (iv) Direct water cooling requires larger fan power than direct hydrogen cooling.

## CHAPTER 3

Main Dimension

3.1   Although as per present trend, commercial designs of transformers and rotating machines are rarely developed from first principles, it is worthwhile for a beginner to consider the fundamental principles of design since whatever remote connection it may have these are the principles on which the design methods are based.

The first step in the design is to obtain main dimensions from the given specifications. For a rotating machine, this means determination of armature diameter D and the effective armature length L_a; and for a transformer, core and window dimensions.

Since the electromagnetic phenomenon in a transformer and rotating machines is concerned with mutual flux in the magnetic circuit and ampere-conductors in the electric circuit, the design principles are based on the consideration of two quantities, namely, the *specific magnetic loading* and the *specific electric loading*.

Specific magnetic loading is related to the specific iron loss (watts per kg.), and the specific electric loading to the specific I^2R-loss. Generally, the choice of these quantities form the starting point of a design.

The determination of main dimensions is based on the following basic equations :-

(1) E.m.f. equation :

Induced voltage in armature,

$$E_0 = 4 \, \phi f_r \, T_a \tag{3.1}$$

$$= \phi \, Z_a n_r \frac{P}{a} \text{volt, for d.c. machine} \tag{3.1a}$$

$$= 4.44 \, \phi_m f_s \, T'_a \text{ volt/phase, for a.c. machine} \tag{3.1b}$$

For a transformer,

induced primary voltage,

$$E_{01} = 4.44 \, \phi_m f_s \, T_1 \text{ volt/phase} \tag{3.1c}$$

and, induced secondary voltage,

$$E_{02} = 4.44 \, \phi_m f_s \, T_2 \text{ volt/phase} \qquad (3.1d)$$

(2) For d.c. machines.

$$\text{Rated kW, } P = E_t \, Ia \cdot 10^{-3} \text{ for generator mode} \qquad (3.2a)$$

$$= \eta \, E_t \, Ia \cdot 10^{-3} \text{ for motor mode} \qquad (3.2b)$$

For a.c. machines of m – phase,

$$\text{Rated kVA, } Q = m \, E_t \, Ia \cdot 10^{-3} \text{ for generator mode} \qquad (3.2c)$$

$$= m\eta \, E_t \, Ia \cdot 10^{-3} \text{ for motor mode} \qquad (3.2d)$$

For transformer,

$$\text{Rated kVA, } Q = m \, E_1 I_1 \cdot 10^{-3} \qquad (3.2e)$$

3.2 MAIN DIMENSIONS

3.2.1 Single-phase Core-type Transformer

Consider, for simplicity, a single-phase core-type transformer.

(a) For a transformer of normal design, resistance-and leakage reactance-drop on full load of both the primary and secondary windings are negligibly small (below 5%) in comparison, to the induced voltage so that,

$$E_1 \approx E_{o1} = 4.44 \, \phi_m f_s T_1 \text{ volt} \qquad (3.3)$$

assuming sinusoidal condition.

Thus,

$$\text{kVA } Q = E_1 \, I_1, \, 10^{-3}$$

$$= 4.44 \, \phi_m f_s \, (I_1 \, T_1) \cdot 10^{-3} \text{ kVA}$$

$$= 4.44 \, B_m f_s \, A_i \, (I_1 \, T_1) \cdot 10^{-3} \text{ kVA} \qquad (3.4)$$

where, A_i = net core area.

(b) The window accomodates the primary and secondary windings including insulation and clearance necessary for cooling purpose. The actual area occupied by conductors $= k_w \cdot A_w$, where, A_w is the window area $= h_w \cdot W_w$ (Figure 3.1), and k_w is the window space-factor.

Assuming mmf-balance, that is, neglecting the no-load mmf,

$$I_1 \, T_1 = I_2 \, T_2 \qquad (3.5)$$

But, the actual area occupied by the primary conductors + that occupied by the secondary conductors

$$= k_w \, A_w$$

Hence,

$$\frac{I_1 \, T_1}{J_1} + \frac{I_2 \, T_2}{J_2} = kw \, A_w \qquad (3.6)$$

where, J_1 and J_1 are primary and secondary current densities (amp. m^{-2}).

[*Note* : It will be shown later (equn. 3.33) that the choice of equal primary and secondary current-densities leads to minimum I^2R loss.]

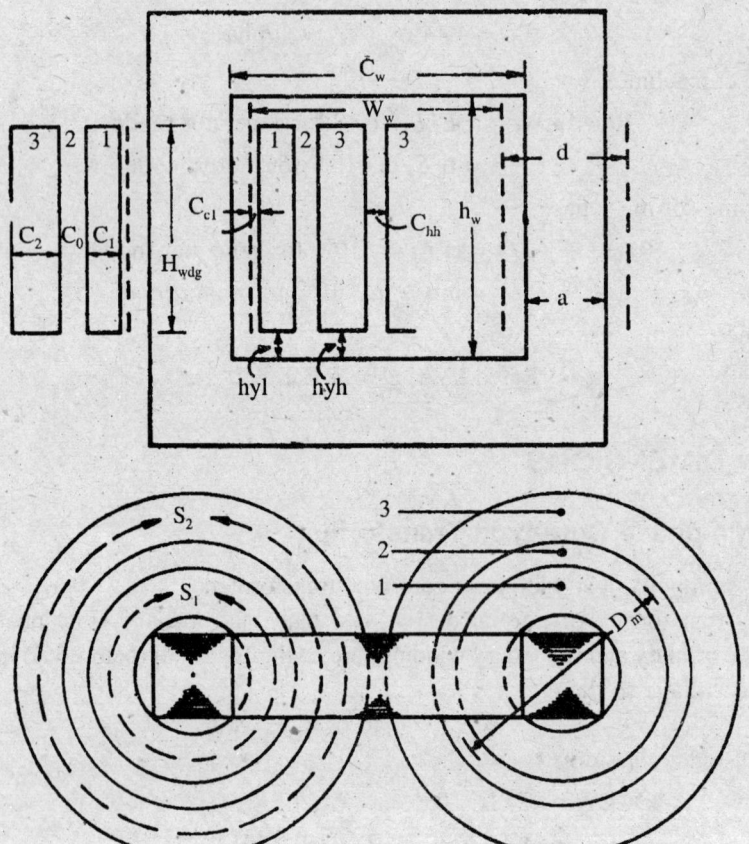

Figure 3.1   Single-phase core-type Transformer : Frame and windings.

1. *l.v.* windings
2. duct between *l.v.* and *h.v.* coils,
3. *h.v.* winding.

$$\frac{I_1 T_1}{J_1} = \frac{I_2 T_2}{J_2} = A_{cond} = \frac{1}{2} k_w A_w \qquad (3.7)$$

(c) From equations (3.4), (3.6) and (3.7),

$$Q = 4.44 \, B_m f_s \, A_i \, A_{cond} \, J. \, 10^{-3} \text{ kVA} \qquad (3.8a)$$

$$= 2.22 \, B_m f_s \, A_i \, k_w \, A_w \, J \, 10^{-3} \text{ kVA} \qquad (3.8b)$$

$$= 2.22 \, B_m f_s \, k_i \, A_c \, k_w \, A_w \, J. \, 10^{-3} \text{ kVA} \qquad (3.8c)$$

where, A_c is the gross core-sectional area, and k_i, the stacking factor.

3.2.2 Single-phase Shell-type Transformer

For shell-type, equations deduced for core-type are applicable with due consideration of the dimensions, as shown in Figure 3.2.

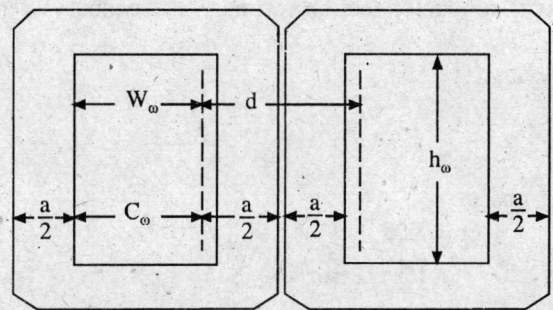

Figure 3.2 Single-phase shell-type transformer frame.

3.2.3 Three-phase Core-type Transformer

(a) From equation (3.4), kVA per phase,

$$Q = 4.44 f_s B_m A_i I_1 T_1 10^{-3}$$

(b) Each window contains two primary and two secondary-phase windings.

That is, the total conductor area in each window (Figure 3.3)

$$= 2\left(\frac{I_1 T_1}{J_1} + \frac{I_2 T_2}{J_2}\right)$$

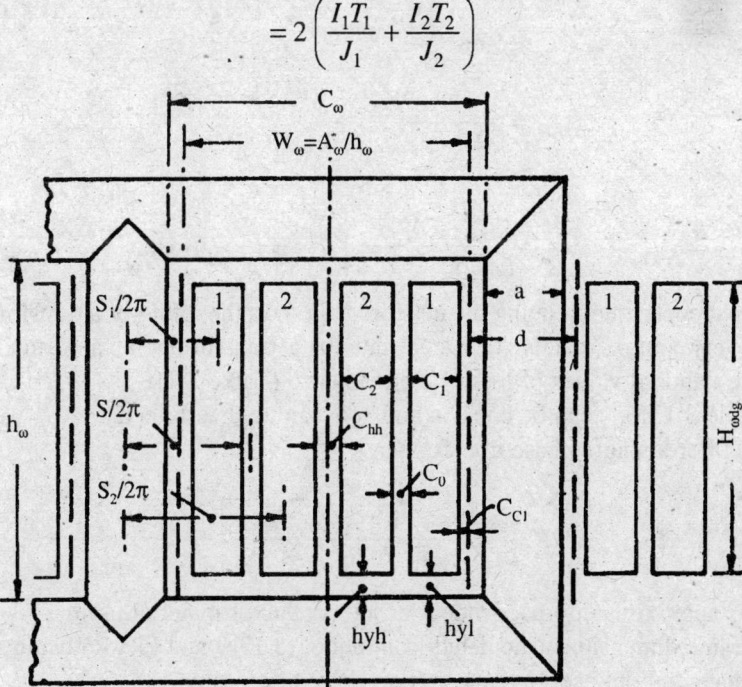

Figure 3.3 Three-phase core-type transformer.

$$= 4 \frac{I_1 T_1}{J} \tag{3.9}$$

$$= 2 A_{con}, \text{ assuming mmf.-balance.} \tag{3.10}$$

where, $2 A_{con}$ is the total conductor section (primary + secondary) per window for a three-phase transformer.

In terms of window area A_w.

$$k_w A_w = 2 A_{con} = 4 \frac{I_1 T_1}{J} \tag{3.11}$$

(c) Thus, kVA per phase,

$$Q = 2.22 f_s B_m A_i A_{con} J \ 10^{-3} \tag{3.12a}$$
$$= 1.11 f_e B_m k_i A_c A_w k_w J \ 10^{-3} \tag{3.12b}$$

3.2.4 Output Coefficient

Re-arranging equation (3.12a).

$$A_i^2 = \frac{Q \ 10^3}{2.22 \ f_s \ B_m J} \frac{A_i}{A_{con}}$$

That is,

$$A_i = \left[\frac{10^3}{2.22 \ f_s \ B_m J} \frac{A_i}{A_{con}} \right]^{1/2} \cdot Q^{1/2} \tag{3.13a}$$

$$= K_{AQ} \cdot Q^{1/2} \ \text{meter}^2 \tag{3.1b}$$

where,

$$K_{AQ} = \left[\frac{10^3}{2.22 \ f_s \ B_m J} \frac{A_i}{A_{con}} \right]^{1/2} \tag{3.14a}$$

$$\approx 3 \left[\frac{A_i}{B_m \ J \ A_{con}} \right]^{1/2} \ \text{for } f_s = 50 \text{ Hz.} \tag{3.1b}$$

K_{AQ} is an output coefficient relating the net core area A_t to the rating of a transformer, and can be regarded, very approximately, as a constant for a transformer of a particular type with approximately standard values of losses and reactance (Table 3.1).

Using Table 3.1, K_{AQ} can be chosen, and A% can be determined.

[Note : (1) For a single-phase transformer,

$$K_{AQ} = \left[\frac{10^3}{4.44 \ f_s \ B_m \ J} \frac{A_i}{A_{cond}} \right]^{1/2} \tag{3.14c}$$

(2) K_{AQ} is approximately independent of the conductor material used.

(3) The frame dimensions calculated from equn. (3.13b) and (3.14c) using Table 3.1 are very approximate and only serves as a starting point of a design.]

Table 3.1 Specific Loadings and K_{AQ} for Core-type Transformers

Class	$Bm(T)$ (average value)	$J(A.mm^{-2})$ (for copper)	K_{AQ}
Generator	1.70	2.5 – 4.0	0.0018 – 0.0024
Transmission :			
Primary	1.55	3.8 – 5.0	0.0014 – 0.0019
Secondary	1.55	2.6 – 3.2	0.0016 – 0.0022
Distribution	1.55	2.5 – 3.0	0.0017 – 0.0019
Rural	1.50	2.0 – 2.4	0.0012 – 0.0014

For Shell-Type Transformers

Three-phase :	K_{AQ}
Power	0.0028 – 0.0044
Distribution	0,0023 – 0.0037
Single-phase	0.0028 – 0.0034

Table 3.2 Window-space Factor

Class	MVA rating	Voltage (kV)	k_w
Generator	340 –144	275	0.21 – 0.17
	144 – 72	132	0.25 – 0.16
Transmission :			
Primary	161 – 156	400	0.14
	125 – 93.6	275	0.15 – 0.14
	120 – 100	275	0.11
	90 – 45	132	0.20 – 0.14
Secondary	20 – 10	33	0.40 – 0.30
Distribution	1 – 0.1	11	0.40 – 0.30
Rural	0.1 – 0.016	11	0.40 – 0.30

3.2.5 Voltage Per Turn

Voltage per turn of primary winding $= \dfrac{E_1}{T_1}$

$\qquad\qquad$ = voltage per turn of secondary winding

$$= \frac{E_2}{T_2}$$

$$= E_T$$

$$= 4.44\, \phi_m f_s$$

That is, $E_T = 4.44\, B_m A_i f_s$ $\qquad\qquad$ (3.15a)

$\qquad\quad = 2.22\, B_m A_i$ for $f_s = 50\text{Hz}.$ $\qquad\qquad$ (3.15b)

$$\text{Putting } A_i = K_{AQ}\, Q^{1/2}$$
$$E_T = 4.44\, B_m f_s\, K_{AQ}\, Q^{1/2}$$
$$= K_{EQ}\, Q^{1/2} \tag{3.15c}$$

Like equn. (3.13b), equn. (3.15c) can also be the starting point of the design, using table 3.3. Also, E_T will yield the number of turns of the primary and secondary windings.

Table 3.3   Voltage Coefficient, K_{EQ}

	Class	KEQ
Core type :		
	Generator	0.68–0.91
	Primary transmission	0.48–0.65
	Secondary transmission	0.55–0.76
	Distribution	0.58–0.65
	Rural	0.40–0.47
Shell type :	Three-phase power	1.06–1.66
	Three-phase distribution	0.77–1.23
	Single-phase	0.84–1.02

3.2.6   Specific Magnetic And Electric Loadings–Transformer

(a) The current-density J is related to the specific I^2R-loss, W_r, through equation (1.3), and the maximum flux-density B_m for a given core material and type of construction is related to the specific iron losses at any frequency through equns. (1.19) and (1.20).

   B_m and J are often termed 'Specific Magnetic Loading' and 'Specific Electric Loading' respectively and their choice (Table 3.1) is usually regarded as the starting point of a design.

(b) When the losses and reactances are not specified, choice of largest acceptable values of B_m and J can be made to start with, as this leads to minimum weight of materials.

(c) The maximum value of current-density is determined by the cooling method employed- the better the method of cooling, the larger is the value of J that can be used. However, a designer must compare the increased cost due to the improved cooling method required with the economy in material due to the choice of increased value of J.

(d) In general, the maximum value of current density for a class A transformer is about 3.2×10^6 A.m^{-2} for distribution transformer, and 5.5×10^6 A.m^{-2} for large transformers with forced cooling.

Example 3.1

   (Using equations 3.1 through 3.14, the preliminary calculations with respect to the design of a transformer could be made, as shown in this example). Specifications of a ON-cooled, class A distribution transformer are :-

   Rating = 1 MVA, 3 phase, 50 Hz.,

   Nominal voltage = 11000 / 433 V, ± 5 %,

   Connections = delta / star.

   Compute core - and window-area, and the h.v. and I.v. turns.

Solution Designer's choice : B_m = 1.55 T, using CROS

$$J = 2.75 \times 10^6 \text{ A.m.}^{-2} \text{ (Average value)}$$
$$K_{AQ} = 0.0018 \text{ for core type.}$$
$$k_i = 0.95$$
$$k_w = 0.35$$

(i) Net core section, A_i (from equn. 3.13b,)

$$= 0.0018 \left(\frac{1000}{3} \right)^{1/2}$$
$$= 0.0329 \text{ m}^2.$$

(ii) From equation 3.12a,

$$A_{con} = \frac{(1000/3).10^3}{2.22 \times 50 \times 1.55 \times 0.0329 \times 2.75 \times 10^6}$$
$$= 0.0250 \text{ m}^2.$$

(iii) Gross core-section, $A_c = \dfrac{0.0329}{0.95} = 0.0346 \; m^2.$

(iv) Window area A_w (from equn. 3.11)

$$= \frac{2 \times 0.0250}{0.35}$$
$$= 0.143 \text{ m}^2.$$

(v) Voltage per turn, E_T (from equn. 3.15a),

$$= 4.44 \times 1.55 \times 0.0329 \times 50$$
$$= 11.32$$

Number of secondary (l.v.) turns, T_2

$$= \frac{433/\sqrt{3}}{11000} \times 971.7$$
$$= 22.08$$

Choose. T_t = 22 turns

and T_1 = 968 turns

Hence, number of h.v. regulating turns = 1.05 × 968 = 48.4
$$= 48 \text{ (choose)}.$$

Thus, steps in h.v. coil : (968 + 48) − 968 − (968 − 48)

: 1016 − 968 − 920

(vii) Revised value of E_T = 11.32 − 971.7 / 968 = 11.36
Revised value of B = 11.36 − 1.55 / 11.32 = 1.556 T

3.2.7 Direct Current Machines

(a) Under rated condition, the armature resistance-drop is negligibly small in comparison to induced armature voltage, so that,

$$E_0 = E_t$$

and thus, rated kW, for generator mode (from equn. 3.2a),

$$P = E_t \cdot Ia \; 10^{-3}$$

$$= (\phi p) \cdot n_r \left(\frac{I_a Z_a}{a} \right) \cdot 10^{-3} \qquad (3.16)$$

where, (ϕp) is the total flux in the machine, often termed as the 'magnetic loading', whereas $(I_a Z_a/a)$ is the total armature ampere-conductors per parallel path, termed as 'electric loading'.

(b) Defining 'Specific Magnetic Loading', Bav, as the average gap flux-density, given by,

$$B_{av} = \frac{\phi P}{\pi D L_a} \; \text{Tesla} \qquad (3.17)$$

and 'Specific Electric Loading', $\overline{ac}$, as the ampere-conductor per metre of armature periphery, i.e.,

$$\overline{ac} = \frac{I_a Z_a}{a \pi D} \qquad (3.18)$$

(c) Thus, for generator mode, rated kW,

$$P = \pi^2 D^2 L_a \, n_r \, B_{av} \, \overline{ac} \; 10^{-3} \; \text{kW} \qquad (3.19a)$$

and for motor mode,

$$P = \eta \pi^2 D^2 L_a n_r B_{av} \, \overline{ac} \; 10^{-3} \; \text{kW} \qquad (3.19b)$$

3.2.8 Alternating Current Machines

(a) As in the case of direct current machines, assuming that the armature resistance and leakage-reactance drops under rated condition are negligibly small in comparison to the induced voltage,

$$E_t = E_0$$

and thus, for a 3-phase machine (generator mode), the rated kVA,

$$Q = 3 \, E_t \, I_a \; 10^{-3}$$
$$= 4.44 \, \phi_m f_s \, (3 \, I_a \, T'_a) \; 10^{-3}$$
$$= 2.22 \, (\phi_m \, p) \, n_s \, (3 \, I_a \, T_a) \, K_w \cdot 10^{-3}$$

where, K_w is the winding factor so that the effective turns per phase

$$T'_a = K_w \, T_a$$

$$= \frac{1}{2} \, K_w \, Z_a \qquad (3.20)$$

T_a and Z_a being the armature turns per phase and armature conductors per phase respectively.

(b) In terms of specific loadings, Bav and ac, knowing,

$$B_{av} \text{ (from equn. 3.17)} = \frac{\phi_m \, P}{\pi \, D \, L_a} \tag{3.21}$$

and
$$\overline{ac} = \frac{3 \, I_a \, Z_a}{\pi \, D} \text{ for a 3-phase machine,} \tag{3.22}$$

the rated kVA, for generator mode,

$$Q = 1.11 \, \pi^2 \, K_w \, D^2 \, L_a \, n_s \, B_{av} \, \overline{ac} \, 10^{-3} \tag{3.23a}$$

and, for motor mode,

$$Q = 1.11 \, \eta \, \pi^2 \, K_w \, D^2 \, L_a \, n_s \, B_{av} \, \overline{ac} \, 10^{-3} \tag{3.23a}$$

3.2.9 Separation of D And L_a

For a given specification, a design procedure starts with the designer's choice of specific loadings, and equations 3.19 and 3.23 enable $D^2 La$ to be evaluated. Separation of D and L_a is made as below : -

(a) For direct current machines, separation is made in accordance with the way in which the poles are proportioned, i.e. according to the ratio (pole arc / pole pitch) or (pole arc / pole length).

For normal design, (pole arc / pole pitch) is 0.65 to 0.70, and (pole arc / pole length) is 0.80 to 1.0.

(b) For alternating current machines of cylindrical rotor type, such as, an induction motor, or a turbo-alternator, or an a.c. commutator motor, D is generally limited by the maximum peripheral velocity (Table 3.4).

Table 3.4

Type of machine	B_{av} (Tesia)	ac (amp. cond. m^{-1})	Maximum peripheral velocity (m. sec^{-1})
1. D.C. machine	0.45 – 0.75	15000 – 51000	45
2. Induction motor	0.30 – 0.60	5000 – 45000	30
3. Synchronous machine :			
(a) salient pole	0.50 – 0.65	20000 – 40000	80
(b) non-salient pole	0.55 – 0.65	50000 – 75000	200
4. A.C. commutator motor	0.30 – 0.65	15000 – 35000	45

3.2.10 Importance of Specific Magnetic and Electric Loadings

(a) The importance of the concept of specific loadings lies in the fact that these do not vary greatly with the size of a particular type of machine.

For example, the specific electric loading $\overline{ac}$ for a direct current machine can lie expressed as,

$$\overline{ac} = \frac{I_c\, Z_a}{\pi\, D} = \frac{I_c}{a_c}\cdot\frac{a_c\, Z_a}{\pi\, D} \tag{3.24}$$

where, $I_c = I_a/a$, the armature current per path ;
and a_c = conductor section, m^2.

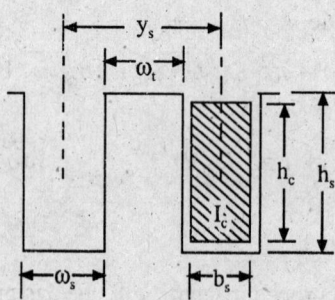

Figure 3.4 Slots and conductors in a rotating machine.

Referring to Figure 3.4, denoting S = number of armature slots,

$$\overline{ac}\, Z_a = S\, b_s\, h_c \tag{3.25a}$$

$$\pi\, D = y_S\, S \tag{3.25b}$$

and

$$y_s = \omega_s + \omega_t \tag{3.25c}$$

Denoting A_{cu}, the copper area in a slot $= b_s\, h_c$ and
A_{sl}, the slot area $= \omega_s\, h_s$,

$$\overline{ac} = J\,\frac{b_s\, h_c}{y_s} \tag{3.26}$$

$$= J\,\frac{b_s\, h_c}{\omega_s\, h_s}\cdot\frac{\omega_s}{w_s + \omega_t}\, h_s$$

$$= J\,\frac{A_{cu}}{A_{sl}}\, h_s\,\frac{1}{1 + (\omega_t\,/\,\omega_s)} \tag{3.27}$$

Thus, specific electric loading is constant provided the current-density, the ratio copper to slot area, ratio of tooth-width to slot width, and the slot height are constants.

(b) Similarly, for the same maximum flux-density and nature of flux-distribution, the specific magnetic loading is constant for all machines of a particular type.

(c) Type of cooling to be employed for the armature depends approximately on the choice of specific electric loading, as shown below -

$I^2 R$ – loss in the active lengths of armature conductors, P_a

$$= I_c^2\, Z_a\,\rho\,\frac{L_a}{a_e},$$

assuming that all armature conductors are in series, and noting that L_a is approximately equal to the length of active part of an armature conductor; a_c, conductor section; and ρ, the resistivity of the conductor material.

or,

$$P_a = \frac{I_c \, Z_a}{\pi \, D} \, \rho \, \frac{I_e}{a_e} \, \pi \, D \, L_a$$

or the i^2R – loss in the active parts of the conductors per unit surface-area of the armature,

$$= \frac{P_a}{\pi \, D \, L_a}$$

$$= \overline{ac} \cdot \rho \cdot J \qquad\qquad (3.28)$$

For any type of cooling, $(P_a / \pi \, D \, L_a)$, the loss per unit surface-area to be dissipated, can be specified. Thus, depending on the choice of the values of $\overline{ac}$, J. and conductor material, type of cooling of armature can be designed.

(d) In practice, however, the assumption of constant specific loadings for a particular type of rotating machine of all sizes is not quite valid.

For example, larger machines have larger slot depth and thus, should have larger specific electric loading provided all other quantities described in equation 3.28 remain unaltered.

(e) The choice of specific magnetic loading depends on excitation and flux condition, the core loss; and power-factor in case of a.c. machines. Ranges of specific loadings for various types of rotating machines are given in Table 3.4.

(f) Selection of larger values of specific loadings generally leads to low-cost design. For example, larger B_{av} and $\overline{ae}$ lead to lower volume (D^2L) and in turn to lower cost of material required. However, such choices lead to increased iron and I^2R losses and demand improved cooling method for effective heat dissipation.

Table 3.5 gives a statement regarding possible major effects due to the choice of large specific loadings.

Table 3.5 Major Effects due to Choice of Large Specific Loadings

---

(1) *Type of machine : Direct Current*

    (a) High B^{av} :   (i)   Leads to high tooth-density (should generally be less than 2.2 Tesla); larger requirement of mmf, increased field I^2R-loss, and iron loss.

                      (ii) For high speed machines, frequency of reversal of armature flux $(= p \, n_r/2)$ is large, and hence higher B_{av} leads to larger iron losses.

    (b) High $\overline{ac}$:   (i)   Leads to increased temperature-rise. Care should be taken in the choice of insulating material in slots to withstand higher temperature-rise.

                      (ii) For high speed machines, ventilation is better, and high value of $\overline{ac}$ can be used.

                    (iii) Results in increased armature mmf, i.e. more distortion of field form on load and reduced effective gap flux. For the maintenance of torque, field flux should thus be increased thereby needing increased field excitation and increased field I^2R-loss.

(iv) High $\overline{ac}$ can be achieved by reduced armature diameter and/or by increased number of armature conductors. Slot depth will be larger which means increased reactance-voltage and inferior commutation.

(v) For high voltage machines, requirement of insulation space is larger. That is, the space-factor will be smaller. To have higher $\overline{ac}$, armature diameter must be larger, and hence larger size.

(2) *Type of machine : Induction motor.*

(a) High B_{av} : (i) Leads to increased magnetising current and poorer power-factor.

(ii) Leads to increased iron losses.

(b) High $\overline{ac}$: (i) Leads to increased armature I^2R-losses and temperature-rise.

(ii) Leads to larger turns per phase and thus increased leakage reactance; reduces diameter of current locus, and lowers overload capacity.

(iii) For high voltage motors, insulation space required is larger. Slot-space factor is thus smaller. To have higher $\overline{ac}$, larger armature diameter is required and hence larger size of the motor.

(3) *Type of machine : Synchronous machine*

(a) High B_{av} : (i) Leads to higher tooth-density (should normally be less than 2.2 Tesla); larger field mmf. is required and hence increased field I^2R-loss; and also increased iron losses due to higher flux-density.

(ii) Leads to increased flux per pole for the same diameter and length. For the same voltage, this means smaller armature turns per phase, lower synchronous reactance, and thus increased stability and improved regulation.

(iii) Leads to reduced leakage reactance, and thus larger initial current on sudden short-circuit.

(iv) For high voltage machines, insulation requirement is large. For the same conductor section this means lower tooth section and thus, higher tooth-density.

(b) High $\overline{ac}$: (i) Leads to increased armature I^2R-loss and higher temperature-rise.

(ii) Means larger armature turns per phase; thus, increased leakage reactance and also increased armature reaction and synchronous reactance. Regulation will be poorer, stability impaired, and reduced short-circuit current.

Example 3.2

Determine suitable values for the armature diameter and the effective core-length of a 10-MVA, 3-phase, 11-kV, star-connected, 50-Hz, 2-pole, turbo-alternator. Take the winding factor as 0.955 and the airgap length as 1.5cm.

Considering all conductors per phase connected in series, estimate the number of stator conductors per phase and choose a suitable value.

Solution Designer's choice : $B_{av} = 0.55$ T

$$\overline{ac} = 50,000 \text{ amp. cond. m}^{-1}$$

Peripheral speed = 120 m. sec^{-1}

From equn. 3.23(a),

$$10,000 = 1.11\,\pi^2\,0.955\,D^2\,L_a\left(\frac{2\times 50}{2}\right)0.55\times 50,000\times 10^{-3}$$

or, $$D^2\,L_a = 0.695\ \text{m}^3.$$

But, $$\pi\,D_r\,n_s \leq 120,\ D_r\ \text{being the rotor diameter.}$$

That is, $$D_r \leq 0.764$$

Choose $$D_r = 0.76\ \text{m.}$$

Hence, the stator diameter, $$D = 0.76 + 2\times 0.15$$
$$= 0.79\ \text{m.}$$

Thus, effective core-length $$L_a = 1.10\ \text{m.}$$

Armature current, $$I_a = \frac{10,000\times 10^3}{\sqrt{3}\times 11000} = 524.86 = 524.86\ \text{amp. per phase}$$

Knowing, $$\overline{ac} = 3\,I_a\,Z_a\,/\,(\pi\,D),$$
Number of stator conductors per phase

$$Z_a = 50000\times \pi\times 0.79\,/\,(3\times 524.86)$$
$$= 79.2$$

Choose stator conductors per phase = 80.

Thus, revised values : $$\overline{ac} = 50,499\ \text{amp. cond. m}^{-1}$$
$$B_{av} = 0.545\ \text{T}$$

Example 3.3
Determine approximate values for the stator bore and the effective core-length of a 55–kW, 415–v, three-phase, star-connected, 50–Hz, four-pole, induction motor. Take the efficiency as 90 percent, the power-factor as 0.91, winding factor as 0.955, and make pole-pitch equal to the core-length.

Estimate the number of stator conductors required for a winding in which all the conductors per phase are connected in series, and choose a suitable number of conductors and slots, limiting the slot-loading to 600 ampere-conductors.

Solution Designer's choice : $$B_{av} = 0.50\ \text{T}$$
$$\overline{ac} = 35,000\ \text{amp. cond. m}^{-1}$$
$$k_w = 0.955$$

From equation 3.23b.

$$\frac{55}{0.91} = 1.11\,\pi^2\,0.9\times D^2\,L_a\left(\frac{2\times 50}{4}\right)\times 0.5\times 35000\times 10^{-3}$$

or, $$D^2\,L_a = 0.014\ \text{m}^3.$$

With pole pitch = core length,

$$\frac{\pi D}{p} = L_a \text{ or, } La = \pi D / 4 .$$

yielding,
$$D^3 = 0.01784 \text{ m}^3,$$
$$D = 0.26 \text{ m},$$
$$L_a = 0.204 \text{ m}.$$

The armature current per phase $I_a = \dfrac{55 \times 10^3}{0.9 \times \sqrt{3} \times 415 \times 0.91}$

$$= 93.4 \text{ A}.$$

Specific electric loading, $\overline{ac} = 35000 = 3\,\dfrac{I_a\,Z_a}{\pi D}$

or, number of stator conductors per phase, Za = 102

Hence, computed total no, of stator conductors = 3 × 102 = 306

But, with slot loading limited to 600 amp. cond per meter,

$$\text{no. of stator slot, } S \geq \frac{35000 \times \pi \times 0.26}{600}$$

$$\geq 47.65$$

Choose S = 48, a multiple of (no. of poles × no. of phases) for integral slot winding (Chapter 5), and revised value of number of stator conductors = 6 × 48 = 288.

Example 3.4

Obtain suitable values of diameter and core-length for a 1500–kVA, 3,300–V three–phase, star–connected, 50–Hz, 10–pole alternator which is to have specific loadings of about 0.51 T and 34000 ampere conductors per metre. The ratio (pole pitch) / (core length) may be taken as 0.8.

Select suitable number of stator conductors and slots, limiting the ampere-conductors per slot to 2000. State the number of turns per coil of a double-layer winding is to be used- (A.R.C.S.T.)

Solution   With designer's choice of k_w as 0.955, we have, from equation 3.23a,

$$1500 = 1.11\,\pi^2 \times 0.955 \times D^2\,L_a \left(\frac{2 \times 50}{10}\right) \times 0.51 \times 3400 \times 10^{-3}$$

or,
$$D^2 L_a = 0.827 \text{ m}^3$$

With
$$\frac{\text{Pole} - \text{pitch}}{\text{core} - \text{length}} = \frac{\pi D}{p\,L_a} = \frac{\pi}{10}\frac{D}{L_a} = 0.8$$

$$D = 2.546\,L_a$$

and thus,
$$D = 1.28 \text{ m, and } L_a = 0.503 \text{ m}.$$

Again, the armature current per phase, $I_a = \dfrac{1500 \times 10^3}{\sqrt{3} \times 3300}$ A.

$$= 262.43 \text{ A.}$$

Specific electric loading, $\overline{ac} = 34000 = 3\,\dfrac{I_a\, Z_a}{\pi\, D}$

Hence, no. of stator conductors per phase, $Z_a = 173.66$

and the total no. of stator conductors $= 3 \times 173.66 = 520.99$

With slot-loading limited to 2000 amp. cond per meter,

$$\text{no. of stator slot, } S \geq \frac{34000 \times \pi\, D}{2000} \geq 68.36$$

Choosing integral-slot winding (Chapter 5), number of stator slots must be a multiple of (no. of phases × no. of poles) i.e., $3 \times 10 = 30$.

$$\text{Choose } S = 90, \text{ and}$$

$$\text{total no. of stator conductors} = 6 \times 90 = 540.$$

For a double-layer winding, no. of coils = no. of slots (Chapter 5). = 90.

$$\text{But, number of turns} = \frac{540}{2} = 270.$$

$$\text{Hence, no. of turns/coil} = \frac{270}{90} = 3.$$

Example 3.5

A 30-kW, 1000-rpm d.c. motor has specific magnetic loading of 0.50 Tesla. Estimate the kW-rating of an 850-rpm d.c. motor which has a specific magnetic loading of 0.55 Tesla, a current-density 10% greater than that of the 30-kW motor and linear dimensions which are all, including those of slots, 20% greater. Assume two motors to have the same efficiency.

Solution From equation 3.28a,

$$\text{specific electric loading, } \overline{ac} = J\,\frac{b\,h_c}{y_s}$$

Denoting the 30-kW motor by subscript '1' and the other motor by '2',

$$\overline{ac}_1 = J_1\,(b_{s1}h_{c1}\,/\,y_{s1})$$

and $\overline{ac}_2 = J_2\,(b_{s2}h_{c2}\,/\,y_{s2})$

But, $J_2 = 1.1\,J_1,\ b_{s2} = 1.2\,b_{s1},\ h_{c2} = 1.2\,h_{c1},\ \text{and}\ y_{s2} = 1.2\,y_{s1}.$

Hence, $\overline{ac}_2 = 1.32\,\overline{ac}_1$

From equation 3.19b,

$$\frac{P_2}{P_1} = \frac{D_2^2 \, L_{a_2}}{D_1^2 \, L_{a_1}} \cdot \frac{n_{r2}}{n_{r_1}} \cdot \frac{B_{av2}}{B_{av1}} \cdot \frac{\overline{ac_2}}{\overline{ac_1}}$$

with $D_2 = 1.2 \, D_1$, $n_{r2} = 0.85 \, n_{r1}$, $B_{av2} = 0.55$, $B_{av1} = 0.50$, and $P_1 = 30$ kW, the required estimated value for the 2nd motor = 64 kW.

3.2.11 Optimum Design

I. Condition for maximum efficiency.

$$\text{Efficiency}, \eta = \frac{\text{Output}}{\text{Output + losses}}$$

Let the value of the current I at which the efficiency is maximum be $(k_s \cdot I_a)$, I_a being the rated current phase.

Thus, output at this current for an m-phase machine $= m \, E_t \, I \cos \phi$

$$I^2 \, r\text{-loss at this current} = I^2 \cdot R = k_s^2 \, I_a^2 \cdot R = k_s^2 \cdot P_{cu}$$

where, P_{cu} is the rated $I^2 R$-loss.

Hence,

$$\eta = \frac{mE_t I \cos \phi}{mE_t I \cos \phi + P_i + I^2 R}$$

P_i being the iron losses, independent of current I.

For maximum efficiency, $(d \, \eta / \, dI) = $ zero. Assuming constant power-factor,

$$[mE_t I \cos \phi + P_i + I^2 R] \, mE_t \cos \phi - mE_t I \cos \phi \, [mE_t \cos \phi + 2IR] = 0$$

or,

$$P_i = I^2 R = k_s^2 \, P_{cu} \qquad (3.29)$$

Equation (3.29) is valid for direct current machine also.

II. Loss- and Mass-ratio for maximum efficiency.

Putting :

$$P_i = W_i \cdot G_i \cdot m_i \qquad (3.30a)$$

and

$$P_{cu} = W_r \cdot G_c \cdot m_c \qquad (3.30b)$$

where, W_i and W_r are specific loss (W· kg^{-1}) for iron and copper respectively ; G and m stand for specific weight (kg. m^{-3}) and mass (m^3) respectively with subscripts i and c standing for iron and copper, equation (3,29) yields,

for maximum efficiency, mass ratio, $\dfrac{m_i}{m_c} = k_s^2 \cdot \dfrac{W_r}{W_i} \cdot \dfrac{G_c}{G_i}$ \qquad (3.31)

For copper, $W_r = 2.41 \, . \, 10^{-12} \, J^2$ W. kg^{-1} with current-density J in amp . m^{-2}; and $G_c = 8890$ kg. m^{-3}.

With CROS at $B_m = 1.55$ T and $f_s = 50$ Hz.,

$$W_i = 1.49 \text{ W. kg}^{-1}, \text{ and } G_i = 7650 \text{ kg. m}^{-3}$$

Hence, mass-ratio for maximum efficiency $= 1.88. \ 10^{-12} \ J^2$.

[Note : For a transformer, approximate values of m_i and m_c can be determined as below :-

Case I Single-phase core-type (Figure 3.1).

Length of iron in core and yoke, $\quad L_i = 2 \ h_w + 2 \ c_w + 4a$ $\hspace{3cm}$ (3.32)

and assuming equal limb and yoke section,.

$$m_i = A_i \ Li \ m^3 \hspace{3cm} (3.33)$$

Again, mass of conducting material per limb

$$= \alpha_1 \frac{T_1}{2} \ s_1 + \alpha_2 \frac{T_2}{2} \ s_2$$

where, α is the cross-sectional area of conductors.

$$= \frac{1}{2} \left[\frac{I_1 \ T_1}{J_1} \ s_1 + \frac{I_2 \ T_2}{J_2} \ s_2 \right]$$

$$= \frac{1}{2} \ A_{con} \ (s_1 + s_2) \text{ using equation 3.7.}$$

With $s_1 \equiv \pi \left(a + \frac{1}{4} \ c_w \right)$, and $s_2 \equiv \pi \left(a + \frac{3}{4} \ \dot{c}_w \right)$, total mass of conducting material,

$$m_c \equiv A_{con.} \ 2\pi \left(a + \frac{1}{2} \ c_w \right) m^3 \hspace{3cm} (3.34)$$

Case II Refer Figure 3.3-three-limb, three-phase core-type. Total length of iron in limb and yoke,

$$L_i = 3 \ h_w + 4 \ C_w + 7 \ a_1 \hspace{3cm} (3.35)$$

and $\hspace{2cm} m_i = A_i : L_i \ m^3$, assuming equal limb and yoke section.

Again, mass of conducting material per phase,

$$= \alpha_1 \ T_1 \ s_1 + \alpha_2 \ T_2 \ s_2 \hspace{3cm} (3.36a)$$

$$= \frac{I_1 \ T_1}{J_1} s_1 + \frac{I_2 \ T_2}{J_2} \ s_2 \hspace{3cm} (3.36b)$$

$$= \frac{1}{2} \ A_{con} (s_1 + s_2) \text{ from equn, (3.11)}$$

But, $\hspace{2cm} s_1 = \pi \left(a_1 + \frac{1}{4} \ C_w \right)$

and $\hspace{2cm} s_2 = \pi \left(a_1 + \frac{3}{4} \ C_w \right) \hspace{3cm}$ (3.37)

Hence, total mass of conducting material,

$$m_c = 3\,A_{con}\,\pi\left(a_1 + \frac{1}{2}\,C_w\right) m^3$$

$$= 3\,A_{con}\cdot s \tag{3.38}$$

where,
$$s = \pi\,D_m = \pi\left(a_1 + \frac{1}{2}\,C_w\right)$$

III. Condition for minimum I^2 R-loss.

The i^2 r-loss per phase for a transformer (or rotating machine),

$$P_R = (I_1^2\,R_1 + I_2^2\,R_2)\,k_c \text{ watt}$$

where, I and R are current and resistance; subscripts 1 and 2 denote l.v. and h.v. (or stator or rotor) respectively; k_c is the eddy-loss factor (art 1.7.2–v) = 1.05 to 1.15.

or,
$$P_R = \left[I_1^2\,\frac{\rho\,s_1\,T_1}{\alpha_1} + I_2^2\,\frac{\rho\,s_2\,T_2}{\alpha_2}\right]k_c$$

$$= \left[J_1^2\,(s_1 T_1 \alpha_1) + J_2^2\,(s_2 T_2 \alpha_2)\right]\rho\cdot k_c$$

For a 3-phase core type transformer,

$$s_1 = \pi\,(D_m - c_o - c_1)$$

and
$$s_2 = \pi\,(D_m + c_o + c_2)$$

Generally, $D_m \gg (c_0 + c_1)$, and also $\gg (c_0 + c_2)$,

so that,
$$s_1 \equiv s_2 \equiv s$$

Hence,
$$P_R = \left[J_1^2\,(T_1\,\alpha_1) + J_2^2\,(T_2\,\alpha_2)\right]\rho\cdot s\cdot k_c \tag{3.39}$$

Writing,
$$T_1\alpha_1 = A_{1cond}, \text{ and } T_2\,\alpha_2 = A_{2cond}, \tag{3.40}$$

where, A_{1cond} and A_{2cond} are respectively the total area occupied by conducting material in l.v. and in h.v. coils per phase, we note, for a 3-phase core type transformer $A_{1cond} + A_{2cond}$

$$= \frac{1}{2}\,k_w\,A_w$$

$$= k_w\,A_w, \text{ for a single phase core-type transformer,}$$

$$= \text{a constant } K_1 \text{ for a particular design of frame.} \tag{3.41}$$

From equations (3.39), (3.40) and (3.41).

$$P_R = \left[J_1^2\,A_{1cond} + J_2^2\,(K_1 - A_{2cond})\right]\rho\cdot s\cdot k_c$$

For minimum loss, $dP_R / d\,A_{1cond} = 0$,

or,
$$J_1 = J_2 = J \tag{3.42}$$

That is, for minimum total I^2R-loss, the current-densities of the l.v. and the h.v. coils are equal, for a transformer of a particular window area.

For rotating machine with wound stator and rotor windings, with small gap length (such as a wound-rotor induction motor), it can be shown that the above condition is also valid.

IV. *Condition for Minimum Total Material Cost*

Defining c_i and c_c as the specific cost (rupee per kg.) for iron and conducting material respectively,

cost of iron
$$C_i = c_i\, G_i\, m_i$$
$$= c_i\, G_i\, L_i\, A_i.$$

and cost of conducting material,
$$C_c = C_c\, G_c\, m_c$$
$$= 3\, c_c\, G_c\, A_{con}\, s \text{ for a 3-phase transformer}$$

For a particular rating of a transformer, $(B_m . J)$ can be assumed to be constant as the dimensions of the design is altered, A_i is proportional to $(1 / A_{con})$ from equation 3.12 (a).

We write,
$$A_i \cdot A_{con} = K$$

Again, from equation (3.14a),

$$\frac{A_i}{A_{con}} = K_{AQ}^2 . 2.22\, f_s\, B_m\, J\, 10^{-3}$$

$$= \gamma \text{ (say)}$$

Thus,
$$A_i = (K\, \gamma)^{1/2};\; A_{con} = (K / \gamma)^{1/2}$$

Hence, total cost
$$C_t = C_i + C_c$$
$$= c_i\, G_i\, L_i\, (K\, \gamma)^{1/2} + 3c_e\, G_c\, s\, (K / \gamma)^{1/2}$$

For minimum total cost $(d\, C_t / d\, \gamma) = 0$, which yields,

$$\gamma = 3\, \frac{c_c}{c_i} \cdot \frac{G_c}{G_i} \cdot \frac{s}{L_i} = \frac{A_i}{A_{con}}$$

or,
$$\frac{A_i\, L_i\, G_i}{3\, A_{con}\, s\, G_c} = \frac{c_c}{c_i}$$

$$\frac{\text{Weight of iron}}{\text{Weight of conducting material}} = \frac{\text{Specific cost of conducting material}}{\text{Specific cost of iron}} \qquad (3.43)$$

[Note : Defining total mass $m = m_i + m_c$
$$= A_i\, Li + 3\, A_{con}\, s$$

and following the above procedure, i.e. letting $A_i\, (K\gamma)^{1/2}$ and $A_{con} = (K/\gamma)^{1/2}$, it can be shown that for minimum total mass.

$$\frac{A_i}{A_{con}} = \frac{s}{L_i}$$

i.e. mass of iron m_i = mass of conducting material m_c \qquad (3.44)]

3.3 STANDARD RATINGS AND FRAMES

Transformers and motors of small and medium sizes are built to standard ratings, as indicated in Tables 3.6 and 3.7.

Table 3.6 Recommended Preferred kVA Ratings of Transformers

(1) 3-phase Power Transformers			
25	250	2500	25000
	315	3150	31500
40	400	4000	40000
	500	5000	50000
63	630	6300	63000
	800	8000	80000
100	1000	10000	100000
125	1250	12500	
160	1600	16000	
200	2000	20000	

(2) 3-phase distribution transformer

16	25		40	50	63	80	100

IEC reviews the list periodically to take into account of technological advances

For small and medium-size transformers, specially distribution transformers which are produced in large scale, standardisation of width of lamination plates for the core results in considerable economy, though the use of such standard lamination plates leads to departure from optimum proportions (chapter 6. Table 6.1).

Similarly, for small and medium size motors, standard frame sizes are available .

Table 3.7 Recommended Preferred kW Rating of Rotating Machines

(1) D.C, and induction motors.

0.06	0.37	2.2	15	45	132	220	335	450	600	800
0.09	0.55	3.7	18.5	55	150	250	355	475	630	850
0.12	0.75	5.5	22	75	160	280	375	500	670	900
0.18	1.1	7.5	30	90	185	300	400	530	710	950
0.25	1.5	11.0	37	110	200	315	425	560	750	1000

(2) Turbo-generators 3-phase, 2-pole, 50 Hz.

(a) Air-cooled

MW	10	12		16		25		40	50
MVA	12.5	15		20		31.25		50	62.5

(b) Hydrogen-cooled.

MW	50		63		80		100		125	
MVA	62.5		78.75		100		125		156.25	

* Shanmugasundaram, A., Gangadharan, G., Palani, R. : Electrical Machine Design Data Book, pp. 168-175

The effect of above mentioned standardisations obviously means imposition of limitations on the designer. Once the dimensions of the standard cores (for transformer) and standard frame size (for motors) have been fixed, it is nearly always cheaper to design the machine on the most appropriate standard, even though this may not be exactly the optimum size. In practice, the existence of standard cores and frames often simplifies the process of obtaining a design by restricting the fields of choice.

QUESTIONS

3.1 Define Specific magnetic loading and Specific electric loading in relation to (i) transformer, (ii) rotating electrical machine.

3.2 Show that the specific electric loading is approximately constant provided the current-density, ratio of conductor to slot-area, ratio of tooth-width to slot-width, and the slot depth in a rotating machine are constants.

3.3 Show that for minimum total I^2R-loss in a transformer, current-densities of primary and secondary windings should be approximately equal.

3.4 A turbo-alternator has smaller diameter and larger length, whereas, a water wheel generator has smaller length and larger diameter. Why?

3.5 Two machines A and B are identical, but A has all linear dimensions $x-$ times those of B. For particular specific loadings, shew that, (i) output of A is x^4-times that of B; (ii) total loss of A is x^3-times that of B; (iii) loss-dissipating surface is x^2-times that of B.

3.6 State advantages and dis-advantages of using standard frame-size in electrical machines.

3.7 State 'true' or 'false';

(a) For the design of transformer where losses and reactances are not specified, the highest acceptable values of flux-density and current-density should be chosen to keep the cost of materials to a minimum.

(b) The value of specific electric loading determines the iron loss in a machine.

(c) The specific magnetic loading is constant for a particular type of machine for the same flux-density and nature of flux-distribution.

(d) In a high speed d.c. machine, smaller value of specific electric loading must be chosen.

(e) In an induction motor, higher value of specific magnetic loading leads to better power-factor.

(f) In a high-voltage induction motor, higher value of specific electric loading means larger size of the motor.

(g) In a synchronous machine, choice of high specific magnetic loading leads to lower synchronous reactance.

(h) Choice of high specific electric loading for synchronous generators leads to improved voltage regulation.

(i) High speed machines have smaller overall size for the same power and voltage ratings.

## CHAPTER 4

Magnetic Circuit Calculations

4.1 MAGNETIC CIRCUIT

Figure 4.1 (a), (b), (c) and (d) shows typical simplified magnetic circuits for a transformer, d.c., induction, and synchronous machines under steady-state conditions. Whereas, the magnetic circuit of a transformer is quite simple, complexity

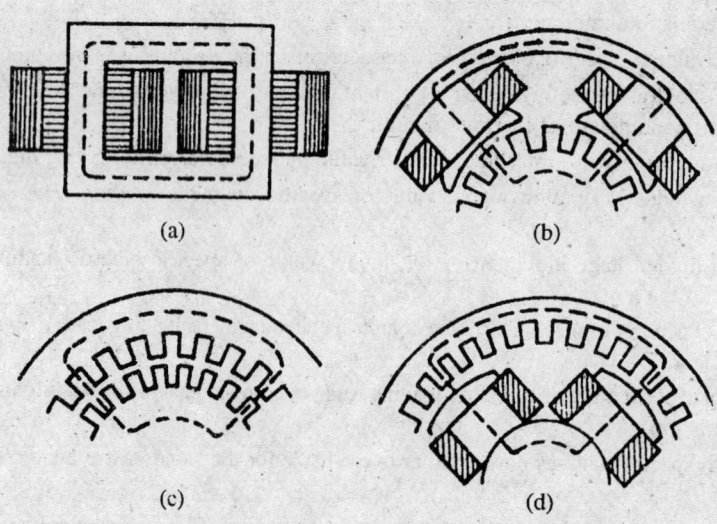

(a)

(b)

(c)

(d)

Figure 4.1 Magnetic circuit of transformer and machines

arises in those of rotating machines due to (i) non-uniform gap length under the poles in d.c. and synchronous machines, (ii) presence of slots and complicated nature of leakage fluxes in both slator and rotor.

In the design of magnetic circuit for the above equipment, the aim is to achieve a reasonably high flux in various parts of the circuit without any increased losses due to forcing of flux in iron into magnetic saturation.

For simplified calculation, the magnetic circuit in transformer and rotating electrical machines may be divided into several parts as shown in Table 4.1.

Tabic 4.1 Magnetic Circuit

Type	Part	Material normally used	Normal maxm. working flux-density (Tesla)
TRANSFORMER	1. Core	HRS/CROS	1.4–1.7*
	2. Yoke	-do-	1.1–1.4
	(* may be increased to 2.0 Tesla for 20 % over-voltage.)		
D.C. MACHINE	1. Yoke	Cast steel	1.5
	2. Pole core	Silicon steel	1.4–1.6
	3. Pole shoe	Silicon steel/ cast steel	1.4–1.6/ 2.0
	4. Gap	Air	0.45–0.75
	5. Armature teeth	Silicon steel	1.8–2.2
	6. Armature core	-do-	0.8–1.2
INDUCTION MOTOR	1. Stator core	Silicon steel	0.8–1.2
	2. Stator teeth	-do-	1.8–2.2
	3. Gap	Air	0.3–0.6
	4. Rotor teeth	Silicon steel	1.8–2.2
	5. Rotor core	-do-	0.8–1.2
SYNCHRONOUS MACHINE	1. Stator core	Silicon Steel†	0.8–1.2
	2. Stator teeth	-do-	1.8–2.2
	3. Gap	Air (in air-cooled machines)	0.5–0.65 (salient-pole) 0.5–0.65 (non-salient pole)
	4. Pole	Silicon steel (for salient pole)	0.8–1.2
		Forged steel (for turbo-alternator)	2.0
	5. Rotor core	-do-	-do-
	† CROS may be used with advantage in large induction motor and water- wheel generators–see chapter 1 art. 1.6.2)		

[*Note* : While the iron parts in the magnetic circuit in an induction motor is completely laminated, the yoke and pole-shoe in d.c. machine and the rotor in turbo-alternator are generally unlaminated. Thus, under varying flux conditions in these parts, eddy-current losses occur. Figure 4.2 gives typical B-H characteristics of important ferro-magnetic materials].

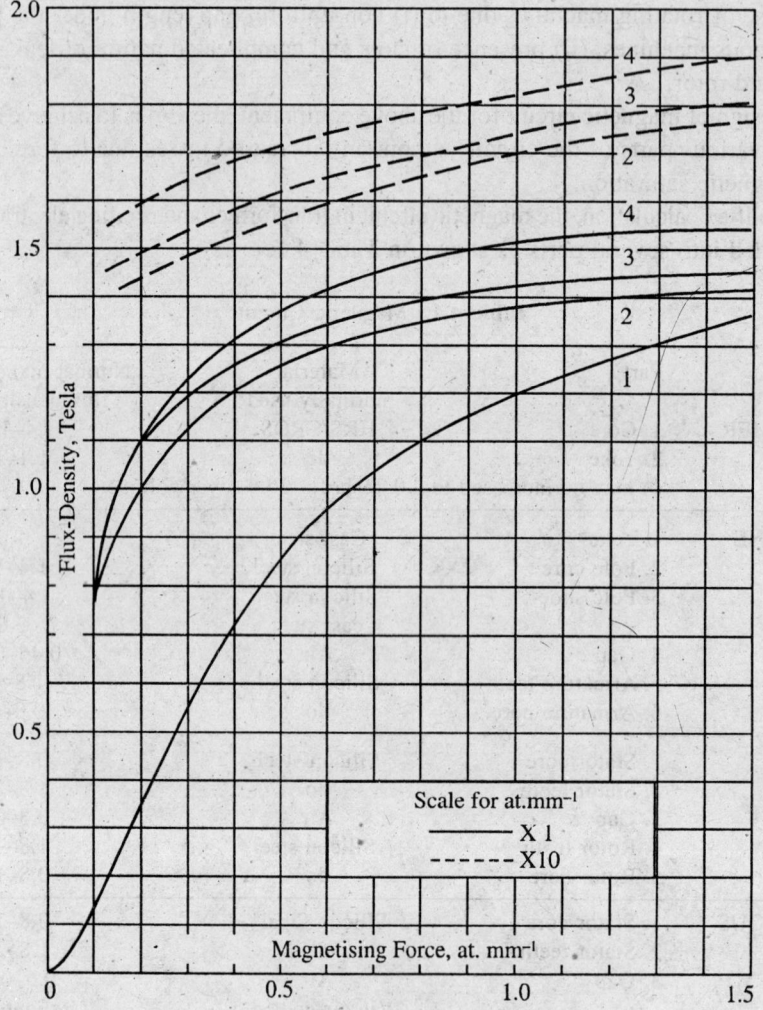

Figure 4.2   B–H characteristics. 1–mild steel, 2-stalloy, 3–42-quality, 4–lohys.

4.2   CALCULATION OF MAGNETISING FORCE

It is evident from Figure 4.1 that the magnetic circuit of a rotating machine can be considered to be a series one, and assuming the flux-density in each part to be constant, the mmf for any part (say 1) is,

$$F_1 = \Phi_1 \, R_1$$

where, Φ and R are magnetic flux and reluctance respectively, and subscript '1' indicates the part number.

or,

$$F_1 = \frac{B_1 A_1}{\mu_0 \, \mu_{r1} \, A_1} l_1 = \frac{B_1}{\mu_0 \, \mu_{r1}} l_1$$

$$= H_1 \cdot l_1$$
$$= at_1 \cdot l_1 \tag{4.1}$$

where, $\mu_0 = 4\pi \times 10^{-7}$,

μ_r = relative permeability of the material of the part,

l = length of flux-path in the part, metre,

H = magnetising force needed to overcome the reluctance of the part

= at, ampere-turn per metre.

The total magnetising force for the magnetic circuit,

$$F = (F_1 + F_2 + F_3 + \ldots\ldots) \text{ amp-turns}$$
$$= \{at_1 l_1 + at_2 \cdot l_2 + \ldots\ldots) \text{ amp-turns} \tag{4.2}$$

Example 4.1

Figure 4.3 shows a magnetic circuit with the following particulars :-

Part	Material	Length of flux-path mm.	Cross-sectional area sq. cm.
(1) Gap	Air	1	25
(2) Pole	Transformer steel plate	150	25
(3) Yoke	Cast steel	600	25

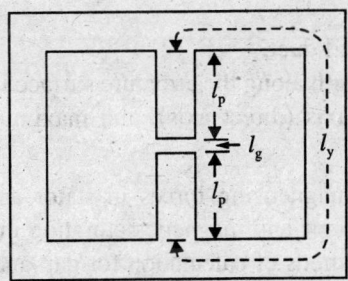

Figure 4.3 Pertaining Example 4.1

Leakage co-efficient for pole and yoke = 1.11.

Neglecting fringing of flux, compute the total magnetising force for a gap flux of 2.25 mWb. B-H characteristics of various part-materials are given in Figure 4.2.

Solution

(i) *For the gap* : density = $\dfrac{2.25 \times 10^{-3}}{25 \times 10^{-4}} = 0.9\ T$

$$at_g = 800,000 \times 0.9 \text{ amp.turn, } m^{-1}$$

Thus, $\qquad F_g = at_g \times 1 \times 10^{-3} = 720 \text{ amp.turn}$

(ii) *For the pole* : density = $\dfrac{2.25 \times 10^{-3}}{25 \times 10^{-4}} \times 1.11 = 1.0\ T$

The corresponding at_p from Figure 4.2 = 447 amp.turn. m^{-1}

[*Note:* In a computer-based design process, at_p is to be obtained. For this, the method as in art 4.2.2 can be used.]

Thus, $\qquad\qquad F_p = at_p \times 2l_p$, since 2 poles in series.

$$= 447 \times 2 \times 15 \times 10^{-2} = 135\ \text{amp.turn}$$

(iii) *For the yoke* : density = $\dfrac{1}{2} \times \dfrac{2.25 \times 10^{-3}}{25 \times 10^{-4}} \times 1.11 = 0.5\ T$

The corresponding at_y from Figure 4.2 = 530 amp.turn. m^{-1}

Thus, $F_y = at_y \times l_y = 530 \times 60 \times 10^{-2}$ since two paths in yoke are in parallel.

$$= 318\ \text{amp.turn}$$

Hence, total magnetising force, $F = F_g + F_p + F_y$

$$= 1173\ \text{amp.turns.}$$

[Note : Example 4.1 shows that for any specified flux per pole, the contribution of gap in the total magnetising force is the largest. In a rotating machine, such contribution of gap under rated condition can be as high as 70–80% of the total mmf. Further, the gap is bounded by iron surfaces on both sides, either slotted or unslotted and the calculation of gap ampere-turns is complicated due to :-

  (i) the presence of slots and duct,
  (ii) the variation of gap-length along the armature surface in a salient-pole machine, being minimum at the pole-axis (direct axis), and maximum along the interpolar or the quadrature axis.

Again, the calculation of magnetising forces in stator and rotor teeth are complicated mainly due to tooth- and slot-shape and magnetic saturation in teeth.]

In the following article, methods of calculation for gap-and tooth-ampere-turns would be discussed. The calculations for the remaining parts of the magnetic circuit are somewhat simpler and are illustrated through Example 4.4 for an induction motor.

4.2.1 Treatment of Magnetic Saturation

The magnetic circuit calculation in an electrical apparatus depends largely on the relations between the mmf and flux in various iron parts of the magnetic circuit, and the gap length. For a particular mmf, the flux depends on the reluctances of the iron parts, and thus magnetic saturation plays an important part in the design analysis of electrical apparatus. The magnetic circuit data essential to the handling of saturation is given by the open-cricuit magnetisation curve or the saturation curve of the apparatus. Unfortunately, ro simple relation exists between the flux and mmf, and for digital computation, the following representations have been useful, the choice depending on the nature of the problem and the operating region on the curve :-

(a) By Froelich equation : $|B| = \dfrac{|H|}{a + b|H|}$ (4.3a)

(b) By transcendental function : $H\ a.\ e^{bB}$ (4.3b)

(c) By power series : $H = a.\ H^n$ (4.3c)

(d) By series polynomial : $H = a_0 + a_1\ B + a_n B_n + a_m\ B^m$ (4.3d)

where, a, b. and n, m are coefficients.

4.2.2 Straight Line Approximation of a Curve

Quite often, results within reasonable accuracy is obtained by replacing the curve by a number of straight lines (11 in Figure 4.4), the larger the number of divisions, the better is the accuracy.

Method: Our aim is to determine X (or H) for any Y (or B).

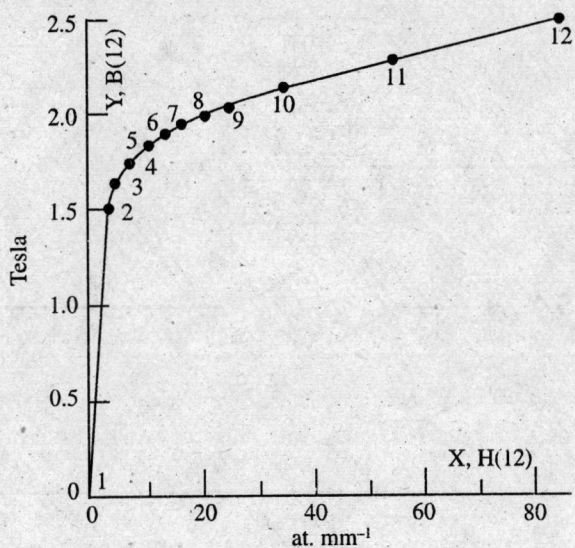

Figure 4.4 Straight line approximation of saturation curve.

For any straight line, the slope S is given by:

$$\text{for line 1 : } S(1) = \frac{H(2) - H(1)}{B(2) - B(1)}$$

where. B(1) and B(2) values of B at nodes 1 and 2, corresponding to values H(1) and H(2) respectively.

Similarly, for

$$\text{line 2 : } S(2) = \frac{H(3) - H(2)}{B(3) - B(2)}$$

$$\cdots \qquad \cdots \qquad \cdots$$

$$\text{for line N : } S(N) = \frac{H(N+1) - H(N)}{B(N+1) - B(N)}$$

For any value of B = Y (in Figure 4.4).

$$H = X = H(2) + [Y - B(2)] \cdot S(2)$$

and in general, $\qquad X = H(N) + [Y - B(N)] \cdot S(N) \qquad\qquad$ (4.3e)

The advantage of using equn. (4.3e) is that it is amenable to translated into a computer program.

These programs are placed in a 'Header File' so that they may be used with any program. The flow chart and the computer program in C++ language (one such case is given below) are applicable for other curve fitting programs with changed data.

File I: Flux-density vs. Magnetising force characteristic of special Lohys.

(B in Tesla, mf in at/m.)

The flow chart and the corresponding computer program (Header File: Mcurlohy.h) are shown in Figures 4.5(a) and (b).

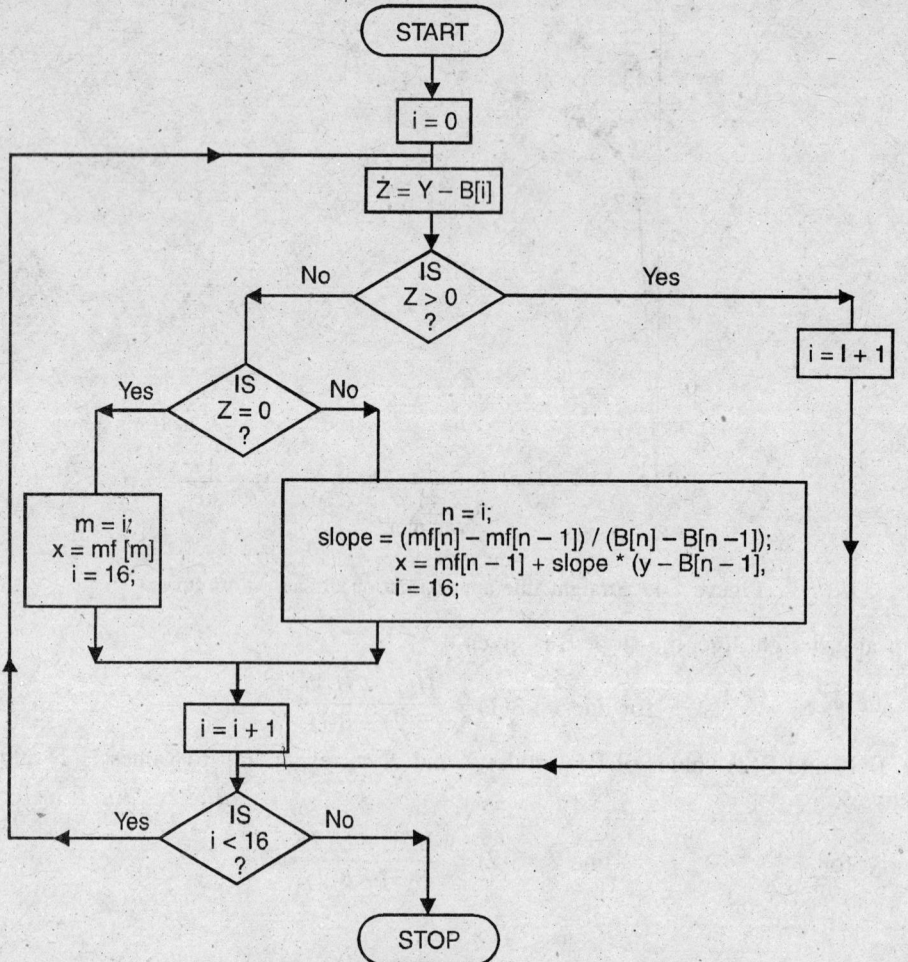

Figure 4.5(a) Flow chart for curve-fitting by straight lines.

Other 'Header Files' used in various design problems with corresponding data for curve-fitting are given below. The students may develop the programs based on the above mentioned flow diagram.

File II: Flux-density vs. Magnetising force characteristic of 42 Quality Stalloy.
    (B in Tesla, mf in at/m.).
    Header file: Mcur42st.h
B [16] = {0.4 0.9 1.0 1.06 1.15 1.20 1.26 1.32 1.35 1.40 1.425 1.45 1.60 1.70 1.80 1.90};
mf[16] = {50 130 160 200 250 315 400 500 600 900 1200 1600 4500 8200 14600 24000};
File III: Flux-density vs. Magnetising force characteristic of Cast Steel.
        (B in Tesla, mf in at/m.).
        Header file : Mcurstl.h

B [9] = {0.5 0.8 1.0 1.1 1.2 1.3 1.4 1.5 1.6};
mf[9] = {200 320 460 550 680 900 1260 2000 3300};

File IV: Flux-density vs. Specific core loss characteristic at 50 Hz, frequency of Machine grade sheet Steel, 0.3 mm. thick.
    (B in Tesla, sl in watt/kg.).
    Header file : sploss3.h

B [18] = {0.4 0.6 0.7 0.8 0.9 1.0 1.1 1.2 1.3 1.4 1.5 1.6 1.7 1.8 1.9 2.0 2.1 2.2};
Sl [18] = {1.0 1.2 1.9 2.85 3.9 4.9 6.0 7.5 8.1 10.0 11.5 13.0 15.0 16.5 17.7 18.7 19.6 20.5};

File V: Flux-density vs. Specific core loss characteristic at 50 Hz. Frequency of Machine grade Sheet Steel, 0.4 mm. thick.

    (B in Tesla, sl in watt/kg.).
    Header file : sploss4.h

B [14] = {0.5 0.6 0.7 0.8 0.9 1.0 1.1 1.2 1.3 1.4 1.5 1.6 1.7 1.8};
Sl [14] = {1.7 2.25 3.35 4.5 5.7 7.0 8.0 9.9 11.3 13.2 15.0 16.6 18.7 21.0};

File VI: Flux-density vs. Specific core loss characteristic at 50 Hz. Frequency of Machine grade sheet Steel, 0.5 mm. thick.

    (B in Tesla, sl in watt/kg.).
    Header file : Sploss5.h

B [14] = {0.5 0.6 0.7 0.8 0.9 1.0 1.1 1.2 1.3 1.4 1.5 1.6 1.7 1.8};
Sl [14] = {4.95 5.65 6.5 7.5 8.7 10.5 12.0 13.5 15.5 17.5 20.5 22.5 25.6 29.0};

File VII: Flux-density vs. Specific core loss characteristic at 50 Hz. Frequency of Transformer grade sheet Steel, cold-rolled, 0.35 mm. thick.

    (B in Tesla, sl in watt/kg.).
    Header file : Loscur_c.h

B [10] = {0.8 0.9 1.0 1.1 1.2 1.3 1.4 1.5 1.6 1.7};

```
float mcurlohy (float y)
{
        int i, n, m ;
        float x, z, slope ;

    // the curve is represented by  15  straight-line segments

5.  float B[15] ={1.20 , 1.25 , 1.31 , 1.37 , 1.40 , 1.43 , 1.45 , 1.48, 1.50, 1.52,
                    1.55,  1.60,  1.64,  1.70,  1.73} ;
6.  float mf[15] ={300 , 350 ,  400,  500 , 570 ,  650 ,  750 ,  900, 1000, 1300,
                   1600, 2400, 3400, 5000, 6500} ;

        i = 0 ;
        do
        {
7.        z = y – B[i] ;
8.        if ( z > 0 )
          {
            i = i + 1 ;
          continue ;
          }
           else
            if ( z = = 0 )
              {
              m = i ;
              x = mf [m] ;
              i = 16 ;
              }
            else
               if ( z < 0 )
                {
                n = i ;
                slope = ( mf[n] – mf[n – 1] ) / ( B[n] – B[n – 1] ) ;
                   x = mf[n – 1] +  slope * ( y – B[n – 1] ) ;
                   i = 16 ;
                }
           i = i + 1 ;
        }
          while ( i < 16 ) ;
        return x ;
}
```

Figure 4.5(b)   Computer program (Header File) for using Flux-density vs. Magnetising. Force characteristic of Special Lohys.

Sl [10] = {0.25 0.38 0.50 0.70 0.82 1.15 1.37 1.60 1.90 2.5};

File VIII: Flux-density vs. Specific core loss characteristic at 50 Hz. Frequency of Transformer grade sheet Steel, hot - rolled, 0.35 mm. thick.

(B in Tesla, sl in watt/kg.).
Header file : loscur_h.h

B [10] = {0.8 0.9 1.0 1.1 1.2 1.3 1.4 1.5 1.6 1.7};
Sl [10] = {0.82 1.05 1.25 1.5 1.92 2.35 2.87 3.55 4.3 5.8};

File IX: Flux-density vs. Magnetising volt-ampere characteristic of hot rolled sheet Steel*.

(B in Tesla, Mgva in VA per kg.)
Header file : Magva_h.H

B [10] = {0.75 1.1 1.15 1.2 1.225 1.25 1.3 1.35 1.4 1.42};
Mgva [10] = {1.3 6.0 6.8 7.5 8.5 10.0 19.0 30.0 50.0 60.0};

File X: Flux-density vs. Magnetising volt-ampere characteristic of cold-rolled sheet Steel*.
(B in Tesla, Mgva in VA per kg.)
Header file : Magva_c.h

B [13] = {0.75 1.0 1.25 1.5 1.6 1.625 1.65 1.675 1.7 1.725 1.75 1.775 1.8};
Mgva [13] = {1.2 2.0 2.5 4.2 5.0 7.5 12.0 16.0 22.0 26.0 32.0 45.0 70.0};

File XI: Carter's co-efficient vs. slot opening / gap length for semi-closed slots (Figure 4.7)

ratio = slot opening / gap length.
(Header File : carter_s.h)

ratio [15] = {0. 0.5 1.0 1.5 2.0 2.5 3.0 3.5 4.0 5.0 6.0 7.5 9.3 10.0 12.0};
kc [15] = {0 .1 .18 .26 .34 .395 .445 .495 .53 .595 .655 .725 .80 .845 .883};

File XII: Carter's co-efficient vs. slot opening / gap length for Open slots (Figure 4.7)
ratio = slot opening / gap length.
Header File : carter_o.h

ratio [15] = {0. 0.5 1.0 1.5 2.0 2.5 3.0 3.5 4.0 5.0 6.0 7.4 8.5 10.0 12.0};
kc [15] = {0.075 .147 .22 .235 .325 .37 .415 .445 .505 .545 .60 .63 .66 .70};

4.2.3 Magnetising Force for the Gap

The following cases are considered :-

Case I   Gap bounded by unslotted (smooth) iron surfaces on both sides and the gap-length is uniform.

The gap-density B_g at any point under a pole is thus constant under steady-state operating condition.

That is,
$$B_g = \frac{\Phi.p}{\pi D L_a}$$
(4.4)

and
$$at_g = \frac{1}{\mu_0} B_g \text{, since } \mu_r = 1 \text{ for the gap.}$$

$$= 800,000\ B_g \text{ amp.turn approx.}$$
(4.5)

Thus, gap ampere-turns, $F_g = at_g l_g$

* Shanmugasundaram, A, *et at*, Electrical Machine design Data Book, page 24

$$= 800,000 \, \mathrm{B}_g \cdot l_g \text{ amp.turn} \tag{4.6}$$

[Note : $\mathrm{B}_g = \mathrm{B}_{av}$, the average flux-density, also known as the specific magnetic loading.]

Case II Gap bounded by smooth surface on one side and slot on the other.

The influence of slot on an otherwise uniform field is shown in Figure 4.6(a)

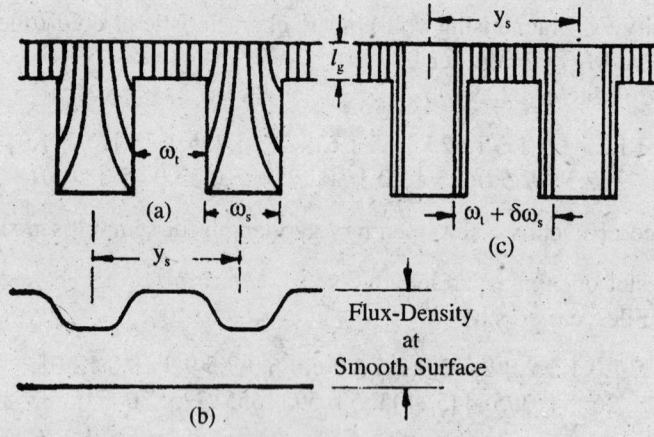

Figure 4.6 (a) Typical flux-distribution due to open slot, (b) approximate distribution, (c) typical flux-density distribution.

for an open-slot, indicating flux concentration at the tooth surface and thinning of the flux as the centre of the slot is approached. Such curves could be obtained by mapping of equipotential lines using numerical methods, such as the finite-difference method, the finite-element method.

The effect is equivalent to far lower density at the centre of the slot and a density larger thar B_{av} (as given by equn. 4.4) at the tooth surface (Figure 4.6b). That is, the effective flux-density $\mathrm{B'}_{av}$ at the tooth surface is larger than B_{av} of case I, which can be measured in terms of an artificial increase in tooth width from w_t to $(w_t + \delta w_s)$ as shown by the aproximiation of the flux-plot in Figure 4.6(c).

From equn. (4.4),
$$B_{av} = \frac{\Phi.p}{(\pi \, D / S)} \cdot \frac{1}{L_a \cdot S}$$

$$= \frac{(\Phi.p / L_a S)}{y_s} \tag{4.7}$$

where, S and y_s. are respectively the number of slot and the slot-pilch (in metre); $y_s = (\pi D / S)$; B_{av} is the specific magnetic loading of the machine i.e. the uniform gap-density neglecting the slot-effect.

With slots present, we may define,
$$y_s = w_t + \delta w_s$$
$$= y_s - kc.w_0 \tag{4.8}$$

where, k_c is the *Carter's coefficient*, a function of the ratio of slot-opening w_o at the gap surface and the gap-length l_g (Figure 4.7).

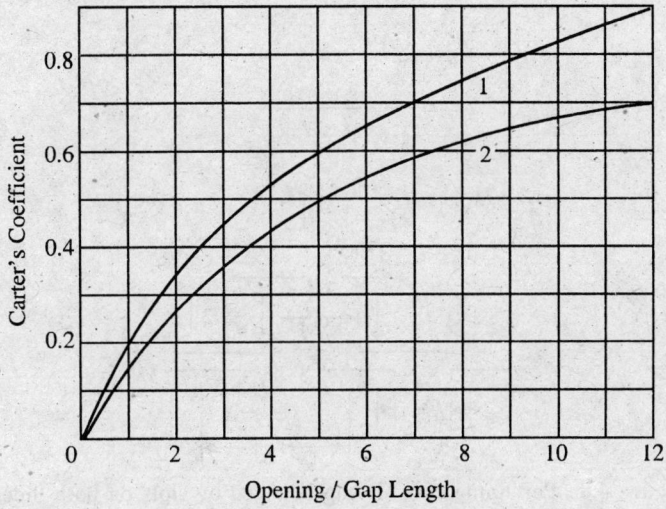

Figure 4.7 Determination of Carter's coefficient. 1–semi-closed slot, 2–open slot.

Defining B_{nv} as the average density of the gap with slotting either on stator or on the rotor,

$$B_{av} = \frac{(\Phi.p / L_a.S)}{y_s - k_c.w_0} = \frac{y_s}{y_s - k_c w_0} B$$

$$= k_g. B \qquad\qquad (4.9)$$

where, $k_g = y_s/(y_s - k_c.w_0)$ is a factor by which the average density is increased in the presence of slots

[Note : (1) It is evident from Figure 4.7 that,

(i) for a large gap length, the effect of slotting is less ;

(ii) k_c for semi-closed slot is larger than that for open-slot since the flux-distribution at the gap surface with the former is more uniform than that with the latter, and thus δ for semi-closed slot is less than that for open slot.

(2) The effect of slotting can be considered as if,

(i) the slol-pilch y_s is 'contracted', or,

(ii) the gap length l_g is enlarged to l'_g, where,

$$l'_g = k_g . l_g$$

$$= \frac{y_s}{y_s - k_c w_0} l_g \qquad\qquad (4.10)$$

with the gap density remaining equal to B_{av}.]

Thus the effect of slotting can be considered in either of the following two ways :-

(1) $F_g = 800,000\ B'_{av}.l_g$ (4.1 la)

(2) $F_g = 800,000\ B_{av} . l'_g$ (4.11b)

Case III   Gap bounded by slotted surfaces on both sides of the gap (Figure 4.8).
Following case II, the gap ampere-turns,

$$F_g = 800,000 \, B_{av}. \, kg. \, lg$$

where, $$k_g = k_{g1} \cdot k_{g2}$$ (4.12a)

and $$k_{g1} = \frac{y_{s1}}{y_{s1} - k_{c1} w_{01}}$$ (4.12b)

$$k_{g2} = \frac{y_{s2}}{y_{s2} - k_{c2} w_{02}}$$

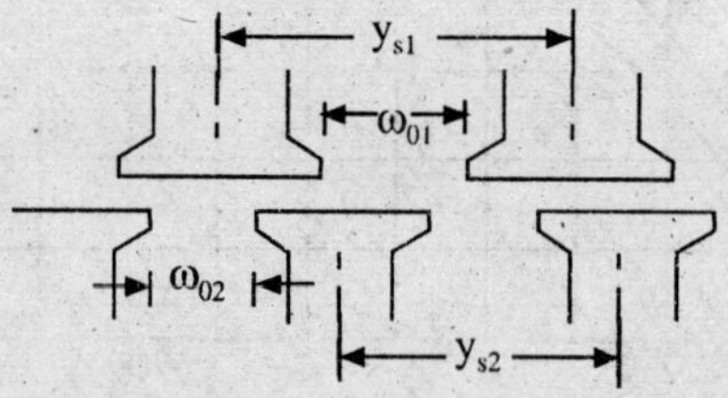

Figure 4.8   Pertaining case III-gap bounded by slots on both Sides.

k_{c1} and k_{c2} are Carter's coefficient for stator and rotor slots respectively and are functions of $(w_{01}/ \, l_g)$ and $(w_{02} / \, l_g)$ as given by Figure 4.7.

Case IV   Effect of radial ventilating ducts.

As discussed in chapter 2, art. 2.4.2.3, radial ventilating ducts are often used in stator and/ or rotor cores to facilitate flow of coolant.

The effect of radial ducts can be handled in the same way as the open slols (cases II & III).

(a) With number of ducts n_d, each of width w_d on either slator or rotor (Figure 4.9a), the core length La can be considered to be reduced to $(L_a - k_d \cdot n_d \cdot w_d)$.

With reference to equns. (4.4) and (4.9),

$$B'_{av} = \frac{\Phi \cdot p}{\pi D \, (L_a - k_d \cdot n_d \cdot w_d)}$$

$$= \frac{\Phi \cdot p}{\pi \, DL_a} \cdot \frac{L_a}{(L_a - k_d \cdot n_d \cdot w_d)}$$

$$= k_{gd} \cdot B_{av}$$ (4.13a)

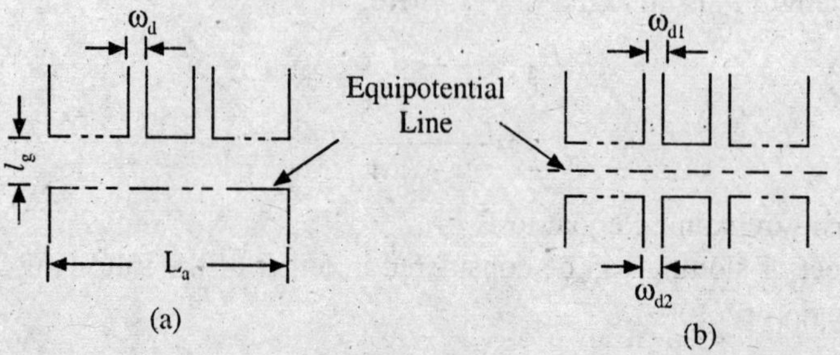

(a)   (b)

Figure 4.9   Pertaining case IV-(a) ducts on one member, (b) ducts on both members.

That is, there is an increase in the effective flux-density by a factor k_{gd}, where,

$$k_{gd} = \frac{L_a}{L_a - k_d^\circ \cdot n_d \cdot w_d} \tag{4.13b}$$

Considering the smooth surface as the equipotential line, k_d is a function of (w_d / l_g) and is given by Figure 4.7 for open slot.

(b) With ducts on both stator and rotor (Figure 4.9b) facing each other, the central plane of the gap can be considered as the equipotential surface, and

$$k_{gd} \text{ (in equn. 4.13a)} = k_{gd1} \cdot k_{gd2} \tag{4.14a}$$

with

$$k_{gd1} = \frac{L_a}{L_a - k_{d1} \cdot n_{d1} \cdot w_{d1}}$$

$$k_{gd2} = \frac{L_a}{L_a - k_{d2}.n_{d2}.w_{d2}} \tag{4.14}$$

k_{d1} and k_{d2} being Carter's coefficients for stator and rotor ducts and are respectively functions

of $\left(w_{d1} / \frac{1}{2} l_g \right)$ and $\left(w_{d2} / \frac{1}{2} l_g \right)$; k_{d1} and k_{d2} could be obtained using Figure 4.7.

CASE V   Effect of saliency.

In salient-pole machines, i.e. direct current and salient-pole synchronous machines, the airgap is minimum along the direct axis increasing as the pole tip is approached. The gap-density distribution is as shown by 1 in Figure 4.10. For the design of electrical machines, it is often convenient to use hypothetical total flux per pole Φ_t instead of the flux per pole Φ ($= B_{av}$ $\pi DL_a/p$) from equn 4.4 using the average gap-density or specific magnetic loading B_p). The hypothetical total flux per pole is obtained by assuming uniform gap-density distribution with its magnitude equal to the peak flux-density B_p. That is,

$$\Phi t = B_p. \pi DL_a/p \tag{4.15}$$

In other words, the gap density distribution is assumed to be rectangular instead of trapezoidal. The ratio of the area under the field form to the area of the hypothetical rectangle (2 in Figure 4.10) is termed as 'Field Form Factor', f_d, so that,

$$\Phi = f_d.\Phi_t \tag{4.16a}$$

and

$$B_{av} = f_d. B_p \tag{4.16b}$$

Thus, the gap ampere-turns per pole for a machine with salient pole on one member and smooth surface on the other is.

$$F_g = 800,000 \ B_p \cdot l_g$$
$$= 800,000 \ (B_{av} / f_d) \ l_g \tag{4.17}$$

l_g being the minimum gap-length along the direct-axis.

Note : (1) For a non-salient pole machine with slots and ducts on both stator and rotor, the gap ampere-turns per pole,

$$F_g = 800,000 \ B_{av}.k_g.k_{gd}.l_g \tag{4.18}$$

k_g and k_{gd} being given by equns. (4.12) and (4.14) respectively.

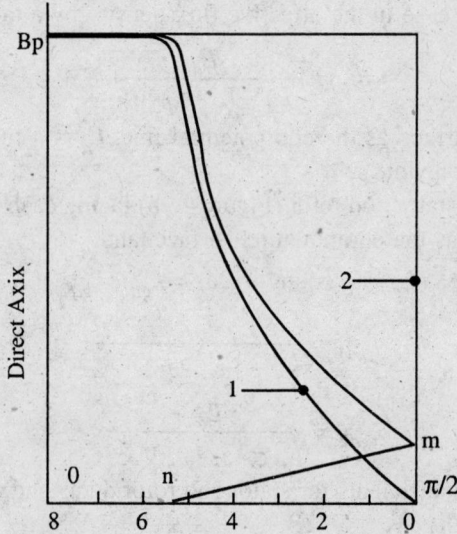

Figure 4.10 Flux-density distribution, salient-pole machine. 1–trapezoidal, 2–rectangular.

(2) For a salient-pole machine with slots and ducts on both stator and rotor.

$$F_g = 800,000 \, (B_{av} / f_d) \cdot k_g \cdot k_{gd} \cdot l_g \tag{4.19}$$

Case VI Induction motor.

A special case arises in induction motors in which, because of very small gap length, tooth saturation flattens the gap-density waveform (Figure 4.11) Such a field-form can be considered equivalent to a fundamental wave with a superposed 3rd harmonic wave. Since, effect of harmonics on the r.m.s. induced voltage due to the gap flux is very small, calculation of ampere-turns for gap and teeth could be based on the density B_{30} at a point 30° electrical from the direct-axis, where the 3rd harmonic is zero. Thus,

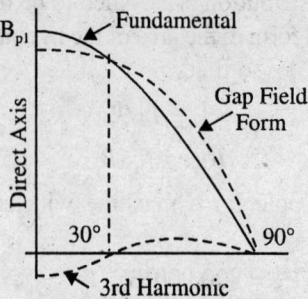

Figure 4.11 Gap density distribution-induction motor.

$$B_{30} = B_{p1} \cos 30°$$

$$= \frac{1}{2} \pi \frac{\Phi P}{\pi D \, L_a'} \times 0.866$$

$$= 1.36 \frac{\Phi}{YL'_a} \qquad (4.20a)$$

Where, Y is the pole-pitch (= $\pi D/p$), metre;

L$'_a$ is the modified armature length due to ducts.

The gap ampere-turns per pole,

$$F_g = 800,000 \ B_{30} \ l'_g \qquad (4.20b)$$

[This stipulation is applicable in computing stator- and rotor-teeth ampere-turns.]

4.2.4 Gap-density Distribution

(A) *Graphical method* The method in which the magnetic field is plotted, is based on the following three important assumptions :-

(1) The iron parts of the magnetic circuit has infinite permeability so that
   (i) the iron boundary surfaces can be considered as equipotentials;
   (ii) the flux lines leave the field surface at right angles to it and enters the armature surface at right angles to it also.

   Thus, the magnetic field in the gap can be considered to be set up by a magnetic potential difference between the two boundary iron surfaces.

(2) Armature core length is large so that the magnetic field of the gap between a pole and the armature surface is two-dimensional in which the space distribution of potential occurs only in one plane and all fluxlines lie in that plane. Such a field in the gap can be expressed by the Laplace's equation

$$\frac{\delta^2 B}{\delta x^2} + \frac{\delta^2 B}{\delta y^2} = 0$$

   and can be expressed completely if the lines of flux and equipotential surfaces are mapped.

(3) The boundary iron surfaces are smooth i.e. the effects due to slots and ducts are absent.

In plotting magnetic fields consideration is given to the fact that the flux in the gap distributes itself in such a way that the reluctance is minimum. The flux-path under the pole is assumed to be divided into tubes of force (Figure 4.12), each tube being of unit width in the direction parallel to the shaft. Considering any tube abed, we have, the permeance of the tube proportional to (b_x / l_x), so that the flux-density B_x at the armature surface is proportional to ($b_x/a_x.l_x$) since the area at the surface = a_x. 1 = a_x for tube of unit axial width. If B_p is the peak density along the direct-axis, we have,

$$B_x : B_p = \frac{b_x}{a_x.l_x} : \frac{1}{l_g}$$

or

$$B_x = \frac{b_x}{a_x} . \frac{l_g}{l_x} . B_p \qquad (4.21)$$

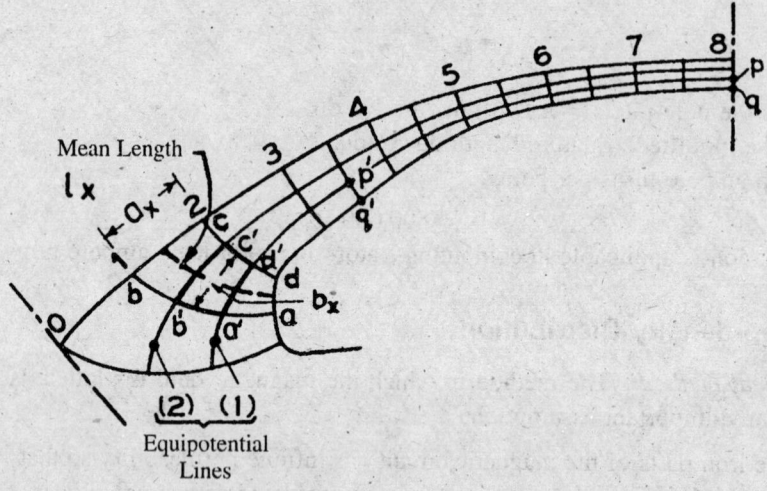

Figure 4.12   Pertaining graphical method of field plotting.

The mapping will be greatly simplified by making $b_x = l_x$. which can be done by drawing equipotential lines (1 and 2 in figure) intersecting the flux lines at right angles at all the four corners a', b', c', and d'. The mesh area a' b' c' d' is thus a curvilinear square.

For simplicity, using a large number of flux and equipotential lines, b_x can be assumed equal to a_x, so that,

$$B_x = \frac{l_g}{l_x} B_p \qquad\qquad (4.22)$$

[*Note :* The method of curvilinear square offers the advantage of easier recognition and checking. The mapping of the field can be easily rectified if each curvilinear square is viewed from all the four sides.]

Example 4.2

Plot the field form of a salient-pole synchronous machine, given layout of pole and armature.

Solution

(1) Lay the outline of the half pole-pitch of the armature surface and half pole-arc of a pole.

(This a sufficient since there is symmetry about the direct-axis.)

(2) Divide the armature surface into several equal divisions (8 in Figure 4.12)

(3) Sketch the tubes of force and the equipotential lines taking care that the equipotential lines interesect the flux lines at right angles.

(It is advantageous to start from the nodes selected on the armature surface and move from the direct-axis towards the quadrature axis.)

(4) Assuming B_p at the direct-axis equal to 100 units, calculate B_x at an other points P' as.

$$B_x = 100 \frac{l_g}{l_x} - = 100 \frac{pq}{p'q'} = 50 \text{ units.}$$

(5) The plot of Bx at the armature surface (Figure 4.10) against the divisions chosen will show a positive value (om) at the quadrature-axis. Since the flux-density at the point 0 on the armature surface should be zero, a correction is necessary. For this draw a straight line mn from point m to a point n where the pole bevel begins. The intercepts between the plot of B_x and the straight line mn gives the actual flux-density.

(B) *The Finite-difference Method.*

By the finite-difference method, the entire two-dimensional space between the armature surface and the pole-face is replaced by a mesh of finite length so that the potentials at the nodes only are determined by solving the Laplace's equation.

$$\frac{\delta^2 U}{\delta x^2} + \frac{\delta^2 U}{\delta y^2} = 0 \tag{4.23}$$

It can be shown (Appendix A.4) that, the potential $U_{i, j}$ at any node O (Figure 4.13) can be written in terms of those at the adjoining nodes A, B, C and D by finite-difference approximation as,

$$U_{i+1, j} + U_{i-1, j} + U_{i, j+1} + U_{i, j-1} - 4U_{i,j} = 0 \tag{4.24}$$

when OA = OB = OC = OD.

For given boundary conditions, a set of linear simultaneous equations of the same number as that of the nodes are obtained, the solution of which yields the nodal potentials,

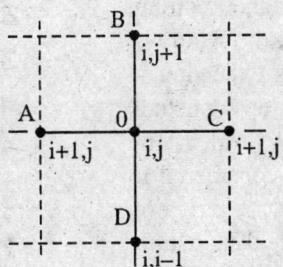

Figure 4.13   Pertaining Finite-differencs method.

For physical problems it is often found that iterative technique is very suitable for determining the nodal potentials, and Liebmann's process, which is well-suited for computer programming, can be used with advantage.

As shown in Appendix A.12 for accelerated convergence of the solution process, equation (4.24) is modified as,

$$U_{i,j}^{k+1} = \alpha \left[U_{i+1,j}^k + U_{i-1,j}^k + U_{i,j+1}^k + U_{i,j-1}^k - \left(4 - \frac{1}{\alpha} \right) U_{i,j}^k \right] \tag{4.25}$$

in which the superscripts show the order of iteration, i.e. the $(k+1)$th iterate utilises the k-th. set of values around the node (i, j). The value of α generally lies between 0.25 and. 0.50.

With irregular boundary, that is when the boundary does not pass through the nodes of the mesh, AO in Figure 4.13 may not be equal to OC, OB not equal to OD. Under such condition, the equation (4.24) is modified as (ref. Fig A.10– Appendix A.4),

$$U_{i,j}^{k+1} = \alpha \left\{ 2\left[\frac{U_{i+1,j}^k}{h_1(h_1+h_2)} + \frac{U_{i-1,j}^k}{h_2(h_1+h_2)} + \frac{U_{i,j+1}^k}{l_1(l_1+l_2)} + \frac{U_{i,j-1}^k}{l_2(l_1+l_2)} \right] \right.$$

$$\left. -\left[2\frac{h_1 h_2 + l_1 l_2}{h_1\,h_2\,l_1\,l_2} - \frac{1}{\alpha} \right] U_{i,j}^k \right\} \tag{4.26}$$

when $OA = h_1$ $OB = l_1$ $OC = h_2$, and $OD = l_2$.

Example 4.3 demonstrates an application of the finite-difference method to a simplified pole-configuration.

Example 4.3

Obtain the equipotential lines in a salient-pole machine by the finite-difference method for the pole configuration as shown in Figure 4.14. In the figure, only the region between the direct- and quadrature-axis is shown, which is sufficient because of symmetry. Further, the curvature and armature slotting are neglected, and iron part of the magnetic circuit is assumed to have infinite permeability. The field winding is distributed along the pole-height as shown.

Solution   Boundary conditions are determined as below :-

(1) Since iron part of the magnetic circuit has infinite permeability, the whole of the magnetic potential is expended in the region between the pole and armature. Thus, the magnetic potential at the armature surface can be considered to be equal to zero.

(2) Because of the nature of field excitation, the entire pole-face will be at constant potential of 1000 units and the inner surface of the yoke at zero potential.

(3) The drop of the magnetic potential along the pole (or direct)-axis is uniform from 1000 units at the pole to zero at the armature.

(4) Symmetry exists about the interpolar (or quadrature) axis. For the development of a computer program,

    (i) the entire region under consideration is replaced by a grid (15 × 35) of equal (for simplicity) mesh length;

    (ii) boundary values of the magnetic potential U (I,J) at any node (i,j) are then fed;

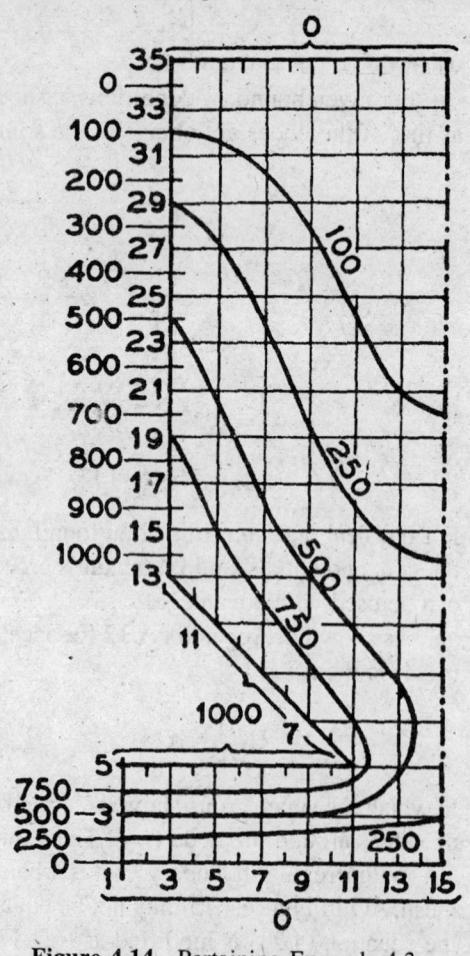

Figure 4.14   Pertaining Example 4.3

(iii) next, the assumed values of tolerance (TOL) and acceleration factor α (A) are fed, and

(iv) during each iteration tolerance is compared with the maximum deviation (DMAX) in any U-value between successive iteration. Thus, TOL limits the number of iteration;

(v) The magnetic potential at all nodes other than those specified by the boundary conditions are next set to zero value.

Figure 4.14 shows a possible solution of the example.

4.2.5 Magnetising Force for Teeth

In normal electrical machines of medium and large sizes, the slots are generally parallel-sided so that the teeth are tapered. Under rated condition, the tooth-density is large at a section where the tooth-width is small and high magnetic saturation may result. This necessitates careful estimation of mmf due to teeth.

4.2.5.1 *True and apparent tooth-densities*

If it is assumed that the whole of the flux per slot-pitch $y_s (= \Phi p / y_s)$ passes through the iron portion of a tooth and none through the adjoining slots, as may happen under unsaturated condition, the flux-density at any section of a tooth,

$$B'_t = \frac{\text{flux per slot pitch, } \Phi_y}{\text{tooth width at the section, } w_t \times \text{net length of armature core, } L_i} \qquad (4.27)$$

With saturation at sections of narrow width, under rated condition, some amount of flux is thrown into the slots (Figure 4.6a) and actual or 'True' density, B_t, is less than B'_t given by equation (4.27). B'_t is generally termed as the 'Apparent tooth-density and a relation between B'_t and B_t can be deduced as below :-

Let under saturated condition, Φ_t and Φ_s denote respectively the flux per slot-pitch in tooth only, and in slot only, so that,

$$\Phi_y = \Phi_t + \Phi_s \qquad (4.28)$$

Dividing by w_t, L_i, $B'_t = \dfrac{\Phi_t}{w_t \, L_i} + \dfrac{\Phi_s}{w_t \, L_i}$

$$= B_t + \frac{\Phi_s}{A_s} \cdot \frac{A_s}{w_t \, L_i} \qquad (4.29)$$

where, A_s is the total area facing the flux Φ_s in slots

$$= L_a \, y_s - w_t \, L_i$$

However, since flux in slots is in the air. $\mu_r = 1$

$$\frac{\Phi_s}{A_s} = 4\pi \times 10^{-7} \, \text{H} \qquad (4.30)$$

where, H is the tooth mmf (amp.turn. metre^{-1}) for the true tooth-density B_t. noting that the magnetic circuit in tooth and in adjoining slots are in parallel. That is, at any section of a tooth,

$$B'_t = B_t + 4\pi \times 10^{-7} H (K - 1) \tag{4.31a}$$

where, $$K = \frac{L_a \cdot y_s}{w_t \, L_i} = \frac{\text{gross area at the section}}{\text{Net iron area at that section}} \tag{4.31b}$$

[*Note* : K may have values between 1.5 for a machine without ducts and 4 for the bottom of a turbo-alternator tooth.]

4.2.5.2   *Determination of B_t and H at any tooth-section.*

Equation (4.31a) involves two unknown quantities, B_t and H, and can be solved for a given B_t-H characteristic of tooth material. Neglecting hysteresis effect and expressing the B_t-H characteristic by Froelich equation (equn. 4.1a).

$$B_t = \frac{H}{a + b\,H} \tag{4.32}$$

where, *a* and *b* are constants*, we have from equation (4.31a),

$$B'_t = \frac{H}{a + bH} + cH$$

where, $c = 4\pi \times 10^{-7} (K - 1)$

or, $$bcH^2 + (1 + ac - b\,B'_t) H - aB'_t = 0 \tag{4.33}$$

For any tooth-section, knowing B'_t from equation (4.27), H and thus B_t can be determined from equation (4.33).

4.2.5.3   *Estimation of tooth–mmf*

(A) *Graphical method*

   (i) For a given flux per slot-pitch, the apparent flux-density at a number of sections from the tip to the bottom of a tooth is calculated using equation (4.27).

   (ii) The correction for true density is to be applied for values of $B'_t \geq 2$ Tesla. As such, for $B'_t < 2$ Tesla, H is determined directly from the given B-H characteristic of the tooth material.

   (iii) For $B_t' \geq 2$ T, the gross area at the concerned section and thus K is calculated, and H is computed from equation (4.33).

   (iv) From a plot of H against the distance from the tooth-tip, the required tooth-mmf is determined by evaluating the area under the curve.

(B) *Three-ordinate method*
    This method is applicable to trapezoidal teeth of moderate taper.

---

* Constants a and b can be determined as below :-
    (i) From equation (4.32),

$$Y = \frac{H}{B_t} = a + bH \,,\ \text{which is the equation of a straight line.}$$

(ii) Plot Y against H and draw a straight line through higher values of H (Figure 4.16) and find *a* and *b*.

If H_1, H_2, and H_3 be the ampere-turn per metre for three equi-distant sections of the taper.

$$\text{mean } H = \frac{1}{6}(H_1 + 4H_2 + H_3) \tag{4.34}$$

by Simpson's rule.

The required tooth-mmf = (mean H × tooth height, m) amp-turn. $\tag{4.35}$

(C) *One-third density method*

This method is applicable to the cases where the taper is slight and the tooth-density at any section is low.

The method is based on the assumption that the average tooth-density over the whole of the tooth-height is equal to the apparent density $B'_{t/3}$ at a section 1/3rd the tooth-height from the bottom of a tooth.

The corresponding $H_{1/3}$ ampere-turn per metre is obtained from the given B-H characteristic of the tooth material and,

the required tooth-mmf = ($H_{1/3}$ × tooth height, m) amp-turn. $\tag{4.36}$

Example 4.4

The data giving rotor tooth dimensions of *a* machine are :-

Distance from top, mm	0	3	6	6	23	40
Tooth-width, mm	15	12	9	14.6	12	9.6

Other data with respect to the machine are :-

flux per slot-pitch	...	3.245 mWb
gross armature length	...	18 cm.;
stacking factor	...	0.90;
armature diameter	...	41.2 cm.;
no. of armature slots	...	48.

Determine the mmf required for teeth by, (A) graphical method, (B) three-ordinate method, (C) one-third density method.

The B-H characteristic of the tooth material is given by,

B(T)	:	1.2	1.3	1.42	1.5	1.6	1.65	1.75	1.8
$H(\overline{at \cdot m^{-1}})$	:	3.4	3.6	3.8	4	4.6	5.2	8	10×10^3

B (T)	:	1.9	1.98		2.05	2.12		2.16	2.22
$H(\overline{at \cdot m^{-1}})$	:	15	19.5		25	31		35.5	42.5×10^3

Solution   Basing on the discussion in art. 4.2.4.1, 4.2.4.2 and 4.2.4.3, Table 4.2 can be completed. The constants *a* and *b* of the Froelich equation are determined (Figure 4.16) as 4000 and 0.355 respectively.

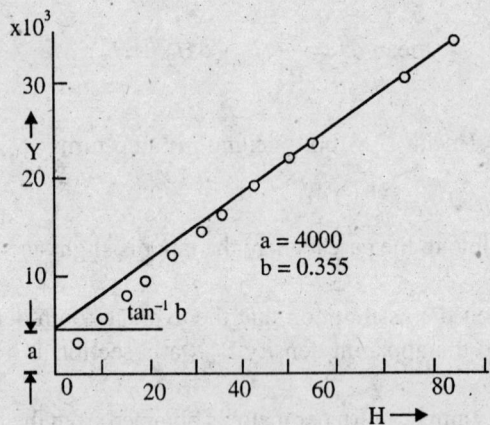

Figure. 4.16 Determination of constants a and b of Froelich equation

In Table 4.2,

Table 4.2

(1) Distance from top mm.	(2) Tooth-width width w_t mm.	(3) Net iron area mm²	(4) Apparent density B_t' (T)	(5) Gross area mm²	(6) K	(7) H at.m⁻¹
0	15	2300	1.41			3800
3	12	1840	1.76			8200
6	9	1458	2.23	4382	3.0	35625
6	14.6	2239	1.44			3900
23	12	1840	1.76			8200
40	9.6	1472	2.20	3530	2.4	35460

(i) column (1) and (2) are given data;

(ii) column (3) = wt. L_t;

(iii) column (4) = flux per slot-pitch/column (3);

(iv) column (5) = gross area = column (3) + ws. La at each tooth-section where $B'_t \geq 2$ Tesla;

(v) column (6) = column (5)/column (3) for sections at which $B'_t \geq 2$ Tesla;

(vi) H from B–H curve for $B'_t < 2$ tesla. from equation (4.33) for $B'_t \geq 2$ Tesla.

The curves for B'_t and H against the distance from tooth-top are plotted in Figure 4.17.

(A) Graphical method : Measuring the area under the curve of H. required tooth amp. turn = 474.

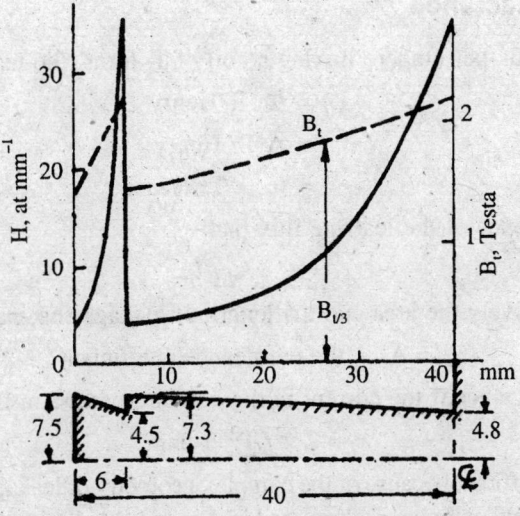

Figure 4.17 Determination of tooth mmf.

(B) Three-ordinate method:

$$\text{Tootli mmf} = \left\{ 0.6 \left[\frac{1}{6}(3800 + 4 \times 8200 + 35625) \right] \right.$$

$$\left. + 34 \left[\frac{1}{6}(3900 + 4 \times 8200 + 35460)] \right] \right\} \times 10^{-2}$$

$$= 481$$

(C) One-third density method :

B'_t at 13.3 mm. from the bottom of the tooth as obtained from Figure 4.17 = 1.84 Tesla. Corresponding H from the B-H characteristic = 11.2 amp-turn. mm^{-1}. Thus, required tooth-mmf = 11.6 × 40 = 464.

4.3 LEAKAGE FLUX

Consideration of leakage flux in electrical machines is often important though, for normal machines, it is about 5 to 10 per cent of the total flux. Leakage flux affects,

   (i) voltage regulation of transformer and rotating machines,

  (ii) forces in and between windings (specially under sudden load and short-circuit conditions),

 (iii) the excitation demands of salient-poles in rotating machines,

  (iv) stray-load losses in transformer and rotating machines

  (v) circulating currents induced due to these fluxes in transformer tank wall.

4.3.1   Leakage Reactance

If Φ_l be the leakage flux per ampere linking a coil of T–turns, the leakage reactance,

$$L_l = \Phi_l. \text{ T Henry} \qquad (4.37\ a)$$

$$= \Lambda T^2 \text{ Henry} \qquad (4.37\ b)$$

$$= \mu_o \lambda T^2 \text{ Henry} \qquad (4.37\ c)$$

where, Λ is the permeance of the leakage flux-path

$$= \mu_o . A_l / l_l,$$

A_l and l_l being respectively the area and the length of leakage flux-path, and

$$\lambda = A_l. l_l \text{ the permeance coefficient} \qquad (4.37d)$$

The corresponding reactance, if the current in the coil alternates at a frequency f Hz. is,

$$x_l = 2 \pi f T^2 \Lambda \text{ ohm} \qquad (4.38)$$

Determination of Λ is difficult because of the complex geometry of leakage flux-paths. Estimation of the flux-pattern by field mapping as discussed in Appendix A-1 with the important assumption of infinitely permeable iron parts of the magnetic circuit is possible, specially using digital computation. However, approximate estimates can be obtained by assuming simplified field geometry and considering the leakage flux as a superposition of straight components.

4.3.1.1   *Transformer*

The following three cases will be considered :-

Case I : Concentric cylindrical coils–primary and secondary of equal length.

Case II : Sandwich coils.

Case III : Concentric cylindrical coils – primary secondary of unequal length. Typical flux-plots for the above three cases are shown in Figure 4.18(a), (b) and (c).

For simplified estimation, the above three flux-plots are replaced by apparent leakage flux assumed to be rectilinear as shown in Figure 4.18 (d), (e), and (f).

The estimation of leakage flux is based on further important assumptions :-

(i) Iron parts of the flux-path is infinitely permeable, i.e. the whole of the winding mmf is expended in the non-magnetic part of the flux-path;

(ii) There is mmf-balance between the primary and the secondary, i.e.

$$I_1 T_1 = I_2 T_2,$$

(iii) One-half of the flux in the duct links with the l.v. and the other half with the h.v. winding;

(iv) Mmf are evenly distributed throughout the winding section.

Case I : Concentric cylindrical coils of equal length.

(a) The length of apparent leakage flux, l_l in Figure 4.18(d) can he expressed by the following empirical equation :-

$$l_l = \frac{H_{wdg}}{K_R} \qquad (4.39a)$$

where,
$$K_R = 1 - \frac{c_o + c_1 + c_2}{\pi H_{wdg}} \qquad \text{... (4.39 b)}$$

is known as Rogowski coefficient; c_o, c_1 and c_2 are defined by C_o, C_1 and C_2 in Figure 4.18(a) respectively.

(b) L. V. winding

Considering an element dx at a distance x from the chosen origin O,

$$F_x = F \cdot \frac{x}{c_1}$$

and the elemental flux in the annular cylinder of thickness dx and length l_l is,

$$d\Phi_x = \mu_o \cdot \frac{F_x}{l_l} \pi (D_1 - c_1 + 2x) dx$$

Total number of turns enclosed by the elemental annulus, $= T_1 \cdot \dfrac{x}{c_1}$ so that the flux-linkage in the l.v winding,

$$\psi'_{l.v} = \int_0^{c_1} T_1 \cdot \frac{x}{c_1} \mu_o \frac{F_x}{l_l} \cdot \frac{x}{c_1} \pi (D_1 - c_1 + 2x) dx$$

$$= \mu_o T_1 \frac{F}{l_l} \cdot \pi \left(D_1 \cdot \frac{c_1}{3} + \frac{c_1^2}{6} \right) \qquad (4.40a)$$

(c) H. V. winding

Following the above procedure, $F_x = F \dfrac{x}{c_2}$

x being measured from the outer surface of the h.v. winding, and,

$$\psi_{h.v.} = \int_0^{c_2} T_2 \cdot \frac{x}{c_2} \cdot \mu_o \frac{F}{l_l} \cdot \frac{x}{c_2} \pi (D_2 + c_2 - 2x) dx$$

$$= \mu_o T_2 \frac{F}{l_l} \cdot \pi \left(D_2 \cdot \frac{c_2}{3} - \frac{c_2^2}{6} \right)$$

(d) Duct.

The mmf F is constant in the duct, and hence, flux in the duct,

$$\Phi_d = \mu_o \frac{F}{l_l} \pi D_m \, c_o T_1 \qquad (4.41)$$

As per assumption, half of this flux will link with l.v. and the other half with h.v. winding. That is, duct flux-linkage with the l.v. finding,

$$\psi_{dlv} = \frac{1}{2}\mu_o \frac{F}{l_l}\pi D_m c_o T_2 \qquad (4.41a)$$

and that with the h.v. winding,

$$\psi_{dhv} = \frac{1}{2}\mu_o \frac{F}{l_l}\pi D_m c_o T_2 \qquad (4.41b)$$

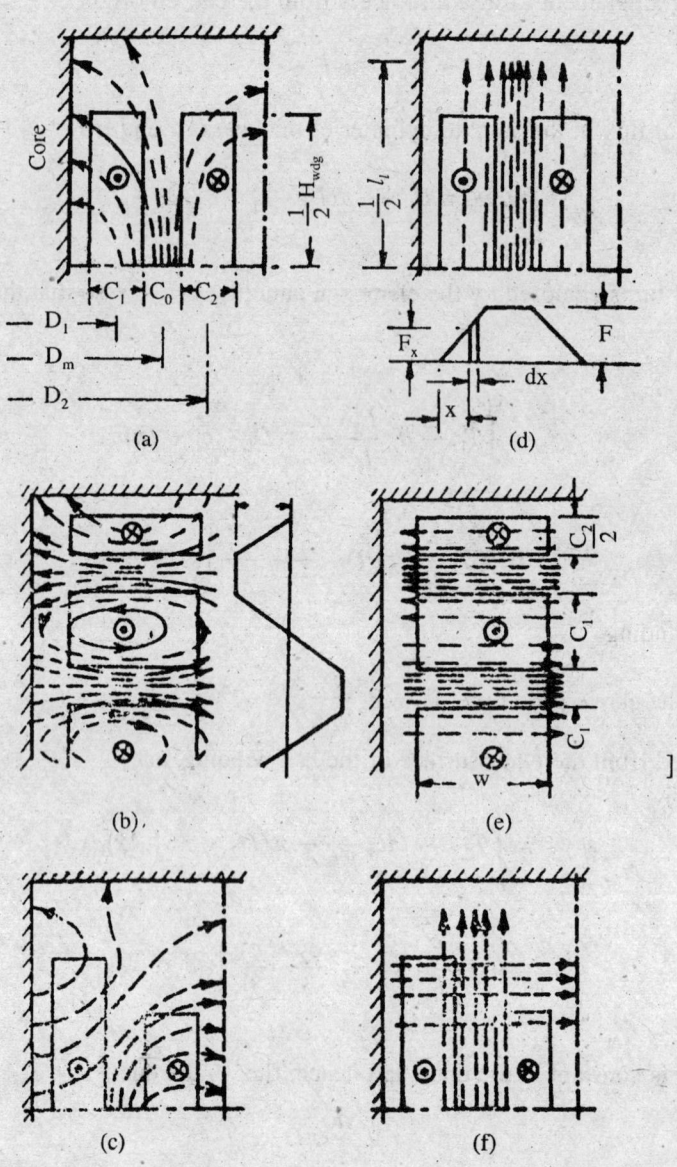

Figure 4.18 Leakage flux in transformers. Core type concentric coils of equal length. (a) actual, (d) assumed, distribution; Shell type–(b) actual, (e) assumed, distribution; Core typs concentric cylindrical coils of unequal length–(c) actual, (f) assumed, distribution.

Hence, from equns. (4.40a) and (4.41a), the total leakage flux linking the l.v. winding,

$$\psi_1 = \psi_{lv} + \psi_{dlv}$$

$$= \mu_o T_1 \frac{F}{l_l} \pi \left(D_m \frac{c_o}{2} + D_1 \frac{c_1}{3} + \frac{c^2_1}{6} \right) \tag{4.42}$$

and that, from equns. (4.40b) and (4.41b), linking the h.v. winding,

$$\psi_2 = \mu_o T_2 \frac{F}{l_l} \pi \left(D_m \frac{c_o}{2} + D_2 \frac{c_2}{3} - \frac{c_2^2}{6} \right) \tag{4.43}$$

But permeance of the leakage flux-path through the l.v. winding,

$$\Lambda_1 = \frac{\text{flux linkage per ampere}}{(\text{no. of l.v. turns})^2} = \frac{\psi_1 / I_1}{T^2_1} \tag{4.44a}$$

and that through the h.v. winding,

$$\Lambda_2 = \frac{\psi_2 / I_2}{T^2_2}. \tag{4.44b}$$

Hence, leakage reactance of the l.v. winding (noting assumption ii)

$$x_1 = 2\pi f T^2_1 \Lambda_1$$

$$= 2\pi f \mu_o T^2_1 \frac{1}{l_l} \pi \left(D_m \frac{c_o}{2} + D_1 \frac{c_1}{3} + \frac{c_1^2}{6} \right) \tag{4.45}$$

and that of the h.v. winding,

$$x_2 = 2\pi f T^2_2 \Lambda_2$$

$$= 2\pi f \mu_o T_2^2 \frac{1}{l_l} \pi \left(D_m \frac{c_o}{2} + D_2 \frac{c_2}{3} - \frac{c_2^2}{6} \right) \tag{4.46}$$

Thus, the total leakage reactance, in l.v. winding terms,

$$X_1 = x_1 + x'_2 = x_1 + x_2 (T_1/T_2)^2$$

$$= 2\pi f \mu_o T_1^2 \frac{1}{l_l} \pi \left[D_m c_o + \frac{1}{3}(D_1 c_1 + D_2 c_2) + \frac{1}{6}(c_1^2 - c_2^2) \right] \text{ ohm.} \tag{4.47}$$

For approximate estimation, noting that for large transformers, $(c_1^2 - c_2^2)$ – term is very small as compared with other terms, and putting,

$$D_1 = D_m - c_o - c_1 \text{ and } D_2 = D_m + c_o + c_2,$$

$$X_1 = 2\pi f \mu_o T_1^2 \frac{1}{l_l} \pi \left[D_m \left\{ c_o + \frac{c_1 + c_2}{3} \right\} - \frac{1}{3} c_o(c_1 - c_2) \right]$$

Again, neglecting $(c_1 - c_2)$-term in comparison with other terms,

$$X_1 = 2\pi f \mu_o T_1^2 \frac{1}{l_1} \pi \, D_m \left(c_o + \frac{c_1 + c_2}{3} \right) \text{ ohm}$$

$$= 2\pi f \mu_o T_1^2 \frac{1}{l_1} s \left(c_o + \frac{c_1 + c_2}{3} \right) \text{ ohm} \tag{4.48a}$$

where, $s = \pi \, D_m$ is the mean circumference of the duct and is approximately equal to the length of mean turn of l.v. or h.v. windings.

[*Note* : (1) Expressed in per unit, the total leakage reactance in l.v.terms, ε_x, is,

$$e_x = \frac{I_1 X_1}{E_{t1}} \text{ p.u.,}$$

$$= 2\pi f \mu_o \frac{F}{E_T} \frac{s}{l_l} \left(c_o + \frac{c_1 + c_2}{3} \right) \text{p.u.} \tag{4.48b}$$

where, $E_T = E_{t1}/T_1$, the voltage per turn.

(2) In terms of per phase kVa-rating of the transformer, knowing $Q = E_{t1}.I_1 \times 10^{-3}$ kVa/phase, and putting

$$c_x \doteq c_o + \frac{1}{3}(c_1 + c_2), \; E_{t1} \cdot I_1 = E_T I_1 T_1 = E_T F,$$

$$e_x = 8\pi^2 \cdot 10^{-7} \cdot f \frac{Q \cdot 10^3}{E_T^2} \frac{s}{l_l} c_x \text{ p.u.}$$

$$\equiv 79 \cdot 10^{-4} f \frac{Q}{E_T^2} \frac{c_x}{(H_{wdg} / s)} \text{p.u.} \tag{4.48c}$$

For $\qquad f = 50$ Hz, $e_x = 0.4 \dfrac{Q}{E_T^2} \dfrac{c_x}{(H_{wdg} / s)}$ p.u.

(3) Normally transformer specification is given in terms of % impedance. From equn. (4.48c),

$$\%X = \frac{I_1 X_1}{E_{t1}} \cdot 100$$

$$\equiv 0.79 f \frac{Q}{E_T^2} \frac{c_x}{(H_{wdg} / s)} \% \tag{4.48d}$$

For $f = 50$ Hz,

$$\% \, X = 39.5 \frac{Q}{E_T^2} \frac{c_x}{(H_{wdg} / s)} \% \tag{4.48e}$$

(4) For better insulation and cooling, quite often, the l.v. and h.v. windings are split up into sections. Figure 4.19 shows a case where the h.v. winding is split into two sections of equal

turns. Obviously, such splitting increases the leakage reactance of the h.v. winding. Following the same procedure as above, the total leakage reactance in terms of I.v. winding is given by equations (4.48 a to d) with,

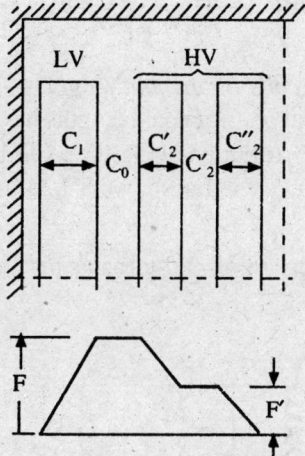

Figure 4.19 Core-type with splitted h.v. coils.

$$c_x = c_o + \frac{1}{3}(c_1 + c_2' + c_2'') + \frac{1}{4}c_o' \qquad (4.49)$$

The increase of leakage reactance is due to the term $(c_o'/4)$ i.e. due to the leakage flux in the duct between the h.v. sections. This is produced by an mmf $= \frac{1}{2} I_2 T_2$ which links $\frac{1}{2} T_2$–turns on one h.v. section.]

Case II : Sandwich coils

With reference to Figure 4.18 (b) and (e), the following are to be noted :-

(a) End l.v. coils have one-half of the turns of interior l.v. coils
(b) Thus, if there are n h.v.coils, there will be $(n-1)$ l.v.coils in addition to 2 half-l.v. coils.
(c) Considering the halves of two adjacent coils and a duct in between forming 'one unit' (Figure 4.18 e), the situation is similar to the case I, and the total leakage reactance for one such unit in l.v. winding terms is,

$$X_n = 2\pi f \mu_o T_n^2 \frac{s}{W}\left[c_o + \frac{1}{6}(c_1 + c_2)\right] \qquad (4.50)$$

where, T_n - no. of turns of each section of l.v. winding.

(d) With n h.v. sections there is a total 2n units and thus, the total leakage reactance, X_1 $= 2 n X_n$.
But, since $T_n = T_1/2n$. where, T_1 is the total l.v. turns,

$$X_1 = 2\pi f \mu_o T_1^2 \frac{s}{2zW}\left[c_o + \frac{1}{6}(c_1 + c_2)\right] \text{ ohm} \qquad (4.51a)$$

$$e_x = \pi f \mu_o \frac{F}{E_T} \frac{s}{nW} \left[c_o + \frac{1}{6}(c_1 + c_2) \right] \text{ p.u.} \qquad (4.51b)$$

$$\% X = 39.5 \times 10^{-4} f \frac{Q}{E_T^2} \frac{1}{(nW/s)} \left[c_o + \frac{1}{6}(c_1 + c_2) \right] \% \qquad (4.51c)$$

Case III : Concentric cylindrical coils of unequal lengths

In case I. concentric cylindrical coils of equal lengths have been considered which in other words considered uniform mmf distribution along the axial length of the winding.

In practice however, either h.v. winding or both h.v. and l.v. windings may have non-uniform mmf-distribution due to :-

(i) the disconnection of a portion of mid-turns as per tap-changing scheme (Figure 4.20 a & b);

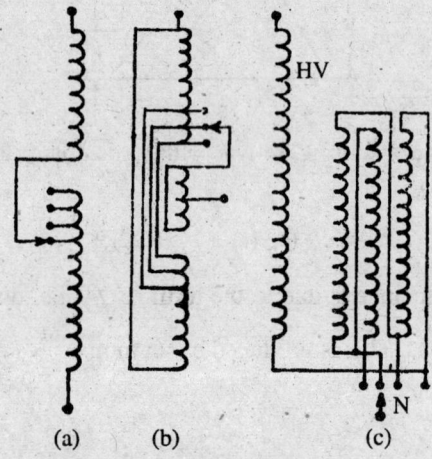

(a)   (b)   (c)

Figure 4.20   Tap coils in h.v. winding.

(ii) use of separate tap-coil at the neutral end of h.v. winding (Figure 4.20 c);

(iii) added insulation of the end coils and provision of wider ducts in h.v helix winding (Figure 4.21).

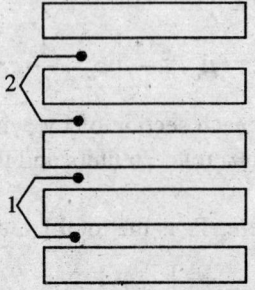

Figure 4.21   Ducts. 1–normal width, 2–enlarged width.

For the determination of leakage flux for the above cases of asymmetry, the mmf distribution can be considered to be a combination of axial (case I) and transverse (case II) fluxes.

For the h.v. winding configuration in Figure 4.22, the non-uniform mmf distribution (is equivalent to a uniform axial distribution) (i) and a transverse distribution (ii). The total leakage reactance in l.v. winding terms is approximately given by the empirical equation.

$$X_1' = X_1 \left[1 + \frac{K}{2}(1 + \frac{\pi}{2} \frac{Kl_l}{c_o + c_1 + c_2}) \right] \qquad (4.52a)$$

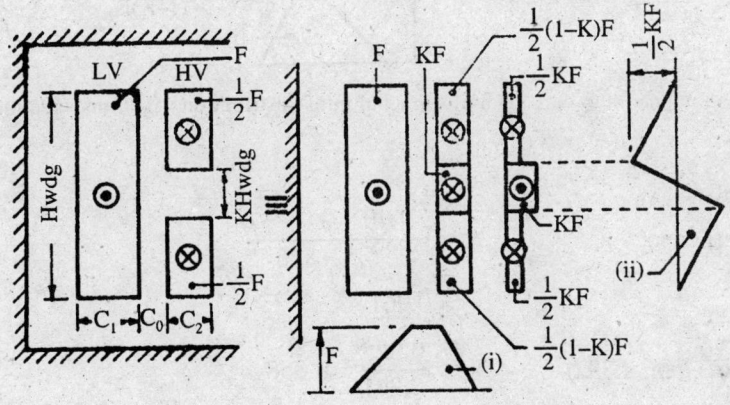

Figure 4.22 Centre–tapped h.v. coil. (i) uniform axial mmf–distribution. (ii) transverse mmf–distribution.

For the h.v. winding configuration in Figure 4.23, the equivalent mmf distribution is due to axial (i), and transverse (ii) mmfs and the total leakage reactance in l.v. terms is,

$$X_1' = X_1 \left[1 + \frac{K}{2}\left(1 + \pi \frac{Kl_l}{c_o + c_1 + c_2}\right) \right] \qquad (4.52)$$

Example 4.5

A 2500-kVa, 35/6.3-kV, Y/Δ, 50-Hz, 3-ph., 3-limb, ON-cooled transformer has the following data :

$c_1 = 24$ mm; $T_1 = 904$; $D_m = 41$ mm.
$c_o = 40$ mm; $T_2 = 282$; $H_{wdg} = 825$ mm;
$c_2 = 37$ mm.

Compute the total leakage reactance in h.v. terms with the windings as concentric coils,

(a) of equal length;
(b) with centre-tapped h.v. coil, as in Figure 4.22;
(c) with end-tapped h.v. coil, as in Figure 4.23;

with K = 0.10.

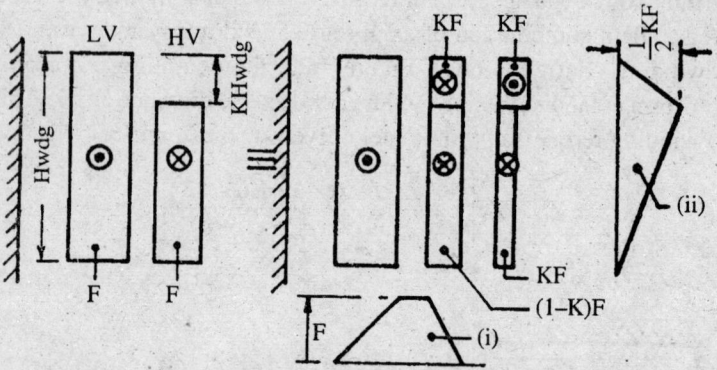

Figure 4.23 End-tapped h.v: coil. (i) uniform axial mmf–distribution, (ii) transverse mmf–distribution.

Solution

Rogowski coefficient
$$K = 1 - \frac{40 + 24 + 37}{\pi \times 825}$$

$$= 0.961$$

Hence, l_l (from equn. 4.39a)
$$= \frac{0.825}{0.961} = 0.858m$$

(a) From equation 4.48a,

$$X_1 = 2\pi \times 50 \times 4\pi \times 10^{-7} \times 904^2 \, \frac{1}{0.858} \pi \times 0.41 \left(40 + \frac{24 + 37}{3}\right) \times 10^{-3}$$

$$= 29.2 \text{ ohm.}$$

$$= 0.06 \text{ p.u.}$$

(b) From equation 4.52 a,

$$X_1' = 0.06\left[1 + \frac{0.10}{2}\left(1 + \frac{\pi}{2}\frac{0.10 \times 0.858}{40 + 37 + 24} \times 10^3\right)\right]\text{p.u.}$$

$$= 0.067 \text{ p.u.}$$

(c) from equation 4.52 b, for configuration of Figure 4.24,

$$X_1' = 0.06\left[1 + \frac{0.10}{2}\left(1 + \pi\frac{0.10 \times 0.858}{40 + 37 + 24} \times 10^3\right)\right]\text{p.u.}$$

$$= 0.071 \text{ p.u}$$

4.3.1.2 *Rotating Machines*

Estimation of leakage flux in rotating machines is more difficult than for transformer, in view of their more complicated configuration. Approximate method leading to simpler equations has been discussed below. For improved estimation, digital computation can be used.

(A) *Salient Field Poles*

Leakage flux due to field excitation can be divided into the following components (Figure 4.24a and b) :-

ϕ_{lx} and Φ_{l2} respectively the component leakage flux per pole between longitudinal surfaces, and between end surfaces of pole-shoes.

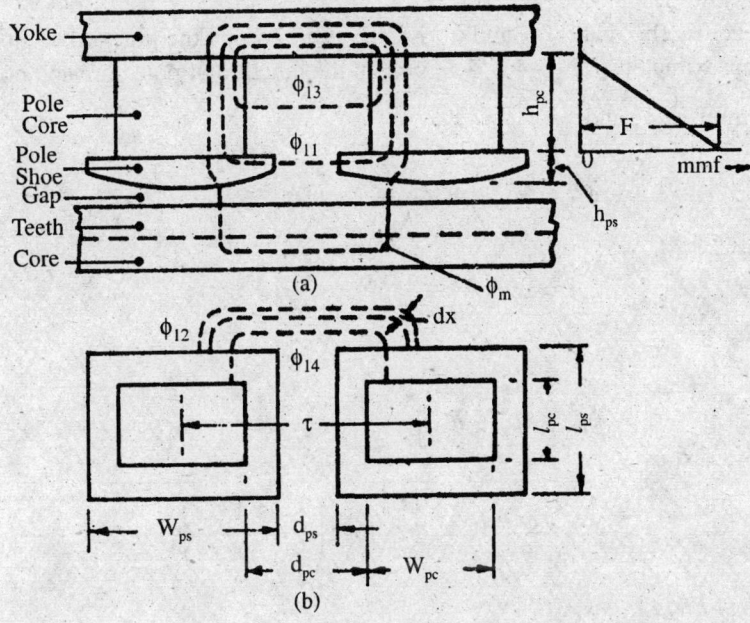

Figure 4.24 Leakage flux-paths in salient field poles.

ϕ_{l3} and ϕ_{l4}: the component leakage flux per pole between longitudinal surfaces, and between end surfaces of pole cores respectively.

The above components can be calculated* as below, by assuming, that

(i) the reluctance of leakage flux-path through iron is negligible;
(ii) the sides of the adjacent poles are parallel (assumption tenable for machines with large diameter as for low-speed salient-pole machines) :
(iii) paths of Φ_{l1} and Φ_{l3} are straight lines between the surfaces;
(iv) paths of Φ_{l2} and Φ_{l4} consists of combination of quarter circles and straight lines.

Noting that the mmf across fluxes ϕ_{l1} and $\phi_{l2} = 2(F_g + F_t + F_c) = F$, where, F_g, F_t, and F_c stand respectively for gap mmf, armature tooth mmf, and armature core mmf, we have,

$$\phi l_1 = 2\mu_o F \frac{l_{ps}\, h_{ps}}{d_{ps}} \tag{4.53a}$$

---

* Gray, A : Electrical Machine Design, McGraw-Hill.

and
$$\phi l_2 = 4\mu_o F \int_o^{\frac{1}{2}W_p/s} \frac{h_{ps}\, dx}{d_{ps} + \pi x}$$

$$= 16 \times 10^{-7} F\, h_{ps} \log_e\left(1 + \frac{\pi\, W_{ps}}{2\, d_{ps}}\right) \tag{4.53b}$$

Again, mmf across flux-paths ϕ_{13} and ϕ_{14} is zero at the root of the pole (where it joins with the yoke) and approximately $= 2\,(F_g + F_t + F_c) = F$ at the junction of pole core and pole shoe. Thus, the average mmf $\frac{1}{2}\, F$, and

$$\phi_{13} = 2\mu_o \frac{1}{2} F \frac{l_{pc}\, h_{pc}}{d_{pc}}$$

$$= \mu_o F \frac{l_{pc}\, h_{pc}}{d_{pc}} \tag{4.53c}$$

$$\phi_{14} = 4\mu_o \frac{1}{2} F \int_o^{\frac{1}{2}W_{pc}} \frac{h_{pc}\, dx}{d_{pc} + \pi x}$$

$$= 8 \times 10^{-7} F\, h_{pc} \log_e\left(1 + \frac{\pi\, W_{pc}}{2\, d_{pc}}\right) \tag{4.53d}$$

(B) *Non-salient Field Poles*

In non-salient pole machines the exciting winding is embedded in slots (Figure 1.9) and field leakage flux exists in combination with the useful flux. Estimation of field leakage is thus to be made step by step with that of the useful flux and cannot be done separately. In such machines, the field overhang leakage is an important component unless non-magnetic (austenitic steel) retaining ring for clamping the overhang is used.

(C) *Armature*

The armature leakage flux due to the current flowing in armature conductors is mainly affected by the winding layout and presence of magnetic material very near to the conductors. It is smaller for distributed winding than for a concentrated winding; smaller for a fractional pitch winding than for a full-pitch winding.

Assuming for simplicity that the armature carries a distributed a.c. winding embedded in slots, the armature leakage flux can be divided into two identifiable components (Table 4.3) :-

(i) leakage flux in the gap region;
and (ii) leakage flux in the overhang region.

As in the case of transformer, the leakage flux is associated with leakage reactance. The procedure for its determination is the same as for transformer, being based on equations 4.37, 4.38.

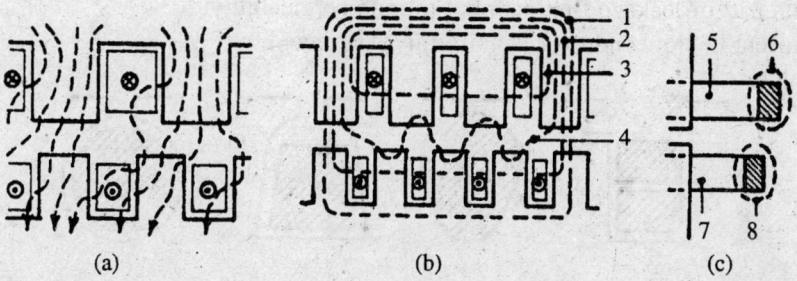

Figure 4.25 (a) Armature leakage flux, (b) approximate components, (c) overhang leakage. 1–mutual flux, 2–tooth leakage, 3–slot leakage, 4–zigzag leakage, 5–stator overhang, 6–stator overhang leakage, 7–rotor overhang, 8–rotor overhang leakage.

Table 4.3

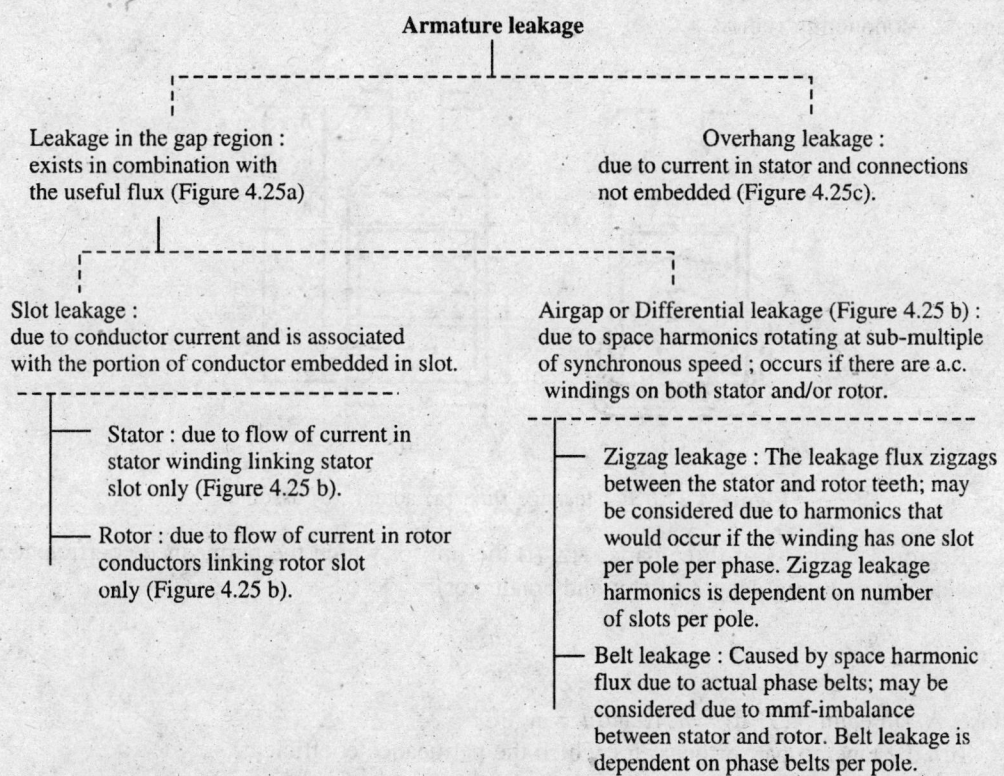

Armature leakage

Leakage in the gap region : exists in combination with the useful flux (Figure 4.25a)

Overhang leakage : due to current in stator and connections not embedded (Figure 4.25c).

Slot leakage : due to conductor current and is associated with the portion of conductor embedded in slot.

Airgap or Differential leakage (Figure 4.25 b) : due to space harmonics rotating at sub-multiple of synchronous speed ; occurs if there are a.c. windings on both stator and/or rotor.

— Stator : due to flow of current in stator winding linking stator slot only (Figure 4.25 b).

— Rotor : due to flow of current in rotor conductors linking rotor slot only (Figure 4.25 b).

— Zigzag leakage : The leakage flux zigzags between the stator and rotor teeth; may be considered due to harmonics that would occur if the winding has one slot per pole per phase. Zigzag leakage harmonics is dependent on number of slots per pole.

— Belt leakage : Caused by space harmonic flux due to actual phase belts; may be considered due to mmf-imbalance between stator and rotor. Belt leakage is dependent on phase belts per pole.

(A) *Slot leakage reactance*

The slot leakage reactance.depends basically on the slot configuration, a few common types being shown in Figure 4.26.

Estimation of the leakage reactance is based on the following assumptions :-

(i) Flux lines across the slot are rectilinear;

. (ii) Iron part of leakage flux-path has infinite permeability;
(iii) Current in slot conductor is uniformly distributed.

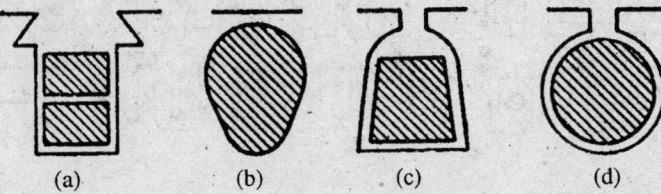

Figure 4.26 Typical rotor slots–(a) parallel–sided, open, (b) closed, (c) tapered, semi-closed, (d) round, semi-closed.

Case I : Semi-closed parallel-sided slot with a single-layer winding

Figure 4.27 shows the actual and the assumed leakage flux-path in a slot. The region of interest can be divided under two heads : region 1–unoccupied by current-carrying conductors; region 2–conducting region.

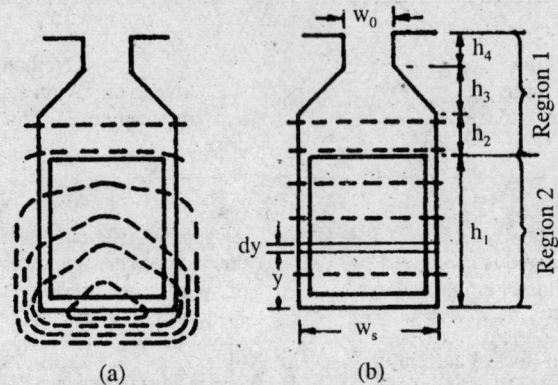

Figure 4.27 Slot leakage flux. (a) actual, (b) assumed.

Region 1 Consists of three parts, viz. (i) the lip, for which the permeance coefficient λ_4 (considering unit axial length of slot and conductor)

$$= \frac{h_4}{w_o}$$

since, A_l (in equn. 4.37 d) $= h_r 1$, and $l_l = w_o$.

(ii) the tapered part or neck, for which the permeance coefficient,

$$\lambda_3 = \frac{2\,h_3}{w_o + w_s}$$

since, $A_l = h_3.\,1$, and mean $l_l = \frac{1}{2}(w_o + w_s)$.

(iii) the part above the conductor, for which,

$$\lambda_2 = \frac{h_2}{w_s}$$

[*Note :* The flux in this region links the whole of the conductor.]

Region 2 Assuming that Z_s conductors in the slot with a slot ampere-conductor I_sZ_s, occupy the full region, that is, the conductor section is $h_1 . w_s$, consider an element dy at y metre from the bottom of the slot.

That is, the permeance coefficient of the element,

$$\lambda_y = \frac{dy}{w_s},$$

and the number of ampere-conductor developing the flux in the element,

$$= I_s Z_s \frac{y w_s}{h_1 w_s} = I_s Z_s \frac{y}{h_1}$$

$$= \text{mmf developing the flux.}$$

Thus, the flux in the element of unity axial length,

$$d\phi_x = \mu_o I_s Z_s \frac{y}{w_s h_1} dy$$

and the flux-linkage in the element,

$$d\Psi_x = \mu_o I_s Z_s \frac{y}{w_s h_1} dy \cdot Z_s \frac{y}{h_1}$$

since the number of turns below the element = no. of conductors below the element

$$= Z_s \frac{y}{h_1} .$$

Hence, total flux-linkage in the height h_1

$$\Psi_1 = \mu_o I_s Z_s^2 \frac{1}{w_s h_1^2} \int_0^{h1} y^2 dy$$

$$= \mu_o I_s Z_s \frac{1}{3w_s} h_1$$

Effective permeance coefficient,

$$\lambda_1 = \frac{\Psi_1}{\mu_o \times \text{total no. of turns} \times \text{total mmf}}$$

$$= \frac{\Psi_1}{\mu_o Z_s I_s \cdot Z_s}$$

$$= \frac{h_1}{3w_s}$$

Hence, total permeance coefficient for the slot,

$$\lambda_s = \lambda_1 + \lambda_2 + \lambda_3 + \lambda_4$$

$$= \frac{h_1}{3w_s} + \frac{h_2}{w_s} + \frac{2h_3}{w_o + w_s} + \frac{h_4}{w_o} \qquad (4.54)$$

Case II : Round Slot (Figure 4.28).

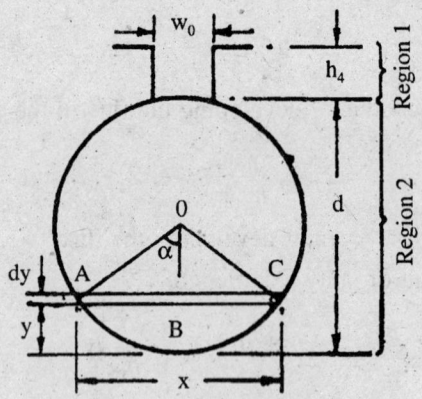

Figure 4.28   Pertaining round slot.

Region I   The lip, for which the permeance coefficient is,

$$\lambda_2 = \frac{h_4}{w_o}$$

Region 2   Considering an element dy at a distance y from the bottom of the slot, the permeance coefficient of the element of unity axial length,

$$\lambda_y = \frac{1}{x}dy = \frac{1}{d \sin \alpha}dy$$

But

$$y = \frac{1}{2}d - \frac{1}{2}d \cos \alpha$$

That is,

$$dy = \frac{1}{2}d \sin \alpha \, d\alpha$$

Hence,

$$\lambda_y = \frac{1}{2}d\alpha$$

No. of ampere-conductors developing the flux in the element

= mmf developing the flux in the element

$$= \frac{\text{area ABC}}{\text{area of the circle}} I_s Z_s$$

Now, area ABC = area OABC–area of triangle OAC

$$= \frac{1}{4}\alpha d^2 - \frac{1}{2} x \cdot \frac{1}{2} d \cos \alpha$$

$$= \frac{1}{4}d^2\left(\alpha - \frac{1}{2}\sin 2\alpha\right)$$

and the area of the circle = $\frac{1}{4}\pi d^2$

Hence, mmf developing the flux in the element,

$$= \frac{1}{\pi}\left(\alpha - \frac{1}{2}\sin 2\alpha\right)I_s Z_s$$

That is, the flux in the element,

$$d\phi_x = \mu_o \frac{1}{\pi}\left(\alpha - \frac{1}{2}\sin 2\alpha\right)I_s Z_s \lambda_y$$

$$= \mu_o \frac{1}{2\pi}\left(\alpha - \frac{1}{2}\sin 2\alpha\right)I_s Z_s d\alpha$$

The above flux links the conductor $= \left(Z_s / \pi\right)\left(\alpha - \frac{1}{2}\sin 2\alpha\right)$ in area ABC.

Hence, elemental flux-linkage,

$$d\Psi_x = \mu_o \frac{1}{2\pi}I_s Z_s\left(\alpha - \frac{1}{2}\sin 2\alpha\right)d\alpha . Z_s \left(Z_s / \pi\right)\left(\alpha - \frac{1}{2}\sin 2\alpha\right)$$

$$= \frac{\mu_o}{2\pi^2}I_s Z_s^2\left(\alpha - \frac{1}{2}\sin 2\alpha\right)^2 d\alpha$$

Hence $\qquad \Psi_1 = \frac{\mu_o}{2\pi^2}I_s Z_s^2 \int_0^{\pi}\left(\alpha - \frac{1}{2}\sin 2\alpha\right)^2 d\alpha$

$$= 0.623\,\mu_o I_s Z_s^2$$

That is, $\qquad \lambda_1 = 0.623$

Hence, total permeance coefficient for the slot,

$$\lambda_s = \lambda_1 + \lambda_2 = 0.623 + \frac{h^4}{w_o} \qquad\qquad (4.55a)$$

In practice,
$$\lambda_s = 0.66 + \frac{h_4}{w_o}$$
(4.55b)

is used considering a correction for actual flux-plot.

Example 4.6

Show that the permeance coefficient for the T-bar slot (Figure 4.29) is given by

$$\lambda_s = \frac{1}{(a+b)^2}\left[\frac{a^2 h_1}{3\,w_s} + \frac{h_2}{w_s}\left(\frac{b^2}{3} + a\,b + a^2\right)\right] + \frac{h_4}{w_o}$$

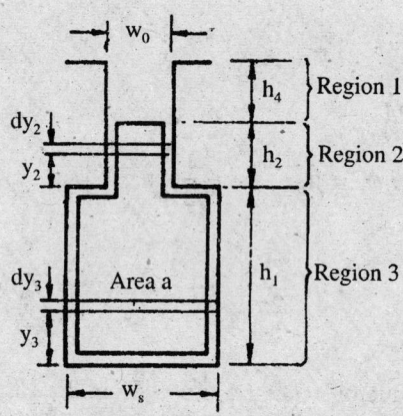

Figure 4.29 Pertaining example 4.6

Solution

Region 1: Following the procedure stated above.

$$\lambda_4 = \frac{h_4}{w_o}$$

Regions 2 & 3:

Consider an element dy_2 at y_2 from the neck of the slot for Region 2, and an element dy_3 at y_3 from the bottom of the slot for Region 3. The Table 4.4 can be completed. In Table 4.4

$$Z_{sa} = Z_s \frac{a}{a+b}, Z_{sb} = Z_s \frac{b}{a+b},$$ a and b being the area of conductors in region 2 and region 3 respectively.

Hence, total permeance for the slot.

$$\lambda_s = \lambda_1 + \lambda_2 + \lambda_4$$

$$= \frac{1}{(a+b)^2}\left[\frac{a^2 h_1}{3w_s} + \frac{h_2}{3w_s}\left(\frac{b^2}{3} + ab + a^2\right)\right] + \frac{h_4}{w_o}$$
(4.56)

Table 4.4 Calculation of component permsance coefficient

	Region 2	Region 3
Permeance coefficient of the element :	$\dfrac{dy_2}{w_o}$	$\dfrac{dy_3}{w_s}$
Mmf developing the flux :	$I_s\left[Z_{sb}\dfrac{y_2}{h_2}+Z_{sa}\right]$	$I_s Z_{sb}\dfrac{y_3}{h_1}$
Flux in the element $d\phi_x$:	$\mu_o\dfrac{I_s}{w_o}\left[Z_{sb}\dfrac{y_2}{h_2}+Z_{sa}\right]dy_2$	$\mu_o I_s Z_{sa}\dfrac{y_3}{h_1\,w_s}\,dy_3$
Flux-linkage $d\psi_x$ in the element :	$\mu_o\dfrac{I_s}{w_o}\left[Z_{sb}\dfrac{y_2}{h_2}+Z_{sa}\right]^2 dy_2$	$\mu_o I_s Z_{sa}^2\dfrac{y_3^2}{h_1^2\,w_s}$
Total flux-linkage :	$\mu_o\dfrac{I_s}{w_o}\left[Z_{sb}^2\dfrac{h_2}{3}+Z_{sa}\,Z_{sb}\,h_2+Z_{sb}^2\,h_2\right]$	$\mu_o\dfrac{I_s}{w_s}Z_{sa}^2\dfrac{h_1}{3}$
Component permeance coefficient	$\lambda_2=\dfrac{1}{(a+b)^2}\left[\dfrac{h_2}{3w_s}\left(\dfrac{b^2}{3}+ab+a^2\right)\right]$	$\lambda_1=\dfrac{a^2}{(a+b)^2}\dfrac{h_1}{3w_s}$

[Note : (1) The slot leakage reactance is higher for a deep narrow slot than for a shallow broad slot.

(2) For a closed slot, the slot leakage reactance is higher because of the presence of iron at the top of the slot. Magnetic saturation in this region influences the slot permeance coefficient].

(B) Other components of leakage flux

The development of equations for other components of leakage flux such as, zigzag, overhang, and belt, are complicated since these components depend on the geometry of the parts of portions of the machine involved, location of adjacent magnetic parts such as, clamps and retaining rings, windings and their layouts, etc.. Generally empirical equations are used for estimation of these components and the following* references are useful.

Airgap leakage reactance In a machine, the airgap field includes a series of harmonic waves with multiples of the fundamental number of poles, revolving at sub-multiples of the synchronous speed. These waves induce rated-frequency voltages in the winding that produce them and thus add to the leakage reactance. The phenomenon may be profound in the case of induction motor.

For convenience, the airgap leakage is divided into two groups :- zigzag, and belt leakages. Whereas the zigzag leakage is dependent on the number of stator and rotor slots and is independent of winding pitch, the belt leakage is dependent on the winding pitch and not on the number of slots.

* Hellmund, R.E. : "Zigzag leakage of Induction Motors", Transaction A.I.E.E., 1907, vol. 26, p. 1505.
Alger;, P. L. : "Calculation of the Armature reactance of synchronous Machines", Transaction A. I. E. E., 1928, vol , 47, p. 493.
Kilgorc L. A. : "Calculation of Synchronous machine constants", Transaction A.I.E.E., 1931, vol., 50, p. 1201.

Zigzag leakage (1) The zigzag leakage reactance may be calculated by the following equation:-

$$x_z = \frac{5}{6} p^2 x_m \left[\frac{1}{S_1^2} + \frac{1}{S_2^2} \right] \qquad (4.57)$$

where, S_1 and S_2 are number of stator and rotor slots; p, the number of poles, and x_m, the magnetising reactance.

(2) The zigzag leakage reactance is practically zero for a polyphase winding with very large number of slots.

(3) For a salient-pole without damper winding, the zigzag leakage does not exist. With dampers, zigzag leakage is present though the gap-length is comparatively long.

(4) For open slots, the zigzag leakage reactance is less in comparison with semi-closed slots.

(5) Skewing increases zigzag leakage because voltage induced by the fundamental wave of flux is displaced in successive elements of conductor length. The resulting decrease in induced voltage can be considered as drop in the main flux-linkage and an increase in leakage.

Belt leakage (1) Belt leakage is due to the actual phase belts which may be several slots wide, and in fractional-slot winding may have varying width (in contrast, the zigzag leakage is as if a phase-belt is formed by one slot). The belt leakage can be attributed to the imbalance in mmf-distribution between the stator and the rotor. Consider the following cases :-

Case I   Reference to Figure 4.30 (a)–the stator and rotor windings are with a turn-ratio 2 : 1 having an identical distribution of mmf.

Net rotor winding linkage = Rotor linkage due to I_1 in stator–rotor winding linkage due to its own current I_2

$$= (6 \times 3I_1 \times 2y_s + 2 \times 2 \times 1 \, I_1 \times y_s) - (3^2 I_2 \times 2y_s + 1^2 \times I_2 \times y_s \times 2)$$
$$= (40I_1 - 20 \, I_2) y_s$$

For mmf-balance, $= I_1 T_1 = I_2 T_2$ or, $I_2 = 2I_1$.

Thus, the net rotor winding linkage = zero.

That is, no belt leakage exists in this case.

Case II   Turn-ratio = 2:1, mmf distribution as in Figure 4.30(b).

Net rotor linkage = $(6 \times 3I_1 \times 2ys + 2 \times 3 \times 1 \, I_1 \times y_s) - (3^2 I_2 \times 2y_s + 2 \times 1^2 \times I_2 \times y_s)$
$$= (42I_1 - 20 \, I_2) y_s$$

Carpenter, C.J. : "Application of the method of Images to machine end-winding fields", Proceding IEE, vol 107 (A), p. 487.

Reece, A. J. B and Pramanik, A : "Calculation of the end-region field of a.c. machines" Proceeding IEE, 1965, vol 112, p. 2083.

Hawley, H, Edwards, I. M., Heaton, J. M. and Stoll, R.L. : "Tubo-geneator end region magnetic fields", Proceeding IEE, 1967, vol 114, p. 1107.

Lawrenson, P.J. : "Calculation of machine end-winding inductances with special reference to Turbogenerators', Proceeding IEE, 1970, vol 117, p 1129.

Alger, P.L. : Induction Machines (Book), Gordon Breach, 1970.

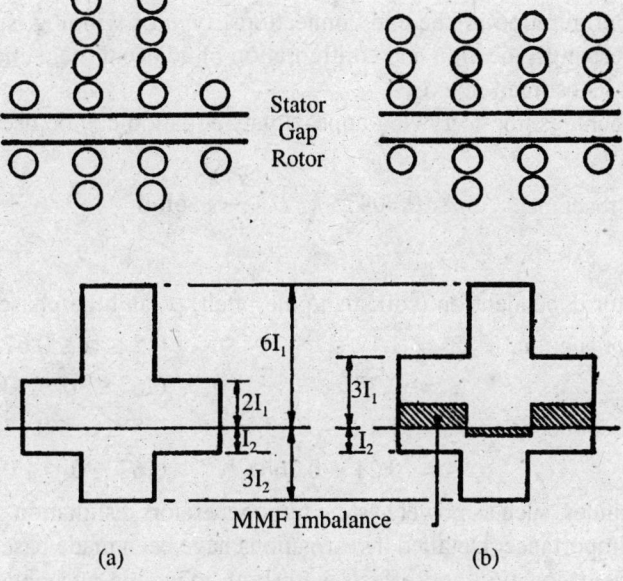

Figure 4.30 Pertaining belt leakage.

For zero rotor linkage, $42 \, I_1 - 20 \, I_2 = 0$ yielding $I_2 = 2.1 \, I_1$

That is, mmf-balance is not possible.

With zero rotor linkage, the net primary linkage responsible for belt leakage flux

$$= (6^2 \, I_1 \times 2y_s + 2 \times 3^2 \, I_1 \times y_s) - 42 \, I_2 \times y_s$$
$$= (90 \, I_1 - 42I_2) \, y_s$$
$$= 1.8 \, I_1 y_s, \text{ putting } I_2 = 2.1 \, I_1$$

(2) The belt leakage reactance, though small may lead to serious handicap to motor performance as additional losses and parasitic torques are produced due to the induced secondary current.

(3) In a machine with relatively short airgap (for example, an induction motor) the magnitude of the airgap leakage reactance is of the same order as the slot-leakage, or overhang (or end-connection) leakage.

(4) Belt leakage reactance is approximately zero for squirrel-cage winding as such winding has no phase-belt.

(5) The belt leakage harmonics of stator winding induce voltage in the rotor winding, and vice versa. This causes opposing circulating currents which reduce the effective reactance. A squirrel-cage rotor thus reduces the belt leakage reactance of the stator winding nearly to zero.

(6) The belt leakage reactance for both wide- and narrow-spread windings are equal for full pitch. As the pitch departs from unity, the wide-spread reactance increases rapidly while the narrow-spread reactance decreases.

(7) The effect of open slots on belt leakage is small since important phase-belt harmonics span several slot-openings.

Overhang (or End-connection) Leakage   The overhang leakage is dependent on the dimensions and configuration of the end connections, type of winding, spacing between the stator and rotor overhangs, location and configuration of adjacent magnetic parts, etc., and no simple method for its estimation exists.

For smaller machines, the following approximate equation can be used :-

Overhang leakage reactance $x_0 = 0.00475 \, k_s \, D \left(\dfrac{T_1}{p} \right)^2$ ohm   (4.58)

at 50 Hz frequency

where, ks, is a factor dependent on (coil span/pole-pitch) α, and the phase-spread.

For wide-spread connection, $ks = 1.12 \, \alpha$ $0.5 \le \alpha \le 0.67$

                               $= 0.75$ $0.67 \le \alpha \le 1.0$

For narrow-spread connection $ks = 1.4 \, \alpha - 0.19$ $0.5 \le \alpha \le 0.67$

                               $= 0.24 + 0.76 \, \alpha$ $0.67 \le \alpha \le 1.0$

For larger machines, such as power system turbogenerators, estimation of overhang leakage is of considerable importance. Detailed investigations have been made based on the solution of two-dimensional field equation and effects with both magnetic and non-magnetic retaining rings, and with and without core screen have been studied. The following conclusions may be drawn :-

(1)   Under steady-state, (a) the core screen increases the overhang leakage reactance; (b) use of magnetic retaining ring results in larger leakage reactance than when non-magnetic ring is used.

(2)   A non-magnetic ring has no steady-state effect. It affects under transient conditions because of eddy-currents.

4.3.2 Mechanical Forces

The wellknown principle of interactive forces (Figure 4.31) when applied to l.v. and h.v. windings of a transformer results in (a) radial, and (b) axial, mechanical forces as shown in Figure 4.32. Whereas the radial forces tend to burst the h.v. winding outwardly, the l.v. winding will tend to crush on the core transmitting the stress to the core.

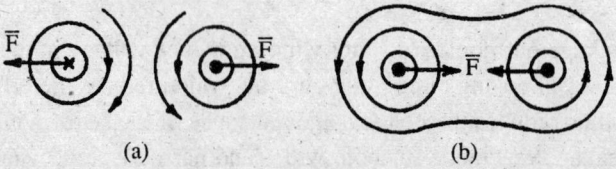

Figure 4.31   Interactive forces, (a) repulsion, (b) attraction.

[*Note* : (1) Under rated conditions, the forces however are small. In cases such as sudden short-circuit, the inrush current may reach a value 20 to 25 limes the rated since, the inrush current is inversely proportional to the per unit impedance. The radial force may reach sufficient value

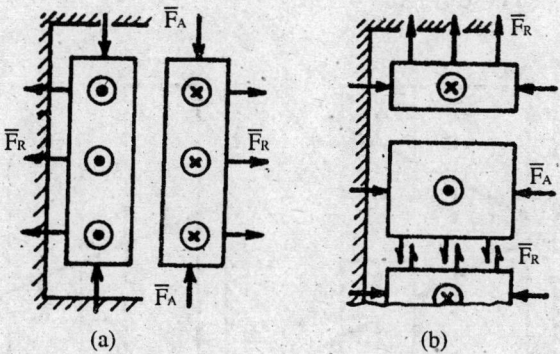

Figure 4.32 Mechanical forces. (a) core type, concentric cylinders of equal length, (b) shell type.

so as to damage the transformer. The windings as such should be suitably braced to withstand the mechanical stresses. Circular coils have a shape best suited to withstand radial stresses.

(2) The axial forces, however are not of any importance even under fault conditions, unless there is asymmetry in the windings.

(3) The maximum axial force is sustained by the turns in the centre of l.v. and h.v. coils, since at this point the forces exerted by turns on each side add up to maximum value.]

(a) *Radial force*

Estimation of radial force can be done by considering the leakage flux only in the duct. For a concentric coil, with reference to Figure 4.18(d), the maximum flux-density in the annular duct (equn. 4.41).

$$= \mu_o \frac{F}{H_{wdg}}$$

where, $F = i_1 T_1 = i_2 T_2$ and i stands for instantaneous current.

The mean radial force per unit axial length and per unit length of the perimeter

$$= \frac{1}{2} \mu_o \frac{F}{H_{wdg}} \cdot \frac{F}{H_{wdg}}$$

For the whole surface area $= \pi\, D_m\, l_l$, the mean radial force,

$$\overline{F}_R = \frac{1}{2} \mu_o F^2 \frac{\pi D_m}{H_{wdg}} \text{ Newton} \tag{4.59}$$

[*Note* : With asymmetry, the mean radial force as given by equation 4.59 should be multiplied by a factor approximately equal to 2 to take into account the 'doubling effect'.]

(b) *Axial force*

Consider the two cases of asymmetry as shown in Figs. 4.22 and 4.23.

Case I : The centre-tapped case (Figure 4.33a).

Axial forces $\overline{F}_A$ in l.v. tend to cancel each other, whereas those in h.v. tend to separate the two coils thereby increasing the asymmetry.

Case II : The end-tapped case (Figure 4.33b).

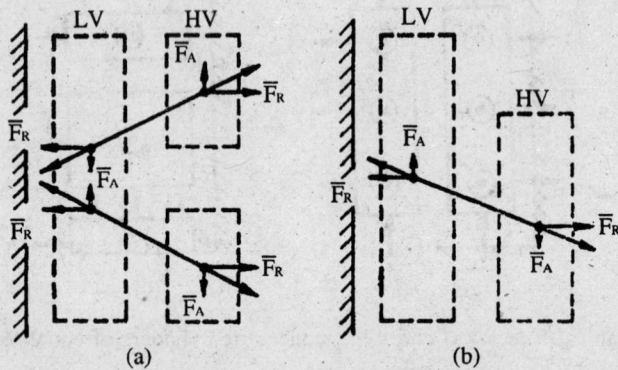

Figure 4.33 Arial forces. (a) centre-tapped, (b) end-tapped, h.v. coil.

$\overline{F}_A$ in l.v. and h.v. coils act in opposition thereby tending to increase the asymmetry.
[*Note* : (1) If the divergence of the coils is allowed to increase, the same fault current applied subsequently will cause a greater axial force.

(2) The increase in asymmetry for the end-tapped case is more than that for the other, and as such, the centre-tapped coils are superior to the end-tapped coils.

The forces of asymmetry can be estimated from the transverse flux developed due to the out of balance mmf KF for the end-tapped case and $\dfrac{1}{2}$ KF for the centre-tapped case.

For the end-tapped case, assuming that the l.v. and h.v. windings are enclosed between two iron surfaces separated by a distance = $2 (c_o + c_1 + c_2)$, the maximum cross-flux density,

$$B_{mA} = \mu_o K \frac{F}{2(c_o + c_1 + c_2)}$$

Thus, the mean axial force developed by the mean cross-flux density $\dfrac{1}{2} B_{mA}$ acting on T_1–l.v. turns or T_2–h.v. turns,

$$\overline{F}_A = \frac{1}{2} \mu_o KF^2 \frac{\pi D_m}{2(c_o + c_1 + c_2)} \tag{4.60}$$

Whereas the inward radial stresses are taken up by the formers and core, the inward axial stresses are taken up by the windings. For the outward axial stresses, suitable end insulation and packings are to be used.

4.4 UNBALANCED MAGNETIC PULL

In a normal rotating machine, magnetic attraction between the stator and the rotor exists as the flux passes from one member to the other. The estimation of the force of attraction may be based on Maxwell stress concept by which a tensile force of magnitude $\dfrac{1}{2}$ BH Joule, metre^{-3}

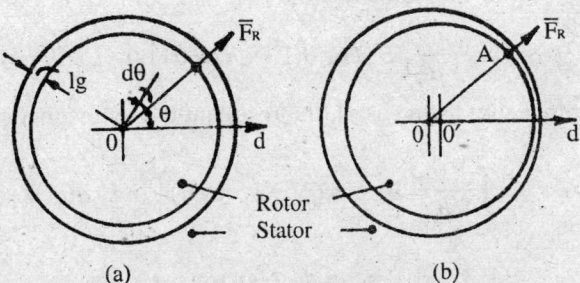

(a)   (b)

Figure 4.34 Unbalanced magnetic pull. (a) stator and rotor concentric, (b) stator and rotor displaced.

along the direction of flux-line exists between the stator and the rotor at any point in the gap, where, B is the gap density at that point and H. the corresponding field intensity (= B/μ_o).

The following two cases may be considered :-

Case I Stator and rotor are concentric (Figure 4.34 a).

Consider a 2-pole machine with gap-flux distributed sinusoidally in space with its peak Bp coinciding with the direct axis.

At any point A at an angle θ from the *d*-axis, the radial density is $B_p \cos \theta$, so that the radial force of attraction over an elemental angle $d\theta$ is,

$$\overline{F}_R = \frac{1}{2\mu_o}(B_p \cos\theta)^2 r \, La \, d\theta$$

where, r and L_a are respectively the rotor radius and its axial length.

The component of $\overline{F}_R$ along the direct axis is $\overline{F}_R \cos\theta$, so that the resultant d-axis magnetic pull

$$= \int_0^{2\pi} \frac{1}{2\mu_o}(B_p \cos\theta)^2 L_a \cos\theta \, d\theta$$

$$= \text{zero.}$$

This is also evident from the physical consideration that since the gap length is uniform, the force of attraction under one pole is equal and opposite to that under the other pole.

Case II Rotor and stator axes are displaced from each other by a small distance e (= 00′) (Figure 4.34b) along the *d*-axis.

The radial gaplength at A is,

$$l_g = l'_g - e \cos \theta = l_g (1 - \in \cos \theta)$$

where, $\in$ is defined as eccentricity = e/l_g.

Assuming that, (1) the peak density is unaltered by the asymmetry, and (2) the flux-density is inversely proportional to the gap length, the flux-density at A

$$= \frac{B_p \cos \theta}{1 - \in \cos \theta}$$

$$= Bp \cos \theta (1 + \in \cos \theta), \text{ since } \in \text{ is small.}$$

The radial force at A over an elemental angle $d\theta$ is,

$$\overline{F}_R = \frac{1}{2\mu_o}\Big[B_p \cos\theta(1+\epsilon\cos\theta)\Big]^2 r \cdot L_\alpha d\theta$$

Following the same procedure as in case I, the resultant d-axis magnetic pull

$$= \int_0^{2\pi} \frac{1}{2\mu_o}\Big[B_p \cos\theta(1+\epsilon\cos\theta)\Big]^2 r\, L_\alpha d\theta$$

$$= 1.25\times10^6 B_p^2\, \epsilon\, rL_a \text{ Newton} \qquad\qquad (4.61)$$

[*Note* : (1) The asymmetry as above may be due to wide tolerances in manufacturing defect in bearing, etc.

(2) Unbalanced magnetic pull may be important in machines with short gap length, such as induction motor, causing vibration and noise.

(3) Experiment* shows that the unbalanced magnetic pull is directly proportional to eccentricty ϵ (as shown by equn. 4.61) for $\epsilon \le 0.1$. With $0.1 \le \epsilon \le 0.3$, saturation in tooth-tips limits B_m and equation 4.61 does not hold good.

(4) In an induction motor, the tooth-tip saturation due to increase in leakage flux under rated load condition restricts the unbalanced magnetic pull. The effects of pull as such, are more likely to be prominent under light-load and no load conditions.]

SHORT QUESTIONS

4.1 Mention 2 points showing why magnetic circuit in rotating machines is more complex than that in transformer.

4.2 Indicate 4 representations of saturation curve.

4.3 Mention 2 effects due to slotting on armature.

4.4 What is Carter's coefficient ?

4.5 Why is Carter's coefficient more for semiclosed slot than for open slot ?

4.6 How does the radial duct on either stator or rotor affect the gap m. m. f. ?

4.7 How are radial ducts on both stator and rotor considered in calculating the gap m: m. f. ?

4.8 Define, Field form factor.

4.9 Mention 3 important assumptions in the graphical method of field plotting.

4.10 What is the basis of Finite-difference method of 2-dimensional field plotting?

4.11 Distinguish between 'Apparent' and 'True' densities.

4.12 State 3 methods of determining tooth m. m. f.

4.13 Mention 5 effects of leakage flux in transformer and rotating machines.

4.14 State the assumptions involved in estimation of leakage reactance in a transformer.

---

* Binns, K. J. and Dye, M : "Identification of the principal factors causing unbalanced magnetic pull in cage induction motors", Proceedings IEE, 1973, vol 120 p 349.

4.15 Sketch typical leakage flux-paths in a core-type transformer with cylindrical coils of (a) equal length, (b) unequal lengths.

4.16 Why is circular coil preferred ?

4.17 Sketch typical leakage flux-paths in a shell type transformer.

4.18 Why are low and/or high voltage windings split into sections?

4.19 Sketch typical leakage flux paths in salient field poles.

4.20 Mention component leakage fluxes of an armature.

4.21 What is airgap leakage ?

4.22 Explain how skewing increases zigzag leakage reactance.

4.23 Discuss the effects of core screen, and magnetic and non-magnetic retaining rings in large power-system turbogenerators.

4.24 Mention the reasons for unbalanced magnetic pull in rotating machines.

4.25 State, 'True' or 'False' :-

(1) The presence of slots on either side of the gap reduces the m.m.f. required.

(2) Open-slot armature needs larger ni. m. f. than the semi-closed slot armature.

(3) The effect of slotting is equivalent to a smaller slot-pitch.

(4) The effect of slotting is more for larger gap-length.

(5) The effect of slotting can be considered as if the gap-length is reduced.

(6) Carter's coefficient for semi-closed slots is larger than for open slots.

(7) The one-third density method is applicable to trapezoidal tooth of moderate taper.

(8) Leakage flux is normally 0.05 to 0.1 per cent of the total flux.

(9) Leakage fluxes do not introduce any additional loss in transformer and rotating machines.

(10) Slot leakage reactance is smaller in deep narrow slots than in shallow broad slots.

(11) Slot leakage reactance is larger in a closed slot than in an open slot.

(12) Zigzag leakage reactance is independent of number of stator and rotor slots.

(13) Belt leakage is dependent on winding pitch.

(14) In polyphase machines with large number of slots, zigzag leakage reactance is large.

(15) Zigzag leakage reactance is larger for open slots than for semi-closed slots.

(16) Belt leakage reactance is zero in squirrel-cage rotor.

(17) Belt leakage reactance of chorded coils for wide-spread winding is larger than for narrow-spread.

(18) Open-slot armature winding has larger belt leakage than semi-closed slot.

(19) Maximum axial force is sustained by the centre-turns of tranformer coils.

(20) Centre-tapped coils are inferior to end-tap coils.

(21) For a winding with no asymmetry, axial forces are of no importance even under fault conditions.

(22) Unbalanced magnetic pull is important in induction motors with large gap length.

4.26 Why are end insulation and packing used in transformers ?

## CHAPTER 5

Electric Circuit Calculations

Design of electric circuit must fulfil the following basic requirements : -

(i) Ability to withstand electric stresses impressed upon it during tests (as per IS 2026–1962). The tests, intended for ensuring trouble-free service are :- (a) power frequency test, for 60 seconds, to prove its margin over its opsrating voltage ; (b) surge test and switching-surge test-to prove its ability to withstand voltage surges due to switching and in some cases, due to atmospheric disturbances.

(ii) Temperature-rise should not be above the specified limit.

(iii) Load loss should be within specified limit.

(iv) For transformer and a.c. machines, per cent impedance should be within specified limit.

(v) Ability to withstand electromagnetic forces (important for large transformer).

(vi) Minimum cost.

5.1 TRANSFORMER

5.1.1 Types of Winding

(a) *Cross-over*: Used for low current requirement (upto about 20 amperes); conductor section-circular; wound in sections connected in series with spacers for horizontal cooling duct (Figure 5.1a): poor space-factor since conductors are round.

(b) *Helix* (Figure 5.1b) : For low voltage and heavy current; rectangular section of conductors wound in the form of helix of constant diameter; can be wound directly on a cylinder or can be separated from it by spacers to provide oil-duct. Spacers can also be arranged between turns to provide axial ducts.

(c) *Disc*: For high voltage aud low current; suitable only for conductors in strip form, wound alternately from inside to outside and outside to inside, and connected in series (Figure 5.1c - i).

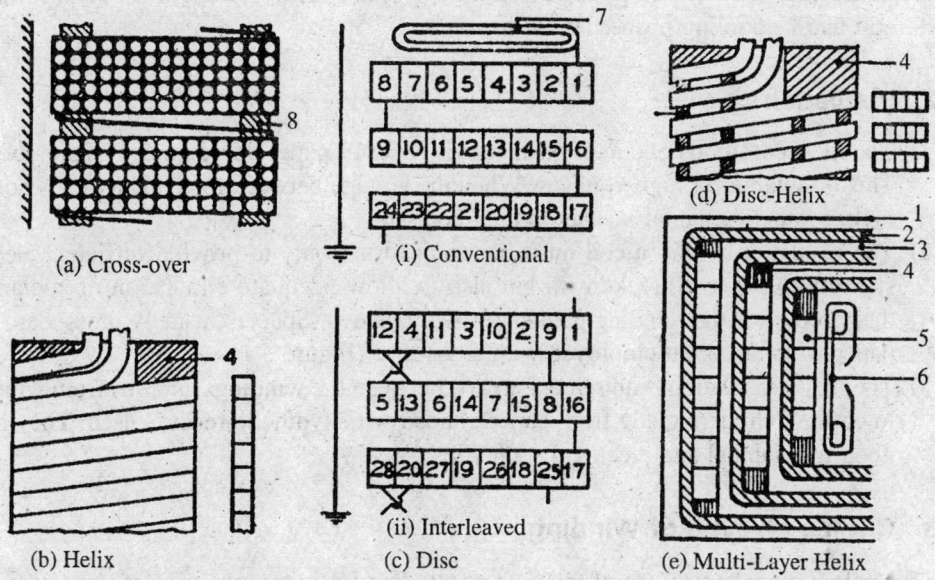

(a) Cross-over   (i) Conventional   (d) Disc-Helix

(b) Helix   (ii) Interleaved   (e) Multi-Layer Helix

(c) Disc

Figure 5.1   Types of transformer winding. 1-earth screen, 2-paper wrap, 3-duct, 4-layer end packing, 5-layer of conductor, 6-line screen, 7-static relay, 8-spacer.

[*Note* : (1) Advantage of disc winding are,

(i) ease with which mmf-balance can be maintained between l.v. and h.v. windings;

(ii) lower axial force ;

(iii) mechanically strong,

(2) A modificition of disc winding, known as interleaved disc (Figure 5.1 c-ii) has the following advantages, though at the expense of a higher voltage between adjacent turns :-

(i) Inductive effect can be almost eliminated since capacitance current flows in opposite directions in adjacent conductors. Thus, initial surge voltage distribution is considerably improved (art 6.8),

(ii) This form of winding does not need any special electric screen.

(d) *Disc-helix*:   Helical type winding with individual strips assembled in radial packs (Figure 5.1d); Strips may be connected in parallel for heavy current, transposed to keep down stray losses.

(e) *Multi–layer Helix* (Figure 5.1e):   Similar to helix, but for high voltage (above 132 kV) range and wide range of current, because it can be of two or more number of layers. Screening and insulation are provided to suit voltage requirement.

[*Note* : Layer nearest the duct between the l.v. and h.v. coils is in a high intensity magnetic field leading to, (a) higher eddy-current loss, and (b) higher radial forces in the layer in comparison with other layers.] .

    (f) Foil Coils made from copper or aluminium sheets or foils; can be used iu distribution transformers, and small power transformers. Space factor is better than other forms of coil and its cooling properties are good.

5.1.2 Insulation

    (1) The most common conductor insulation is paper or paper-board known as pressboard. This is suitable for high voltages. When the voltage between the conductors is not too high, synthetic enamel insulation may be used.

    (2) The insulation within a coil must be arranged not only to provide sufficient electric strength to prevent breakdown, but also to allow adequate circulation of coolant so that no part of the winding gets excessively heated. Spacers, usually pressboard and paper-wrap are often employed with advantage (Figure 5.1).

    (3) The major insulation requirement are, (i) between the windings, and (it) from winding to earth. Cylinders made from paper bonded with synthetic resin is used. They have good mechanical and electrical properties.

5.1.3 Choice of Type of Winding

Table 5.1 below gives the choice of the type of winding for various classes of transformer.

Table 5.1

Class of Transformer	MVA range	High voltage winding kV.	Type	Low voltage winding kV.	Type
1. Generator	72–340	132–275	Disc or multi-layer helix	11, 22	Disc-helix
2. Primary transmission	45–161	132–440	Disc or multi-layer helix	11, 33, 66	Disc or Disc-helix
3. Secondary transmission	10–20	33–66	Disc	11	Disc or helix
4. Distribution	0.1–1	11–33	Foil, Cross-over, Multi-layer helix, Disc.	0.43	Helix or Multi-layer helix
5. Rural	0.016–0.1	3.3–11	Foil, Cross-over	0.43	Helix

5.2 ROTATING MACHINES

5.2.1 Salient Field Poles

Figure 5.2(a) shows a typical salient-pole for a d.c. and synchronous machine. The conductors are generally placed 'strip on edge'. Normal insulation between conductors can be treated asbestos paper. Pole end-plates are used so that the sharp corners are converted into round, to avoid damage of conductors at the bends. For larger machines, some of the conductors may be protruded to have larger cooling surface, or ducts may be provided between coils (Figure 5.2b).

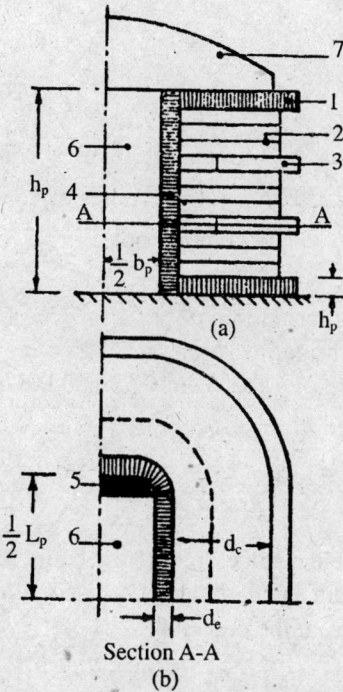

(a)

Section A-A
(b)

Figure 5.2 Salient field pole. 1-coil end insulation, 2-interturn insulation, 3-conductor, strip on edge protruding for better cooling, 4-pole body insulation, 5-pole end plate, 6-pole core, 7-pole shoe.

For the design of field winding, the data generally furnished are :- (i) voltage per coil*, v_f, (ii) field mmf per pole, F_f, and (iii) mechanical dimensions such as, armature -diameter, D; length, breadth and height of the pole. From equation (1.2),

$$R_f = \rho T_f \frac{s}{\alpha_f}$$

where, ρ = 0.0214 ohm-mm^2, m^{-1} for copper at 75°C;

s = the mean length of turn, metre (Figure 5.2a);

α = conductor section, mm^2,

and subscript 'f' stands for the field winding.

Putting, $v_f = R_f I_f$, we have,

$$\alpha_f = \rho \, s \, T_f \frac{I_f}{v_f} = \rho \, s \, \frac{F_f}{v_f} \qquad (5.1)$$

---

* For d.c. self-excited generator, v_f = 80% of the full-load terminal voltage E_t ; 20% being kept for field control; for motor, $v_f = E_t$.

The design steps can be

Step 1   Compute : -

  (i) voltage per coil $V_f = E_f/p$ (5.2)

  (ii) peripheral speed $v_r = \pi D n_r$ (5.3)

  (iii) heat transfer coefficient,

$$c_f = (6.25 \text{ to } 7.1)\,(1 + 0.07\,v_r)\ \text{W.m}^{-2}.\ {}^{\circ}\text{C}^{-1}$$

for stationary poles   (5.4a)

$$= (12.5 \text{ to } 16)\,(1 + 0.1\ vr)\ \text{W.m}^{-2}.\ {}^{\circ}\text{C}^{-1}$$

for rotating poles   (5.4b)

Step 2   Assume, coil depth, d_c;

               width of pole body insulation, d_e ;

               width of coil end insulation, h_e;

               winding space factor, k_w (between 0.55 and 0.67)

Step 3   Compute length of mean turn for one pole (Figure 5.2a),

$$s = 2(l_p + b_p) + \pi\,(2d_e + d_c)m \qquad\qquad (5.5)$$

Step 4–7   Compute

  (i) cross-sectional area of field conductor, α_f (from equn. 5.1). Choose standard conductor section and SWG no. from standard table corresponding to the computed value of α_f

  (ii) field turns per pole, T_f from the equation,

$$\alpha_f T_f = k_w\,d_c\,(h_p - 2h_e) \qquad\qquad (5.6)$$

and choose nearest whole number.

  (iii) field current $I_f = F_f/T_f$ (5.7)

  (iv) field resistance per pole, $R_f = v_f/I_f$ (5.8)

Step 8   Compute approximate cooling surface,

$$S_f = (s + d_c)\,(h_p - 2\,h_e)\,m^2 \qquad\qquad (5.9)$$

Step 9 & 10   Compute

  (i) I^2R-loss, $\dot{W}_{Lf} = I_f^2\,R_f$ Watt.   (5.10)

  (ii) temperature - rise $\theta = W_{Lf}\,(c_f\,S_f)\,{}^{\circ}\text{C}$ (5.11)

Check whether θ is satisfactorily near the specified maximum temperature-rise of 65°C. If not, modify the assumed value of dc and repeat above procedure.

The procedure is illustrated in Example 5.1.

Example 5.1

    A 15-kW, 220-V, 1150-rpm, 4-pole d.c. shunt motor has field ampere-turns per pole $F_f = 2000$, armature diameter $D = 25.5$ cm., $b_p = 8.0$ cm., $l_p = 14.0$ cm, $h_p = 10.0$ cm.

    Design a field winding for a maximum temperature-rise of 65°C, using copper as the conducting material.

Solution

(1) $v_f = \dfrac{220}{4} = 55v$ per pole.

$$v_r = \pi \times 25.5 \times 10^{-2} \left(\dfrac{1150}{60}\right) = 15.4\ m.\sec^{-1}$$

and $c_f = 6.5\ (1 + 0.07 \times 15.4) = 13.5$ W. $m^{-2}.\ ^\circ C^{-1}$

(2) Assume : $d_c = 1.5$ cm.,
$ d_e = 0.3$ cm.
$ h_e = 0.5$ cm.
$ k_w = 0.6$

(3) Length of mean turn, $s = 2(0.14 + 0.08) + \pi\ (2 \times 0.003 + 0.015)$
$$= 0.506 \text{ m.}$$

(4) Conductor section, $\alpha_f = 0.0214 \times 0.506\ (2000/55)$
$$= 0.3937 \text{ mm}^2.$$

Choose 22 SWG round copper wire.
Thus, modified bare conductor-section = 0.3973 mm^2,
enamel insulated section = 0.4550 mm^2.

(5) Number of turns, T_f

$$= \dfrac{0.6 \times 0.015\ (0.10 - 2 \times 0.005)}{0.3973 \times 10^{-6}}$$

$$= 2038.7$$

Choose $T_f = 2039$

(6) Field current, $I_f = \dfrac{2000}{2039} = 0.981$ A.

(7) Field resistance per pole,

$$R_f = \dfrac{55}{0.981} = 56.07\ \Omega$$

(8) Approximate cooling surface,
$$S_f = 0.506\ (0.10 - 2 \times 0.005)$$
$$= 0.0456 \text{ m}^2$$

(9) Copper-loss per pole, $W_{Lf} = 55 \times 0.981 = 53.96$ W

(10) Volume of copper, per pole = $0.506 \times 0.3973 \times 10^{-6} \times 2039$
$$= 409.9 \times 10^{-6}\ m^3$$

Weight of copper, per pole = $409.9 \times 10^{-6} \times 8890$
$$= 3.644 \text{ kg.}$$

(11) Temperature-rise, $\theta = \dfrac{53.96}{13.5 \times 0.0456} = 87.65^\circ C$

The computed temperature-rise is more than the permissible value of 65°C. Hence, de is to be increased suitably. This can be done by trial, which can be performed by a simple computer program as indicated in Figure 5.3.

A computer program

The design calculation as discussed in art. 5.2.1 and Example 5.1 have been translated into a computer program as shown in Figure 5.3. A switch (o) enables choice of heat transfer coefficient for either stationary poles or rotating poles (equn. 5.4 a and b). For optimised design against a given maximum temperature rise, program 'optm' have been written based on the flow chart in Figure 5.3.

```
// Design of field coil for a given maxmimun temperature-rise


#include <iostream.h>
#include <iomanip.h>
#include <math.h>
#include <conio.h>


class spec
{   private:
     float wsf;
     public:
     int n, at;
     float d;
     void getdata(int n, int at, float d, float wsf );
     void display() ;
     void fieldcoil(int n, int at, float d, float wsf) ;
} ;

void spec :: getdata( int rpm, int amtur, float dia, float wsfac )
{
      n = rpm ; at = amtur ; d = dia ; wsf= wsfac ;
}

void spec :: display()
{   cout <<"speed=    "<< n <<" r.p.m, Field ampere-turn= "<<at
        <<" ,armature dia.=  "<< d <<"winding Space factor=  "<< wsf;
}


void spec :: fieldcoil (int n, int at, float d, float wsf)
{
  float he,hp,k1,k2,k3;

  cout<<"Enter the values of pole height(hp) and width of coil-end"
       " insulation(he)" ;
  cin >> hp >> he ;

// Assumed copper as winding material:Sp.Resis=0.0214x10 pow(-6);
  k1 = ( hp - 2.* he) * 1000.;
  k2 = at * 0.0214 * at / ( wsf * k1 * k1 );

  int o;
  float vr,cf,v;
  const float PI = 3.14159;
```

```
//          Computation of Heat transfer coefficient, cf
  vr = PI * d * n / 60.;                    //Peripheral speed
  cout<< endl <<"Enter the value is o" ;
  cin >> o;
  switch(o)        // o=1 for stationary poles; =2 for rotating poles
  {
  case 1: cout << endl << setw(30) <<"Stationary Poles" ;
        cf = 6.60 * ( 1. + 0.07 * vr );
        cout << endl << setw(50) <<"cf =  "<< cf ;
        break;
  case 2: cout << endl << setw(30) <<"Rotating Poles" ;
        cf = 14.25 * ( 1. + 0.10 * vr );
        cout << endl << setw(50) <<"cf =  "<<cf ;
        break;
  }
  k3 = k2 / cf;
  cout << endl <<"k3=   "<< k3 ;
  void optm (float,float&,float&,float&,float&,float&,float&);
  int k5;
  float lp,bp,de,dc,fs,k4,C1,s,tmax,theta,Dc ;
  float af,maf,swg,vf,p,Tf,mTf,If,Wcp,Cvo,Cpwt ;

  cout <<"Enter the values of lp,bp,de" ;        // see Fig. 5.2
  cin >> lp >> bp >> de ;

  k4 = 2. * ((lp + bp) + PI * de) ;

  cout <<endl <<"Enter the value of Assumed Coil-depth, dc" ;
  cin >> dc ;                //start with a chosen value of dc,metre.

  s = k4 + PI * dc ;
  cout << endl << setw(40) <<"s =  "<< s ;
  fs = s / ((s + de) * dc) ;
  theta = k3 * fs ;
  cout << endl << setw(30) <<"Computed temp-rise=  "<< theta ;

  cin >> rpm >> amtur >> dia >> wsfac ;
  s.getdata( rpm, amtur, dia, wsfac );
  s.display() ;
  s.fieldcoil( rpm, amtur, dia, wsfac ) ;
}

void optm(float dc,float& tmax,float& k3,float& k4,float& de,float& theta,float& Dc)
{                            // optimisation program
  float fs,s;
  const float PI = 3.14159 ;
  if (theta >= tmax)
  {
  do
  {
  dc = dc + 0.0001 ;
  s = k4 + PI * dc ;
  fs = s / ((s + de) * dc) ;
  Dc = dc ;
  theta = k3 * fs ;
  }
```

```
while(theta>tmax) ;
cout << endl << setw(30) <<"Dc:  "<< Dc << setw(30) <<"temp.rise:  "<< theta ;
}
else if (theta<tmax)
 {

do
{
dc = dc - 0.0001 ;
s = k4 + PI * dc;
fs = s / ((s + de) * dc);
theta = k3 * fs;
}
while(theta<tmax);
dc = dc + 0.0001 ;
s = k4 + PI * dc ;
fs = s / ((s + de) * dc) ;
theta = k3 * fs ;
Dc = dc ;
 }
}
```

Figure 5.3 A computer program for the design of field coil for a given maximum temperature-rise.

RESULT

Run 1

Enter(E)   1150   2000   0.255   0.6
              Enter pole height(hp) and width of coil-end insulation(he)
(E)   0.1   0.005
              Enter the value is o
(E)   o = 1
Output(o)   stationary poles
              cf = 13.694
              k3 = 1.286
(E)   lp = 0.14 bp = 0.08 de = 0.003
(E)   dc = 0.015

              temp. rise = 85.24
(E)   allowable maxm. temp. rise = 65
(o)   Dc = 0.0197
              temp. rise = 64.92
              Enter vp, p.
(E)   55   4
(o)   af = 1.629
(E)   modified bare cond. section (from standard table), maf = 1.70354,
              16 3/4 swg
(o)   Temperature-rise = 64.92

Total Field turns = 624
Bare conductor section = 1.70354 mm. sq.
Conductor guage No. = 16 ¾
Field current = 3.205 amp
Total copper loss = 176.3 watts
Total weight of copper = 19.68 kg.

Run 2

Enter(E)	1150 2000 0.255 0.6
	Enter pole height(hp) and width of coil-end insulation(he)
(E)	0.1 0.005
	Enter the value is o
(E)	o = 2
output(o)	rotating poles
	cf = 36.13
	k3 = 0.4875
(E)	lp = 0.14 bp = 0.08 de = 0.003
(E)	dc = 0.01
(o)	s = 0.773
	temp. rise = 4.856
(E)	allowable maxm. temp. rise = 65
(o)	Dc = 0.0075
	temp. rise = 64.61
(E)	vp = 55 at = 2000 p = 4 hp = 0.075 he = 0.005 wsf = 0.6
(o)	af = 1.502
(E)	modified bare cond. section (from standard table), maf = 1.589 17 swg
(o)	Temperature-rise = 64.61
	Total Field turns = 252
	Bare conductor secion = 1.581 mm. sq.
	Conductor guage No. = 17
	Field current = 7.94 amp
	Total copper loss = 436.5 watts
	Total weight of copper = 6.87 kg.

The above design with rotating poles needs modification-coil-depth is very small, field current and copper losses are very high. As such, computation is done with height of poles reduced to 0.075 m. The result is as below:

Temperature-rise = 64.97
Total Field turns = 3396

Bare conductor secion = 1.6417 mm. sq.

Conductor gauge No. = 26

Field current = 0.059 amp

Total copper loss = 3.24 watts

Total weight of copper = 9.99 kg.

Example 5.2

For the data in example 5.1, obtain the design for minimum copper weight.

Solution   It can be seen in example 5.1 that the temperature-rise θ attains a value less than the permissible 65°C with $d_c = 0.02$ metre and the corresponding copper weight is 5.01 kg.

The procedure used in example 5.1 can be used to determine dc for the minimum copper weight knowing that θ will always be less than the permissible 65°C for $d_c > 0.02$ metre. Hence, by increasing d_c in certain steps (say 0.001 m.) the complete set of calculation is performed and the design for minimum copper-weight is chosen.

It will be seen that the design with $d_c = 0.02$ metre gives the minimum copper-weight. With $d_c < 0.02$ m., copper weight is lesser but the temperature-rise is above its acceptable value.

5.2.2   Non-Salient Field Poles

Non-salient field poles form the rotor of turbo-generator, where the centrifugal force due to high peripheral speed necessitates the use of rotor made of massive steel forging. The field winding is distributed in slots which are deep and are milled from the rotor cylinder (Figure 1.9). The design of such winding can be made on the same lines as for salient-pole field winding distributing the turns in the slots and with special attention to the mode of heat-dissipation (Figure 2.17).

5.2.3   Armature

In a rotating machine, armature winding consists of coils uniformly distributed in slots along the armature periphery, and can be broadly divided under two heads : -

(a) Closed type,

(b) Open type.

The former, as the name signifies, is closed in itself, and is accessible through commutator. The open type, on the other hand, has end terminals which can be connected as star or delta in three-phase machines.

5.2.3.1   *Winding terms*

(i) A coil in a winding (Figure 5.5) has two sides known as COILSIDES and two OVERHANGS. The coilsides are placed in slots–its parts lying within slots are *active* parts of the winding.

---

* The readers are advised to formulate the program.

(ii) The distance between the coilsides (y_w) of a coil in slots is the coilspan. Coilspan is generally indicated in terms of number of teeth enclosed by the coil.

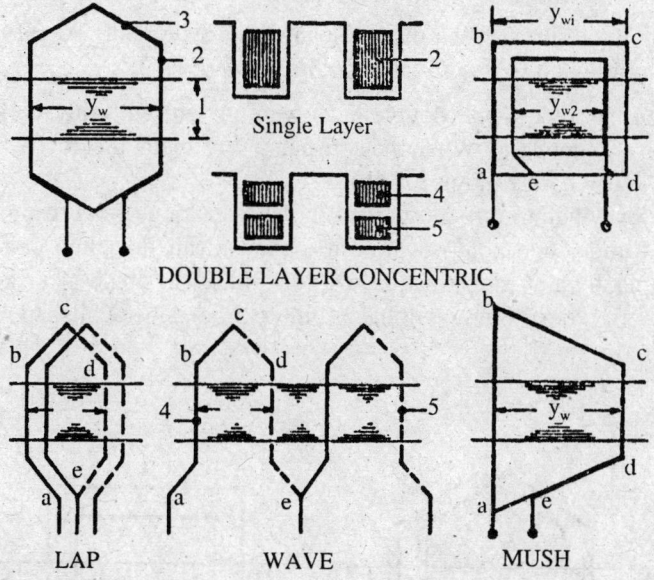

Figure 5.5   Type of armature winding. 1—active part, 2—coilside, 3—overhang, 4—top layer, 5—bottom layer.

(iii) An armature winding may be of one layer, called *single layer*, or of two layers, called *double layer*. In a single layer winding each coilside occupies a whole slot, whereas in a double-layer winding there are one or more coilsides per layer in a slot. For the end connections in a double-layer winding, one coilside of a coil is placed on the top layer of one slot and the other coilside in the bottom layer of another slot.

In a d.c. machine, the armature winding is invariably of double layer whereas in a.c. machines, single layer windings are used upto a few *kw* rating; otherwise, armature winding is always double-layer.

If S be the number of armature slots, for a single-layer winding, number of armature coils

$$C = \frac{1}{2}S \tag{5.13a}$$

since, the slot is fully occupied by one coilside; and for a double-layer winding, with n coilsides per layer, number of armature coils $C = n.\ S$ (5.13b)

(iv) Depending on how the coilsides are connected, an armature winding can be classified as,

   (a)   Concentric ⎫
                    ⎬ single - layer
   (b)   Mush   ⎭

   (c)   Lap   ⎫
              ⎬ double-layer
   (d)   Wave ⎭

In concentric windings, coils have different coilspans, coilspan (Figure 5.5) of the inner coil (y_{w2}) being smaller than y_{w1} coilspan of the outer coil. On the otherhand, coils in each of mush, lap and wave windings have the same coilspan.

(v) A coilside is *full-pitch* when the coilspan is one pole pitch. A coil having span other than one pole-pitch is *fractional-pitch or chorded.*

For a 24-slot armature, a 4-pole double-layer winding with full-pitch coils will have a span of (24/4 =) 6 teeth (Figure 5.6a). When the coilspan y_w = 5 teeth, the double-layer winding has fractional pitch = 5/6 = 83.3% (Figure 5.6b).

In a single layer winding, difficulty arises in the formation of overhang, and in one form, pitch of each coil under one pole is varied to obtain mean full-pitch or fractional-pitch as desired. For example, for a 24-slot armature, a 4-pole, fullpitch, single-layer, concentric winding with 2 coil-sides per pole per phase will have (Figure 5.6c), since coilspan for full-pitch = 24/

$$4 = 6 \text{ teeth} = \frac{1}{2}(y_{w_1} + y_{w_2}).$$

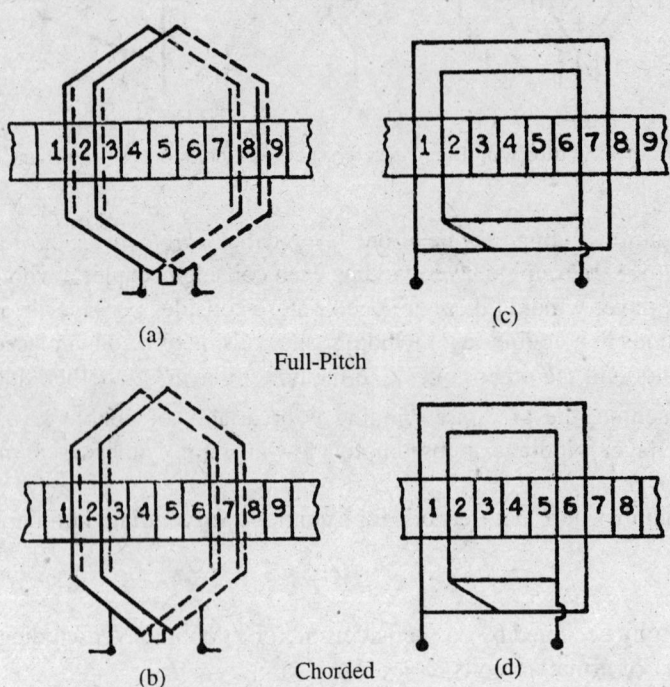

(a)

(c)

Full-Pitch

(b)   Chorded   (d)

Figure 5.6 Full-pitch and chorded coils. (a) & (b)—double-layer, (c) & (d)–single-layer.

$$y_{w1} = 7 \text{ teeth and } y_{w2} = 5 \text{ teeth,}$$

Again, the same armature with 4-pole, fractional-pitch, single-layer, concentric winding with 2 coils per pole (as in a single-phase motor) will have (Figure 5.6d),

$$y_{w1} = 6 \text{ teeth and } y_{w2} = 4 \text{ teeth,}$$

giving a pitch of (5/6) i.e. 83.3%, since, $\frac{1}{2}(y_{w_1} + y_{w_2}) = 5$ teeth and coilspan for full-pitch is 6 teeth.

[*Note* : (1) Fractional-pitch winding has a reduced overhang length and its pitch may be so chosen as to suppress or reduce certain harmonics. However, chording reduces the induced voltage (which can, however, be compensated by an equivalent increase in number of turns) and is not generally made below 2/3rd.]

5.2.3.2 *Closed Winding*

Closed winding, suitable for commutator machines (both d.c. and a.c.) are composed of lap and wave *winding elements*. A winding element may be defined as that part of winding which spans two successive (according to winding scheme) commutator segments.

Simple form of lap and wave winding elements (to distinguish it from more camplicated types) are *simple lap* and *simple wave*. Figure 5.7 shows (a) simple lap-

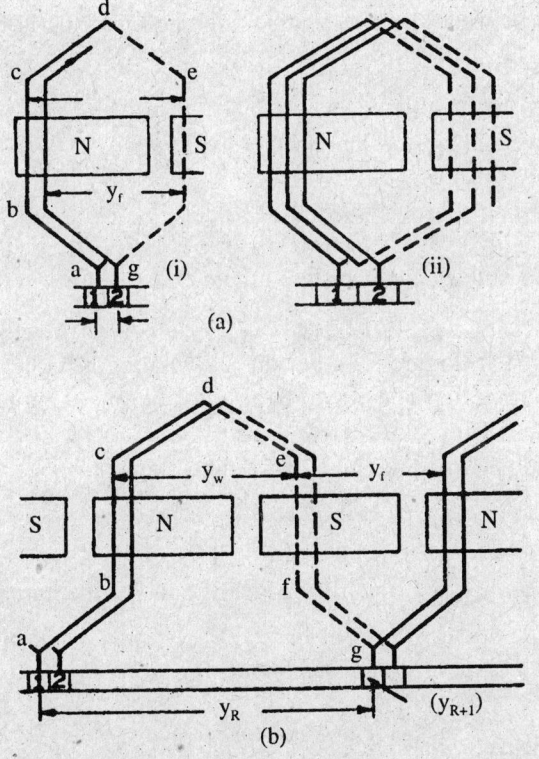

Figure 5.7 Winding elements, (a) simple lap, (b) simple wave

(i) single turn, and (ii) multi-turn; and (b) simple wave winding elements. Multiturn wave is not generally used.

For simple lap, a winding element a-b-c-d-e-f-g starting from say, segment 1 covers the coilsides under two consecutive poles and returns to segment 2.

For simple wave, winding element a-b-c-d-e-f-g starts from segment 1, covers coilsides under two consecutive poles, and moves ahead to segment $(1 + y_R)$, where, y_R is called the winding pitch.

In an armature winding, the winding elements are joined *in series* till the complete armature periphery is traversed.

5.2.3.3 *Winding terms*

(i) An armature winding may be either *progressive* or *retrogressive*. For the lap type, the winding is progressive when the winding element terminates in a segment (2 in Figure 5.7) to the right of the segment from where it has started (1 in Figure), and retrogressive, when it terminates in a segment to the left.

For the wave type, the winding elements in series after traversing the armature periphery may return to the right (progressive) or to the left (retrogressive) of the first segment from where the first winding element has started.

(ii) The distance between the consecutive winding elements is called the *winding pitch* (y_R).

For simple lap, $y_R = \pm 1$ (5.14)

'+' sign for progressive and '–' for the retrogressive.

For simple wave, $y_R = \dfrac{2}{p}(C \pm 1)$ (5.15)

C being the number of coils = number of commutator segments.

Again, in terms of coilspan y_w, $y_R = y_w + y_f$ (5.16)

where, y_f is the front pitch (Figure 5.7). For the lap winding, y_f is negative.

[*Note* : (1) For simple lap and wave, there must be one commutator segment for each winding element. That is,

No. of commutator segment, C = no. of winding elements

$$= \frac{1}{2}(Z/T_w)$$ (5.17)

where, Z is the total number of armature conductors and T_w, the number of turns per winding element.

(2) For symmetrical lap, *total number of coils must be divisible by the number of pole-pairs*.

For symmetrical wave, *total number of coils must not be divisible by the number of pole-pairs*.

For example, an armature with 21 coils is unsuitable for 4-pole simple lap but suitable for 4-pole simple wave, and also for 6-pole simple lap.]

Design of simple lap and wave windings are illustrated through examples 5.3 and 5.4.

Example 5.3

Design a simple lap winding—4-pole, double–layer, one coilside per layer and progressive, for a 12-slot armature.

Solution   From equation (5.13b), for $n = 1$, no. of coils = 12.
Hence, no. of commutator segments = 12.

$$\text{Pole-pitch} = S/p = 3 \text{ teeth.}$$

Choose coilspan $y_w = 3$ teeth for full-pitch coil.

Hence,   front pitch $y_f = -2$ teeth.

Figure 5.8(a) shows a simple lap winding for the given data.

Alternatively, the winding can be depicted through *winding table* as below : -

(1) Identification of coilsides : -

	Slot no.	1	2	3	4	5	6	7	8	9	10	11	12
Coilside	Top layer	1	3	5	7	9	11	13	15	17	19	21	23
	Bottom layer	2	4	6	8	10	12	14	16	18	20	22	24

(2) Identification of coils in terms of coil-throw : -

With $y = 3$ teeth, a lop-layer coilside 1 in slot 1 should be connected to the bottom-layer coilside 8 in slot 4 and so on, to form full-pitch coils. Thus, we have,

Coil no.	1	2	3	4	5	6	7	8	9	10	11	12
Coil throw	1–8	3–10	5–12	7–14	9–16	11–18	13–20	15–22	17–24	19–2	21–4	23–6

(3) Since the front pitch $y_f = -2$ teeth, the end connection between two consecutive coils is known. Thus, end of the bottom-layer coilside 8 in slot 4 should be connected to the top-layer coilside 3 in slot 2 (two teeth behind slot no. 4), and so on.

Thus the winding table is,

$$\text{Coil no. } 1 + 2 + 3 + 4 + 5 + 6 + 7 + 8 + 9 + 10 + 11 + 12 + 1 -$$
$$\text{or, in terms of coil throw,}$$

$$(1 - 8) + (3 - 10) + (5 - 12) + (7 - 14) + (9 - 16) + (11 - 18) +$$
$$(13 - 20) + (15 - 22) + (17 - 24) + (19 - 2) + (21 - 4) + (23 - 6) + (1 -$$

[*Note* : (1) *Brush position*

With pole-pitch (= 3 teeth) indicated on the winding diagram, the instantaneous directions of induced e.in.f.s (and hence currents) in the coilsides can bo marked. It can be seen (following the instantaneous directions of currents) that the complete winding is divided into 4 parallel paths (Figure 5.8b) and thus the positions of 4 brushes can be fixed up.

If the brushes c and d are omitted, the winding will have two parallel paths consisting of 3 coils (coils 1 to 3) in one path and the remaining in another. In the latter path however, 6 coils would be ineffective as the e.m.f.s in coils 7, 8 & 9 can cancel those in coils 10, 11 & 12, and thus the power rating will be reduced by 50%. Four brushes are thus essential to utilise the full winding.

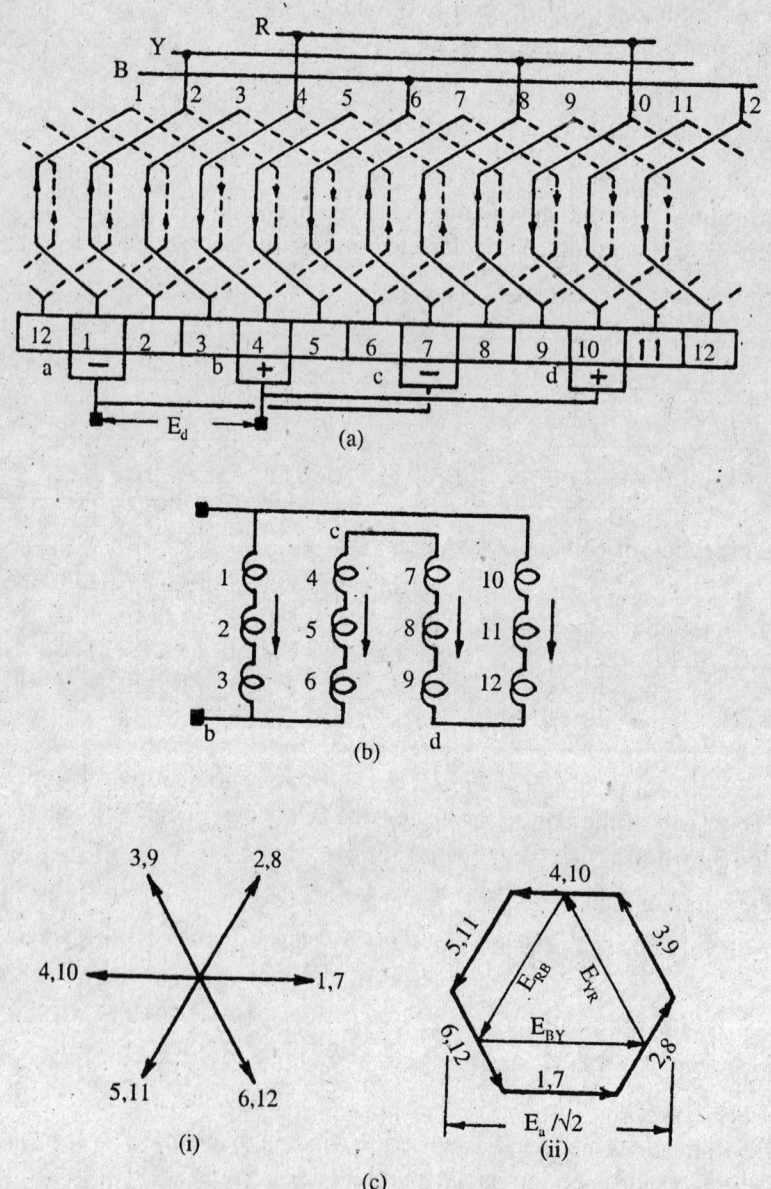

Figure 5.8 Simple lap winding–12-slot, 4-pole (a) winding diagram, (b) parallel paths, (c)–(i) coil emf star, (ii) emf polygon.

(2) E.m.f. polygon

A knowledge about the coil e.m.f. and the resultant voltage can be obtained from the e.m.f. polygon. The procedure is as below : -

(a) Compute electrical angle between any two consecutive coil e.m.f.s. The required angle is given by the winding pitch y_R and for simple lap, such angle

$$\gamma = 1 \text{ slot pitch}$$

$$= \frac{p}{S} \times 180° \text{ elec.} \tag{5.18}$$

where, p is the number of poles and S, the number of slots.

In example 5.3, $\gamma = \frac{4}{12} \times 180 = 60°$ electrical

(b) With the induced voltage (r.m.s. value) in any coil (say, coil 1) as reference phasor, induced voltage phasors of coils 2, 3, 4, etc. are drawn to form the coil-emf star (Figure 5.8 c-i), Basing on the coil-e.m.f. star, the e.m.f. polygon (Figure 5.8 c-ii) can be drawn.

It is of interest to note that,

(i) for a complete winding, the polygon is traversed twice in the example 5.3. In fact, *in a p-pole machine the number of such traversal is equal to the pole-pair.*

(ii) with the brushes placed as discussed under Note (i) above, the voltage across the brushes,

$$E_4 = \sqrt{2} \, E_{ab} \tag{5.19}$$

since the phasors in the e.m.f. polygon denote r.m.s. values.

(iii) the polygon once traversed, contains two parallel paths. That is, the total number of parallel paths equals the number of poles, as shown by Figure 5.8(b).

(3) Coilside per layer

In a simple lap winding, the voltage per coil is limited by the permissible voltage of about 12 V between two consecutive commutator segments. The turns per coil is thus limited and generally the required number of coils is large.

In view of above and to avoid unreasonably small slots, double-layer winding with more than one coilside per layer is used. The number of coilsides per slot must be even and is not generally more than 10.

Figure 5.9 shows connections to coilsides for double-layer winding with 2 coilsides per layer.

(4) Connections for a.c. supply

With direct voltage appearing across the brushes, alternating voltages are available at points inside of the commutator. The closed winding can bo thus,

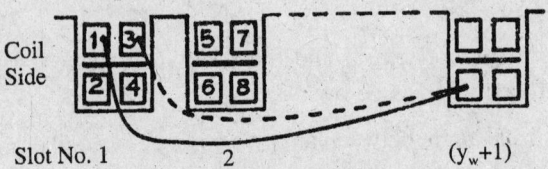

Figure 5.9   Dauble-layer winding with 2 coilsides per layer.

tapped suitably at the non-commutator end to provide for a symmetrical m-phase supply. Figure 5.8(d) shows tappings for a symmetrical 3-phase supply in which each section of the winding extending over two pole-pitches is tapped at 3 equally-spaced points measured in terms of phase-pitch,

$$y_{ph} = \frac{2C}{3p} \text{ for 3-phase,}$$

and

$$= \frac{2\,C}{m\,p} \text{ for m-phase supply.} \tag{5.20}$$

In example $y_{ph} = \dfrac{2 \times 12}{3 \times 4} = 2$ coils. That is, every second coil must be tapped and connected to alternate sliprings (Figure 5.8d). The corresponding line voltages E_{RY}, E_{YB} and E_{RR} are shown on the e.m.f. polygon.

Such a.c. winding viewed from the slip-ring end is a closed delta-connected winding. It is evident from above that for a symmetrical simple lap winding the total number of coils must not only be divisible by the number of pole-pairs, but also by the number of phases.

(5)   Equilising rings

Since a simple lap winding consists of several parallel branches it would appear permissible to assume that the branch voltages are equal. Such assumption is not necessarily valid in a *multipolar* machine since the reluctance of various branch-magnetic circuits may differ due to the variation of gap-length under different poles and iron parts may also have variations.

Such differences in branch voltages would result in circulating currents which may be of considerable magnitude even though the voltage differences are small (art 1.7.2-ii) since, the currents are only limited by small resistances of the armature circuits. Circulating currents would complete their paths through brushes, and under load conditions, may overload the brushes resulting in excessive sparking.

Circulating currents may be prevented from flowing through the brushes by using *equilising rings*. An equilising ring connects equipotential points on the winding. In a multipolar lap winding it generally suffices to use 10 to 20 such rings, each ring connecting a number of equipotential points equal to the pole-pair.

Figure 5.10 shows the non-commutator end of a 8-pole lap winding having 2 coilsides per layer and 16 slots. 4 equilising rings have been used so that the pitch of tapping points

$$y_{eq} = \frac{2\,C}{n_e\,p} \tag{5.21}$$

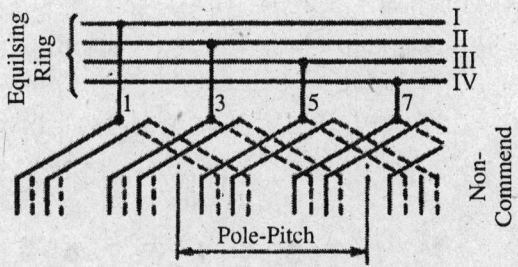

Figure 5.10 Equilising ring–4 coilsides per slot.

where, n_e is the number of equilising rings. For the example under consideration,

$$y_{eq} = \frac{2 \times 2 \times 16}{4 \times 8} = 2 \text{ coils.}$$

Example 5.4

For a 15-slot armature, design a 4-pole, double-layer, progressive, wave winding with one coilside per layer*.

Solution   From equation (5.13b),

no. of coils = 15

no. of commutator segments = 15.

For progressive wave $y_R = 2\dfrac{15+1}{4} = 8$ coils = 8 teeth.

$$\text{Pole-pitch} = \frac{15}{4} = 3\frac{3}{4} \text{ teeth.}$$

Choosing, coilspan $= y_w = 4$ teeth,

the front pitch $y_f = -y_w = 4$ teeth.

Figure 5.11(a) shows the winding diagram.

Alternatively, identifying the coilsides in the same manner as in example 5.3, the throw of each coil is

Coil no.	1	2	3	4	5	6	7
Throw	1–10	3–12	5–14	7–16	9–18	11–20	13–22

	8	9	10	11	12	13	14	15
	15–24	17–26	19–28	21–30	23–2	25–4	27–6	29–8

* See note (2) below, $n = 1$ has been chosen for simplicity.

The winding table is (with $y_R = 8$ teeth), $\leftarrow y_R \rightarrow$

Coil $1 + 9 + 2 + 10 + 3 + 11 + 4 + 12 + 5 + 13 + 6 + 14 + 7 + 15 + 8 + 1-$

In terms of coil throw,

$(1 - 10) + (17 - 26) + (3 - 12) + (19 - 28) + (5 - 14) + (21 - 30) + (7 - 16) + (23 - 2) +$
$(9 - 18) + (25 - 4) + (11 - 20) + (27 - 6) + (13 - 22) + (29 - 8) + (15 - 24) + (1 -$

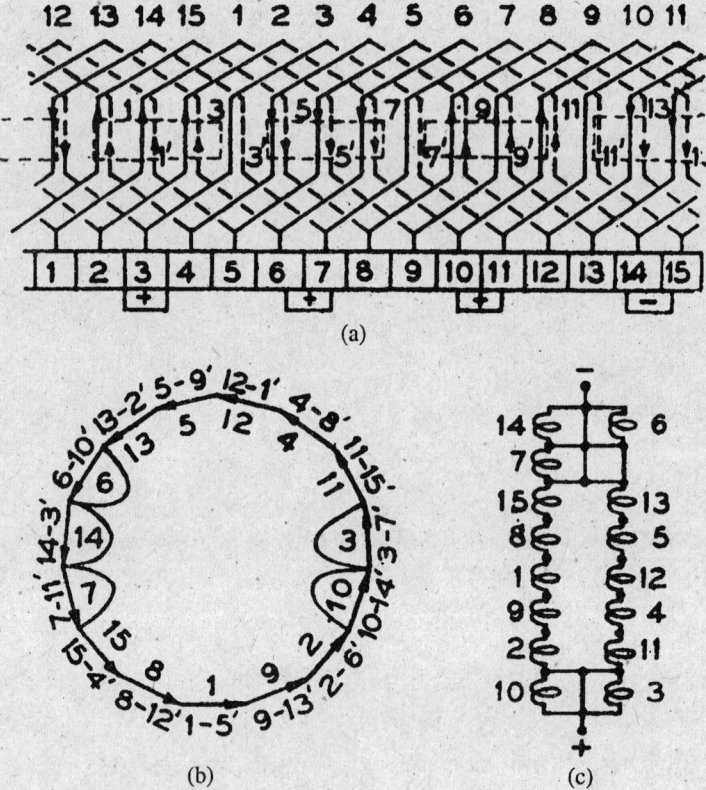

Figure 5.11 Simple wave winding- 15-slot, 4-poie, (w) winding diagram, (b) emf polygon, (c) parallel paths.

[*Note :* (1) In simple wave winding, since the number of slots per pole-pair is not an integer, the pole-pitch in terms of number of teeth is fractional. The coilspan cannot be an exact* pole-pitch and the number of teeth embraced by a coil is made as close to the pole-pitch as possible. Example 5.5 illustrates a simple wave winding when the number of coilsides per layer is more than 1.

---

* The number of coils per pole can be made an integer if one or more *dummy coils* are used. Dummy coils, as the name signifies, are coils which fill the slot-spaces but are not connected in the circuit. Generally this is avoided, but use of standard armature stampings often forces a designer to use dummy coils.

In general, the choice of coilspan in machines having coilsides per layer n more than 1 is given by the equation,

$$y_w = (\text{an integer} \times n) + 1 \tag{5.22}$$

if the coilgroups are bound together.

(2) *Coilside per layer*

In a simple wave winding, I coilside per layer would rarely occur. The number of coilsides per layer must be odd for even number of pole-pairs and even for odd pole-pairs. That is, for 4, 8, and 12 poles, n is 3, 5, etc; and for 6 and 10 poles, n is 2, 4, etc.

(3) *E.m.f. polygon*

The electrical angle between coil 1 and coil 2 is (by equn. 5.18),

$$\gamma = \frac{4 \times 180}{15} = 48°$$

and that between coil 1 and 9 is,

$$y_R \cdot \gamma = 8 \times 48 = 384° \approx 24°$$

The coil-emf star and the emf polygon is shown in Figure 5.11(b). It can be seen that,
  (i) whatever be the number of poles, the polygon is traversed once.
  (ii) the winding is equivalent to a simple lap winding of two poles having 15 coils.

(4) *Brush position*

Since a simple wave winding is equivalent to a two-pole lap winding, only two brushes would suffice, thereby dividing the winding into 2 parallel paths (independent of the number of poles). In example 5.4 brushes on segments 3 and 6/7 would suffice. However, such practice for a large machine normally leads to larger commutator and brush length.

On the other hand, if the number of brushes is made equal to the number of poles (as in a simple lap winding), current per brush is reduced, thereby reducing the required lengths of brushes and commutator segments.

The effect of placement of 4 brushes at 1 pole-pitch apart is shown in Figure 5.11(c). Coils 6 and 14 are shorted by brush b, 10 and 3 by brush c, and, 14 and 7 by brush d. That is, – ve brushes short-circuit 3 coils and the others 2 coils.

The winding is still divided into 2 parallel paths of equal number of coils (5 in the example).

(5) *Use in a.c.*

Wave winding is not in common use in polyphase commutator machines, though frequently used in single phase motors. Lap winding is preferred as it results in larger number of parallel paths. The current per brush arm is low, and thus the voltage across consecutive commutator segments is low.

Example 5.5

For an 11-slot armature, design a 4-pole, doublelayer progressive winding having 3 coilsides per layer.

Solution

From equation (5.13b), for $n = 3$,
no. of coil = $3 \times 11 = 33$.
no. of commutator segments = 33.

Number of coils per pole-pair $= \dfrac{33}{4}$ is not an integer.

Hence, simple wave winding is suitable.
For progressive winding,

$$\text{winding pitch, } y_R = 2\,\frac{33+1}{4} = 17 \text{ coils;}$$

$$\text{pole-pitch } \frac{11}{4} = 2\,\frac{3}{4} \text{ teeth.}$$

$$\text{Choose } y_w = 3 \text{ teeth.}$$

(1)   *Identification of coilsides*

Slot no.	1			2			3			4			5		
Top coilside	1	3	5	7	9	11	13	15	17	19	21	23	25	27	29
Bottom coilside	2	4	6	8	10	12	14	16	18	20	22	24	26	28	30

6			7			8			9			10			11		
31	33	35	37	39	41	43	45	47	49	51	53	55	57	59	61	63	65
32	34	36	38	40	42	44	46	48	50	52	54	56	58	60	62	64	66

(2)   *Coil throw*

Coil no	1	2	3	4	5	6
Throw	1–20	3–22	5–24	7–26	9–28	11–30

7	8	9	10	11	12	13
13–32	15–34	17–36	19–38	21–40	23–42	25–44

14	15	16	17	18	19	20
27–46	29–48	31–50	33–52	35–54	37–56	39–58

21	22	23	24	25	26	27
41–60	43–62	45–64	47–66	49–2	51–4	53–6

28	29	30	31	32	33
55–8	57–10	59–12	61–14	63–16	65–18

(3)   *Winding table*

$$y_r = 17 \text{ coils}$$

Coil no.   1 + 18 + 2 + 19 + 3 + 20 + 4 + 21 + 5 + 22 + 6 + 23 + 7 + 24 + 8 + 25 + 9 + 26
               + 10 + 27 + 11 + 28 + 12 + 29 + 13 + 30 + 14 + 31 + 15 + 32 + 16 + 33 + 1−

or, in terms of coil-throw,

(1 − 20) + (35 − 54) + (3 − 22) + (37 − 56) + (5 − 24) + (39 − 58) + (7 − 26) + (41 − 60) +
(9 − 28) + (43 − 62) + (11 − 30) + (45 − 64) + (13 − 32) + (47 − 66) + (15 − 34) + (49 − 2) +
(17 − 36) + (51 − 4) + (19 − 38) + (53 − 6) + (21 − 40) + (55 − 8) + (23 − 42) + (57 − 10) +
(25 − 44) + (59 − 12) + (27 − 44) + (61 − 14) + (29 − 46) + (63 − 16) + (31 − 48) + (65 − 18)
+ (33 − 50) + (1−

5.2.3.4   *Comparison*

Important points of distinction between simple lap and simple wave windin' are given in
Table 5.2.

Table 5.2   Comparison between Simple lap and Simple wave windings

Item	Simple lap	Simple wave
1. E.m.f.	Independent of number of poles.	Dependent on number of poles.
2. Permissible current in armature conductor	Dependent on number of poles.	Independent of number of poles
3. Winding pitch	± 1	$\dfrac{2}{p}(c \pm 1)$
4. Front pitch	Negative	Positive
5. Number of coils per pole-pair.	Integer	Not an integer
6. Number of parallel paths	Number of poles	2 (Independent of number of of poles).

p Number of poles; *c*.. Number of coils. ....

5.2.3.5   *Multiplex winding*

It is sometimes necessary to have the number of parallel paths greater than that given by
simple lap winding. Multiplex winding can be useful under the above condition.

In a lap winding, if the winding pitch is made equal to 2 (instead of 1 as in simple lap), the
coils can be arranged into twice the number of poles. Such winding is known as *Duplex* lap
winding. The brush-width is equal to or more than the width of two commutator segments and
the number of coils in series is one-half the total number of coils per pole. The following cases
may be considered : -

Case I :   Duplex doubly-closed lap winding

When the number of armature coils is even, two separate closed circuits (*doubly-closed*)
are formed—one consisting of odd numbered coils in series and the other of even numbered in

series. The two circuits, adjacent to each other, are connected in parallel by the brushes.

Figure 5.12 illustrates a duplex doubly-closed winding for the same data of example 5.3, and the corresponding e.m.f. polygons.

Case II :   Duplex singly-closed lap winding

With odd number of armature coils, the winding will not be closed after the armature periphery is traversed once. Instead, the coils left during the first traversal are now incorporated and a singly closed winding is obtained (Figure 5.13).

[*Note* : The multiplex lap winding need not be limited to duplex winding. If y_R is greater than 2, higher-multiple lap winding is obtained. For duplex lap, the number of parallel circuits = $2p$. In general, for $y_R = b$, the number of parallel circuits in a multiplex lap winding = $b. p$].

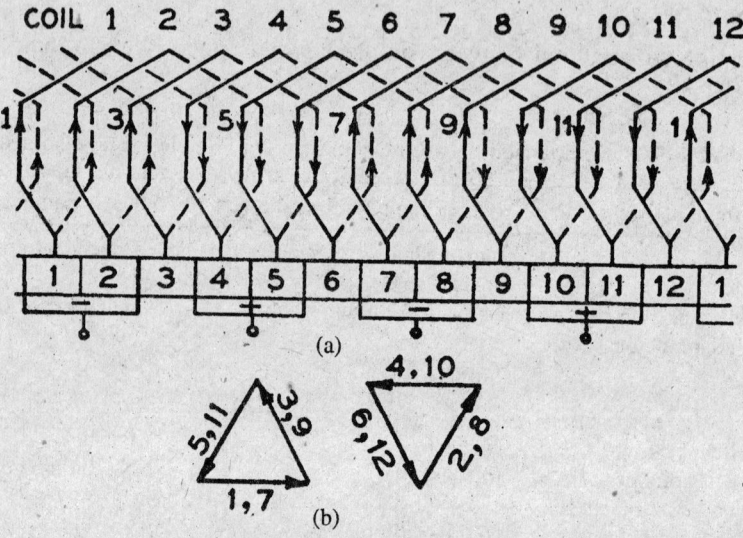

Figure 5.12   Duplex lap winding–doubly-closed, 12-slot, 4-pole. (a) winding diagram, (b) emf polygon.

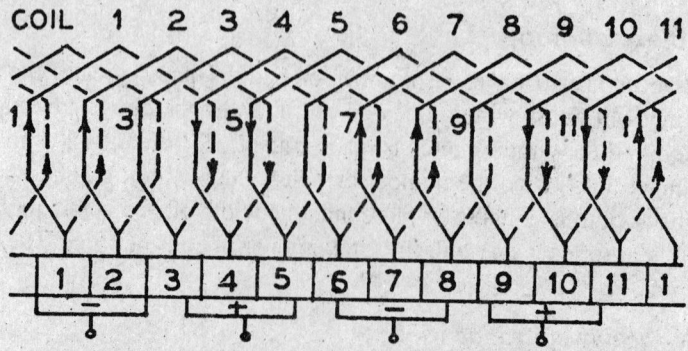

Figure 5.13   Duplex wave winding-singly-closed, 11-slot, 4-pole.

5.2.3.5 *Open winding*

With armature winding assumed equivalent to a current sheet, the e.m.f. polygon (Figure 5.8 c–ii) becomes a circle. As discussed in art. 5.2.3.4–note (4) under example 5.3, the winding is converted into a closed 3-phase delta-connected winding by tapping it at 3 equally-spaced points (120° electrical apart) as in Figure 5.14(a).

On the other hand, if the winding is opened at the 3 tapping points and terminals marked in proper sequence (such as, S(start) and F(finish) as in Figure 5.14b), a 3-phase starconnected winding can be formed by connecting 3 like-terminals as star point and the other 3 as winding terminals.

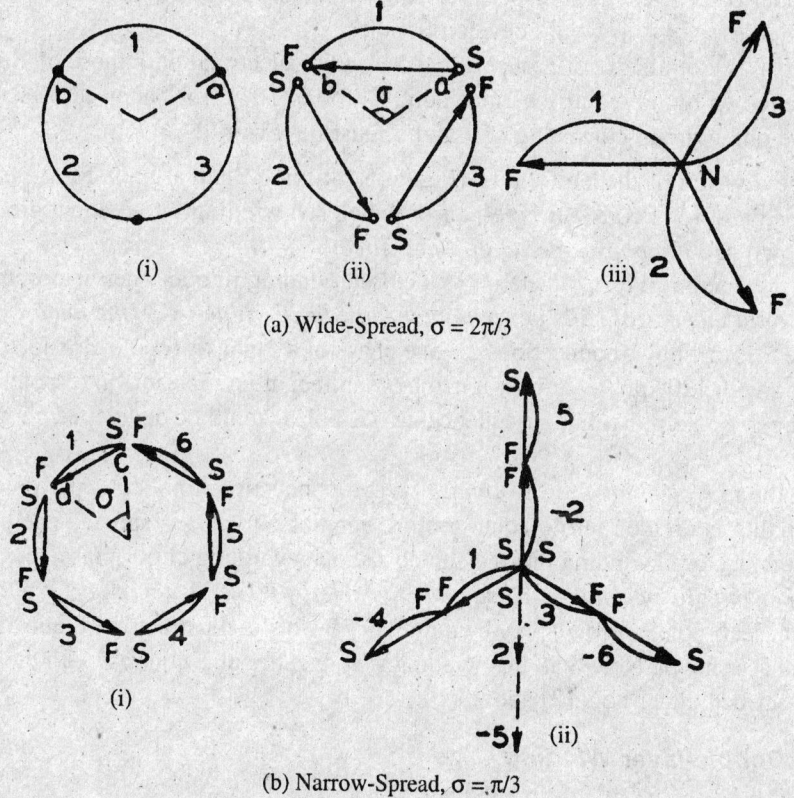

(a) Wide-Spread, $\sigma = 2\pi/3$

(b) Narrow-Spread, $\sigma = \pi/3$

Figure 5.14 Open winding.

For both delta- and star-connected windings formulated above, the ratio between the phase emf (chord ab) and the algebraic sum of phasors between tappings a & b (arc ab) gives the *distribution factor or breadth factor k_b*, with reference to Figure 5.14 (a–i & ii).

$$k_b = \frac{\text{chord ab}}{\text{arc ab}} = \frac{\text{radius} \times 2 \sin (\pi/3)}{(1/3)\, 2\pi \times \text{radius}} = 0.827 \qquad (5.20)$$

This shows that for a 3-phase winding as above, the phase emf is 82.7% of the algebraic sum of coil emfs in each phase-group.

The breadth-factor can be improved upon if the emf-polygon is divided by 6 equally-spaced tappings per pole pair (Figure 5.14 b-i) as calculated below :-

$$k_b = \frac{\text{chord cd}}{\text{arc cd}} = \frac{\text{radius} \times 2 \sin (\pi / 6)}{(1/6) \times 2\pi \times \text{radius}} = 0.955 \qquad (5.20)$$

[*Note* : (1) For connection of the 6 parts per pole-pair into a 3-phase star or delta, it is to be noted that,

   (i) the coils under parts 4, 5 and 6 must be connected in opposite sequence with the coils under parts 1, 2, and 3 respectively, so that, at any instant if coils 1, 2, and 3 develop N–poles, 2, 4 & 6 would develop S–poles.

   (ii) the axis of emfs of coilgroups 2 and 5 is at 60° electrical with those of 1 & 4 and 3 & 6 (shown by dotted arrows in Figure 5.14b-ii). Thus, for balanced 3-phase winding, the phase group consisting of 2 and 5 must be reversed.

   (2) The division of the 'emf circle' (Figure 5.14a) into 3 equal parts for 3-phase winding means division of total coils per pole-pair of the closed winding into 3 equal groups of coils. This is known as 120° *phase-spread* or *Wide-spread*.

Thus, the condition for conversion of a closed winding into an open widespread winding is that *the total number of coils per pole-pair must be divisible by 3 (the number of phases).*

Figure 5.15(a) shows connections for one phase of a 3-phase, 6-pole, 36-slot, double-layer lap winding with 120° phase-spread. It can be seen that at any instant 'like'-poles (say N) are developed per 2 pole-pairs. In consequence, two S-poles would be developed between the two N-poles*.

   (3) In the *narrow-spread* or 60° *phase-spread* connection, total coils per pole-pair of the closed winding is divided into 6 equal groups, each of 60° phase- spread. The condition for conversion of a closed winding into a symmetrical open winding of 60' phase-spread is that *the total number of coils per pole-pair must be divisible by 6 (i.e. 2 × number of phases).*

A 3-phase, 6-pole, narrow-spread winding in 36 slots is illustrated in Figure 5.15(b), with only one phase shown. Note that, this winding is conventional is contrast with the consequent-pole as described under note (2) above.]

5.2.3.6 *Double-layer Winding*

A double-layer winding in an a.c. machine may be,

   (a) an integral-slot winding, when the number of slots per pole is an integer;
   (b) a fractional-slot winding, when the number of slots per pole is not an integer.

   [*Note* : (1) For a symmetrical fractional-slot winding, the number of slots per phase must be an integer.

---

* The development of consequent poles has been utilised in a method of speed control of induction motors by changing number of poles, known as the 'consequent-pole technique'. See, Sen, S. K. Electrical Machinery, Third Edition, pp. 340.2

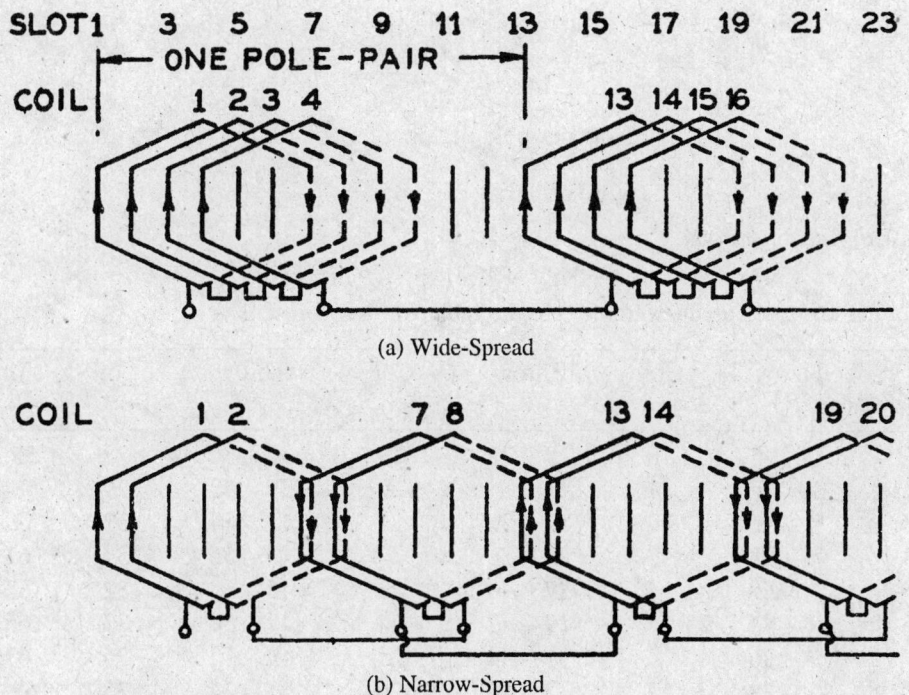

Figure 5.15 Open winding–3-phase, double-layer lap, 36-slot, 6-pole.
— top layer, – – – – bottom layer.

(2) Advantages of fractional-slot winding are :-

   (a) The winding is useful for low-speed machines, since the number of poles is large, and in comparison with fractional-slot winding an integral-slot winding will need a larger number of slots in order to have a normal breadth-factor;

   (b) Much freedom to the designer with regard to the choice of number of slots and standard armature stampings;

   (c) Minimisation of slot harmonics. In an integral-slot winding, harmonic emfs induced in the coils of a particular phase add up, since the coilsides occupy similar positions under the N-and S-poles. In fractional-slot winding, on the other hand, coilsides occupy dissimilar positions and suitable selection of displacements of coilsides from the pole-axes enables reduction of certain space harmonics.]

Design of integral-slot and fractional-slot windings are discussed in Examples 5.6 and 5.7.

Example 5.6

For a 36 -slot armature, obtain 3-phase, 6-pole, integral-slot, double-layer windings of (a) 120- phase-spread, (b) 60- phase-spread, each with (i) full-pitch coils, (ii) chorded coils.

Solution

$$\text{Slots per pole} = \frac{36}{6} = 6$$

$$\text{Slots per pole per phase} = \frac{6}{3} = 2$$

$$\text{Total number of coils} = 36$$

(A)   *With fall-pitch coils*

$$\text{Coil span } y_w = 6 \text{ teeth.}$$

Identifying coilsides as in example 5.3, the table for coil-throw is :-

Coil no.	Throw	Coil no.	Throw	Coil no.	Throw	Coil no.	Throw
1	1 – 14	10	19 – 32	19	37 – 50	28	55 – 68
2	3 – 16	11	21 – 34	20	39 – 52	29	57 – 70
3	5 – 18	12	23 – 36	21	41 – 54	30	59 – 72
4	7 – 20	13	25 – 38	22	43 – 56	31	61 – 2
5	9 – 22	14	27 – 40	23	45 – 58	32	63 – 4
6	11 – 24	15	29 – 42	24	47 – 60	33	65 – 6
7	13 – 26	16	31 – 44	25	49 – 62	34	67 – 8
8	15 – 28	17	33 – 46	26	51 – 64	35	69 – 10
9	17 – 30	18	35 – 48	27	53 – 66	36	71 – 12

The winding table for closed lap winding is :-

Coil   1 + 2 + 3 + 4 + 5 + 6 + 7 + 8 + 9 + 10 + 11 + 12 + 13 + 14 +15 + 16 + 17 + 18 + 19 + 20 + 21 + 22 + 23 + 24 + 25 + 26 + 27 + 28 + 29 + 30 + 31 + 32 + 33 + 34 + 35 + 36 + 1 –

Dividing the above into number of phase-groups = no. of pole-pairs
$$= 3 \text{ (in this example)}$$

Group I   :   1 + 2+ 3+ 4 + 5 + 6 + 7 + 8 + 9 + 10 + 11 + 12
Group II   :   13 + 14 + 15 + 16 + 17 + 18 + 19 + 20 + 21 + 22 + 23 + 24
Group III :   25 + 26 + 27 + 28 + 29 + 30 + 31 + 32 + 33 + 34 + 35 + 36

(a) To obtain 120° -phase spread, each of the above groups is to be opened at 3 equidistant points. That is,

in Group I :   after coils 4, 8, and 12;
   Group II :   after coils 16, 20 and 24;
   Group III :   after coils 28, 32, and 36.

The winding table is, thus,

Phase R :   (S) coil   1 + 2 + 3 + 4 ⌐   ... group I
                        ⌐ –13 + 14 + 15 + 16 ⌐   ... group II
                        ⌐ –25 + 26 + 27 + 28   (F)   ... group III

Phase B :   (S) coil   5 + 6 + 7 + 8 ⌐   ... group I

     ⌐ 17 + 18 + 19 + 20 ⌐   ... group II

     ⌐ 29 + 30 + 31 + 32   (F)   ... group III

Phase Y :   (S) coil   9 + 10 + 11 + 12 ⌐   ... group I

     ⌐ 21 + 22 + 23 + 24 ⌐   ... group II

     ⌐ 33 + 34 + 35 + 36   (F)   ... group III

[*Note* : (1) (S) standing for 'start', (F) for 'finish', indicate polarities of terminals in each case.

(2) For wide-spread, phase-spread = number of slots per pole-pair per phase = 4, which checks with above.

(3) For wide-spread, in phase R, coil 4 is connected to 13. and coil 16 to 25 developing 3 like poles. Similarly for the other phases.]

(b) To obtain 60° phase-spread, each group is to be opened at 6 equidistant points, i.e. in

Group I   :   after coils 2, 4, 6 8, 10, and 12 ;

Group II   :   after coils 14, 16, 18, 20, 22, and 24;

Group III   :   after coils 26, 28, 30, 32, 34, and 36.

The winding table is, thus,

Phase R :   (S) coil   1 + 2 ⌐   ....   group I

     ⌐ 7 + 8 ⌐   ...   I

     ⌐ 13 + 14 ⌐   ...   II

     ⌐ 19 + 20 ⌐   ...   II

     ⌐ 25 + 26   ...   III

     (F)   31 + 32   ...   III

Phase B :   (F) coil   3 + 4 ⌐   ...   group I

     ⌐ 9 + 10 ⌐   ...   I

     ⌐ 15 + 16 ⌐   ...   II

     ⌐ 21 + 22 ⌐   ...   III

     ⌐ 27 + 28   ...   III

     (S)   33 + 34   ...   III

Phase Y :   (S) coil   5 + 6 ⌐   ...   group I

     ⌐ 11 + 12 ⌐   ...   I

     ⌐ 17 + 18 ⌐   ...   II

     ⌐ 23 + 24 ⌐   ...   II

     ⌐ 29 + 30 ⌐   ...   III

     (F)   35 + 36 ⌐   ...   III

[*Note* : (1) For narrow-spread

phase-spread = number of slots per pole per phase $= \dfrac{36}{6 \times 3} = 2$ with above.

(2) An alternative system of depicting coilgroupings for the above two cases is in the form of a box as shown in Figure 5.16. The box shows slot no. 1 to 12 covering one pole-pair, which will be repeated for slot no. 13 to 24, and for 25 to 36.

$\longleftarrow$ ——————— ONE POLE-PAIR ——————— $\longrightarrow$

Slot no.	1	2	3	4	5	6	7	8	9	10	11	12
Top layer	R	R	R	R	B	B	B	B	Y	Y	Y	Y
Bottom layer	B'	B'	Y'	Y'	Y'	Y'	R'	R'	R'	R'	B'	B'

PHASE
$\longleftarrow$ SPREAD $\longrightarrow$
120°

(a) Integral-slot, full-pitch, 3-phase, wide-spread winding-6 slots per pole.

$\longleftarrow$ ——————— ONE POLE-PAIR ——————— $\longrightarrow$

Slot no.	1	2	3	4	5	6	7	8	9	10	11	12
Top layer	R	R	B'	B'	Y	Y	R'	R'	B	B	Y'	Y'
Bottom layer	R	R	B'	B'	Y	Y	R'	R'	B	B	Y'	Y'

PHASE
$\longrightarrow$ SPREAD $\longleftarrow$
60°.

(b) Integral-slot, full-pitch, 3-phase, narrow-spread winding-6 slots per pole.

Figure 5.16 Representation of double-layer winding.

The box is formed as below : -

(i) Knowing the phase-spread (= 4 teeth), the top layer of slots 1 to 4 are filled up by coilside (R) of R-phase, followed by coilside (B) of B-phase in slots 5 to 8, and (Y) in slot 9 to 12.

(ii) For full-pitch winding, the coilspan (=6 teeth) is computed. Hence, coilside (R') of coil 1 is placed in the bottom layer of slot 7. The procedure in (i) above is then followed.

(iii) For narrow-spread with 6 slots per pole, the phase-spread is 2 slots. Hence, top layer of slots 1 & 2 are filled up by coilsides (R). The top layer of the next two slots should be by (B') instead of (B) since, for symmetrical winding, this phase is to be reversed. Top layer of slots 5 & 6 is filled up by Y-coilsides. For slots 7 to 12, the same procedure is followed but all the coils are to be reversed to get opposite poles.

(iv) Since the coilspan is 6 teeth, the bottom layers of slot 7 and 8 are filled up with (R'), and the procedure as above is repeated.

(3) E.m.f. Diagram

$$\text{Slot pitch } \gamma = \frac{6 \times 180}{36} = 30° \text{ electrical.}$$

Figure 5.17 (a) shows the coil e.m.f.-star for the 36-slot winding. Knowing the coil-grouping under the respective phases, the emf-diagrams for both wide- and narrow-spread windings are drawn (Figure 5.17b).

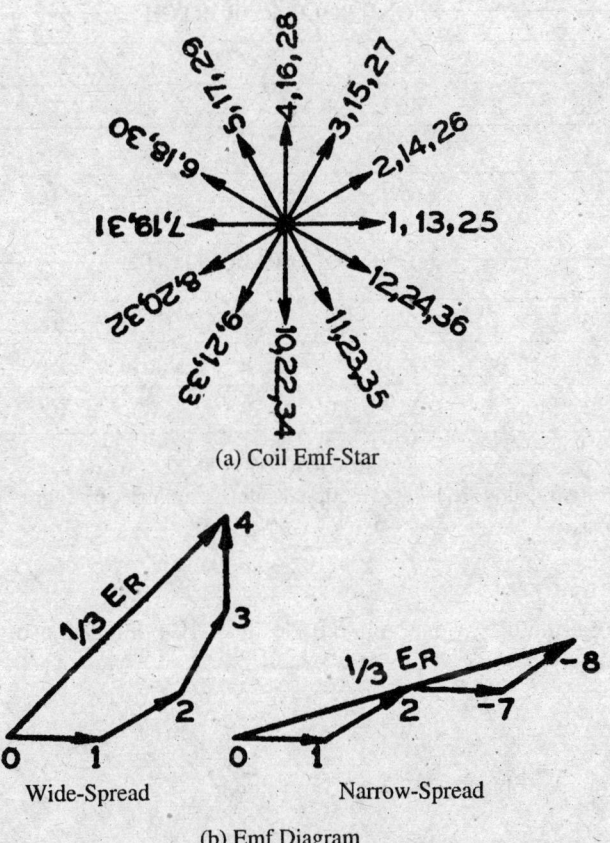

(a) Coil Emf-Star

(b) Emf Diagram

Figure 5.17 Pertaining integral slot, full-pitch, 3-phase winding-6 slots per pole.

It can be seen that the phase-sequence for the narrow-spread is opposite to that for wide-spread.]

(B) *With Chorded coils*

Chording in this winding can be chosen as 6/7 or 5/7 (chording is not generally made below 2/3).

With chording of 6/7, the coil-throw table for the full-pitch coils is modified as :-

Coil no.	1	2	3	4	5	6	7	8
Throw	1–12	3–14	5–16	7–18	9–20	11–22	13–24	15–26

and so on.

The winding table is the same as for the full-pitch coils, and so are the coil emf-star and the emf-polygon.

The representation in the form of a box is however, altered as shown in Figure 5.18.

←——————————————— ONE POLE–PAIR PITCH ———————————————→

1	2	3	4	5	6	7	8	9	10	11	12
R	R	R	R	B	B	B	B	Y	Y	Y	Y
B′	Y′	Y′	Y′	Y′	R′	R′	R′	R′	B′	B′	B′

(a) Wide-spread.

←——————————————— ONE POLE–PAIR PITCH ———————————————→

1	2	3	4	5	6	7	8	9	10	11	12
R	R	B′	B′	Y	Y	R′	R′	B	B	Y′	Y′
R	B′	B′	Y	Y	R′	R′	B	B	Y′	Y′	R

(b) Narrow-spread.

Figure 5.18 Integral-slot, double-layer, fractional-pitch winding - 6 slots per pole, 2 slots per pole per phase.

Example 5.7

For the 36-slot armature of example 5.6, design a 10-pole, 3-phase, double-layer winding with 60° phase-spread.

Solution

(i) No. of slots per phase $= \dfrac{36}{3} = 12$, an integer,

Hence symmetrical 3-phase winding is possible.

(ii) No. of coils per pole $= \dfrac{36}{10} = \dfrac{18}{5} = 3\dfrac{3}{5}$

The winding must be fractional-slot.

We choose a coil-span = 3 teeth.

Thus the winding is chorded. The table for coil-throw is :-

Coil no.	Throw	Coil no.	Throw	Coil no.	Throw	Coil no.	Throw	Coil no.	Throw	Coil no.	Throw
1	1–8	7	13–20	13	25–32	19	37–44	25	49–56	31	61–68
2	3–10	8	15–22	14	27–34	20	39–46	26	51–58	32	63–70
3	5–12	9	17–24	15	29–36	21	41–48	27	53–60	33	65–72
4	7–14	10	19–26	16	31–38	22	43–50	28	55–62	34	67–2
5	9–16	11	21–28	17	33–40	23	45–52	29	57–64	35	69–4
6	11–18	12	23–30	18	35–42	24	47–54	30	59–66	36	71–6

(iii) No. of slots per pole per phase $= \dfrac{36}{10 \times 3} = \dfrac{6}{5} = \dfrac{X}{Y}$

Since the number of slots per pole per phase is fractional, it can be seen that there are X coils over a distancs of Y pole-pitches. It is thus sufficient to examine only Y pole-pitches of the winding. The coil-grouping per phase can be determined as below :-

(1) Form a table of X columns per phase, 3X columns for 3 phases, and Y rows, as shown in Table 5.3.

Table 5.3 Coil Group for Fractional-Slot Winding

Pole	R-phase					(–) B-phase					Y-phase				
1	1				2					3			4		
2		5					6				7				8
3				9				10					11		
4		12				13				14				15	
5			16				17					18			
6	1														

(2) Mark top left hand box (under pole 1, R-phase) by coil no. 1. Count Y (= 5 in this example) from the next box under pole 1 and mark the (Y + l)th box by coil no. 2, (2Y + 1)th box by coil no. 3, and so on, till the table is fully covered.

[Check : If the marking is correct, the first box under pole no. 0 will have a coil in it.]

(3) From table 5.3 it can be ssen that under Y number of poles, coils 1, 2, 5, 9, 12 and 16 are under R-phase, coils 3, 6, 10, 13, 14 and 17 are under B-phase, and coils 4, 7, 8, 11, 15, and 18 are under Y-phase.

(iv) Knowing that for 60° phase-spread, the coils under phase B must be reversed, wo have the following winding table :

Phase R	*Phase B*	*Phasy Y*

(S)   coil   1 + 2 ⌐   (F)   coil   3 ⌐   (S)   coil   4 ⌐

⌐ 5 ⌐   ⌐ 6 ⌐   ⌐ 7 + 8 ⌐

9 ⌐   10 ⌐   11 ⌐

⌐ 12 ⌐   ⌐ 13 + 14 ⌐   ⌐ 15 ⌐

16 ⌐   17 ⌐   18 ⌐

⌐ 19 + 20 ⌐   ⌐ 21 ⌐   22 ⌐

23 ⌐   24 ⌐   ⌐ 25 + 26 ⌐

⌐ 27 ⌐   28 ⌐   ⌐ 29 ⌐

30 ⌐   ⌐ 31 + 32 ⌐   33 ⌐

(F) 34 ⌐   (S) 35 ⌐   (F) 36 ⌐

(v) Alternatively, in the form of a box, the winding can be illustrated as in Figure 5.19(a).

←————————————————— **5 Pole–Pitch** —————————————————→

R	R	B′	Y	R′	B	Y′	Y′	R	B′	Y	R′	B	B	Y′	R	B′	Y
R′	B	Y′	R′	R′	B	Y′	R	B′	Y	Y	R′	B	Y′	R	B′	B′	Y

(a) Fractional-slot winding diagram.

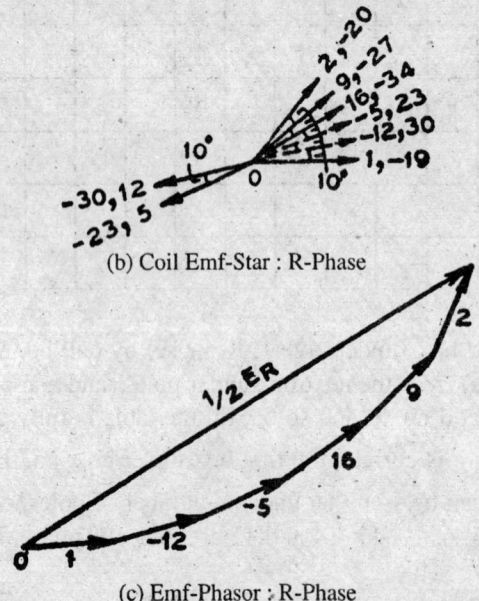

(b) Coil Emf-Star : R-Phase

(c) Emf-Phasor : R-Phase

Figure 5.19   Symmetrical fractional-slot winding-3-phase, double-layer, 36-slot, 10-pole.

(vi) *E.m.f. Diagram*

$$\text{Slot pitch} = \gamma = \frac{10 \times 180}{36} = 50° \text{ electrical.}$$

The coil einf-star and the corresponding emf phasor diagram for the R-phase is shown in Figure 5.19(b) and (c).

[*Note* : (1) A study of the coii-emf star and the phasor diagram would indicate that the 36-slot, 10-pole, 3-phase, fractional-slot winding is equivalent to an 180-slot, 10-pole, 3-phase, inlegral-slot winding giving a value of $\gamma = 10°$ electrical.

In general, a fractional-slot winding with slots per pole per phase (X/Y) reduced to its lowest term is equivalent to an integral-slot winding of X -slots per pole per phase.

(2) With slots per phase = an integer, and slots per pole per phase (X/Y) = a fraction, balanced winding is only possible if Y of the fraction reduced to its lowest term is not a multiple of the number of phases, since coils would then be unequally distributed in the phases. For example, with 42 slots, 3-phase, 12-pole armature winding will have,

$$\text{slols per phase} = \frac{42}{3} = 14 \text{ an integer;}$$

$$\text{slots per pole} = \frac{42}{12} = \frac{7}{2}$$

$$\text{slots per pole per phase} = \frac{42}{12 \times 3} = \frac{7}{6} \text{ a fraction whose}$$

denominator Y is a multiple of the number of phases 3. The coil-groupings for 6 repetitive poles would be as in Table 5.4.

Table 5.4 Coil-grouping for 7 slots per phase for 6 poles

Pole	R-phase						(−) B-phase						Y-phase				
1	1				2						3					4	
2				5				6						7			
3	8				9						10					11	
4				12				13						14			
5	15				16						17					18	
6				19			20						21				

Table 5.4 shows that for 6 repetitive poles, there are 9 coils under phase R. 6 under phase B and Y each. The winding is thus unbalanced.]

5.2.3.7 *Single-layer Winding*

A single-layer winding can be conceived from a double-layer winding by ommitting certain phase-groups of the double-layer full-pitch winding. In a double-layer winding, both full-pitch and chorded arrangements are possible, whereas in single-layer winding, pitch (or mean pitch) of the coils are always 100%. For example, if the alternate phase-groups in the 6-pole, 36-slot, narrow-spread, full-pitch winding (example 5.6) are omitted, we, have the winding table as.

	R – phase	B – phase	Y – phase
(S)	1 + 2	9 + 10 (F)	(S) 5 + 6
	13 + 14	21 + 22	17 + 18
	25 + 26 (F)	(S) 33 + 34	29 + 30 (F)

with the coil-throw as.

Coil No.		Throw	Coil No.		Throw
As in example 5.6	Renumbered for single layer		As in example 5.6	Renumbered for single-layer	
1	1	1 – 7	18	10	18 – 24
2	2	2 – 8	21	11	21 – 27
5	3	5 – 11	22	12	22 – 38
6	4	6 – 12	25	13	25 – 31
9	5	9 – 15	26	14	26 – 32
10	6	10 – 16	29	15	29 – 35
13	7	13 – 19	30	16	30 – 36
14	8	14 – 20	33	17	33 – 3
17	9	17 – 23	34	18	34 – 4

The winding arrangement for the above case together with the coil-emf star and the emf-diagram are shown in Figure 5.20 (a) to (c).

[*Note* : In a single-layer winding, a coilside completely fills a slot. The number of turns per coil can be twice that of corresponding double-layer coil. The induced emf per phase can be the same in both types of winding.]

The winding shown in Figure 5.20 (a), known as 'Lattice' winding is not generally usable as difficulty is experienced in arranging overhangs which cross each other. The difficulty can be overcome by using (A) Concentric, and (B) Mush connections.

Important features of a single-layer winding are : -

(i) the number of coils per phase must be an integer for symmetrical winding (as in the case of double-layer winding);

(ii) No. of slots per pole per phase must be an integer;

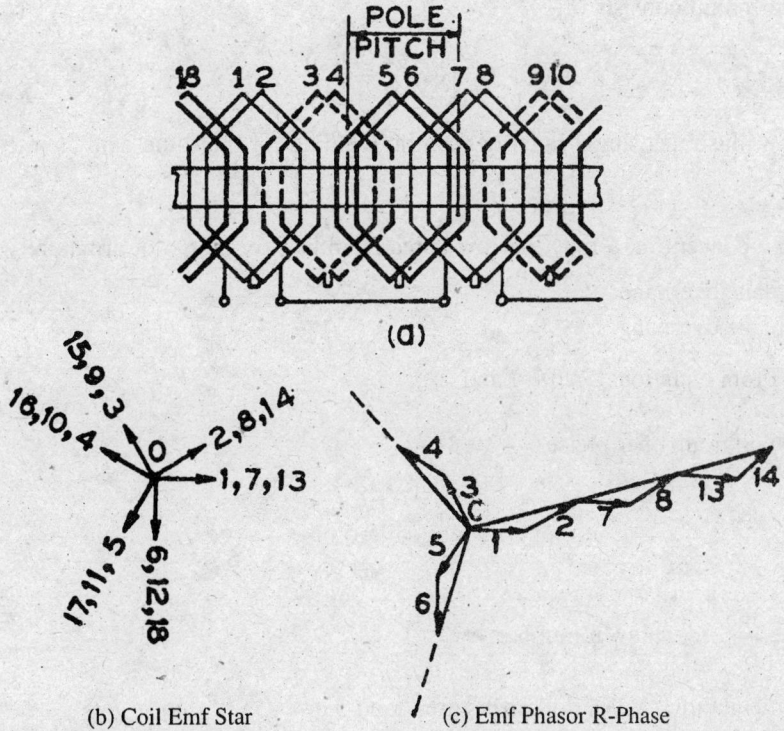

(b) Coil Emf Star   (c) Emf Phasor R-Phase

Figure 5.20   Single-layer winding-3-phase, 36-slot, 6-pole.

(iii) The coilspan must be odd.

(A) In concentric winding,

(a) end-connections are negotiated in 3- or 2- planes;

(b) when slots per pole (S/p) is even,

$$\text{coilspan } y_w = \frac{S}{P} \pm 1 \tag{5.23}$$

[e.g. For a 24-slot, 4-pole winding.

$$y_w = 6 \pm 1 = 7 \text{ and } 5]$$

(c) when (S/p) is odd,

$$y_w = \frac{S}{p} \pm 2 \text{ and } \frac{S}{p} \tag{5.24}$$

[e.g. S = 36, p = 4,

$$y_w = 11, 7, \text{ and } 9]$$

(d) number of coilgroups *per pole-pair and phase* = 1

i.e. No. of coilgroups per phase = $\frac{P}{2}$ \hfill (5.25) or, Each coilgroup consists of,

$$\text{for 3-phase,} \quad \frac{2C}{3p} = \frac{S}{3p} \text{ coils} \qquad (5.25b)$$

Example 5.8 illustrates single-layer concentric winding for the same data as in example 5.6.

Example 5.8

Design a concentric, 6-pole, narrow-spread winding for a 36-slot armature with,

(i) 3-plane overhang;

(ii) 2-plane overhang.

Solution   From equations 5.25 (a) and (b),

$$\text{No. of coilgroups per phase} = \frac{6}{2} = 3$$

and $$\text{no. of coils per coilgroup} = \frac{36}{3 \times 6} = 2$$

Again, $\dfrac{S}{p} = \dfrac{36}{6} = 6$, an even number;

hence, from equation (5.23), coil-spans are 7 and 5.

(i) 3-plane overhang, and (ii) 2-plane overhang based on the above data are shown in Figure 5.21 (a) and (b) respectively.

[*Note* : (1) For 3-plane overhang, the pitch between consecutive coil-groups is 60° electrical. Hence, for symmetrical 3-phase connection, the middle phase is to be reversed.

On the otherhand, for 2-plane overhang, the pitch between consecutive coilgroups is 120° electrical.

(2) For 2-plane overhang with even number of pole-pairs, coilgroups per phase is even. Half of the coilgroups per phase will be on the 2nd plane and the other half will be on the 1st plane.

(3) For 2-pole overhang with odd number of pole-pairs (as in the above example), number of coilgroups per phase is odd. End connections can only be arranged by cranking (Figure 5.21b) the overhang of one of the groups.

(4) *Split-concentric winding*

If each coilgroup of concentric 2-plane winding is splitted up into two sets of concentric coils so that each set shares its return coilsides with those of another group under the same phase, a split-concentric winding with 3-plane overhang will result (Figure 5.22). There will thus be one group of concentric coils per pole and phase, and the number of coilgroups per phase will always be even. That is, for the number of coils per group to be constant the number of slots per pole and phase must be even.]

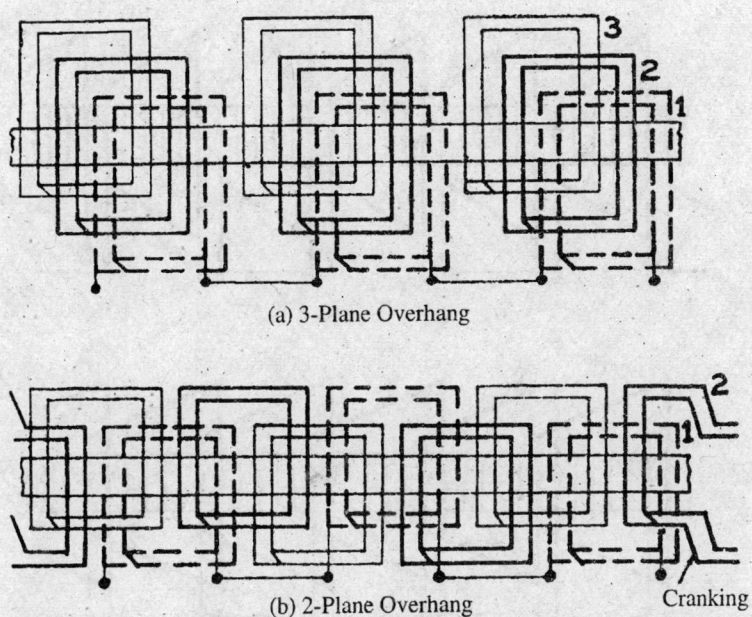

(a) 3-Plane Overhang

(b) 2-Plane Overhang   Cranking

Figure 5.21 single-layer concentric winding-2 slot per pole per phase. 1, 2, 3-1st, 2nd and 3rd plane respectively.

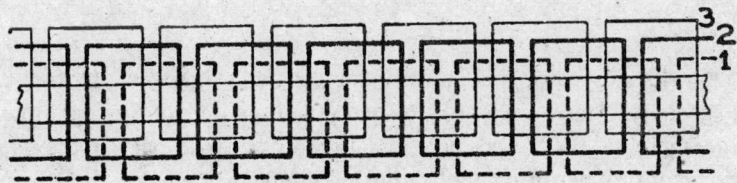

Figure 5.22 Split-concentric winding-2 slot per pole and phase. 1, 2, 3-1st, 2nd and 3rd plane respectively.

(B) *In Mush winding*

  (i) coilspan in terms of number of teeth should be odd.

  (e.g. For y_w - 6, coilspan can be either 5 or 7 — equn. 5.23)

  (ii) the coils have the same span, and are trapezoidal (Figure 5.23) with one coilside shorter than the other. The winding is done with the shorter coilsides placed in the slots first.

  (iii) short and long coilsides .should occupy alternate slots.

  (iv) coilsides in slots are connected to the right and to the left alternately.

  (v) coils can be dropped conductor by conductor into the slots and are therefore, suitable for semi-closed slots.

  Figure 5.23 shows 4-pole, 3rphase, mush winding for a 36-slot armature.

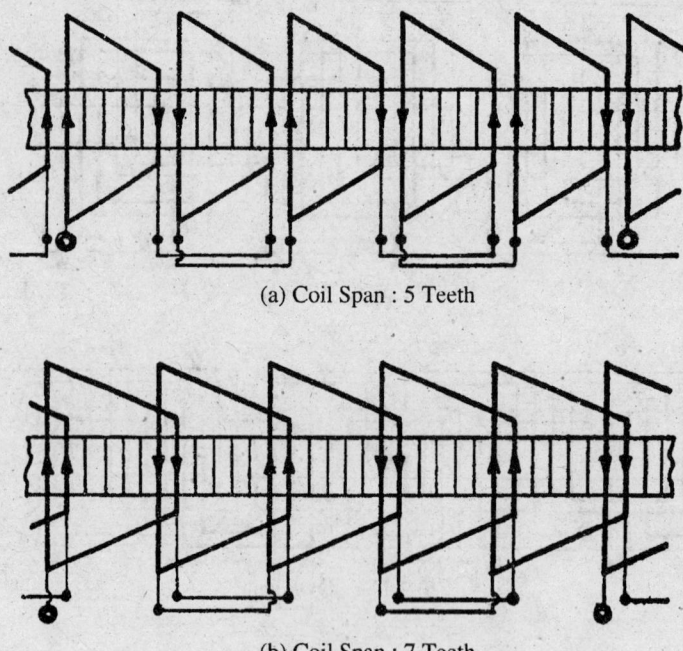

(a) Coil Span : 5 Teeth

(b) Coil Span : 7 Teeth

Figure 5.23   Mush winding-3-phase, 36-slot, 6-pole.

5.2.4 Skewing

The presence of slots in stator and/or rotor may generate gap-flux harmonics in the flux-density wave specially, if the number of slots per pole is an integer. For example, in a 36 – slot, 4 pole machine, there are 9 slots per pole and 18 slots per pole-pair, In the gap, there will be an 18-th harmonic of flux-density modulated in amplitude by the poles, the effect being constant amplitude flux-harmonics of order 18 ± 1; that is, a 17th and a 19th. Figure 5.24 shows a typical gap-flux density waveform in a salient-pole machine.

Such harmonics can be taken care of by (i) using fractional slot winding or chording (art. 5.2.3.6), (ii) skewing, (iii) shaping the pole shoes in a salient-pole machine so that the maximum gap-length at the pole-tip is about twice the minimum at the pole-centre, and (iv) using composite steel-bronze wedges in the rotor slots of a turbo-alternator.

Skewing is the method of providing an angular twist of the slot so that the conductors in a slot, instead of being parallel to the axis of the stator or rotor, are at an angle γ with it (Figure 5.25b). A comparison with a unskewed rotor coilside or conductor (Figure 5 25a) positioned under a harmonic flux-density wave at any instant shows that,

   (i) the induced voltage in the unskewed coilside (or bar) is proportional to the peak density B_h,
   (ii) the induced harmonic voltage in one element of length ab of the skewed coilside (Figure 5.25b) is cancelled by an equal and opposite voltage in another element of length bc.

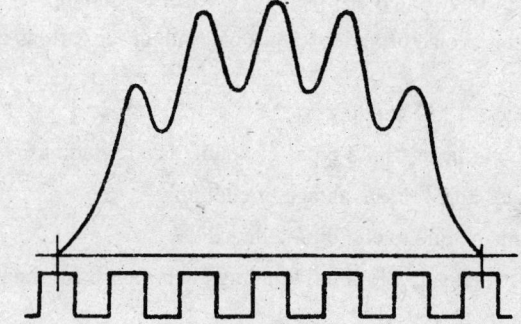

Figure 5.24 Typical Field form, salient-pole machine.

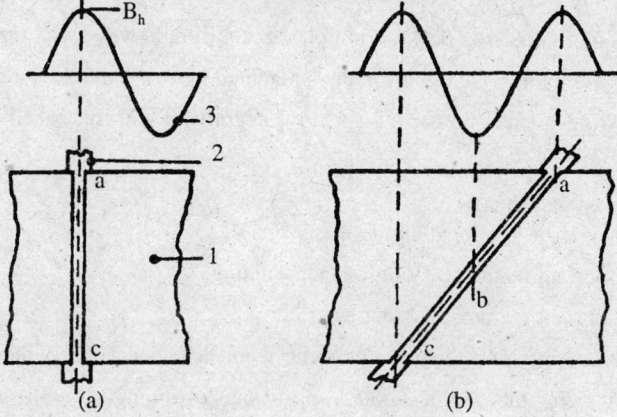

Figure 5.25 Skewing. 1-stator or rotor body, 2-Coilside, 3-harmonic flux-density wave.

This conclusion applies regardless of the instantaneous relative position between the coilside and the harmonic flux-wave.

Ganerally the amount of skewing used is of the order of magnitude of one slot-pitch.

[*Note* : (1) The most important effects of flux-and mmf-harmonics are : - (i) introduction of harmonics in the induced voltage (as in synchronous generators), (ii) production of parasitic torques, and (iii) production of vibration and noise.

(2) The last two effects are of special importance in squirrel cage induction motors since the stator mmf-harmonics is intensified by the presence of rotor slots.]

SHORT QUESTIONS

5.1 State 6 basic requirements which a design of electric circuit must fulfil.

5.2 Describe and compare important types of winding of a transformer.

5.3 State 3 advantages of disc winding.

5.4   State 2 advantages of interleaved disc winding over disc winding.

5.5   Form a table showing choice of various types of winding for various classes and voltage-ratings of transformers.

5.6   Mention 2 advantages of foil winding.

5.7   Name 2 types of single-layer, and 2 types of double-layer windings.

5.8   Mention 2 advantages and 1 disadvantage of chording.

5.9   How is the disadvantage due to chording reduced ?

5.10   Define the terms :– Coilspan, Chording, Winding pitch, Equilising ring, Dummy coil, consequent pole, Phase-spread.

5.11   Why is it advantageous to have stator number of coils per pole and phase, an integer ?

5.12   Compute the phase-spread and coil-span for a 54 slot, 6 pole, 3 phase full-pitch simple lap winding.

5.13   Why is equilising ring used in armature winding ?

5.14   Give 6 points of comparison between simple lap and simple wave windings.

5.15   Compute the distribution factors for wide-spread and narrow-spread windings.

5.16   Compare :– Wide-spread and narrow-spread; Integral-slot and fractional-slot; 2-plane overhang and 3-plane overhang.

5.17   What is 'coil e.m.f. star' ?

5.18   State 3 advantages of fractional-slot winding.

5.19   Mention 3 important features of single-layer winding.

5.20   Mention 5 important features of Mush winding.

5.21   What is the basic difference between concentric and split-concentric windings ?

5.22   Develop the winding table for a 4-pole, 3-phase mush winding for a 36-slot armature.

5.23   Form the winding table for a 3-phase, 16-pole, fractional-slot, narrow-spread double-layer winding for a 108-slot armature with a coil-span of 6 slots. How many poles are there in a repeatable section of the winding ? Show that the above winding is equivalent to a 3-phase, 16-pole, double-layer, integral-slot winding in 432 slots.

5.24   Mention 4 methods of negotiating the gap-flux harmonics due to the presence of slots.

5.25   What is meant by Skewing ?

## CHAPTER 6

Design of Transformer

6.1 SOME IMPORTANT CONSIDERATIONS

6.1.1 Core and Shell Types

(i) In core type, cooling of winding is better since the winding encloses the core. In shell type, on the other hand, the cooling of core is better.

(ii) Shell type gives better support to windings against electromagnetic forces since the windings are enclosed by core.

(iii) Leakage reactance in shell type is smaller than that in core type since, in sandwich type winding, the linkage between the primary and secondary is better.

(iv) In core type, repair of winding is easier.

(v) Shell type is more suitable for large transformers as strip-conductor, one turn per layer, can be used with advantage.

6.1.2 Distribution Transformers

(i) Distribution transformers are used to supply power, for general purposes, at final distribution voltage level from a high voltage distribution system. The minimum rating is about 5 kVa and the maximum 1 MVa. For industrial substations, the rating may be as high as 3 MVa above which the percent impedance to limit fault current is impractically high.

(ii) Distribution transformers have low percent reactance leading to superior voltage regulation. For transformers of 100–1000 kVa rating, the percent reactance is between 4 to 6%. For smaller ratings, lower values may be employed since an actual fault current will fall in proportion to the rating at constant reactance.

(iii) In distribution transformers, core material is invariably CROS except in very small units. Three-phase units are generally of core type. Single phase units would be cheaper if of core type.

Mitre corner joints are usually used to reduce corner losses. Alternatively strip-wound core is used, specially in rural transformers upto 100 kVa as this results in low loss (about 22 W on a 5-kVa transformer).

(iv) Flux-density with CROS is of nominal value 1.65 Tesla. For larger sizes, lower values are used to limit noise and switching inrush current. For rural transformers, lower density is used so that iron loss is low. The density can be as low as 1.0 Tesla.

(v) Distribution transformers are designed for maximum efficiency at about 50% rated load since these transformers are always connected to the line and have iron losses throughout 24-hours irrespective of whether it supplies load or not. Quite often, for rural transformers, with low load factor, the ratio of iron to I^2R losses is made low for economic considerations.

(vi) In view of item (v) above, distributiou transformers are designed for low iron losses. The iron mass and thus A_i / A_{cond} are restricted, Since inherent reactance is very small, low value of A_i / A_{cond} can he used.

(vii) For distribution transformers, calculation over wide ranges have shown that H_{wdg}/s between 0.3 and 1.0 for most economic utilisation of active materials.

(viii) Distribution transformers are mostly ON-cooled whether indoor or outdoor type, pole-mounted or ground-mounted.

Increasing use is being made of insulating and cooling media other than oil, either AN-cooled with class C insulation or synthetic liquid (SN) cooled.

6.1.3 Generator Transformers

(i) The function of generator transformers is to step up the generated voltage to the transmission voltage and are generally of delta-star connection. Such connection facilitates use of h.v. tapchanger at relatively lower voltage, electrically adjacent to the neutral point which is generally directly earthed. When a three-phase on-load tap-changer is used, the neutral point is often made within the tap changer and is brought out through a bushing for earthing.

(ii) To suit above, winding arrangement (Figure 6.1) alternative to that shown in Figure 4.23 has been used. The h.v. winding is in three parts (the innermost one carrying the taps) with the I.v. winding in between two h.v. parts. The percent of turns apportioned to various parts depends on the value of reactance desired.

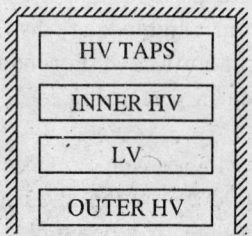

Figure 6.1 Alternative coil-arrangement in generator transformer

[Note : (1) For the winding arrangement of Figure 6.1, care must be taken to provide correct insulation for the various windings, particularly in relation to the stresses resulting from surges.

(2) This winding arrangement has higher winding losses but mechanical forces due to short-circuit are reduced. Further, in comparison with the conventional arrangements (Figure 4.23) it has lesser radial stray flux entering the core. The circulating eddy-currents in the core and corresponding heating are thereby reduced.]

(iii) The range of variation of flux-density in a generator transformer is dependent on the voltage adjustment provided on the alternator. The UK prictice is to restrict the adjustment to + 5% whereas US practice allows wider variation so that tap-changer is avoided.

(iv) When tap-changer is used, quite often it is required to have off-set tapping range, for example, + 2 to −16%.

(v) % reactance of generator transformers are generally dependent on stability considerations, for which the alternator and the transformer must be considered together. Normally, large alternators have a transient reactance of 25 to 32%. Thus the reactance of the generator transformer should be less than 18% since, the maximum total alternator and transformer reactance should be about 50% for stability reasons.

(vi) A_i/A_{cond} for nearly all classes of power transformers lie between 1.0 and above 3.5.

(vii) For power transformers of normal losses and reactance, the practical range of H_{wdg}/s is 0.55 to 0.74.

For ratio of specific cost* of copper to iron = 3, the design based on, minimum cost of materials gives $H_{wdg}/s = 0.6$ to 0.7.

(viii) Generator transformer; operate primarily on full load. Thus two-stage cooling is not required. It is usual to employ OFAF or OFW cooliiig. The latter method is popular because, (a) it requires smaller space for the oil-water heat-exchangers as compared with those for oil-air units; (b) there is an abundant and reliable supply of cooling water in a power station.

6.1.4 Transmission Transformers

(i) Transmission transformers may be auto or two-winding. There is no hard and fast rule regarding the use of auto in preference to two-winding. Various considerations such as system earthing, ratio, tapping range, availability of tap changers, transport facilities, etc. are to be taken.

(ii) Almost all auto transformers used on large power systems employ star-star connection. With such connection, there is no phase-shift between the primary and the secondary voltage. This can be achieved also by using a two-wmding star-star transformer, in which case there is no electric connection between the networks.

(iii) If h.v. and l.v. networks are in phase and the networks are solidly earthed at their neutral points, it is possible to use auto-transformers with graded insulation.

In many substations, two transformers are used instead of one, the rating being such that one can support the entire substation load if the other is out of circuit for

---

* Cost per unit mass

maintenance or repair. In such case, it is common to employ transformers of dual rating, for example, natural (ONAN) and forced (OFAN or OFAF) cooling, thus avoiding the running of cooling plant during lighter load.

(iv) With regard to A_i/A_{cond} and H_{wdg}/s for transmission transformers discussions in (vi) and (vii) art 6.1.3 on generator transformers are applicable.

6.2 CORE SECTION

It has been indicated in Chapter 3 – art 3.2.4, how the net core section Ai can be calculated for a given specification. Assuming a suitable value for the stacking factor (Table 1.4), the gross core-section $A_c = A_i/k_i$.

It is advantageous to have the primary and the secondary coils wound on round bobbins, since this lends to minimum length of mean turn and is more rugged and easy to manufacture. However, such practice leads to non-utilisation of the full circular area, the maximum utilisation of which is desirable. With I – laminations or plates such utilisation can only be effected to by increasing the number of steps. Since, increased number of steps means increased cost of assembly and inventory, choice of steps obviously depends on the size of the transformer.

Figure 1.3 illustrates square, two-stepped (or cruciform) and three-stepped core sections of a transformer. Computation of dimensions for optimum fill for the above three cases is given below.

I. *Square section*
Reference : Figure 1.3 (a),

$$\text{gross core-section A}_c = a^2;$$
$$\text{and } a = 0.707\, d \tag{6.1a}$$

$$\text{Optimum fill} = \frac{(0.707\, d)^2}{\pi\, d^2/4} \times 100\%$$

$$= 63.6\,\% \tag{6.1b}$$

II. *Two-stepped or Cruciform*
With reference to Figure 1.3(b),

$$a_1 = d\cos\theta = b_1 + 2b_2$$
$$a_2 = d\sin\theta = b_1$$

Gross core-section, $A_c = a_1 b_1 + 2a_2 b_2$

$$= d^2\sin\theta\cdot\cos\theta + 2d\sin\theta\cdot\frac{d}{2}(\cos\theta - \sin\theta)$$

$$= d^2(\sin 2\theta - \sin^2\theta)$$

For optimum fill, $\dfrac{d\,A_c}{d\,\theta} = 0$.

or, $2\cos 2\theta - \sin 2\theta = 0$; giving $\theta = 31.72°$.

That is,
$$a_1 = 0.85d$$
$$a_2 = b_1 = 0.526d$$
$$b_2 = 0.112d$$

and
$$\% \text{ fill} = \frac{A_c}{\pi d^2/4} \times 100\% = 78.6\% \tag{6.2b}$$

III. Three-stepped section

Reference to Figure 1.3 (c),

$$a_1 = d \cos \theta = b_1 + 2(b_2 + b_3) = a_2 + 2b_3$$
$$a_2 = 0.707\, d = b_1 + 2b_2$$
$$a_3 = d \sin \theta = b_1$$

Gross core-section, $A_c = a_1 b_1 + 2a_2 b_2 + 2a_3 b_3$

$$= d^2 \sin \theta . \cos \theta + 0.707\, d\,(0.707\, d - d \sin \theta)$$
$$+ d \sin \theta \cdot (d \cos \theta - 0.707\, d)$$
$$= d^2 (\sin 2\theta - 1.414 \sin\theta + 0.5)$$

Putting $(d\, A_c / d\, \theta) = 0$, for optimum fill,

$$2 \cos 2\theta - 1.414 \cos \theta = 0$$

Solving, $\cos \theta = 0.906$, and we have,

$$a_1 = 0.906\, d,\ a_2 = 0.707d,\ a_3 = 0.423d;$$
$$b_1, = 0.423d,\ b_2 = 0.142d,\ b_3 = 0.10d \tag{6.3a}$$

and % fill = 85.1% (6.3b)

For steps 4 and above, the above procedure will lead to a number of partial differential equations equal to the number of steps, the solution of which will give the plate dimensions and % fill under optimum condition. Table 6.1 gives the step width for optimum fill. It is evident that with 11 steps, about 96% of fill is obtained, beyond which increase in number of steps does not generally justify the increased assembly and inventory costs. However, it must be noted that,

Table 6.1 Strip width for optimum fill

Plate width/diameter d*

No of Steps	% fill	a_1	a_2	a_3	a_4	a_5	a_6	a_7	a_8	a_9	a_{10}	a_{11}
4	88.5	.935	.80	.60	.355							
5	90.8	.950	.85	.71	.53	.31						
6	92.3	.96	.89	.78	.63	.47	.28					
7	93.4	.97	.90	.82	.71	.58	.43	.25				
9	94.8	.98	.93	.87	.80	.71	.61	.50	.37	.22		
11	95.8	.98	.94	.89	.83	.76	.71	.65	.56	.45	.33	.19

* Refer to Figure 6.2

(i) clamping of laminations is usually done by bolts passing through holes punched in the laminations;

(ii) For number of steps 6 and above, longitudinal and transverse ducts in core are used for cooling purposes.

Both the above factors reduce the % fill. Further, clamping of bolts introduces extra iron losses. Use of high-strength resin glass bands[†] applied to the periphery of the core-section of large power transformers with a view to eliminate clamping bolts has been reported.

[Note : *Standardisation* : In practice, departure from the strip sizes as in Table 6.1 is made so that,

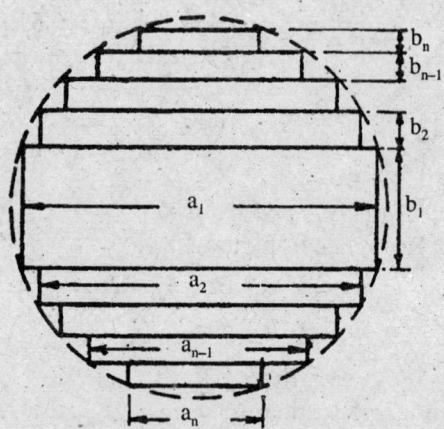

Figure 6.2   Pertaining strip width.

(i) laminations can be manufactured in standard sizes to avoid an excessively wide assortment of lamination sizes and to reduce wastage to a minimum ;

(ii) space can be provided for clamping the laminations.

Such standardisation is advantageous in the case of distribution transformers which are produced in bulk.]

As a starting point the choice of number of steps for various values of gross core-section may be made from Table 6.2.

Table 6.2   Choice of number of steps

Gross core-section, $m^2 \times 10^{-3}$	No. of steps	No. of ducts
< 3	1	–
3 – 5	2	–
5 – 7	3	–
7 – 15	4	–
15 – 45	5	–
45 – 80	6	1
80 – 200	7	1
200 – 400	9	2
400 – 750	11	2

†   Kerr, H.W. and Palner, S : "Developments in the design of large power transformer" Proceedings IEE, vol 111. 1964, pp. 823.

6.3 YOKE SECTION

With HRS, the usual practice is to make the cross-sectional area of the yoke greater than that of the limbs by increasing plate widths. The vertical height is thus increased. There is a consequent reduction in flux-density of yoke (and thus reduction in iron losses) which however, is not in direct proportion to the increase in area, due to non-uniformity of flux-distribution.

Usual practice is to use, (a) rectangular yoke-section equal to 1.1 to 1.15 times the limb section in small transformers; (b) two-stepped yoke section, 1.05 times the limb cross-section in medium-sized transformers.

With CROS, gain made by increasing yoke section is generally outweighed by the increased cost of extra plates and it is now normal practice in three-phase, three-limb cores to make yoke section and plate widths exactly equal to those of the limbs.

For very large power transformers, use of 5 limbs reduces yoke depth. Cooling is improved without the use of ducts. Relatively low iron loss is obtained if the unwound limb is made about 40–50 % and the main portion of yoke about 58% of the total wound-limb cross-section.

Example 6.1

Calculation of core and yoke sections. Specification of the transformer is the same as in Example 3.1.

Solution

Designer's choice :   No. of steps in limb = 5

(From Table 6.2, for gross core-section of 0.0346 m^2)

No. of steps in yoke = 2.

For 5 steps, optimum fill (from Table 6.1) is 90.8%

Hence, diameter of the core-circumscribing circle,

$$d = \left[\frac{0.0346}{0.908} \frac{4}{\pi} \right]^{1/2} = 0.22 \text{ m}.$$

Computation of core-section

The width of laminations and the depths of stacks are computed as below : -

Calculated lamination width mm.	Designer's choice	Stack depth mm.	Designer's choice
$a_1 = 0.95 \times 220$ $= 209$	210	$b_1 = (220^2 - 210^2)^{1/2}$ $= 65.57$	65
$a_2 = 0.85 \times 220$ $= 187$	190	$b_2 = \frac{1}{2}\left[(220^2 - 190^2)^{\frac{1}{2}} - b_1 \right]$ $= 22.95$	23

$$a_3 = 0.71 \times 220 \qquad 155 \qquad b_3 = \frac{1}{2}[(220^2 - 155^2)^{\frac{1}{2}} \qquad 22$$
$$- (b_1 + 2b_2)]$$
$$= 156.2 \qquad\qquad = 22.56$$

$$a_4 = 0.53 \times 220 \qquad 115 \qquad b_4 = \frac{1}{2}[(220^2 - 155^2)^{\frac{1}{2}} \qquad 16$$
$$= 116.6 \qquad\qquad - (b_1 + 2b_2 + 2b_3)]$$
$$= 16.27$$

$$a_5 = 0.31 \times 220 \qquad 70 \qquad b_5 = \frac{1}{2}[(220^2 - 70^2)^{\frac{1}{2}} \qquad 11$$
$$= 68.2 \qquad\qquad - (b_1 + 2b_2 + 2b_3 + 2b_4)]$$
$$= 10.78$$

Hence, gross core-section, A_c, is

stack   1 : $a_1 \cdot b_1 = 0.01365$ m^2

        2 : $2a_2 \cdot b_2 = 0.00874$

        3 : $2a_3 . b_3 = 0.00682$

        4 : $2a_4 \cdot b_4 = 0.00368$

        5 : $2a_5 \cdot b_5 = 0.00154$

                  $A_C = 0.0344$ m^2

Net core-section, $A_i = 0.95 \times 0.0344 = 0.0327$ m^2

Computation of yoke-section

The yoke section is taken as 2-stepped and equal to the core-section. Reference to Figure 6.3,

$$b_{y1} = b_1 + 2b_2 \qquad = 0.111 \text{ m.}$$
$$b_{y2} = b_3 + b_4 + b_5 = 0.049 \text{ m.}$$
$$b = b_{y1} + 2\, b_{y2} \qquad = 0.209 \text{ m.}$$

$$h_1 = \frac{0.0344}{0.209} \qquad = 0.165 \text{ m.}$$

$$h_2 = 0.8 \times 0.165 = 0.132 \text{ m.}$$

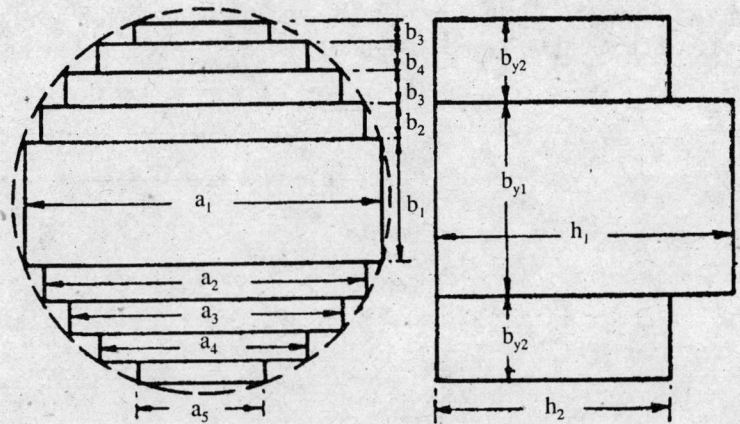

Figure 6.3 Core-and yoke-section.

6.4 CLEARANCES

Clearances between core and windings, between the windings, etc. are chosen so as to guard the windings, taps, leads, etc against short-circuit to earth or short-circuit between windings at both normal service voltage and possible over-voltages. Clearance is governed by mutual position of windings, core, tank, and other parts. Main clearances are : -

For core type (Figure 6.4)

    (i) Between core and l.v. winding, c_{c1};

    (ii) Between l.v. and h.v. windings, c_o;

    (iii) Between h.v. windings on two consecutive limbs, c_{hh};

    (iv) Between yoke and l.v., h_{y1}. and yoke and h.v., h_{yh}.

For shell type

    (i) Between h.v. and l.v. windings c_o;

    (ii) Between l.v. and core, b_{ci}, and between h.v. and core, h_{ch};

    (iii) Between h.v. and l.v. coils and tank wall;

    (iv) Between yoke and l.v., h_{y1} and, yoke and h.v., h_{yh}.

End clearances between yoke and l.v. winding is not so much determined by insulation requirement than to make provision for mechanical support at the ends of l.v. coil.

Clearance between l.v. and h.v. coils depend on,

    (i) voltages of the windings,

    (ii) type of insulation used, their dielectric property, and mechanical strength,

    (iii) the specified reactance-value,

    (iv) width of duct (about 5 mm. for small capacity transformers, and 8 to 13 mm. in large capacity h.v. transformers) for easy circulation of cooling oil.

It is impossible to formulate rules regarding clearances covering wide range of transformer kVa and voltage-ratings, range of insulating materials and operating conditions. Table 6.3 gives typical clearances which could be a guide to a designer.

Table 6.3 Clearances

Type	MVa	volt-ratio	c_o	c_{c1}	mm c_{hh}	h_{yl}, h_{yh}*
Rural	.016–.1	3300/433				20
		6600/433	9	5	10	25
		11000/433				30
Distribution	.1–1	11000/433	12	5	12	30
		33000/433	20		25	65
Secondary transmission	.10–20	33kV/11kV	20	15	25	65
		66kV/11kV	30		40	75
Primary transmission	45–161	132kV/11,33, 66kV	60	15	80	70
		220kV/11,33, 66kV	80	20	100	90
		400kV/11,33, 66kV	120	30	160	140
Generator	72–340	11,22kV/132kV	60	15	80	70
		11,22kV/220kV	80	18	100	90

* Refer to Figure 3.1

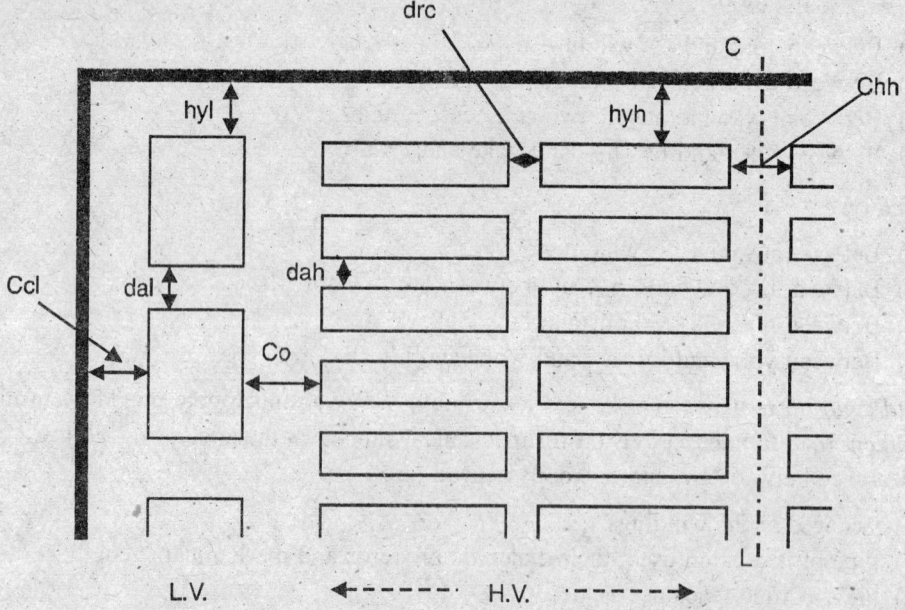

Figure 6.4 Clearances in a three-phase core-type transformer.

6.5 INSULATION

Of all the electrical apparatus, transformer presents the most serious problems in insulation, because voltage reaches a much higher value (400 kV, for example) than in rotating machines, where voltage is limited to about 10 to 20 kV. The insulation must correspond to voltage conditions which exists either in the steady-state or in transient over-voltage condition such as faults in the system operation of circuit-breakers, etc.

The question of insulation accounts for most of the constructional arrangements of the windings, such as, thickness of paper insulation, spacing insulating tubes, etc.

(i) Between core and winding, and between l.v. and h.v. windings : moulded cylinders or tubes, or a number of wraps of pressboard, of high dielectric strength, not affected by insulating oil may be used with a set of insulating spacers as and when required.

(ii) For smaller ratings, non-supporting coils can be made with h.v. coil wound directly on the l.v. and insulation. On larger ratings, a considerable space is required between coils to obtain specified reactance. On the smallest srzes clearance required to obtain the reactance may be too small to allow for duct.

[*Note* : In power transformers, oil is the weakest insulation, its strength varies with the size of the duct, wider the duct, the lower is the strength. Under quite ideal condition, oil in ducts, 20 mm wide, withstands stress of about 6 kv. mm^{-1} but such condition is rare in practice.]

6.6 WINDING DESIGN

A description of various types of transformer windings, insulation requirements, and choice have been discussed in art, 5.1. For the design of transformer winding, it is often necessary to try alternative designs so that the most suitable one is selected.

Choice of the type of winding is dependent on given voltage and current ratings, as shown in Table 5.1. Once a type is chosen it is now required to design the winding so that it fits with the computed window dimensions with proper provisions for ducts, insulation, and clearances. It does not follow however that there is a unique design which meets all requirements. There is of course an overlap between the ranges covered by any two types of winding, but in addition external factors such as manufacturing requirements cannot often be neglected.

Once a design procedure is established it is possible to optimise the design as per customer's requirements using digital computation.

Some important design considerations for various types of winding are discussed below. The required specifications for the design of winding are :– Phase voltage; Phase current; Turns per phase; % tap, if any; Winding height, H_{wdg} : Internal diameter of coil, D_{ico}.

6.6.1 Cross-over

Transformers with low current requirement use circular conductors instead of strips. The space-factor is poor. Winding is generally done in several coil-sections in series with spacers for axial cooling ducts.

(i) For a given winding height H_{wdg} (= window height, h_w minus the axial end clearances at top and bottom), the total axial height of insulated conductors,

$$H_{sic} = 1.03\, H_{wdg} - \text{total clearances of axial ducts} \qquad (6.4)$$

The factor 1.03 accounts for non-uniform arrangement of turns.

(ii) Assuming suitable voltage par coil-section (generally 1300–1400 volts per coil-section), number of coil-sections nc can be computed which should be an even number. Thus, the total clearance

$$= (n_c - 1) \times \text{chosen axial duct width, } b_d \qquad (6.5)$$

(iii) The values of given current per phase and assumed current-density enable choice of circular conductors with insulation, from standard table, so that the number of axial strands per coil-section,

$$A_{str} = \frac{H_{sic}}{n_c \times \text{diameter of inslated conductor}} \qquad (6.6)$$

(iv) Knowing the turns per coil-section, the number of radial strands, R_{str}, can be determined which however should be odd so that the coil ends on the side other than where it has started.

Computerised Design

Governing Equations

Enter the given specifications.

Step 1: Choose voltage per coil, current density (based on whether copper or aluminium is used as the conductor material the clearances (Figure 6.4), and compute the number of coils and corresponding integer value.

Step 2: Compute:
maximum winding height Hwdgo = specified window height 'h' - 2. hyl -
                                                    (no. of coils 'nco'- 1.). dal.
Maximum height of each insulated coil, hico = 1.03 H_{wdgo}/nco.
Computed bare conductor section, cso = phase current/assumed current density.
   Choose the conductor data from standard table for ROUND conductors, modify the current density accordingly, compute number of axial and radial strands in the coil converting them into nearest integer values and compute modified number of turns per phase.

Step 3: Compute :
   winding height and width; length of mean turn = ([internal diameter of coil 'dic' + 2. Ccl + width of winding 'Wwdgo']; and hence,
   resistance per phase, conductor loss, and volume and weight of conductor material.

```
//
//Design of cross-over winding
//given specification:voltage per ph,vo;current per ph ,ao.
//  turns per phase,to;percent tap,tapo;height of winding,ho
//  internal dia. of coil,dico.

#include<iostream.h>
#include<iomanip.h>
#include<conio.h>
#include<math.h>

class spec
    {
      public:
        int vo,to,ho,dico ;
        float ao,tapo;
       spec()
          { vo=0.0; ao=0.0; to=0.0; ho=0.0; dico=0.0; tapo=0.0; }
       spec( int v, float a, int t, int h, int dic, float tap )
          { vo=v ; ao=a ; to=t ; ho=h ; dico=dic ; tapo=tap ; }
       void getvalue(void)
          { cout << endl <<"Enter the values of: vo,ao,to,ho,dicoo,tapo" ;
            cin >> vo >> ao >> to >> ho >> dico >> tapo ;
          }
       void display()
          {
           cout <<"volt/ph=  "<< vo << setw(20) <<"amp/ph.= "<< ao
              << setw(20) <<"turns/ph.=  "<< to << setw(20)
              <<"ht of window.=  "<< ho << setw(20)
              <<"internal dia.of coil=  "<< dico
              << setw(20) <<"%tap= "<< tapo ;
          }
        int getv()  { return vo ; }
       float geta()  { return ao ; }
        int gett()  { return to ; }
        int geth()  { return ho ; }
        int getdic()  { return dico;}
       float gettap()  { return tapo;}
    };

class winding : public spec

{
  public:
   float hyl ;      // Axial clearance between yoke & coil at each end,mm.
   float dal ;       //Axial duct width between two consecutive coils,mm.
   float cdo ;         //Assumed current densityamp per mm-sq.
   float Hwdgo,nco,cso,hico,cbdo,cido,cbso,ciso,azho,azwo,Wwdgo,doco,dmco ;
   float lmto,reso,wcpo,vcoo,cpwto,cwto,k1,roe,spwt ;
    int anc, turns ;

   winding()
   { cdo=0.0; Hwdgo=0.0;  nco=0.0;  hico=0.0; cbdo=0.0; cido=0.0; cbso=0.0;
     ciso=0.0; Wwdgo=0.0; azho=0.0; azwo=0.0; doco=0.0; dmco=0.0; lmto=0.0;
     reso=0.0;  wcpo=0.0; vcoo=0.0; cpwto=0.0; cwto=0.0; }
```

```
winding
( float cd,float Hwdg,  float nc,  float hic,  float cbd, float cid ,
  float cbs,  float cis, float azh, float azw, float Wwdg, float doc ,
  float dmc, float lmt, float res,  float wcp,  float vco, float cpwt,
  float cwt
)
{ cdo=cd ; Hwdgo=Hwdg; nco=nc ; hico=hic ; cbdo=cbd; cido=cid;cbso=cbs;
 ciso=cis; Wwdgo=Wwdg;azho=azh; azwo=azw ; doco=doc; dmco=dmc;lmto=lmt;
 reso=res;  wcpo=wcp; vcoo=vco;cpwto=cpwt; cwto=cwt;
}
winding ( spec x )
{

   int vpc ;
  float nco, k1, k2 ;

    int v = x.getv() ;
   float a = x.geta() ;
    int t = x.gett() ;
    int h = x.geth() ;
  float dic = x.getdic();
  float tap = x.gettap();

   cout << endl <<"Enter the value of voltage per coil" ;
   cin >> vpc ;
   cout << endl <<"Enter the values of hyl & dal, in mm " ;
   cin >> hyl >> dal ;
   cout << endl <<"Enter the value of assumed current-density, amp/mm.sq." ;
   cin >> cdo ;
     nco = v / vpc ;                // number of coils
     cout << endl <<"nco=  "<< nco ;
     cout <<"Enter the integer value of number of coils" ;
     cin >> anc ;
      k2 = 2. * hyl + ( anc - 1. ) * dal ;
     Hwdgo = h - k2 ;
      hico = Hwdgo * 1.03 / anc ;        //ht. of each insul.coil
      cso  = a / cdo ;                // bare conductor section
     cout << endl <<"bare conductor section =  "<< cso ;

     cout << endl <<"Enter data for round conductor from Standard Table:"
               "bare dia (cbdo)-mm., insul. dia (cido)-mm.,"
               "bare section(cbso)-mm-sq.,insulated section (ciso)-mm-sq." ;
     cin >> cbdo >> cido >> cbso >> ciso ;

      cdo = cdo * cso / cbso ;          // modified current density
      azho = hico / ciso ;           // no. of axial strands
      azwo = t /( anc * azho ) ;        // no. of radial strands/coil
      cout << endl <<"azho=  "<< azho << setw(20) <<"azwo=  "<<azwo ;
      cout << endl <<"Enter the modified values of axial (azho) and "
               " radial (azwo) strands" ;

     cin >> azho >> azwo ;
     turns = anc * azho * azwo ;          //modified turns/ph.
     Hwdgo = cido * azho * anc / 1.03 ;    //Winding height,mm
     Wwdgo = cido * azwo ;             //Winding width,mm

//       Compute conductor loss and weight
```

```
        float roe, spwt ;
    const float PI = 3.14159 ;
            k1 = pow(10,-3) ;
    cout << endl <<"Enter the values of specific resis,- ohm.mm.sq/m(roe) &"
            <<"specific weight- kg per mm-cubed(spwt)";
    cin >> roe >> spwt ;

    dmco = dic + Wwdgo ;          // mean dia. of coil,mm.
    lmto = PI * dmco ;                     // length of mean turn,mm.
    reso = roe * lmto * turns * k1 / cbso ;     // resistance,ohm.
    wcpo = a * a * reso ;              // copper loss, watt.
    vcoo = turns * lmto * cbso ;          //volume of copper- mm.cubed.
    cpwto = spwt * vcoo * k1 ;            //copper weight, g.

}
void display()
{
cout << endl << setw(40) <<"DESIGN SHEET" ;
cout << endl <<"clearances:" ;
cout << endl <<"Axial between yoke and coil,at each end= "<< hyl <<" mm.";
cout << endl <<"Axial duct between two consecutive coils= "<< dal <<" mm.";
cout << endl <<"modified current-density=    "<< cdo <<" amp/mm.sq";
cout << endl << setw(15) <<"Dia.of round bare conductor=    "<< cbdo <<"mm";
cout << endl << setw(25) <<"Dia.of round insul. conductor= "<< cido <<"mm";
cout << endl << setw(35) <<"Number of axial coils=    " << anc ;
cout << endl << setw(45) <<"No. of radial strands in coil=  "<< azho ;
cout << endl << setw(55) <<"Number of axial strands/ coil=  "<< azwo ;
cout << endl << setw(55) <<"Number of turns/ phase=  "<< turns ;
cout << endl << setw(45) <<"winding height=    "<< Hwdgo <<" mm" ;
cout << endl << setw(35) <<"winding width =    "<< Wwdgo <<" mm" ;
cout << endl << setw(25) <<"conductor-loss=    "<< wcpo  <<" watt" ;
cout << endl << setw(15) <<"conductor weight=  "<< cpwto <<" g" ;
}
};


void main()
{
 clrscr();
 spec m;
 winding n;
 m.getvalue();
 cout <<"\n"; m.display() ;


 n = m ;
//  n.getvalue();
 cout <<"\n"; n.display() ;
}
```

Figure 6.5(a)   A computer program for the design of Cross-over winding.

RESULT

(E) 11000   6.06   2303   396   210   0

(E) 1350

(E) 20   8

(E) 2.75

(o) nco = 8

(E) 8

(o) bare conductor section = 2.203636

(E) 1.78   1.96   2.488   3.0

(o) azho = 12.875   azwo = 22.36

(E) 13   22

(E) 0.0341 0.0027

(o)   DESIGN SHEET

Clearances:

Axial between yoke and coil, at each end = 20 mm.

Axial duct width between consecutive coils = 8 mm.

Modified current-density = 2.44 amp/sq. mm.

Dia. of round bare conductor = 1.78 mm.

Dia. of round insulated conductor = 1.96 mm.

Number of axial coils = 8

Number of radial strands = 22

Number of axial strands = 13

Number of turns/phase = 2280

Winding height = 198 mm.

Winding width = 43.1 mm.

Conductor loss = 915.8 watt.

Conductor weight = 12.2 gram.

6.6.2 Helix

For low voltage high current requirement the most common form of winding is helix of constant diameter. Rectangular strips are wound on a former progressing turn by turn from one end to the other.

For a given winding height H_{wdg}, the axial height of conductors H_{stc} plus ducts (if used) width

$$= \frac{1.03\, H_{wdg}}{T + 1} \qquad (6.7a)$$

T being the turns per phase.

[Note : (1) As per requirement of helix winding, a space equivalent to the axial dimension of one extra turn must be allowed.

(2) If transposition is used, space for 3 groups of transposed conductors must be provided for so that equn. 6.7a is modified as,

$$H_{stc} + \text{axial duct width} = \frac{1.03\,H_{wdg}}{T+4} \qquad (6.7b)$$

(3) For minimisation of short-circuit forces, axial packing between turns at the centre of the winding is often desirable. For large transformers, axial width of such packing may be 15 mm.]

From standard table of rectangular conductors the dimensions of the strip-conductor can be chosen.

Computerised design

Governing equations

Enter the given specifications.

Step 1: Choose number of axial coil and the clearances (art. 6.4 and Figure 6.4). Assume current density (based on whether copper or aluminium is used as the conductor material).

Step 2: Compute:
maximum winding height, H_{wdgo} = specified window height 'h' - 2. hyl + (no. of coils 'nco' - 1.). dal.
Maximum height of each insulated coil, hico = 1.03 H_{wdgo}/[(turns + 1.) * nco.].

Step 3: Compute:
bare conductor section, cso = phase current/asumed current density
Choose the conductor date from standard table for RECTANGULAR STRAP conductors, modify the current density accordingly.

Step 4: Compute:
Winding height = (number of turns + 1.). height of insulated conductor, and winding width = width of insulated conductor;

Step 5: length of mean turn = ([internal diameter of coil 'dic' + width of winding 'W_{wdgo}']; and hence, resistance per phase, conductor loss, and volume and weight of conductor material.

```
// Design of Helix winding
// Given specification: current per phase(ao), turns per phase (to),height of
// window(ho- mm.), internal diameter of coil(dico-mm.), % tap (tapo)
#include <iostream.h>
#include <iomanip.h>

#include <conio.h>
#include <math.h>

class spec
  {
    public:
     int to,ho,dico ;
     float ao,tapo;
    SPEC()
```

```
         { ao=0.0; to=0.0; ho=0.0; dico=0.0; tapo=0.0; }
      SPEC( float a, int t, int h, int dic, float tap )
         { ao=a ; to=t ; ho=h ; dico=dic ; tapo=tap ; }
      void getvalue(void)
         { cout << endl <<"Enter the values of: ao,to,ho,dico,tapo" ;
           cin >> ao >> to >> ho >> dico >> tapo ;
         }
      void display()
         {
          cout <<"amp/ph.= "<< ao << setw(20) <<"turns/ph.=  "<< to
             << setw(20) <<"ht of window.=  "<< ho << setw(20)
             <<"internal dia.of coil= "<< dico << setw(20)
             <<"%tap= "<< tapo ;
         }
      float geta() { return ao ; }
       int gett() { return to ; }
       int geth() { return ho ; }
      int getdic() { return dico; }
      float gettap() { return tapo; }
   };


class winding : public spec
{
    public:
     float hyl ;        // Clearance between yoke & coil at each end,mm
     float dal ;        //Axial duct width between two consecutive coils,mm
     float cdo ;           //Assumed current density

   float Hwdgo,nco,cso,hico,cbho,ciho,cbwo,ciwo,cbso,ciso ;
   float Wwdgo,doco,dmco,lmto,reso,wcpo,vcoo,cpwto,cwto ;
   float k1, k2, roe, spwt ;

winding()
{ cdo=0.0; Hwdgo=0.0;  nco=0.0;  hico=0.0; cbho=0.0; ciho=0.0; cbwo=0.0;
 ciwo=0.0;  cbso=0.0; ciso=0.0; Wwdgo=0.0; doco=0.0; dmco=0.0; lmto=0.0;
 reso=0.0;  wcpo=0.0; vcoo=0.0; cpwto=0.0; cwto=0.0; }

winding( float cd, float Hwdg,  float nc,  float hic,  float cbh,
        float cih,  float cbw,  float ciw,  float cbs,  float cis,
        float Wwdg,  float doc,  float dmc,  float lmt,  float res,
        float wcp,  float vco,  float cpwt,  float cwt)
   { cdo=cd; Hwdgo=Hwdg;  nco=nc ; hico=hic; cbho=cbh ; ciho=cih ;
     cbwo=cbw; ciwo=ciw ; cbso=cbs; ciso=cis; Wwdgo=Wwdg; doco=doc ;
     dmco=dmc; lmto=lmt ; reso=res; wcpo=wcp; vcoo=vco ; cpwto=cpwt;
     cwto=cwt;
   }

   winding ( spec x )
   {
     cout << endl <<"Enter the no. of axial coils" ;
      cin >> nco ;
     cout << endl <<"Enter the values of hyl (mm.) & dal (mm.) " ;
      cin >> hyl >> dal ;
     cout << endl <<"Enter the value of assumed current-density" ;
      cin >> cdo ;
```

```
    float a = x.geta()   ;
      int t = x.gett()   ;
      int h = x.geth()   ;
    float dic = x.getdic() ;
   float tap  = x.gettap() ;
    k1 = pow(10,-3) ;
    k2 = 2. * hyl + ( nco - 1. ) * dal ;
   Hwdgo = h - k2 ;
   hico = Hwdgo * 1.03 / (( t + 1.) * nco );   //ht. of each insul.conductor
   cso  = a / cdo ;                //conductor section, mm-sq.
    cout << endl <<"hico =  "<< hico << setw(20) <<" cso=  "<< cso ;
// Since high current, rectangular strap conductor is used
    cout << endl <<"Enter the values of cbho(bare ht- mm),ciho(insul.ht.- mm)"
      <<"cbwo(bare width- mm.),ciwo(insu.width- mm.),cbso(bare section),"
      <<"- sq.mm.,ciso(insulated section- sq.mm.) from Standard Table" ;
    cin >> cbho >> ciho >> cbwo >> ciwo >> cbso >> ciso ;

      cdo = cdo * cso / cbso ;
    Hwdgo = ciho *( t + 1.)/ 1.03;
    Wwdgo = ciwo ;

    const float PI = 3.14159 ;
    cout << endl <<"Enter the values of roe(sp.resis.ohm.mm-sq/metre) &"
          "spwt(sp.weight, kg per mm-cubed" ;
    cin >> roe >> spwt ;
    cout<<endl<<"roe=  "<<roe<<setw(20)<<"spwt=  "<<spwt;
    doco = dic + 2. * Wwdgo ;           //outside dia. of coil,mm.
    dmco = 0.5 * (dic + doco) ;         //mean diameter of coil,mm.
    lmto = PI * dmco ;                  //length of mean turn, mm.
    reso = roe * lmto * t * k1 / cbso ;   // coil resistance, ohm.
    wcpo = a * a * reso ;
    vcoo = t * lmto * cbso ;           // conductor voulme,mm-cubed.
    cpwto= spwt * vcoo * k1 ;
    }
  void display()
  {
    cout << endl << setw(40) <<"DESIGN DATA" ;
    cout << endl <<"clearances:" ;
    cout << endl <<"axial between yoke and coil,at both ends=  "<< hyl<<"mm.";
    cout << endl <<"duct width-axial, between consecutve coils= "<< dal<<"mm";
    cout << endl <<"modified current-density=  "<< cdo <<" amp/mm.-sq";
    cout << endl << setw(15) <<"no. of axial coils=  " << nco ;
    cout << endl << setw(25) <<"Height of rect. bare conduc.= "<< cbho <<"mm";
    cout << endl << setw(45) <<"Width of rect. bare conduc.= "<< cbwo <<"mm";
    cout << endl << setw(50) <<"Height of rect.insul conduc.= "<< ciho <<"mm";
    cout << endl << setw(65) <<"Width of rect.insul. conduc.= "<< ciwo <<"mm";

  cout << endl << setw(45) <<"Radial width of coil=  "<< Wwdgo <<"mm";
  cout << endl << setw(30) <<"winding height= "<< Hwdgo <<"mm";
  cout << endl << setw(20) <<"conductor loss= "<< wcpo <<"watt";
  cout << endl << setw(10) <<"conductor volume= "<< vcoo <<"mm-cubed";
   cout << endl << "conductor weight=  "<< cpwto <<" g";
  }
};
```

```
void main()
{
 clrscr() ;
 spec m ;
 winding n ;
 m.getvalue() ;

 n = m ;
 cout <<"\n"; n.display() ;
}
```

Figure 6.5(b)   A computer program for the design of Helix winding.

RESULT

(E)   166.7   22   325   140   0
(E)   1
(E)   10   5
(E)   2.75
(o)   hico = 13.66   cso = 60.62
(E)   14.0   14.49   4.0   4.49   55.1   63.9
(E)   0.0341   0.0027

(o)   <u>DESIGN SHEET</u>

Clearances:

Axial between yoke and coil, at each end = 10 mm.

Duct width - axial, between consecutive coils = 5 mm.

Modified current-density = 3.025 amp/sq. mm.

Height. of rectangular bare conductor = 14.0 mm.

Height. of rectangular insulated conductor = 14.49 mm.

Width. of rectangular bare conductor = 4.0 mm.

Width. of rectangular insulated conductor = 4.49 mm.

Radial width of coil = 4.49 mm.

Winding height = 323.6 mm.

Conductor loss = 171.7 watt.

Conductor volume = 550251 mm. cubed.

Conductor weight = 1.49 grams.

6.6.3 Disc

Disc type is generally used for winding where helix is uneconomical as the voltage increases. It consists of a number of disc sections wound alternately from outside to inside and inside to outside. The sections are connected is series which can either be done as a separate manufacturing operation or can be wound continuously. For use where the current is very high, disc sections can be connected in parallel. In such cases, the low-voltage winding is placed outside to facilitate connections to the terminals.

The design of disc coils start with the selection of number of disc-sections. For winding without tap, such selection is by trial and error, basing on the choice of suitable conductor section, insulation and duct width. For winding with tap, limitation is imposed on the selection. For convenience, one may assume that,

(i) the tapping points are on the outer surface of the winding. Thus at the tapping points, the number of turns in a pair of disc-sections equals the number of turns in a pair of disc-sections equals the number of turns between tappings;

(ii) the number of turns per section throughout the winding is equal.

To comply with the above assumptions, it will be often found that the computed (turns per section × number of disc-sections) may be larger than the required total turns on maximum tap. Under this case, the excess of turns over the requirement must be omitted. The gaps left due to this omission must be distributed evenly throughout the winding.

(iii) With d_s disc-sections all in series, the number of axial ducts = $(d_s - 1)$, and the height of insulated conductor,

$$H_{stc} = H_{wdg} - d_s \cdot b_t - (d_s - 1) \, w_d \qquad (6.8a)$$

b_i and w_d being width of insulating paper and width of axial duct respectively.

With 2 disc-sections in parallel,

$$H_{stc} = H_{wdg} - \frac{1}{2} d_s \cdot b_i - (\frac{1}{2} d_s - 1) \, w_d \qquad (6.8b)$$

Computerised design

Governing equations

Enter the given specifications.

Step 1: Choose the values of axial clearances - hyl, between yoke and winding, and dal, duct width between two consecutive coils (art. 6.4 and Figure 6.4).

Assume current density (based on whether copper or aluminium is used as the conductor material.

Step 2: Compute,

(i) turns at maximum tap = (1 + tap). number of turns t;

(ii) turns on minimum tap = (1 − tap). number of turns t;

and enter in each case the designer's choice.

Step 3 : Divide the tap into suitable 'tapping step' - say, in two halves, i.e. ts = 0.50 and enter the same. Thus compute,

(i) turns between taps, tbt = tapping step x tap x turns per phase;

(ii) turns per section, tps = tapping step x tbt;

(iii) number of disc sections = turns on maximum tap x tps.

Enter integer values for each of above, and modify the total turns on maximum tap = modified turns per section x modified number of disc sections.

Step 4 : We can now compute the number of turns to be omitted, tsub = turns on maximum tap - modified total turns on maximum tap.

tsub should have a negative value. If not, increase the number of disc sections by 2. In the computer program this has been achieved through an 'if-statement'.

As stated in Example 6.2, omission of turns will lead to a gap in the winding which should be evenly distributed through the coil.

Step 5 : We are now in a position to compute the total height of insulated conductor, hico. With reference to Example 6.2,

  (i) available winding height = window height 'h' - 2. hyl.
 (ii) we take discs in pair with insulated washer of thickness 0.2 mm. in between, so that available height for conductor is reduced by, 0.5 x number of disc sections x 0.2 (washer thickness);
(iii) the available height is further reduced by the axial ducts between the disc sections = (0.5 x number of disc sections - 1) x duct width 'dal'.
       The required height of insulated conductor, hico = [(i) - (ii) - (iii)]/ number of disc sections.

Step 6: Computation of conductor sections.
  (i) Compute bare conductor section = current per phase / assumed current-density.
 (ii) From the computed value of the height of insulated conductor, hico, and computed bare conductor section, choose the STANDARD conductor using standard table for rectangular strap conductors, and enter the values.
(iii) Knowing bare conductor section and current, modify the assumed value of current-density.

Step 7 : Winding height = number of disc sections x height of insulated conductor
          as chosen from standard table
          + [(ii) + (iii)] computed in step 5 (= k2 in computer program).

Coil width, Wwdgo = width of insulated conductor, ciwo x turns per disc section, tps x 1.03.

Step 8: Compute, lenght of mean turn = ([internal diameter of coil 'dic' +
         width of winding 'Wwdgo'];

and hence, resistance per phase, conductor loss, and volume and weight of conductor material.

```
//Design of disc winding
// Given specification: winding current per phase, ap; winding turns, tp;
// height of window, ht; internal dia.ofcoil, dic; % tap, tap.
#include<iostream.h>
#include<iomanip.h>
#include<conio.h>
#include<math.h>

class spec
{
  public:
  int c, d, e ;
  float a, b;
  void getdata(float a, float b, int c, int d, int e) ;
  void display() ;
  void disc( float a, float b, int c, int d, int e ) ;
} ;
```

```
void spec :: getdata(float ap, float tap, int tp, int ht, int dic)
        { a = ap; b = tap; c = tp; d = ht; e = dic; }

void spec :: display()
{
  cout <<"amp/ph.= "<< a << setw(20) <<"%tap= "<< b << setw(20)
      <<"turns/ph.= "<< c << setw(20) <<"height of window.= "<< d
      << setw(20) <<"internal diameter of coil= "<< e ;
}

void spec :: disc ( float a, float b, int c, int d, int e )
{

float f ;      //Axial clearance between yoke and coil,at each end.
float g ;      //Axial duct width between two consecutive coils,mm.
float h ;       //Assumed current density,amp per mm.sq.

float l, m, n, o, q, cbh, cih, cbw, ciw, cbs, cis ;
float r, s, t, u, v, w, x, y, z, ts, aa, bb, cc, dd, ee, ff ;
float k1, k2, roe, spwt ;

cout << endl <<"Enter the values of axial clearance(f, mm) between "
    "yoke and winding; & ( g,mm)-that between two consecutive coils " ;
cin >> f >> g ;
cout << endl <<"Enter the value of assumed current-density,amp/mm.sq" ;
cin >> h ;

l = (1. + b) * c ;             //turns at maximum tap.
m = (1. - b) * c ;             //turns at minimum tap.
cout << endl <<"turn on maxm. tap= "<< l << setw(20) <<"turn on minm.tap"
    <<" = "<< m ;

cout << endl <<"Enter designer's choice of l and m" ;
cin >> l >> m ;
cout << endl <<"Turns on Maximum tap = "<< l ;
cout << endl <<"Turns on Minimumm tap= "<< m ;

cout << endl <<"Enter per cent tapping step, ts " ;
cin >> ts ;

  aa = ts * b * c ;                   //turns between taps
  bb = ts * aa ;                      //turns per section
  dd = l / bb ;                       //no. of disc sections
  cout << endl <<"turns bet. tap: "<< aa <<setw(20) <<"turns per"
    " section: "<< bb << setw(20) <<"No.of disc section: "<< dd ;

cout << endl <<"Enter modified values aa,bb,dd:" ;
  cin >> aa ;                     //modified turns between taps
  cin >> bb ;                     //modified turns per section
  cin >> dd ;                     //modified no.of disc sections
  ee = bb * dd ;                  //modified total turns on maxm.tap
  ff = l - ee;                    //no.of turns to be omitted

if (ff<=0)
{
```

```
        cout << endl <<"no.of turns to be omitted: "<< ff ;
        }
        else
        dd = dd + 2.;
        cout << endl <<"Turns bet.taps= "<< aa << setw(20) << "No. of disc"
             "sections=  "<< dd << setw(20) <<"turns/section= "<< bb ;
//               Conductor section
//        the gap left by omitted turns should be evenly
//             distributed throughout the winding


        o = a / h ;              // computed conduc.section, mm-sq.
        k2 =  0.5 * dd * 0.2 + (0.5 * dd - 1.) * g ;
        q = (d - 2 * f - k2) / dd ;    // height of insulated conductor.
     cout << endl <<"height of insulated conductor= "<< q << setw(20)
             <<" bare conductor section= "<< o ;
// Choose rectangular strap conductor
     cout << endl <<" Enter the values of bare height(cbh,mm), width(cbw,mm),&"
             " rounded-of section(cbs,mm.sq); insulated height(cih,mm),"
             " width(ciw,mm), & rounded-of section(cis,mm.sq)"
             " from standard table" ;
        cin >> cbh >> cbw >> cbs >> cih >> ciw >> cis ;

        h  = a / cbs;           //modified current density
        n  = dd * cih + k2 ;           //axial winidng height,mm.
        r = ciw * bb * 1.03;       //coil width,mm

//    compute conductorr-loss per phase wcp watts,conductor weight cwt kg

        const float PI = 3.14159 ;
               k1 = pow(10,-3) ;
     cout << endl <<"Enter the values of specific resistance(roe-ohm.mm.sq/m)"
        "& specific weight( g/mmcubed),spwt" ;
      cin >> roe >> spwt ;
      s = e + 2. * r ;          // outer dia. of coil,mm.
      t = 0.5 * ( e + s ) ;          // mean dia. of coil,mm.
      u = PI * t ;              // length of mean turn,mm.
      v = roe * u * c * k1 / cbs ;     // resistance,ohm.
      w = a * a * v ;           // conductor loss,watt.
      x = t * u * cbs ;            //conductor volume,mm-cubed.
      y = spwt * x * k1;              //conductor weight, kg.

     cout << endl << setw(40) <<"DESIGN SHEET" ;
     cout << endl <<"modified current density= "<< h <<"amp/mm.sq";
     cout << endl <<"clearances:" ;
     cout << endl <<"Axial- between yoke and winding= "<< f <<"mm.";
     cout << endl <<"axial between consecutive coils= "<< g <<"mm.";
     cout << endl << setw(20) <<"ht.of rect.bare conductor= "<< cbh <<"mm";
     cout << endl << setw(30) <<"width of rect.bare conductor= "<< cbw <<"mm";
     cout << endl << setw(45) <<"ht. of rect.insul.conductor= "<< cih <<"mm";
     cout << endl << setw(60) <<"width of insulated conduc.= "<< ciw <<"mm";
     cout<< endl << setw(60) <<"winding height= "<< n <<" mm.";
     cout << endl << setw(45) <<"radial width of coil= "<< r <<" mm.";
     cout << endl << setw(38) <<"resis. per phase=  "<< v <<"ohm";
     cout << endl << setw(20) <<"conductor loss=  "<< w <<" watt" ;
     cout << endl << setw(10) <<"conductor volume=  "<< x <<"mm.-cubed" ;
     cout << endl << "conductor weight=   "<< y <<" kg.";
```

```
void main()
{
  clrscr() ;
  spec S ;
  float ap, tap ;
    int tp, ht, dic ;
  cout << endl <<"Enter the values of: current/ph, %tap, turns/ph,"
          " height of window, internal diameter of coil" ;
        cin >> ap >> tap >> tp >> ht >> dic ;
  S. getdata(ap, tap, tp, ht, dic) ;
  S.display() ;
  S.disc(ap, tap, tp, ht, dic) ;
}
```

Figure 6.5(c) A computer program for the design of Disc winding.

RESULT

(E)   30.3   0.05   968   805   280

(E)   10   4

(E)   2.75

(o)   tmxo = 1016.4   tmno = 919.6

(E)   1016   920

(o) · Turns on maximum tap = 1016

       Turns on maximum tap = 920

(E)   per cent tapping step = 0.50

(o)   tbt = 24.2   tps = 12.1   ds = 83.9

(E)   Turns between taps = 24 Turns per section = 12 No. of disc sections = 84

(o)   height of insulated coil = 7.07   bare conductor section = 11.02

(E)   6.5   1.7   10.7   6.9   2.1   13.8

(E)   0.0341 0.0027

(o)   DESIGN SHEET

Modified current-density = 2.83 amp / sq. mm.

Clearances:

Axial between yoke and coil, at each end = 10 mm.

Duct width - axial, between consecutive coils = 4 mm.

Height. of rectangular bare conductor = 6.5 mm.

  Height. of rectangular insulated conductor = 6.9 mm.

      Width. of rectangular bare conductor = 1.7 mm.

        Width. of rectangular insulated conductor = 2.1 mm.

          Winding height = 770.0 mm.

        Radial width of coil = 26 mm.

      Resistance per phase = 2.97 ohm.

    Conductor loss = 2722 watt.

  Conductor volume = 3146670 mm. cubed.

Conductor weight = 8.49 grams.

6.6.4 Disc-helix

The computation of winding details is similar to the helix type. Disc-helix can be simplex when all the disc-sections are in series, or duplex, for 2 disc-sections in parallel.

With the provision of 3 groups of transposed conductors, height of insulated conductor, for simplex type,

$$H_{stc} = \frac{1.02\, H_{wdq}}{T + 4} - wd \tag{6.9a}$$

For duplex disc-helix,

$$H_{stc} = \frac{1}{2} \frac{1.02\, H_{wdq}}{T + 1} - wd \tag{6-9b}$$

since in this type, transposed conductors do not need any additional space.

Computerised design

Governing equations

Enter the given specifications.

Step 1: Choose the values of axial clearances - hyl, between yoke and winding, dal, duct width between two consecutive coils and radial duct between conductors, drc (art. 6.4 and Figure 9.6).

Assume current density (based on whether copper or aluminium is used as the conductor material).

Step 2: Compute:
maximum winding height, Hwdg = specified window height 'h' - 2. hyl.

With the above computed value of the winding height, and knowing the turns per phase 't' and the axial duct width between consecutive coils, the height of the insulated conductor will be determined. The computer at this stage uses switch to identify the simplex or the duplex winding and computes accordingly.

Step 3: For the Simplex type, the height of the insulated conductor is given by,
cih = [1.02 Hwdg/(turns per phase + 4.)] - dal.
For the Duplex type, cih = 0.5 [1.02 Hwdg/(turns per phase + 1.)] - dal.
Step 4: From the standard table for RECTANGULAR STRAP conductors, choose the nearest dimension for the height of insulated conductor 'cih', corresponding height of bare conductor 'cbh' and modify the value of Hwdg computed earlier as below:
For Simplex: Hwdg = [cih. (turns per phase + 4.) + dal. (turns per phase + 3.]/1.02;
For Duplex: Hwdg = [2. (cih + dal). (turns per phase + 1.)]/1.02.

Step 5: Compute:
(i) bare conductor section, cs = phase current/asumed current density;
(ii) total width of bare conductor, twbc = cs/cbh;
and using the values of cs, cbh and cih, choose the standard dimensions as well as the number of radial strands 'azw'.

Modify the current density accordingly.

<u>Step 6:</u> Compute:

length of mean turn = [internal diameter of coil 'dic' + width of winding 'Wwdg'];
and hence, resistance per phase, conductor loss, and volume and weight of conductor material.

```
//Design of Disc-helix winding
//Given specification:current per phase(ap),% tap (tapo),turns per phase (tp,),
//  height of window(hp- mm.), internal diameter of coil(dic-mm.),

#include <iostream.h>
#include <iomanip.h>
#include <conio.h>
#include <math.h>

class spec
{
  public:
    int c, d, e ;
    float a, b ;
    void getdata(float a, float b, int c, int d, int e) ;
    void display() ;
    void dischelix(float a, float b, int c, int d, int e) ;
} ;

void spec :: getdata(float ap, float tap, int tp, int ht, int dic)
    { a=ap; b=tap; c=tp; d=ht; e=dic; }
void spec :: display()
{
  cout << endl << setw(20) <<"amp/ph.=      "<< a ;
  cout << endl << setw(30) <<"% tap=  "<< b ;
  cout << endl << setw(40) <<"turns/ph.=   "<< c ;
  cout << endl << setw(50) <<"window ht.=  "<< d ;
  cout << endl << setw(60) <<"internal dia of coil= "<< e ;
}

void spec :: dischelix( float ap, float tap, int tp, int ht, int dic )
{
  float cd ;           //assumed current density, amp per mm-sq.
  float c0 ;           // axial clearance betweenlimb & coil,  mm.
  float c1 ;           //axial clearance between yoke & coil, mm
  float c2 ;           // axial clearance between layers, mm.
  float c3 ;           // radial duct between coils, mm.
  float roe ;          //specific resis. of conduc.material(Al),ohm/mm-sq
  float spwt ;         //specific weight of conductor material,kg/mm-cubed

  cout << endl <<"Enter the values of : cd, hyl, drc, roe, spwt" ;

  cin >> cd >> c0 >> c1 >> c2 >> c3 >> roe >> spwt ;

  float f, g, h, i, n, o, q, r, s, t, u, w ;
  long v ;
  float cbh,cih,cbw,ciw,cbs,cis ;
  int O ;
  const float PI = 3.14159 ;
  float k1 = pow(10,-3) ;

  f = d - 2 * c1 ;
```

```
        cout << endl <<"ENTER THE VALUE OF O = 1 for SIMPLEX, = 2 for DUPLEX" ;
        cin >> O ;
        if ( O == 1.)
        {
           cout << endl << setw(20) <<"simplex" ;
           float simplex(float,int,float&,float&) ;
           {
            simplex(f,c,c2,cih) ;
           }
        }
         else
         if (O == 2.)
         {
           cout << endl << setw(20) <<"duplex" ;
           float duplex(float,int,float&,float&) ;
           {
            duplex(f,c,c2,cih) ;
           }
         }

    cout << endl << "Height of insulated conductor=   "<< cih ;

    cout << endl <<"Enter the values of cbh ( height of bare conductor) "
                 "from Standard Table" ;
     cin >> cbh ;
     g = a / cd ;            // Total bare conductor section
     cout << endl <<" total bare conductor section: "<< g ;
     cbw = g / cbh ;         //total width of bare conductor
     cout << endl <<" total width of bare conductor =    "<< cbw ;

     cout << endl <<" Enter the values of cbw (bare width-mm.),"

                 " ciw (insulated width -mm.),cbs (rounded off bare "
                 " section- mm.-sq),cis (rounded off insulated"
                 " section-mm-sq),from Standard Table"
                         " & "
                 " number of radial strands (l)"
             " so that cbw x l equals total bare conductor section cbw approx." ;

      cin >> cbw >> ciw >> cbs >> cis >> l;

     n = l * cbs ;           //modified total cond. section
     o = l * ciw + (l - 1.) * c3 ;    //radial width of coil
     cd = a / n ;                     //modified current-density

 //    compute conductor-loss per phase wcp watts,conductor weight cpwt kg

     q = e + 2. * c0 + 2. * o ;        // outer dia. of coil,mm.
     r = 0.5 * ( e + q) ;              // mean dia. of coil,mm.
     s = PI * r ;                      // length of mean turn,mm.
     t = roe * s * c * kl / n ;        // resistance,ohm.
     u = a * a * t ;           // conductor loss,watt.
     v = c * s * n ;           // conductor volume,mm-cubed
     w = spwt * v * kl ;        // conductor weight, kg.
```

```
    cout << endl << setw(45) <<"DESIGN  SHEET";
    cout << endl << "Modified Current density: "<< cd <<" amp/mm-sq";
    cout << endl <<"clearances";
    cout << endl <<"axial between limb and coil= "<< c0 <<" mm";
    cout << endl <<"axial between yoke and coil= "<< c1 <<" mm";
    cout << endl <<"duct- axial between layers=  "<< c2 <<" mm";
    cout << endl <<"duct- radial between coils=  "<< c3 <<" mm";
    cout << endl << setw(10) <<"Height of rect.bare conduc= "<< cbh <<" mm";
    cout << endl << setw(30) <<"Width of rect.bare conduc=  "<< cbw <<" mm";
    cout << endl << setw(40) <<"Height of rect.insul.conduc= "<< cih<<" mm";
    cout << endl << setw(50) <<"Width of rect.insul.conducc= "<< ciw<<" mm";
    cout << endl << setw(70) <<"No. of Radial strands=  "<<l ;
    cout << endl << setw(55) <<"Winding height=  "<< f <<" mm.";
    cout << endl << setw(40) <<"Radial width of coil= "<< o <<" mm.";
    cout << endl << setw(30) <<"coil resistance=   "<< t <<" ohm";
    cout << endl << setw(20) <<"conductor volume=   "<< v <<" mm-cubed";
    cout << endl << setw(10) <<"conductor loss=   "<< u <<" watt";
    cout << endl <<"conductor weight= "<< w <<" kg";
} ;

float simplex(float f, int c, float& c2, float& cih)
{
        float hic ;     // height of insulated conductor

//              Choose rectangular strap conductor
    hic = 1.02 * f / (c + 4.) ;  // total ht. of insul conduc.
//                      and axial duct between conduc.
    cih = hic - c2;             // height of insulated conductor
    cout << endl <<"ht.of insul. conductor= "<< cih ;
    cout << endl <<"Enter the value of cih(ht.of insul. conductor from Standard
//                              Table" ;
    cin >> cih ;
    f = (cih *(c + 4.) + c2 * (c + 3.))/1.02 ;  //winding height
} ;


float  duplex(float f,int c,float& c2,float& cih)
{
        float hic ;

//              Choose rectangular strap conductor
    hic = 1.02 * f / (c + 1.) ;      // total ht. of insul conductor
//                      and axial duct between conduc.
    cih = 0.5 * (hic - 2.* c2) ;  // height of insulated conductor
    cout << endl <<"ht.of insul. conductor= "<< cih ;
    cout << endl <<"Enter the value of cih(ht.of insul. conductor)"
                                "from Standard Table" ;
    cin >> cih ;
    f = 2. * (cih + c2) * (c+1.) / 1.02 ;          //winding height
} ;

void main()
{
    clrscr() ;
    spec S;
```

```
    float ap, tap ;
    int tp, ht, dic ;
    cout << endl <<"Enter the data: armature current / ph(ap),%tap(tap),"
        "turns / ph.(tp), window height(ht),internal dia.of coil(dic)" ;
    cin >>  ap >> tap >> tp >> ht >> dic ;

    S.getdata(ap, tap, tp, ht, dic) ;
    S.display() ;
    S.dischelix( ap, tap, tp, ht, dic) ;
  } ;
```

Figure 6.5(d)   A computer program for the design of Disc-helix winding.

RESULT

Run 1

(E)   1333   0   22   740   200

(E)   5   5   5   2.75

(E)   o = 1

(o)   Simplex.   ht. of insulated conductor = 23.64

(E)   Enter the value of cih (ht. of insulated cond) from standard table 23.9

(o)   height of insulated conductor = 23.9

(E)   Enter the value of cbh (ht. of bare cond) from standard table 23.5

(o)   bare conductor section = 484.73
      width of bare conductor = 20.63

(E)   4.0   4.4   91.8   103.4   5

(E)   0.0341   0.0027

(o)   DESIGN SHEET

Modified current-density = 2.9 amp/sq. mm.

Clearances:

Axial between yoke and coil, at each end = 5 mm.

Duct width - axial, between consecutive coils = 5 mm.

Duct width - radial between conductors = 5 mm.

Height. of rectangular bare conductor = 23.5 mm.

Width. of rectangular bare conductor = 4.0 mm.

Height. of rectangular insulated conductor = 23.9 mm.

Width. of rectangular insulated conductor = 4.4 mm.

Number of radial strands =5

Winding height = 730 mm.

Radial width of coil = 42 mm.

Resistance per phase = 0.00124 ohm.

Conductor loss = 2207.95 watt.

Conductor volume = 7677154 mm. cubed.

Conductor weight = 20.73 grams.

Run 2

(E) 1333   0   22   740   200

(E) 5   5   5   2.75

(E)   o = 2

(o) Duplex.   ht. of insulated conductor = 11.187

(E) Enter the value of cih (ht. of insulated cond) from standard table 11.4

(o) height of insulated conductor = 11.4

(E) Enter the value of cbh (ht. of bare cond) from standard table 11.0

(o) bare conductor section = 484.73

width of bare conductor = 44.07

(E) 2.5   2.9   26.6   32.9   18

(E) 0.0341   0.0027

(o)   DESIGN SHEET

Modified current-density = 2.78 amp/sq. mm.

Clearances:

Axial between yoke and coil, at each end = 5 mm.

Duct width - axial, between consecutive coils = 5 mm.

Duct width - radial between conductors = 5 mm.

Height. of rectangular bare conductor = 11.0 mm.

Width. of rectangular bare conductor = 2.5 mm.

Height. of rectangular insulated conductor = 11.4 mm.

Width. of rectangular insulated conductor = 2.9 mm.

Number of radial strands = 18

Winding height = 730 mm.

Radial width of coil = 137.2 mm.

Resistance per phase = 0.00166 ohm.

Conductor loss = 2949.3 watt.

Conductor volume = 11158708 mm. cubed.

Conductor weight = 30.13 grams.

6.6.5 Multilayer-Helix

This winding is similar to helix, but can be designed for high voltage on which the choice of number of layers depends. Once the number of layers has been chosen, turns per layer can be decided upon. The number of inner-layer turns must be larger than the successive outer-layer turns to make room for end packings (Figure 5.1e). The height of insulated conductors can then be computed using equn. 6.7 a.

Computerised design

Governing equations

Enter the given specifications.

[*Note:* The program as in Figure 6.5(e) includes values of assumed current density of 2.75 amps per mm. sq. - taking aluminium as the conducting material and also the values of axial clearance 'hyl' and radial duct width between layers 'drc'. These are to be altered as per designer's requirement, as and when necessary.]

<u>Step 1:</u>   Choose the number of layers = two, three or four, through a 'switch', and with specified turns per phase 't' at the input,

compute-for two-layers: turns per phase per layer which enables the designer to select the number of turns per phase for the inner layer 'til', and that for the outer layer, 'tol';

-    for three-layers: turns per phase per layer which enables the designer to select the number of turns per phase for the inner layer 'til', and those for the middle 'tml' and the outer layer, 'tol';

-    for four layers: turns per phase per layer which enables the designer to select the number of turns per phase for the inner layer 'til', and those for the 2nd layer 't21', 3rd layer 't31' and the outer layer, 'tol';

<u>Step 2:</u> Compute,
(i) maximum winding height, Hwdgo = specified window height 'h' - 2. hyl;
(ii) height of insulated coil = 1.03 Hwdg / (turns per phase, inner layer, 'til' + 1).

<u>Step 3:</u> Compute, bare conductor section, cs = phase current / asumed current density.

Choose the conductor data from standard table for RECTANGULAR STRAP conductors, modify the assumed current density accordingly.

<u>Step 4:</u> Compute: modified winding height = 1.03 (number of turns of inner layer + 1.) x height of insulated conductor, and winding width, Wwdg = no. of layers x width of insulated conductor + (no. of layers - 1.) x width of radial duct, drc.

<u>Step 5:</u> Compute length of mean turn = [internal diameter of coil 'dic' + width of winding 'Wwdg'];
and hence, resistance per phase, conductor loss, and volume and weight of conductor material.

```
// Design of multilayer helix winding
//Given specification: current per phase(ap),% tap (tap),turns per phase (tp),
// height of window(ht- mm.), internal diameter of coil(dic-mm.),

#include<iostream.h>
#include<iomanip.h>
#include<conio.h>
#include<math.h>

class spec
{
    public:
     int c, d, e ;
     float a, b ;
    void getdata(float a, float b,int c, int d, int e) ;
    void display() ;
    void multilayerhelix(float a, float b,int c, int d, int e) ;
} ;

void spec :: getdata(float ap, float tap, int tp, int ht, int dic)
    { a=ap; b=tap; c=tp; d=ht; e=dic; }
void spec :: display()
{
```

```
    cout << endl << setw(20) <<"amp/ph.=    "<< a ;
    cout << endl << setw(30) <<"% tap=  "<< b ;
    cout << endl << setw(40) <<"turns/ph.=  "<< c ;
    cout << endl << setw(50) <<"window ht.=  "<< d ;
    cout << endl << setw(60) <<"internal dia.of coil=  "<< e ;
}

void spec :: multilayerhelix( float ap, float tap, int tp, int ht, int dic )
{
    float cd ;              //Assumed current density, amp per mm-sq.
    float hyl ;             // Axial clearance between yoke & coil,mm.
    float drc ;             // radial width of duct between layers, mm.
    float roe ;             //specific resis. of conduc.material(Al),ohm/mm-sq
    float spwt ;            //specific weight of conductor material,kg/mm-cubed

    cout << endl <<"Enter the values of : cd, hyl, drc, roe, spwt" ;
    cin >> cd >> hyl >> drc >> roe >> spwt ;

float f, g, h, k, l, m, n, o, q, r, u ;
float til, cbh,cih,cbw,ciw,cbs,cis ;
 int O, M ;
const float PI = 3.14159 ;
float k1 = pow(10,-3) ;
    cout << endl <<"Enter the O = 1 for TWO layer, = 2 for THREE layer,"
            "= 3 for FOUR layer" ;
    cin >> O ;

    if (O==1)
        {                               // No. of Layers = m
        cout << endl << setw(45) <<"no. of layers, m = 2" ;
            void twolayer(int,int&) ;
            {
                twolayer(c, til) ;
            }
        } else
        if (O==2)
        {
        cout << endl << setw(45) <<"no. of layers, m = 3" ;
            void threelayer(int,int&) ;
            {
                threelayer(c, til) ;
            }
        } else
        if (O==3)
        {
        cout << endl << setw(45) <<"no. of layers, m = 4" ;
            void fourlayer(int,int&);
            {
                fourlayer(c, til);
            }
        }
    getch() ;

f = d - (2 * hyl) ;
h = f * 1.03 / (til + 1.) ;      //height of insulated coil
cout << endl << setw(12) <<"height of insulated coil:  "<< h ;
```

```
    g = a / cd ;                    //conductor section, mm-sq.
    cout << endl <<" conductor section=  "<< g <<"mm-sq." ;
    getch() ;

// Rectangular strip conductors are chosen
    cout << endl <<"Enter the values of ht.of bare(cbh-mm.), ht.of insul"
            "(cih-mm.),width of bare(cbw-mm.),width of insul.(ciw-mm.)"
            "conductor;  bare conduc.section(cbs- mm-sq) &"
            "insul.conduc.section(cis- mm-sq) from standard table" ;
    cin>> cbh >> cih >> cbw >> ciw >> cbs >> cis ;

    cd = a / cbs ;                  //modified currnt density
     f = cih * (til + 1.) / 1.03 ;          //winding height,mm

    cout << endl <<"Enter the value of No. of Layer, M" ;
     cin >> M ;                     // number of layers
    k = M * ciw + (M - 1.) * drc ;          //radial width of coil,mm

//          Compute copper loss,watt; and copper weight,kg

     l = e + 2.* k ;            //outside dia. of coil,mm
     m = 0.5 * ( e + l ) ;          //mean dia. of coil,mm
     n = PI * m ;
     o = roe * n * c * kl / cbs ;
     q = a * a * o ;            //conductor loss,watt
     r = c * n * cbs ;          //conductor volume,mm-cubed
     u = spwt * r * kl;         //conductor weight,kg

    cout << endl << setw(45) <<"DESIGN SHEET" ;
    cout << endl <<"clearances:" ;
    cout << endl <<"axial- between yoke and coil=  "<< hyl <<"mm";
    cout << endl <<"radial width of duct between layers =  "<< drc <<"mm";
    cout << endl <<"modified current-density=  "<< cd <<"amp/mm-sq";
    cout << endl << setw(25) <<"Height of rect.bare conduc.=  "<< cbh <<" mm";
    cout << endl << setw(45) <<"Width of rect. bare conduc.=  "<< cbw <<" mm";
    cout << endl << setw(50) <<"Height of rect.insul conduc.= "<< cih <<" mm";
    cout << endl << setw(65) <<"Width of rect.insul.conduc.=  "<< ciw <<" mm";
    cout << endl << setw(45) <<"Radial width of coil=  "<< k <<" mm";
    cout << endl << setw(30) <<"winding height=  "<< f <<" mm";
    cout << endl << setw(20) <<"conductor-loss=  "<< q  <<" ohm";
    cout << endl << setw(20) <<"conductor volume= "<< r  <<" mm-cubed";
    cout << endl << "conductor weight=   "<< u <<" kg";
};


//twolayer()
void twolayer(int c,int& til)

{
 int tol,M;
 M = 2.;                    //number of layers
 float tpl;
 tpl = c / M;                   //turns/phase/layer
 cout << endl << setw(30) <<"tpl=   "<< tpl ;

 cout << endl <<"Enter the values of til & tol" ;
  cin >> til >> tol ;               //turns/ph-inner/outer layer
```

```
    cout << endl <<"turns/ph.-inner layer= "<< til << setw(30)
            <<"turns/ph.-outer layer= "<< tol ;
}


//threelayer()
void threelayer(int c,int& til)
{
  int M,tml,tol;
  float tpl;                          //turns/ph./layer
  M = 3;                              // no. of layers

  tpl = c / M;
  cout << endl << setw(12) <<"tpl=  "<< tpl ;

  cout << endl <<"Enter the values of til, tml & tol" ;
  cin >> til >> tml >> tol ;          //turns-inner/middle/outer layer
  cout << endl <<"turns/ph.-inner layer=  "<< til << setw(30)
            <<"turns/ph.-middle layer=  "<< tml  << setw(30)
            <<"turns/ph.-outer layer=   "<< tol ;
}

//fourlayer()
void fourlayer(int c,int& til)
{
  int M,t2l,t3l,tol;                  //t2l:2nd layer;t3l:3rd layer
  float tpl;                          //turns/ph./layer
  M = 4.;                             // no. of layers

  tpl = c / M;
  cout << endl << setw(12) <<"tpl=   "<< tpl ;

  cout << endl <<"Enter the values of til,t2l,t3l & tol" ;
  cin >> til  >> t2l >> t3l >> tol ;  //turns/ph.-inner/2nd/3rd/outer layer
  cout << endl <<"turns/ph.-inner layer=   "<< til << setw(30)
            <<"turns/ph.-second layer=   "<< t2l << setw(30)
            <<"turns/ph.-third layer=   "<< t3l << setw(30)
            <<"turns/ph.-outer layer=   "<< tol ;
}

void main()
{
  spec S ;
  float ap, tap ;
  int tp, ht, dic ;
  clrscr() ;
  cout << endl <<"Enter the data:  a ,b, c, d, e" ;
  cin >> ap >> tap >> tp >> ht >> dic ;
  S.getdata( ap, tap, tp, ht, dic ) ;
  S.display() ;
  S.multilayerhelix( ap, tap, tp, ht, dic ) ;
}
```

Figure 6.5(e)   A computer program for the design of Multilater-helix winding.

RESULT

Run 1

(E) 66.7   0   292   740   220

(E) 2.75   10   10   0.0341   0.0027

(E) O = 1

(o) TWO LAYER   tpl = 146

(E) turns per phase-inner 152   outer 140

(o) height of insulated coil = 4.84 conductor section = 24.25 mm. sq.

(E) 4.5   4.9   5.5   5.9   23.9   27.8

(E) No. of layers = 2

(o)   DESIGN SHEET

Clearances:

Axial between yoke and coil, at each end = 10 mm.

Duct width - radial, between layers = 10 mm.

Modified current-density = 2.79 amp/sq. mm.

Height. of rectangular bare conductor = 4.5 mm.

Width. of rectangular bare conductor = 5.5 mm.

Height. of rectangular insulated conductor = 4.9 mm.

Width. of rectangular insulated conductor = 5.9 mm.

Radial width of coil = 16.8 mm.

Winding height = 728 mm.

Conductor loss = 1378.9 watt.

Conductor volume = 5191728 mm. cubed.

Conductor weight = 14 kg.

Run 2

(E) 66.7   0   292   740   220

(E) 2.75   10   10   .0341   .0027

(E) O = 2

(o) THREELAYER   tpl = 97

(E) turns per phase-inner 107   middle 97   outer 88

(o) height of insulated coil = 6.87   conductor section = 24.33 mm. sq.

(E) 6.5   6.9   4.0   4.4   25.1   29.4

(E) No. of layers = 3

(o)   DESIGN SHEET

Clearances:

Axial between yoke and coil, at each end = 10 mm.

Duct width - radial, between layers = 10 mm.

Modified current-density = 2.67 amp/sq. mm.

Height. of rectangular bare conductor = 6.5 mm.

Width. of rectangular bare conductor = 4.0 mm.

Height. of rectangular insulated conductor = 6.9 mm.

Width. of rectangular insulated conductor = 4.4 mm.

Radial width of coil = 23.2 mm.

Winding height = 723.9 mm.

Conductor loss = 1348.4 watt.

Conductor volume = 5599764 mm. cubed.

Conductor weight = 15.1 kg.

Run 3

(E) 66.7   0   292   740   220

(E) 2.75   10   10   0.0341   0.0027

(E) O = 3

(o) FOUR LAYER   tpl = 73

(E) turns per phase-inner 83   2nd 73   3rd 73   outer 63

(o) height of insulated coil = 8.83   conductor section = 24.25 mm. sq.

(E) 8.0   8.4   3.0   3.4   23.3   27.5

(E) No. of layers = 4

(o)   **DESIGN SHEET**

Clearances:

Axial between yoke and coil, at each end = 10 mm.

Duct width - radial, between layers = 10 mm.

Modified current-density = 2.86 amp/sq. mm.

Height. of rectangular bare conductor = 8.0 mm.

Width. of rectangular bare conductor = 3.0 mm.

Height. of rectangular insulated conductor = 8.4 mm.

Example 6.2

For the 1 MVA, 11000 ± 5% / 433–V, delta/star distribution transformer of Example 3.1, design the l.v. and h.v. windings, given,

$$\text{rated load loss at } 75°C = 11880 \text{ Watts;}$$

$$\% \text{ impedance} = 6.50 \%$$

$$\text{no load loss at rated voltage and frequency} = 1800 \text{ Watts.}$$

Solution

From example 3.1 : $E_T = 11.30, J = 2.75$ A. mm^{-2},

$$A_w = 0.143 \text{ m}^2, T_1 = 22,$$

$$T_2 = 1016 - 968 - 920$$

(i) % resistance $= \dfrac{\text{Load loss}}{\text{volt. amp rating}} \times 100$

$$= \frac{11880}{1 \times 10^6} \times 100 = 1.19 \%$$

% rectance $= (6.50^2 - 1.19^2)^{1/2} = 6.4\%$

$$= 39.5 \frac{Q}{E_T^2} \frac{c_x}{(H_{wdg}/s)} \quad \text{from equn. 4.48 (e)}$$

That is, $\dfrac{c_x}{(H_{wdg}/s)} = 0.063$

(ii) $c_x = c_o + \dfrac{c_1 + c_2}{3}$

Assume $H_{wdg}/s = 0.6$ & $c = 0.01$ m (from Table 6.3)

With $c_1 - c_2$, we have,

$c_x = 0.038$ m ; $c_1 = c_2 = 0.039$ m

(iii) Load loss = $3\,P_R$ + stray loss

where, P_R is the I^2R loss per phase.

Assuming stray loss as 5% of the load loss,

$P_R = \dfrac{1}{3} \times 0.95 \times 11880 = 3762 \ W/ph.$

$= \dfrac{2\,k_e\,Q\,J\,\rho}{E_T}\,s \times 10^3$ from equn. 3.39 with $J_1 = J_2$

$= 2 \times 1.15 \times (1000/3) \times 2.75 \times 10^8 \times 0.0214 \times 10^{-6} \times 10^3 \times \dfrac{s}{11.36}$

or, $s = 0.947$ m.

or, $H_{wdg} = 0.6 \times 0.947 = 0.568$ m.

$$\text{Window height, } h_w = H_{wdg} + 2\,h_{yl}$$
$$= 0.568 + 2 \times 0.03 = 0.628 \text{ m.}$$

(iv) From Figure 3.1

$$C_w = (d - a_1) + 2(c_{cl} + c_1 + c_0 + c_2 + \tfrac{1}{2}\,c_{hh})$$

$$= (0.22 - 0.21) + (0.005 + 0.039 + 0.012 + 0.039 + \tfrac{1}{2}\,0.012)$$

$$= 0.212 \text{ m.}$$
$$A_w = C_w \cdot h_w = 0.212 \times 0.628 = 0.133 \text{ m}^2$$

Modified window space-factor, $k_w = 0.35 \times \dfrac{0.143}{0.133} = 0.375$ which is within limit.

(v) *Winding design*

(a) Low voltage winding :

$$\text{Current/phase} = \dfrac{1000 \times 10^3}{\sqrt{3} \times 433} = 1335 \text{ Amp per phase.}$$

$$\text{Sectional area of l.v. winding} = \dfrac{1335}{2.75 \times 10^6} = 486 \times 10^{-6} \text{ m}^2.$$

(1) We choose helix winding. Keeping an extra axial space for one turn, total l.v. turns = 22 + 1 = 23.

[*Note* : If transposition of conductors is needed, additional space for 3 turns must be provided for.]

Single-layer helix	*Two-layer helix*
	We choose :
No. of turns = 23	No. of turns-
	inner layer = 12
	-outer layer =11
	Radial gap between
	two layers = 5 mm.

Axial height of insulated conductor
assuming 2 mm. axial spacer between

$$= \frac{568 - (22 - 1) \times 2}{23} \qquad\qquad = \frac{568 - (11 - 1) \times 2}{12}$$

$$= 22.98 \text{ mm} \qquad\qquad\qquad = 45.7 \text{ mm.}$$

(3) From standard table for strip conductor.

		Single-layer	Two-layer
we choose, bare conductor		11 mm. × 2.8 mm.	11 mm. × 2.8 mm.
sectional area taking into account rounding off of edges	:	30.3 mm^2	30.3 mm^2
number of axial strands	:	2	4
number of radial strands	:	8	4

giving total conductor-section of
l.v. winding

$$= 30.3 \times 16 \qquad\qquad = 30.3 \times 16$$
$$484.8 \times 10^{-6} \text{ m}^2 \qquad\quad 484.6 \times 10^{-6} \text{m}^2.$$

Modified current density,   :

$$: \quad 2.75 \times 10^6 \times \frac{485}{484.6}$$

J_1 $\qquad\qquad\quad = 2.76 \times 10^6 \text{ m}^2 \qquad 2.76 \times 10^6 \text{ m}^2.$

Axial height of coil   : $\quad (11 + 0.4) \times 2 \times 23 \qquad (11 + 0.4) \times 4 \times 12 + 2 \times$

(assuming total insulation thickness $\quad + 2 \times 21 = 566.4$ mm $10 = 567$ mm.

thicness/of .4 mm per l.v.
stand axially)

With 0.4 mm total insulation thickness of each l.v. strand radially,
radial width of coil :-

duct between cil and core	5.0 mm	5.0 mm
radial width– coil	= 8×(2.8+.4) = 25.6	inner layer
duct between inner and uter layers		= 4 × (2.8+.4) = 12.8
		5.0
		outer layer 12.8
Total radial width, c_1	30.6 mm.	35.6 mm.

(b) High voltage winding :

$$\text{Current per phase} = \frac{1000 \times 10^3}{3 \times 11000} = 30.3 \text{ amps.}$$

Sectional area of h.v. winding with current-density of 2.75 A.mm^{-2}

$$= \frac{30.3}{2.75 \times 10^6} = 11.1 \times 10^{-6} \; m^2$$

(1) We choose disc type, assuming (i) an axial duct width of 4 mm. between disc-sections, and an insulating washer of 0.2 mm. between each section of a pair; (ii) tapping points are on the outer surface. That is, at the tapping points, no. of turns in a pair of sections = no. of turns between tappings; (iii) number of turns per section throughout the winding be equal; (iv) insulation on conductors 0.25 mm thick.
Since, number of turns at the principal tapping = 968,
with 2½% tapping steps, turns between taps,

$$= \frac{2.5 \times 968}{100} = 24.2$$

We choose 24.

Turns per section $= \dfrac{1}{2} \times 24 = 12$

Number of disc-sections $= \dfrac{1016}{12} = 84.7$

We choose 86.
Total turns on maximum tap = 86 × 12 = 1032
Thus to get the required 1016 turns, 16 turns must be omitted. The gaps left by these omitted turns should be evenly distributed throughout the winding.

(2) Height of insulated conductor, H_{stc}

$$= \frac{1}{86}\left[h_w - 30 \times 2 - \frac{1}{2} \times 86 \times 0.2 - \left(\frac{1}{2} \times 86 - 1\right) \times 4 \right]$$

$$= -\frac{1}{86}(628 - 60 - 8.6 - 168) = 4.55 \text{ mm.}$$

From standard table for strip conductors, we choose copper conductor, bare :
4 mm × 2.8 mm; rounded-off section : 10.7 mm^2.
Axial height of h.v. winding

$$= 86 \times (4.0 + 0.5) + 43 \times 0.2 + 42 \times 4 + 4 = 568 \text{ mm.}$$

Radial width, $c_2 = 12 \times (2.8 + 0.5) = 39.6$ mm.
(c) Taking $c_{cl} = 5$ mm., and $c_{hh} = 12$ mm.,
(1) for l.v. winding :-

inside diameter $= d + 2\,c_{cl} = 220 + 10 = 230$ mm.

outside diameter = 230 + 2 cl = 230 + 71.2 = 301.2 mm.

(2) for h.v. winding :-

inside diameter = 301.2 + 2 c_0 = 321.2 mm.

outside diameter = 321.2 + 2 c_2 = 400.4 mm.

(d) No-load loss :

Length of iron : core, L_c = 3 h_w = 3 × 0.628 = 1.88 m.

yoke, Ly = 4 C_w + 7 α_1 = 4 × 0.207 + 7 × 0.21 = 2.30 m.

Volume of iron : core = $A_i L_c$ = 0.0327 × 1.88 = 0.062 m^3

yoke = $A_i L_y$ = 0.0327 × 2.30 = 0.075 m^3

Mass of iron = 7650 × (0.062 + 0.075) = 1043.0 kg.

At B_m = 1.556 T specific iron loss at 50 Hz frequency = 1.71 W. kg^{-1}.

Hence the no-load loss at rated voltage and frequency = 1043.0 × 1.71 = 1795 W.

Summary of winding design

Winding	Low voltage	High voltage
Phase voltage (volts)	250	11550–11000–10450
Phase current (amp)	1335	30.3–(31.9)
Turns per phase	22	1016–968–920
Type	2-layer helix	Disc
Turns per layer	inner 12	No. of disc section
	outer 11	(in pairs) = 85
		Turns/section = 12
Conductors		
Current density (A. mm-8)	2.76	2.83
dimension : bare	11 mm × 2.8 mm	4 mm × 2.8 mm.
insulated	1.4 mm × 3.2 mm.	4.5 mm × 3.3 mm.
Strands : axial	4	
radial	4	
Turn dimensions	45.6 mm × 12.8 mm.	4.5 mm × 3.3 mm
Turn area	484.6 mm^2	10.7 mm^2
Redial width	35.6 mm	39.6
Duct width between l.v. & h.v.	10 mm.	
Inside diameter	220 mm.	311.2 mm.
Outside diameter	291.2 mm.	390.4 mm.
End insulation	2 × 30 mm.	2 × 30 mm.
		duct 42 × 4 mm.
Axial spacers	22 × 2 mm.	washer 41 × 0.2 mm.
		at centre 4 mm.
Height	567 mm.	568 mm.

6.7 DESIGN OF THREE-PHASE CORE-TYPE TRANSFORMER

A typical computer program written in C++ - language is given in Figure 6.7. The program uses a single class SPEC with a friend function TRAN to facilitate data encapsulation and use of sub-programs. In the program, three such sub-programs have been used as 'header files' and one for the computation of temperature-rise [Figure 6.6(a)]. The use of friend function enables calling of data directly from the computer representation of characteristics of core-magnetic material.

Various other alternatives are possible using multiple and hybrid inheritancy, and one such is shown in Figure 6.6(b).

[*Note:* Students may develop computer program on the basis of Figure 6.6(b).]

In Figure 6.7, a computer program is written for a 25 kVa, 3-phase, 50-Hz, 11,000/433 star-delta, core-type, natural oil-cooled transformer for maximum mean top-oil temperature-rise of 40 degree celsius, on the basis of an 'implementation diagram' in Figure 6.6(a). For any other specification, the data-input are to be changed. Further, in this program, equations for the design of lv. and hv. windings are the same since disc windings have been used for both l.v. and h.v. windings, and as such, a sub-program 'void winding (int, float, float and,...)' has been called twice.

Governing Equations

Step 1: Enter the assumed values of voltage co-efficient 'keq', the maximum flux density 'Bm' (Tesla), current density 'cd' (amp per mm. sq.), window space-factor 'wsf', stacking factor 'fst', core space-factor 'fsp',

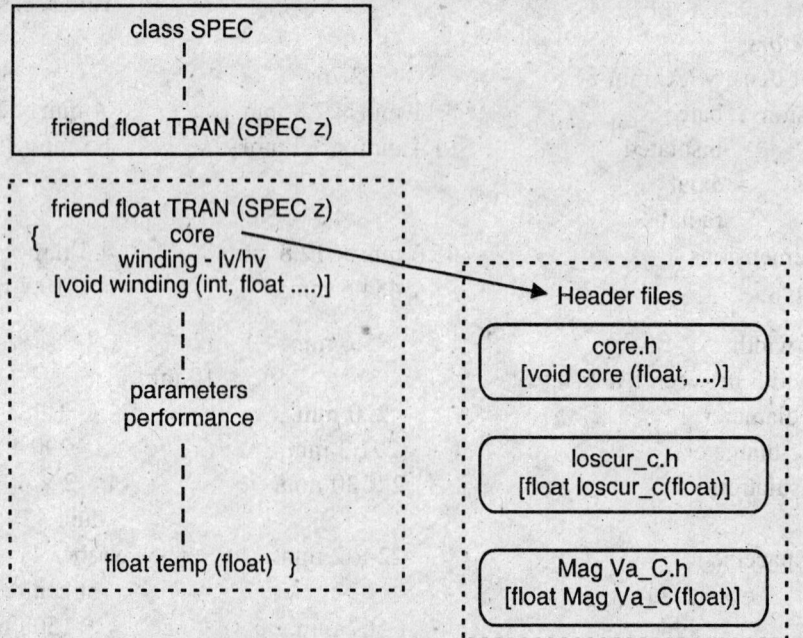

Figure 6.6(a)   Transformer Design — class with friend function.

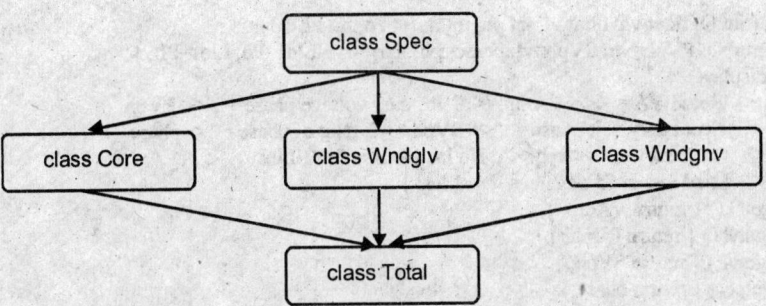

Figure 6.6(b)   Transformer design-multiple Inheritance.

Compute:
-   (i)   voltage per turn = keq [rated VA . 10^{-3}./ no. of phase];
-   (ii)   net core-section area 'ai' = voltage per turn. 10^6 / (4.44 Bm. Frequency 'fr');
-   (iii) gross core-section area 'ac' = net core-section area/stacking factor.

Step 2:   Compute number of steps and width of the limb using the header file "core. h" developed on the basis of discussion in art. 6.2.

Step 3:   Modify 'ai' and 'Bm' and compute the window area 'Aw' (mm. sq.) as, Aw = rated VA . 10^6 / (1.11 x no. of phase. ai. wsf. Bm. cd. fr).

[*Note:* Aw = window width x window height.]

Choose the ratio of window height and window width (generally between 1.5 and 2.5 for 3-phase core-type transformer) and compute width and height. Hence, compute height of limb 'HA', choosing suitable vertical width of the yoke (art. 6.3). In the program vertical width of yoke has been assumed equal to core width 'cw'.

Step 4: The computer program then computes core losses using the header file - "loscur_c.h" which gives the specific loss of the core material (in this case cold rolled grain-oriented sheet steel of 0.35 mm. in thickness.

```
// Design of a Three phase Core type Transformer :
#include <iostream.h>
#include <iomanip.h>
#include <math.h>
#include <conio.h>
#include "core.h"
#include "loscur_c.h"
#include "MagVa_C.h"


class SPEC
  {
  private:
    int NLlos, Llos;
  public:
    int VAmp, phase, freq;
   float PVph, SVph ;
   SPEC()
   { VAmp=0.0; PVph=0.0; SVph=0.0; phase=0.0; freq=0.0; NLlos=0.0; Llos=0.0; }
```

```
    SPEC( int Q, float vh, float vl, int ph, int fr, int Po, int PL )
    { VAmp=Q; PVph=vh; SVph=vl; phase=ph; freq=fr; NLlos=Po; Llos=PL; }
    void display()
    { cout<<"Rated VA=  "<< VAmp <<", Primary voltage/phase=  "<< PVph
        <<" , Secondary volt/phase=  "<< SVph <<" ,No. of phase= "<<phase
        <<", Frequency=  "<<freq<<", No load loss=  "<<NLlos
        <<", Load loss at 75 deg.cel= "<<Llos; }
    int getQ () { return VAmp; }
    float getvh() { return PVph; }
    float getvl() { return SVph; }
    int getph() { return phase;}
    int getfr() { return freq; }
    int getPo() { return NLlos;}
    int getPL() { return Llos; }

    friend float TRAN( SPEC z ) ;
};


float TRAN ( SPEC z )
{
    int Q, ph, fr, Po, PL ;
    float vh,vl,q,Et,ac,ai,Bm,dia,cw,Aw,w,h,Ic,Im,Io ;
    float keq, cd, wsf, fst, fsp ;

    float k1 = pow(10,-3) ;
    cout << endl <<"Enter assumed values of keq, Bmo, cd, wsf, fst, fsp" ;
    cin >> keq >> Bm >> cd >> wsf >> fst >> fsp ;

    Q  = z.getQ() ;
    vh = z.getvh() ;
    vl = z.getvl() ;
    ph = z.getph() ;
    fr = z.getfr() ;
    Po = z.getPo() ;
    PL = z.getPL() ;

    q = Q * k1 / ph ;              //kVA per phase
    Et = keq * sqrt( q ) ;         // Voltage per turn
    cout << "Voltage / turn =  "<< Et << endl ;

//              Design of core and winding
// CRGO steel sheet, 0.35 mm. in thickness, and mitred 45 degree will be used.

    float k2 = pow(10,6) ;
    ai = Et * k2 / ( 4.44 * Bm * fr ) ;   //Net core section area,mm.sq
    cout << endl <<"Computed net core section=  "<< ai <<" mm.sq" ;
    ac = ai / fst ;                //Gross core sectiom,mm.sq.
    cout << endl <<"gross core section=  " << ac <<" mm.sq" ;

//Compute number of step and width of limb

    float st, opf, k3, k4, ratio, HA ;
    const float PI = 3.14159 ;
```

```
void core(float, float&, float&, float&, float&) ;
core(ac,st,cw,dia,opf) ;
cout << endl <<"no. of step=  "<< st << setw(20) << "core width =  "
        << cw <<" mm."<< setw(25) <<"dia=  "<< dia <<" mm."
        << setw(30) <<"optimum fill=  "<< opf ;

ai = PI * fst * fsp * dia * dia / 4. ;      // Modified ai.
Bm = Et * k2 / ( 4.44 * ai * fr ) ;         // Modified Bm.

Aw = Q * k2 / ( ph * 1.11 * ai * wsf * Bm * cd * fr ) ; //Window area
```

// Aw is the product of window-width(w) and window-height(b). The ratio
// (=h / w) is generally between 1.5 and 2.5, depending on customers choice.

```
cout << " Enter ratio of window height and width" << endl ;
cin >> ratio ;
k3 = Aw / ratio ;
w = sqrt(k3) ;                      // Window width
h = ratio * w ;                     // Window height
HA = h + 2. * cw ;                  // height of limb
```

//   Core losses and No-load current
```
k4 = pow(10,-6) ;

float Crwt, Crlos, y, x1, x2, Mgva, k5 ;
Crwt = 7.85 * k4* opf * ai * ( 3.* HA + 2. * w ); //Total weight of
```
//   core(kg),assuming yoke section equal to limb section
```
getch();

y = Bm ;
float loscur_c(float) ;
x1 = loscur_c( y );
Crlos = x1 * Crwt ;
        cout<<endl<<" Core loss =  "<< Crlos <<" watts" ;
```
// Core loss
```
getch();

y = Bm ;
float MagVa_C(float) ;
x2 = MagVa_C(y);
Mgva = x2 * Crwt ;
        cout<<endl<<"Magnetising volt-ampere =  "<< Mgva ;
getch() ;

Ic = Crlos / (ph * vh) ;            // Core-loss current
Im = Mgva / (ph * vh) ;            // Magnetising current
Io = sqrt( Ic * Ic + Im * Im ) ;   // No load current
k5 = Im * 100./ Io ;               // Magnetising current as per cent
```
//   of no-load current
```
cout << endl << setw(50) <<" DESIGN SHEET : CORE " ;

cout << endl << setw(25) <<"gross core section=  "<< ac <<"mm-sq." ;
cout << endl << setw(40) <<"modified net iron section=  "<< ai <<"mm-sq";
cout << endl << setw(55) <<"modified flux-density=  "<< Bm <<" Tesla" ;
cout << endl << setw(60) <<"window area=  "<< Aw <<" mm-sq" ;
cout << endl << setw(55) <<"window width=  "<< w <<" mm." ;
cout << endl << setw(45) <<"window height=  "<< h <<" mm." ;
```

```cpp
    cout << endl << setw(40) <<"core weight=   "<< Crwt <<" kg" ;
    cout << endl << setw(35) <<"Core loss=   "<< Crlos <<" watts" ;
    cout << endl << setw(30) <<"Core-loss comp.of no-load curr.= . "<<Ic<<" amp";
    cout << endl << setw(20) <<"Magentising current=   "<< Im <<" amp" ;
    cout << endl << setw(10) <<"No load current=   " << Io <<" amp" ;
    cout << endl <<"Magnetising current as % of no-load current=   "<< k5 ;

    cout << endl << setw(50) <<"DESIGN OF LOW-VOLTAGE WINDING" ;
     int hyl,hy,dal,da,Cc1,C ;
    float v,cdm,cdl,tl,t,ncl,nc,cbhl,cbh,cihl,cih,cbwl,cbw,ciwl,ciw,cbsl,cbs;
    float cisl,cis,zhl,zh,zwl,zw,dmcl,dmc,Hwdgl,Hwdg ;
    float Wwdgl,Wwdg,resl,res,cpwtl,cpwt,wcp,wcpl,vco,vcol,odlv ;

     v = vl ;

    void winding(int,float,float&,float&,float&,float&,float&,float&,float&,
           float&,float&,float&,float&,float&,float&,float&,int&,int&,
           int&,float&,float&,float&,float&,float&,float&,float&,
           float&,float&) ;
    winding(ph,q,Et,v,cd,h,dia,cdm,t,nc,cbh,cih,cbw,ciw,cbs,cis,hy,da,C,zh,
       zw,dmc,Hwdg,Wwdg,res,wcp,vco,cpwt) ;
cdl=cdm; tl=t; ncl=nc; cbhl=cbh; cihl=cih; cbwl=cbw; ciwl=ciw; cbsl=cbs ;
cisl=cis; hyl=hy; dal=da; Cc1=C; zhl=zh; zwl=zw; dmcl=dmc; Hwdgl=Hwdg;
Wwdgl=Wwdg; resl=res; wcpl=wcp; vcol=vco; cpwtl=cpwt ;

    cout << endl << setw(30) <<"L.V.WINDING DESIGN SHEET :" ;

    cout << endl <<"L.V.winding current density=   "<< cdl <<" amp/mm.sq" ;
    cout << endl << setw(10) << "Number of turns=   "<< tl ;
    cout << endl << setw(10) <<"Number of coils=   " << ncl ;
    cout << endl <<"Winding Details:   Rectangular Strap" ;
    cout << endl << setw(31) <<"Bare"<< setw(43) <<"Insulated" ;
    cout << endl << "Height, mm.="<< setw(18) << cbhl << setw(40) << cihl ;
    cout << endl << "Width, mm. ="<< setw(18) << cbwl << setw(40) <<ciwl ;
    cout << endl << "Section, mm-sq="<< setw(15) << cbsl << setw(40) << cisl ;
    cout << endl <<"Clearances: " ;
    cout << endl <<"Axial-between yoke & winding=   "<< hyl <<" mm" ;
    cout << endl <<"Axial-between two consecutive l.v.coils=   "<< dal <<" mm";
    cout << endl <<"Radial-between limb & lv winding =   "<< Cc1 <<" mm" ;
    cout << endl << setw(10) <<"Number of axial strands=   "<< zhl ;
    cout << endl << setw(20) <<"Number of radial strands=   "<< zwl ;
    cout << endl << setw(40) <<"Winding height=   "<< Hwdgl <<" mm" ;
    cout << endl << setw(50) <<"Winding width=   " << Wwdgl <<" mm" ;
    cout << endl << setw(40) <<"Resistance / ph =   "<< resl <<" ohm" ;
    cout << endl << setw(30) <<"Conductor loss=   "<< wcpl <<" watt" ;
    cout << endl << setw(20) <<"Conductor volume=   "<< vcol <<" mm-cubed" ;
    cout << endl << setw(10) <<"Conductor weight =   "<< cpwt <<" kg" ;

odlv = dmcl + Wwdgl ;
    cout << endl << "Outside diameter of l.v.coil=   "<< odlv <<" mm" ;

cout << endl << setw(50) <<"DESIGN OF HIGH VOLTAGE WINDING" ;

 int hyh,dah,Co ;
float cdh,th,nch,cbdh,cidh,cbwh,ciwh,cbsh,cish,zhh,zwh,dmch ;
```

```
float Hwdgh,Wwdgh,resh,cpwth,wcph,vcoh ;

 v = vh ;

void winding(int,float,float&,float&,float&,float&,float&,float&,float&,
        float&,float&,float&,float&,float&,float&,float&,int&,int&,
        int&,float&,float&,float&,float&,float&,float&,float&,
        float&,float&) ;
winding(ph,q,Et,v,cd,h,dia,cdm,t,nc,cbh,cih,cbw,ciw,cbs,cis,hy,da,C,zh,
     zw,dmc,Hwdg,Wwdg,res,wcp,vco,cpwt) ;

cdh=cdm; th=t; nch=nc; cbdh=cbh; cidh=cih; cbwh=cbw; ciwh=ciw; cbsh=cbs ;
cish=cis; hyh=hy; dah=da; Co=C; zhh=zh; zwh=zw; dmch=dmc; Hwdgh=Hwdg;
Wwdgh=Wwdg; resh=res; wcph=wcp; vcoh=vco; cpwth=cpwt ;


 cout << endl << setw(50) <<"H.V.WINDING DESIGN SHEET :" ;

 cout << endl <<"H.V.winding.current density=   "<< cdh <<" amp/mm.sq" ;
 cout << endl << setw(10) << "Number of turns=   "<< th ;
 cout << endl << setw(10) <<"Number of coils=   " << nch ;
 cout << endl <<"Winding Details:   Round" ;
 cout << endl << setw(28) <<"Bare"<< setw(44) <<"Insulated" ;
 cout << endl << "Diameter,mm.="<< setw(15) << cbdh <<setw(40) <<cidh;
 cout << endl << "Section,mm-sq="<< setw(15) <<cbsh <<setw(40) << cish ;
 cout << endl <<"Clearances: " ;
 cout << endl <<"Axial-between yoke & winding=  "<< hyh <<" mm" ;
 cout << endl <<"Axial-between two consecutive h.v.coils=  "<< dah <<" mm";
 cout << endl <<"Radial-between lv & hv windings =  "<< Co <<" mm" ;
 cout << endl << setw(10) <<"Number of axial strands=  "<< zhh ;
 cout << endl << setw(20) <<"Number of radial strands=  "<< zwh ;
 cout << endl << setw(40) <<"Winding height=  "<< Hwdgh <<" mm" ;
 cout << endl << setw(50) <<"Winding width=  " << Wwdgh <<" mm" ;
 cout << endl << setw(40) <<"Resistance / ph =  "<< resh <<"ohm" ;
 cout << endl << setw(30) <<"Conductor loss= "<< wcph <<"watt" ;
 cout << endl << setw(20) <<"Conductor volume=  "<< vcoh <<" mm-cubed" ;
 cout << endl <<"weight of Conductor material=  "<< cpwth <<" kg" ;

//           Computation of per cent impedance

 float ah,k9,Dm,avHwdg,tWwdg,cx,KR,LL,X,req,r,R,Z ;

 k9 = Q /(ph * ph * Et * Et) ;
 float k6 = pow(10,3) ;
     ah = q * k6 / vh ;          // High volt.winding current
   Dm = 0.5 * ( dmcl + dmch ) ;   //Average of mean dia.of HV & LV coils
avHwdg = 0.5 * ( Hwdgl + Hwdgh ) ;   // Average height of H.V. & L.V.coils
tWwdg = Wwdgl + Wwdgh ;           // total width of H.V. & L.V.coils
   cx = Co + tWwdg / 3.;
   KR = 1.- (( Co + tWwdg )/(PI * avHwdg)); // Ragawosky coefficient
   LL = avHwdg / KR ;
    X = 0.0395 * k9 * k1 * cx * PI * Dm / LL ; // % Reactance

 r = vh / vl ;       // the primary to secondary phase-voltage ratio
 req = resl * r * r ;    // l.v.resis.per ph. in h.v.(primary) terms
```

```
          R = ( req + resh ) * ah * 100. / vh ;      // % resistance in hv.terms

          Z = sqrt( R * R + X * X ) ;               // % impedance

      getch() ;
//              Losses in windings

      float tcpwt, cnloss, Strloss, FLloss, Eff, k10, Reg, theta ;

        tcpwt = cpwtl + cpwth ;          // Total conductor weight
        cnloss = ph * vh * ah * R  / 100. ;   // Total conductor loss at u.p.f.
        Strloss = 0.07 * cnloss ;   //considering stray loss = 7% conductor loss

        FLloss = Crlos + cnloss + Strloss ;   // Full load loss at u.p.f.

//              Per cent Efficiency
        Eff = Q * 100. / ( Q + FLloss ) ;

//              Per cent Regulation on full-load at u.p.f.
        k10 = ( 100. + R )*( 100. + R ) + ( 100. + X )*( 100. + X ) ;
        Reg = sqrt(k10) / 100. ;          //Per cent regulation

        cout << endl << setw(50) <<"PERFORMANCE SHEET :" ;
        cout << endl << setw(20) <<"Per ecnt resistance=   "<< R ;
        cout << endl << setw(35) <<"Percent reactance=     "<< X ;
        cout << endl << setw(50) <<"Per cent impedance=    "<< Z ;
        cout << endl <<"Per cent Regulation on full-load at u.p.f.= "<< Reg ;
        cout << endl <<"Total cond.weight on full-load at u.p.f.= "<< tcpwt<<"kg";
        cout << endl <<"Total cond.loss on full-load at u.p.f= "<< cnloss<<"watt";
        cout << endl <<"Full load loss at u.p.f.= "<< FLloss <<" watt" ;
        cout << endl <<"Per cent Efficiency on full-load at u.p.f.=  "<< Eff ;

//            Overall Dimensons of Core-Winding Assembly
        float Chh, k11, k12 ;
        float Wc, LA, WA ;

        cout << endl <<"Enter Chh - mm.,the radial clearance between two"
                "h.v.windings of consecutive phases" ;
        cin >> Chh ;
        Wc = Cc1 + tWwdg + Co + 0.5 * Chh ;   //Total width of windings & radial
//                        clearances at any limb

        k11 = 0.5 * w - Wc ;     //to check whether the total winding&clearance-
//                  assembly is accomodated in the half window space
        cout << endl <<" k11=  "<< k11 ;
//NOTE: k11 should be ideally zero or should have a small positive value;
//    If negative, increase `ratio` suitably.
        k12 = 2. * ( Cc1 + tWwdg + Co ) ;
        LA = 2. *  w + k12  + 3. * dia ;          // length of assembly
        WA = k12 + dia ;                          //  width of assembly

        cout << endl <<"length of assembly=  "<< LA <<" mm."<< setw(20)
              <<"width of assembly=   "<< WA <<" mm."<< setw(20)
              <<"height of assembly=  "<< HA <<" mm." ;
      getch() ;
```

```
//              Winding dimensions
//    Assign values of clearances : hy and da
//       hy = axial clearance between yoke & winding at top, and at bottom
//       da = axial duct between two consecutive coils


        cout << endl <<"Enter the values of axial clearence between yoke & "
               "winding(hy- mm.) & the axial duct width between two"
               " consecutive coils(dal - mm.)" ;
        cin >> hy >> da ;

        k7 = 2. * hy + (nc - 1.) * da ;
        hco = ( h - k7 ) / nc ;        // height of each coil
        azh = hco / cih ;        // no.of axial strands,should be an integer
        azw = tc / azh ;         // no of radial strands,should be an integer
        cout << endl <<"no. of axial strands,azh=  "<< azh << setw(20)
               <<"no. of radial strands,azw=  "<< azw ;

    getch() ;
    cout << endl <<"Enter Designor's choice of azh and azw ( both should"
           " be integer )" ;
    cin >> zh >> zw ;
    Hwdg = (zh * cih) * nc + ( nc - 1.) * da ;  //Winding height
    Wwdg = zw * cih ;                           //Winding width

//              Computed Parameters
    float roe ;              //specific resistance(ohm.mm)
    float spwt ;             //specific weight kg per mm.cubed
    cout << endl <<"Enter the values of sp.resis (roe-ohm.mm) & sp.weight"
           " (spwt- kg/mm-cubed)" ;
    cin >> roe >> spwt ;

    cout << endl <<"Enter the value of the radial clearance between limb and"
           " coil FOR L.V.winding /  between l.v and h.v.coils for"
           "the design of H.V.winding(C - mm.)" ;
    cin >> C ;

    dic = dia + 2. * C ;        // internal diameter of coil,mm
    dmc = dic + Wwdg ;          // mean diameter of coil,mm
    lmt = PI * dmc ;            // length of mean turn,mm
    res = roe * lmt * t * k1 / cbs ; // resistance per phase
    wcp = a * a * res ;         // conductor loss, watt
    vco = t  * lmt * cbs ;      // volume of conduc.material,mm-cube
    cpwt = spwt * vco * k1 ;    // weight of conduc.material,kg
}


float temp(float k12)
{
    float z, k, theta ;

    z = log (k12) ;
    k = z / 1.25 ;
    theta = exp(k) ;
    return(theta) ;
}
```

```
void main()
{

    clrscr() ;
    SPEC s(25000.,11000.,250.,3.,50.,200.,1000.) ;

    cout << "\n" ; s.display() ;

    float Ntube ;
    Ntube = TRAN (s) ;
    cout << endl << setw(10) <<"Number of tubes required =   "<< Ntube ;
}
```

Figure 6.7 A computer program for the design of Three-phase core-type Transformer.

RESULT

Output (E)   25000   11000   250   3   50   200   1000
Enter (E)   0.54   1.50   2.75   0.35   0.95   0.84
Output (o)

    Voltage per turn = 1.559

    Computed net core section = 4388.6 mm. sq.

    Gross core section = 4619 mm. sq.

    Computed width of limb, mm. = 73.53

    Enter modified value of cw

(E)   73

(o) no. of steps = 2   core width = 73 mm. dia. = 85.9 mm. optimum fill = 0.786

    Enter ratio of window height and width.

(E) 1.75

(o) core loss = 75.81 watt

    magnetising volt-ampere = 199.2

DESIGN SHEET : CORE

    gross core section = 4619.6 mm. sq.

    modified net iron section = 4567.7 mm. sq.

    modified flux-density = 1.537 Tesla

    window area = 42766.7 mm. sq.

    window width = 156.3 mm.

    window height = 273.6 mm.

    core weight = 273.6 kg.

    core loss = 75.8 watt.

    core loss component of no-load current = 0.0023 amp.

    Magnetising current = 0.006 amp.

No load current = 0.0065 amp.

Magnetising current as % of no-load current = 93.46

DESIGN OF LOW VOLTAGE WINDING

(o)   conductor section, mm. sq. = 13.33
      Enter conductor data:
      for RECTANGULAR STRAP: bare ht. (cbh-mm.), bare width (cbw-mm.),
      and bare section (cs-mm.-sq.), insulated height (cih-mm.), insulated width (ciw-mm.),
      and insulated section (cis-mm-sq.);
      for ROUND conductor: bare dia. (cbh-mm.), bare section (cbs-mm.-sq),
      insulated dia. (cih-mm.), insulated section (cis-mm.-sq.)
      [PUTTING cbw = ciw = 0], FROM STANDARD TABLE.

(E)   5.5   2.5   12.7   6.   3.   17.1

(o)   modified l.v. current density = 2.625 amp./mm.-sq.
      Enter voltage / coil

(E)   250

(o)   number of coils = 1 number of turns per phase = 160.38 No. of turns
      per coil = 160.38
      Enter integer values of no. of coils (nc), no. of turns per phase (t), and
      no. of turns per coil/phase (tc).

(E)   1   160   160

(o)   Enter the values of axial clearance between yoke and winding (hy-mm.) and
      the axial duct between two consecutive coils (dal)

(E)   10   5

(o)   no. of axial strands, azh = 42.26   no. of radial strands, azw = 3.79
      Enter Designer's choice of azh and azw (both should be integer).

(E)   42   4

(o)   Enter the values of sp. resis (roe-ohm.mm.) and sp. weight (spwt-kg per mm. cubed).

(E)   0.0341   0.0027

(o)   Enter the value of the axial clearance between LIMB and COIL for the
      design of L.V. winding/between L.V. and H.V. COILS for the design of
      H.V. winding.

(E)   10

(o)   L.V. WINDING DESIGN SHEET

      L.V. winding current density = 2.625 amp/mm. sq.
         Number of turns per phase = 160
         Number of coils = 1

Winding Details: Rectangular Strap

	Bare	Insulated
Height, mm.	5.5	6.0
Width, mm.	2.5	3.0
Section, mm. sq.	12.7	17.1

Clearances:
Axial - between yoke and winding = 10 mm.
Axial - between two consecutive l.v. coils = 5 mm.

Radial - between limb and winding = 10 mm.
Number of axial strands = 42
Number of radial strands = 4
Winding height = 252 mm.
Winding width = 24 mm.
Resistance / ph. = 0.175 ohm.
Conductor loss = 194.8 watts
Conductor volume = 829134 mm. cubed.
Conductor weight = 2.24 kg.
Outside diameter of l.v. coils = 154 mm.

DESIGN OF HIGH VOLTAGE WINDING

(o)   conductor section, mm. sq. = 0.303
Enter conductor data:
for RECTANGULAR STRAP: bare ht. (cbh-mm.), bare width (cbw-mm.),
and bare section (cbs-mm. sq.), insulated height (cih-mm.), insulated width
(ciw-mm.), and insulated section (cis-mm-sq.);
for ROUND conductor: bare dia. (cbh-mm.), bare section (cbs-mm.-sq),
insulated dia.(cih-mm.), insulated section(cis-mm.-sq.)
[PUTTING cbw = ciw = 0], FROM STANDARD TABLE.

(E)   0.635   0   0.31669   0.865   0.   0.5877.

(o)   Modified l.v. current density - 2.39 amp./mm. sq.
Enter voltage / coil

(E)   1350

(o)   number of coils = 8.148 number of turns per phase = 7056.5 No. of turns
per coil per phase - 866.025
Enter integer values of no. of coils (nc), no. of turns per phase (t), and
no. of turns per coil/phase (tc).

(E)   8   7056   882

(o)   Enter the values of axial clearance between yoke and winding (hy-mm.) and the axial
duct between two consecutive coils (dal-mm.)

(E)   15   5

(o)   no. of axial strands, azh = 30.14   no. of radial strands, azw = 29.26
Enter Designer's choice of azh and azw (both should be integer).

(E)   30   29

(o)   Enter the values of sp. resis (roe-ohm. mm.) and sp. weight (spwt-kg per
mm. cubed).

(E)   0.0341   0.0027

(o)   Enter the value of the axial clearance between LIMB and COIL for the
design of L.V. winding/between L.V. and H.V. COILS for the design of
H.V. winding.

(E)   12

(o)   H.V. WINDING DESIGN SHEET

H.V. winding current density - 2.39 amp/mm.sq.
   Number of turns per phase = 7056
   Number of coils - 8

Winding Details:	Rectangular Strap	
	Bare	Insulated
diameter, mm.	0.635	0.865
Section, mm. sq.	0.31669	0.5877

Clearances:
Axial - between yoke and winding - 15 mm.
Axial - between two consecutive l.v. coils = 5 mm.
Radial - between limb and winding - 12 mm.
   Number of axial strands = 30
       Number of radial strands = 4
          Winding height - 252 mm.
           Winding width = 24 mm.
           Resistance/ph. - 0.175 ohm.
        Conductor loss = 194.8 watts
      Conductor volume - 829134 mm. cubed.
    Conductor weight = 2.24 kg.
Outside diameter of l.v. coils - 154 mm.

PERFORMANCE SHEET

Per cent resistance = 4.56
Per cent reactance = 1.985
Per cent impedance = 4.97
Per cent regulation on full load at u.p.f. = 1.46
Total cond. weight = 4.8 kg.
Total cond. loss on full load at u.p.f. = 1139 watt.
Full-load loss at u.p.f. = 1295 watts
Per cent Efficiency = 95.1

(o)   Enter Chh - mm., the radial clearance between two h.v. windings of consecutive phases
(E)   12
(o)   k11 = 1.078
   Length of assembly - 712.5 mm. Width of assembly - 228 mm. height of Assembly = 420 mm.
   Enter the values of clearances between the core winding assembly and tank walls:
   Lengthwise on eachside: ct1 - mm.; widthwise on each side: ct2 - mm.; at top - upto Oil level, ct3 - mm.; between oil level and lid ct4 - mm.; at base: ct5 - mm.
(E)   30   40   170   170   30
(o)   HT - 789.6 mm.   WT - 300 mm.   LT - 772.5 mm.

AT = 1.706   qet = 758.7
Mean top-oil Temperature-rise - 54.4
        Enter perimeter of tube, Atube - metre; and effective length of each tube, Ltube-metre
(E)   0.1   4.5
(o)   Minimum number of tubes required - 6

6.8 SURGE VOLTAGE

6.8.1   The reliable operation ofa power transformer depends on its insulation level which
determines its capability to withstand continuously applied rated frequency service voltage and
impulse over-voltages. Over-voltage is defined as the abnormal voltage much higher than the
service voltage of the transformer generally caused by atmospheric disturbances such as lighting
as well as by routine switching operations in power transmission lines. Also faults such as
breaks and shot-circuits in the transmission line may give rise to breakdown overvoltages
higher than switching surges. The switching surges may be 3 to 4 times the phase voltage
persisting only for a small fraction of a second whereas lightning surges may produce over-
voltages of the order of ten times the phase voltage with duration not more than a few micro-
seconds.

        Appropriate choice of quality of insulating materials, selection of proper clearances between
h.v. and l.v. windings, between windings and limb and yoke ensures maintenance of insulation
level in a transformer. However it must be remembered that too large margins on insulation
strength may result in heavy expenditure on insulating materials, increase in size of the transformer
and its total cost.

        It is thus necessary to devise the most rational insulation having sufficient electric strength
without appreciable increase in the transformer size and / or other means of protection of the
winding against surge voltages and a study of the effects of surge voltage is thus imperative.

6.8.2 Model for the Study

Experiments have shown that the surge voltage phenomenon is fairly complicated in the sense
that

   (a) the waveform of the lightning surge impulses are exceedingly diverse;
   (b) the effect of impulse wave apart from its amplitude is grealty dependent on the
       steepness if its wavefront (Figure 6.9). Surges may have a very steep wavefront
       with a rate of rise one-thousand times the peak rate of normal voltage at rated
       frequency. Such suges may impose intense and rapidly changing electric stresses
       within the transformer.

The development of an appropriate model may be based on representing the transformer
by a capacitance network, the capacitance-value being too small to affect rated frequency
behaviour. The basis of such representation are :

   (i) a steep wavefront is equivalent to a part of a high-frequency sine-wave (Figure 6.7).
       Thus, with rapid changes in voltage at high frequencies the initial potential gradient is
       determined mainly by the inter-turn (or inter-coil) series capacitance of h.v. winding

and the shunt capacitances between h.v. winding, tank and core, between h.v. and l.v. windings (i.e. in effect between h.v. and earth in view of proximity of l.v. winding to the earthed core-limb).

(ii) the h.v. winding has distributed series inductance and resistance, which cannot affect the initial potential gradient since, the reactance value at high frequencies is large.

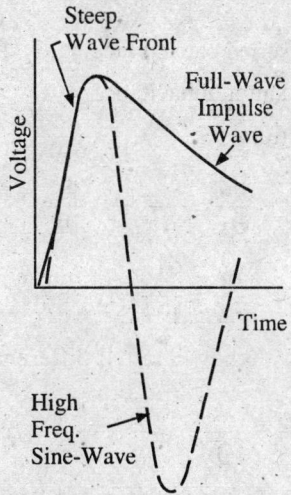

Figure 6.7   Waveform of lightning surge impulse.

Figure 6.8(b) shows a model in which C_s is the total distributed series capacitance of the h.v. winding and C_g, the total distributed shunt capacitance to ground. The h.v. winding has a total distributed series inductance L and resistance R, the length of the winding being l.

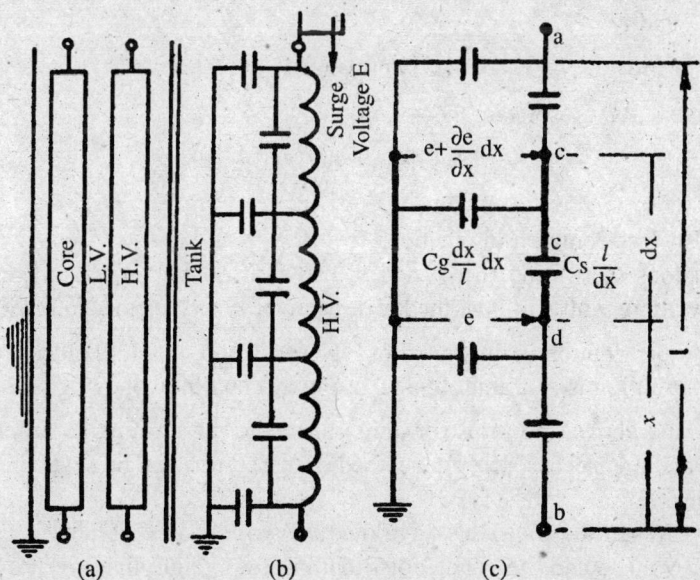

Figure 6.8   (a) Core-type transformer, (b) simplified model, (c) Model at high frequencies.

6.8.3 Analysis

In view of the large winding reactance at higher frequencies the model in Figure 6.8(b) may be approximated to that in Figure 6.8(c).

Consider that the network is subjected to unit function surge voltage of magnitude E at terminal a. Choosing an elemental length cd $(= dx)$ of the network at a distance x from the end b, cd has

$$a \text{ series capacitance} = C_s. \ (l/ \ dx),$$
$$\text{and a shunt capacitance} = C_g \ . \ (dx \ /l).$$

We may write the following equations :

$$\frac{\partial q}{\partial x} = \frac{C_g}{l} \cdot dx \cdot \frac{\partial e}{\partial t}$$

$$\frac{\partial e}{\partial x} = \frac{1}{C_s l} \cdot dx \cdot \frac{\partial q}{\partial t} \qquad (6.11)$$

The above equations lead to the wellknown partial differential equation,

or

$$\frac{\partial^2 e}{\partial x^2} = \frac{\alpha^2}{l^2} \cdot \frac{\partial^2 e}{\partial t^2} \qquad (6.12)$$

where,

$$\alpha = [C_g / C_s]^{1/2} \qquad (6.13)$$

the solution of which is of the form

$$e = Ae^{\alpha x/l} + Be^{-\alpha x/l}$$

Case I : If the end b is earthed, then at $x = 0$, $e = 0$.

i.e. $$A = - B.$$

At end a, $x = l$, and $e = E$.

Solving, $$e = E \ [\sinh \alpha \ (x/l)/\sinh \alpha] \qquad (6.14)$$

Case II : If the end b is isolated, the solution is,

$$e = E \ [\cosh \beta(x/l)/\cosh \beta] \qquad (6.15)$$

where, $$\beta = [2 \ C_g/C_s \ ]^{1/2} \qquad (6.16)$$

[*Note* : Case II is less common in practice.]

From equations (6.12) and (6.13), the distribution of the surge voltage over the entire length of the winding with end b earthed, can be determined for the. following three states:

I. Initial : Distribution is illustrated in Figure 6.12. in which the ideal case is when $\alpha = 0$, when the distribution is a straight line (1 in Figure) and the voltage gradient is uniform.

For $\alpha = 4$ and above, the turns (or coils) nearer to the terminal a suffer most from the impulse effect and the electric stress developed in these turns can be several hundred times of the normal value.

II. Final : When the transients have died away, the surge voltage is equivalent to the direct sustained voltage and the voltage distribution is a straight line given by,

$$e = E \frac{x}{l} \tag{6.17}$$

The voltage distribution characteristic will be identical with the ideal case ($\alpha = 0$) in Figure 6.9.

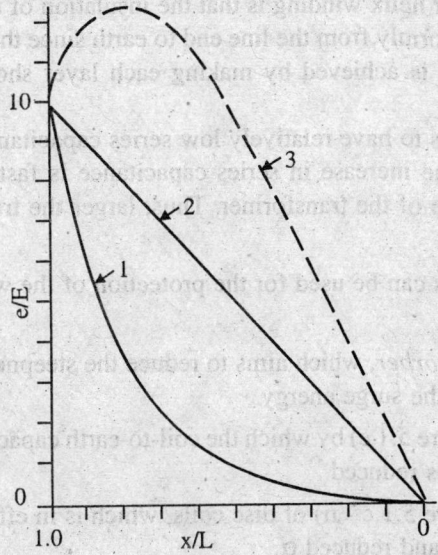

Figure 6.9 Surge voltage distribution and oscillation envelope. 1-initial with $\alpha = 5$, 2 final, 3-envelope of maximum voltage to earth.

III. Intermediate : Between the initial and the final states will occur an oscillation which may result in voltages in excess of the applied voltage appearing on the winding. A typical envelope of the maximum voltage to earth is shown in Figure 6.12- curve 3. This condition demands serious attention since,

    (i) the voltage-rise does not take place in zero time;

    (ii) The distance from the end b at which the voltage-rise will occur and the magnitude of the rise depend on the value of a;

    (iii) the voltage is oscillating when the applied surge voltage is reducing;

    (iv) there is some attenuation of voltage within the winding itself.

6.8.4 Designer's Aim

From the above analysis, it is clear that the designer should aim to obtain an initial distribution as near to the ideal case as possible with minimum oscillations. Various types of h.v. winding discussed in art. 5.1.1 can be studied in the light of their capacity to withstand the surge phenomenon.

    (i) Helix winding is poor because its capacitance network results in high stress. However, helix winding is not generally used for high voltage winding (Table 5.1).

    (ii) In the multilayer helix winding, screens provided inside and outside tend to produce equipotential lines parallel to themselves. Inter-turn capacitances are somewhat cancelled, and inter-layer capacitances present lead to high value of series capacitance C_s. Again, since only the layer-ends are exposed to earth, the shunt capacitance C_g is low. Thus α is low, and surge voltage distribution can be said to be very good. An added advantage with the multilayer helix winding is that the insulation of the winding to earth can be graded nearly uniformly from the line end to earth since the surge voltage distribution is near-ideal. This is achieved by making each layer shorter than the one inside it (Figure 5.1e).

(iii) Disc winding tends to have relatively low series capacitance and high capacitance to earth. However, the increase in series capacitance is faster than the shunt with the increase in the size of the transformer. Thus, larger the transformer, the better is the surge response.

The following methods can be used for the protection of the winding and to improve the surge response : -

(a) *External surge absorber*, which aims to reduce the steepness of the wavefront and to dissipate some of the surge energy.

(b) *Static shields* (Figure 5.1 e) by which the coil-to-earth capacitance C_g can be somewhat neutralised and α is reduced.

(c) *Interleaving* (Figure 5.1 c - ii) of disc coils, which is in effect increases the inter-turn series capacitance and reduced α.

6.9 COMPUTERISED DESIGN OF LARGE TRANSFORMER

6.9.1   The computerised design of transformers uses the same philosophy as that followed by an experienced designer. Once the input informations, such as, kVA-rating, % load loss, % impedance etc., has been fed into the machine, the computer initially selects or assumes certain design parameters, such as maximum flux-density, maximum current density, space-factors etc. necessary to calculate design data and performance characteristics. These characteristics are then compared with the specified guarantees. If the calculated values exceed the guaranteed, the initial assumptions are revised and the process 'repeated till a favourable design is produced.

Figure 6.13 illustrates a typical flow-diagram for a computer-aided transformer design program, specially applicable to large machines, which consists of 15 subroutines.

I. Subroutine SR 1 includes all items generally specified by the customer, such as,

(i) Continuous maximum rating of the transformer.

(ii) Number of phases.

(iii) Frequency.

(iv) L.v. and h.v. line voltages.

(v) L.v. and h.v. winding connections.

(vi) Arrangement of windings with respect to core limbs,, including tapping windings.

(vii) % l.v. and h.v. tappings,, if an.

(viii) Maximum temperature-rise of oil.

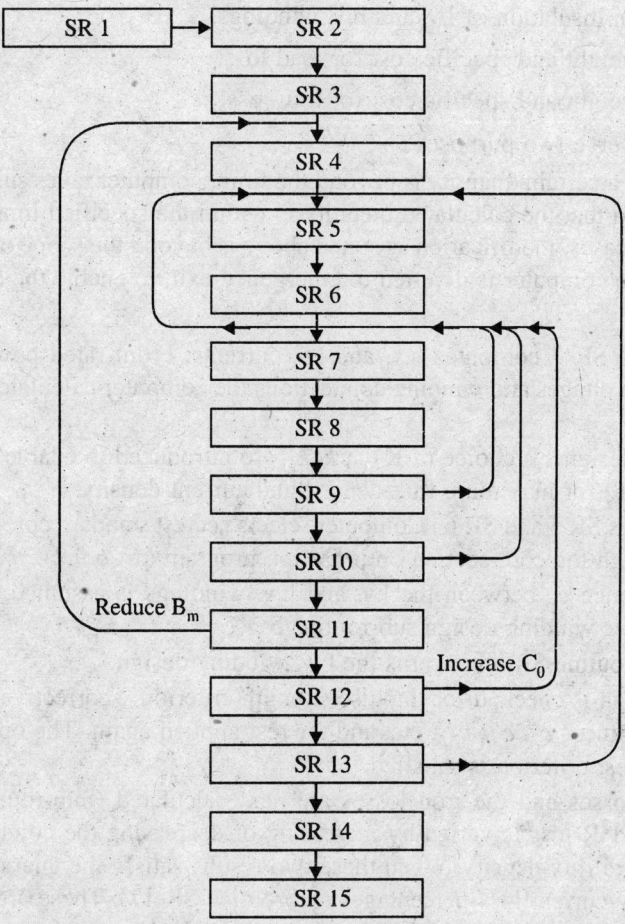

Figure 6.10 Flow-diagram for designing transformers.

(ix) % reactance at continuous maximum rating.

(x) % load less at continous maximum rating.

Apart from above, quite often for large power transformers noise level is specified by the customer.

II. In subroutine SR2 the designer sets maximum and/or minimum limits of some quantities, as below.

(i) Maximum core limb flux-density.

(ii) Maximum l.v. and h.v. current-densities.

(iii) Radial distance between l.v. and h.v. windings (c_o).

(iv) Maximum core height.

(v) Maximum and minimum % l.v. and/or h.v. tappings, if any.

(vi) Maximum % l.v. and h.v. eddy-current losses.

(vii) Breakdown insulation of l.v. and h.v. windings.

(viii) Specific weight and specific cost for load loss.

(ix) Specific weight and specific cost for iron loss.

Such settings serve two purposes : –

(1) Whenever a certain quantity is beyond the limit, computer takes suitable modification program so that the calculated quantity is within the specified limit.

(2) In certain cases, modification as stated above is beyond the scope of the computer and as such the computer is diverted to emergency exit, to enable the designer to modify the limit.

III. Subroutine SR 3 computes h.v. and l.v. currents. From rated power, the number of limbs wound, line voltages and winding connections the computer calculates the limb currents and voltages.

IV. In SR 4, designer's choice of K_{AQ}, k_f k_w are introduced to enable computation of A_t A_{con}, A_w, E_T, T_1, T_2 for maximum flux-density and current-density.

V. Subroutines SR 5 and SR 6 : Computer selects nearest standard core-sections (modified A_t) and goes through the core-section optimisation program (art 6.9.2).

VI. The clearance c_o between the l.v. and h.v. windings is assumed, and the computer goes through the l.v. winding design sub-routine SR 7.

VII. The sub-routine SR 8 concerns the h.v. wiuding design.

VIII. The design is checked for impulse strength criterion*, corrective changes are made mainly in the clearances c_o, c_{ei}, c_{hh}, etc. and the test applied again. The operation is repeated until the impulse test criterion is satisfied.

IX. The I^2R-losses and the iron losses are next calculated (Sub-routine SR 10 and 11 respectively). The I^2R-loss is varied by increasing or decreasing the conductor-size, and the iron loss by the core flux-density. When these two results satisfy the guarantee, the computer calculates the eddy-current loss-percentages (subroutine SR 12). These are tested against the maximum values set by the designer. If they are smaller, the computer proceeds to subroutine 13; otherwise, the only alternative is to increase the inter-winding clearance c_0 Increase in c_0 results in increase in the winding height. Thus, the radial depths c_1 and c_2 are reduced to maintain the same reactance.

X. Subroutine SR 13 computes per-cent reactance; SR 14, the cooling system; and SR 15, the weights of core and conducting materials and cost.

SHORT QUESTIONS

6.1 Compare : Core type & Shell type transformers;
Distribution & Power transformers.

6.2 What are the functions of :
Distribution transformer, Generator transformer,
Transmission transformer ?

6.3 What is the advantage of using two-stage cooling in transmission transformers ?

6.4 State the advantages of using standard strip sizes in transformers.

6.5 Mention 4 factors on which the clearance between l.v. and h.v. coils depends.

6.6 Why is axial packing betwen turns at the centre of a helix winding desirable ?

6.7 Why are distribution transformers designed to develop maximum efficiency at loads somewhat lower than the rated ?

6.8 Why are the first few turns of high voltage coils specially well insulated ?

6.9 What is meant by, over-voltage in a power transformer ?

6.10 How is insulation level in a transformer maintained ?

6.11 What are basis of representing a transformer by a capacitance network for studying surge-voltage phenomenon ?

6.12 Discuss the helix, multi-layer helix, and disc windings in the light of their capacity to withstand the surge phenomenon,

6.13 Mention 3 methods of protection of windings against the surge, and for improvement of surge response.

6.14 How is impulse strength level determined ?

6.15 On what factors does the loss-cost depend ?

6.16 Indicate, 'True' or 'False' :–

    (1) Distribution transformers have low percent reactance.

    (2) In a core type transformer, repair of winding is easier.

    (3) Core type has better support to windings against electromagnetic forces.

    (4) Core type is more suitable for large transformers.

    (5) In rural transformers, low magnetic density is normally used.

    (6) Distribution transformers are designed for low iron losses.

    (7) Power transformers are designed to operate at maximum efficiency when it is delivering rated load.

    (8) In helix winding, a space equivalent to the axial dimension of one extra turn must be provided.

    (9) In disc-helix, transposed conductors do not need any additional space.

    (10) Interleaving of disc coils reduces inter-turn series capacitance.

## CHAPTER 7

Design of Rotating Machines

7.1 Design of rotating electrical machines is more complicated than that of transformers, because,

   (i) in case of latter, relatively few variables are to be handled than those essential for a rotating machine;

  (ii) such variables in the case of transformer have simpler inter-relations between them; and

 (iii) problems may arise with discrete variables in machines, such as, the number of stator slots, number of rotor slots, number of stator turns in series per phase, etc.

Consequently it is easier to have fully synthetic design programs for transformers. In the case of rotating electrical machines, synthetic design procedures for induction motors* and complete design procedure for salient-pole synchronous machine incorporating both synthesis and analysis have been developed, but such procedures need computer which can handle large data and variables.

As has been mentioned in Chapter 1, the most promising approach seems to be in the combination of human designer with the analytic support of a computer. A number of smaller analytic programs on various topics, such as, stator coils, magnetic circuit, reactance, etc. are often helpful to a designer which give him insight into the effects of variations of important electrical and mechanical parameters, and quite often, a minicomputer will serve his purpose.

---

* Chalmers, B.J. & Bennington, B.J : "Digital Computer Program for design synthesis of Large Squirrel-cage Induction Motors", Proceedings IEE, 1967, vol 114, no. 2, p. 261.

+ Toader Stefan : "Optimum Design of Electrical Machines using Non-linear Programming". Technical Report no. 102, Chalmers University of Technology, Goteburg, Sweden, Nov., 1979.

Such small programs can be made near-optimum by incorporating non-linear effects and by checking their validity and accuracy with test data on many individual machines.

Figure 7.1 gives a typical flow chart for the design of electrical machines. Customer's specifications consist of kW (or kVA)-rating, voltage, efficiency, etc.; Standard specifications mean the constraints as per Indian and other Standards; and manufacturer's Specifications include tooling constraints, standard materials and components, insulation thickness, etc. Based on the design experience and insight, the designer next chooses the basic quantities such as, specific electric and magnetic loadings, and carries on design calculations and performance calculations. The calculated performances are then compared with customer's requirements. If not upto satisfaction, one or more basic assumptions are modified suitably. The computer next tests whether the design is optimum. If not, the basic assumptions are again reviewed and modified.

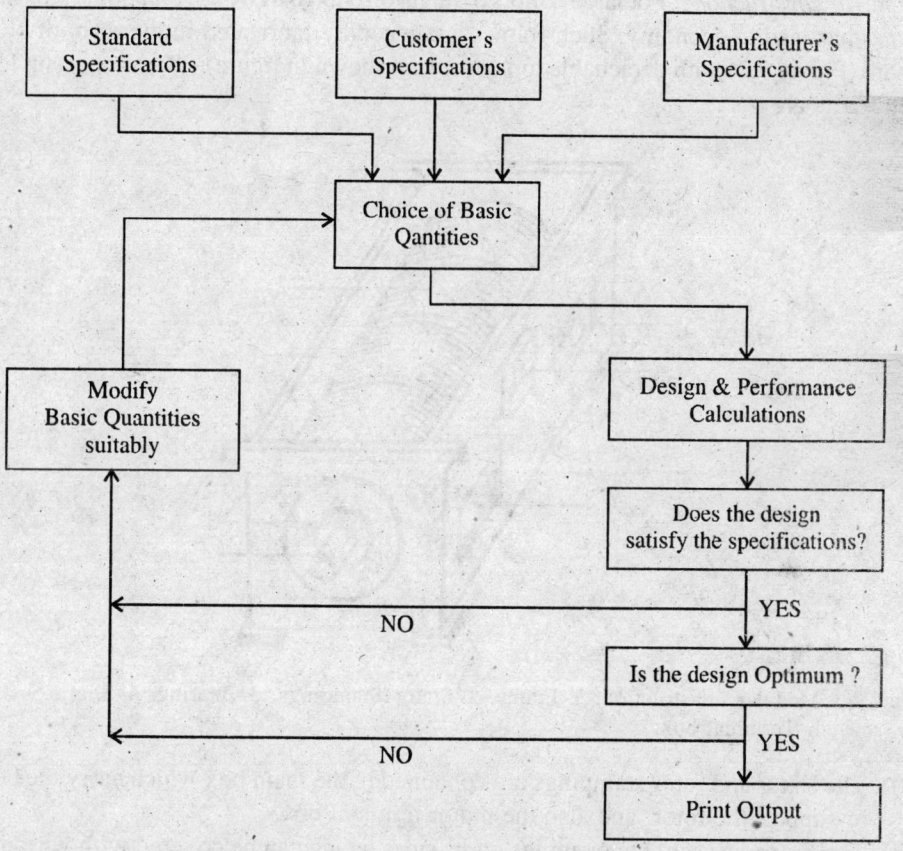

Figure 7.1 Flow Chart for design function.

In this chapter, no effort has been made to give the complete design procedures of various rotating electrical machines. Instead, important design considerations have been enumerated, followed by brief discussion on some computerised design procedure.

7.2 THREE-PHASE INDUCTION MOTOR

7.2.1 Important Design Considerations

(a) *Standard Frames and Stampings*

Three-phase induction motors constitute most widely used motors, and the products, specially low- and medium-size, are highly competitive. Thus, for the reason of economy, it is necessary to use standard frames and stampings.

IS 1231–1967 gives dimensions of standard frames for three-phase foot-mounted motors upto 200 kW. Standard frames for larger sizes are given in IS 8223-1976. Frames are designated by a number followed by letters such as S, M or L (e.g. 132 M), the number giving the height of the shaft-centre in mm. from the base of the frame, and the letter indicating frame length (S : Short, M : Medium, and L : Long).

Modular construction   For large motors of rating 0.25 to 10 MW, use of modular construction of frames has lead to economy. Such frames are generally fabricated in the form of box-type enclosure (Figure 7.2) with detachable top cover (not shown in figure). The important features of such frame are :

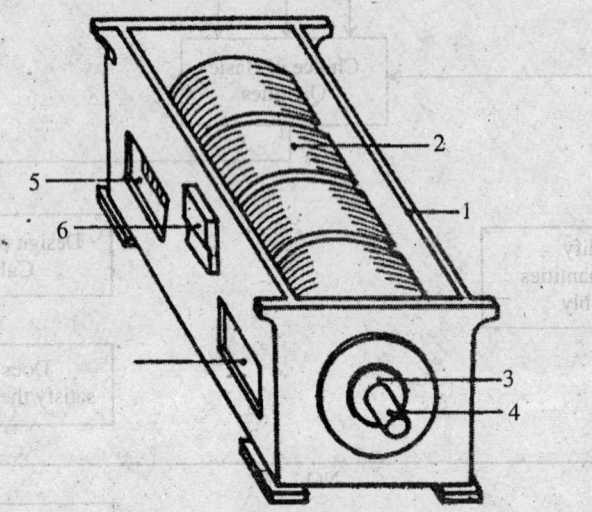

Figure 7.2   Modular construction, 1–Frame, 2–Stator Stampings, 3–Bearing, 4–Shaft, 5–Opening, 6–Terminal box.

  (i) The stator and rotor stampings are contained in the main box which carry the bearings to support the rotor, and also the motor terminal box.

  (ii) The frame has suitable openings on its sides which can be covered with wire-mesh or perforated plates, or which can be connected to air-duct, as demanded by required cooling system.

  (iii) The top cover is suitably designed to satisfy the requirements of enclosure such as drip-proof, flame-proof, or totally-enclosed.

(b) *Gap length*

Of the various parts of the magnetic circuit of a rotating electrical machine the contribution of the gap in relation to the mmf for the rated magnetic flux is generally maximum. The chief design difficulty lies in the necessity of having a small gap length so that the reactive power consumption is minimum. But mechanical considerations stand in the way and a compromise between the electrical and mechanical considerations is needed. However, it must be noted that minimum gap-length attainable from mechanical considerations does not necessarily result in best electrical performance.* In general, with the increased gap-length, the temperature-rise in stator drops, efficiency slightly increases, and power-factor is reduced. Figure 7.3 shows some important characteristics of a 100 kW, 8-pole, slip-ring induction motor with gap-length varied.

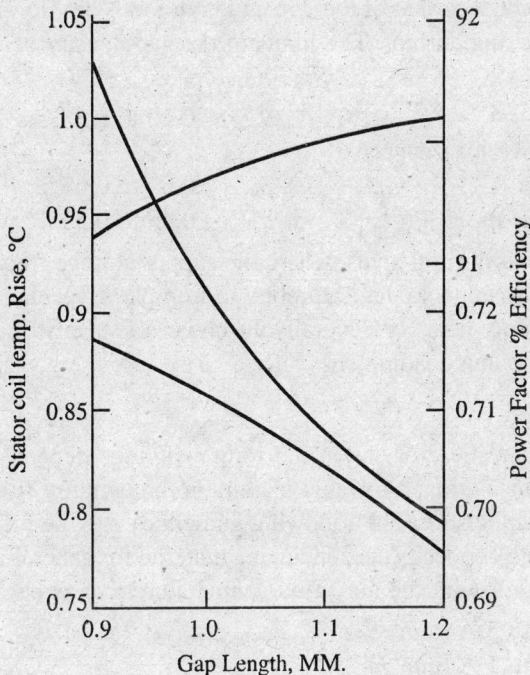

Figure 7.3 Characteristics of a 100-kW, 8-pole slip-ring motor.

For computation of gap-length, empirical equations have been presented by various authors from time to time. Based on an investigation on a large number of three-phase general-purpose induction motors of class A, E, and B insulation, the following equation has been found useful for low and medium ranges :-

---

* Guha, A. K. & Sen, S. K. : "Effect of varying air gap on the performance of squirrel-cage induction motor", Journal of Technology, vol. IV, 1959.

Mitra, A. : "Computer-aided Optimal Design of Induction Motor", M.E. Thesis, University of Calcutta, 1978.

Gap length $l_g = 0.000554 \; x. \; y$ mm   (7.1)

where, $x = 7.6 + p,$ for $p \leq 6$

                                   $= 12.4 + 0.2 \, p,$ for $p > 6$

and $y = 47.5 + P_0,$ for $P_0 \leq 150$

                                   $= 137.5 + 0.4 \, P_0,$ for $P_0 > 150$

P_0 being the power output, kW.

(c) Flux-density

Even for very short gap-length, the mmf required for the gap is larger than the total mmf of the remaining parts of the magnetic circuit, unless there is excessive tooth-saturation. To limit the magnetising current, moderate gap-density is therefore used. The density of stator tooth is directly proportional to that of the gap and high tooth-density produce excessive core losses and large magnetising current. The limits of the gap-density of induction motor are as below :-

General purpose   ...   ...   0.30 to 0.55 T;

Crane duty where overload capacity
       should be large   ...   ...   0.60 to 0.65 T;

Large capacity, high speed   ...   ...   0.45 to 0.55 T.

For motors to be used in rural areas, where the supply voltage frequently drops, the flux-density is to be chosen at value so that saturation in iron parts, specially teeth, is within limit at the top limit of the supply voltage. Generally the choice of specific magnetic loading 1.05 to 1.18 times that for continuous loading will suffice.

(d) Current-density

Choice of current-densities for stator and rotor windings depends upon the conducting material used, class of insulation, allowable maximum temperature-rise of each winding and the type of cooling and enclosure. The following guidelines may be followed in the choice of stator current-density with copper as the conducting material for general-purpose, continuously-rated, protected type motor with the maximum temperature-rise as per I.S. :-

Class A   ....   4 to 5 A. mm^{-2};

Class E   ....   6 to 7 A. mm^{-2};

Class B   ....   7 to 8 A. mm^{-2}.

For the rotor, the current-density is higher since cooling of rotor is better than the stator. For squirrel-cage rotors, bar and end-ring current density with copper as conducting material is 7 to 8.5 A. mm^{-2}, and with aluminium, 4.5 to 5.5 A. mm^{-2}.

For short-time rating upto 5 mins. and enclosed type, the above values of current-density should be multiplied by a factor of 1.8 to 2.0. That is, the value of the specific electric loading is approximately 1.8 to 2.0 times that for continuous rating.

For motors used in rural areas, lower value of current-density should be chosen so that the temperature-rise at the lowest limit of voltage when the current is highest, is within the specified limit.

(e) *Power-factor*

Full-load power-factor generally varies between 0.82 and 0.92, High power factor is generally obtained with high speed motors.

For good power-factor, one may use the following equations :-

$$\tau = 0.44 \sqrt{L_a}, \text{ for small capacity,}$$

$$= 0.41 \sqrt{L_a}, \text{ for medium-size.} \tag{7.2}$$

where, τ is the pole-pitch and L_a, the gross armature length.

(f) *Efficiency.*

The full-load efficiency varies generally between 0.82 and 0.93.

(g) *Slot combination*

   (i) (Number of rotor slots/number of stator slots) must not be an integer;

   (ii) For low reactance, number of rotor slots must be larger than the number of stator slots;

  (iii) For quiet motor, number of rotor slots should differ from the number of stator slots by 20% or more.

(h) *Winding*

   (i) The stator winding for three-phase squirrel-cage and wound-rotor motors are generally lap-connected.

   (ii) For slip-ring motors the rotor winding is generally wave-wound and star-connected. Standstill voltage at the slip-rings on open-circuit varies from 100 volts in small motors to 1 kV for large motors.

   (i) *For Rural use*, such as with pumps, frequent drop in supply voltage results in lower torque and larger current. Motors designed for continuous rated operation may be used in rural areas after suitable de-rating so that the temperature-rise with the lowest voltage limit is within specified value.

Alternatively, such motors should be designed for lower specific electric loading and higher specific magnetic loading as discussed earlier.

(j) *Design for non-sinusoidal supply voltage*

   (i) The method of wide speed control by varying frequency necessitates reduction of supply voltage with frequency. At reduced speeds, say 50% of the rated, voltage must be reduced to approximately 50% of rated, and under this condition, harmonics in the voltage and therefore in the current are large. The induced high harmonic frequency currents produced in the cage of a squirrel-cage induction motor will create high losses and increased motor temperature, so that special provisions must be made for the ventilation of the motor. Also, in order to hold the current at half speed down to a low enough value, it is necessary that the full-load slip be of the order of 10% making motor efficiency much lower than normal.

     Improvement can be obtained by using Alnico bars in rotor, leading to a flat speed-torque curve and low starting current. Half speed can he obtained at a higher value (about 60% of rated) of voltage so that harmonics in the voltage and induced high frequency rotor currents are smaller. Further, at lower speeds, cooling by fan mounted on the shaft is not quite so effective. Unless such motors are separately cooled, it should be de-rated suitably for use at lower speeds.

(ii) Theoretical investigation on a 4-pole 50-Hz squirrel-cage motor showed loss increased by 15 % in stator and by 89% in rotor for six-step square-wave excitation, and 57% in stator and 267% in rotor for pulse-width modulated excitation. Efficiency dropped by 1.4% in six-step square-wave and 5.1% in PWM.

(k)   *Inverter-controlled motor*

(i) The operation of such motors are generally based on constant flux in the machine. However, in the voltage-inverter (VI) type, voltage harmonics are present, the amplitude decreasing with increasing frequency; and in the pulse-inverter (PI) type, there are currents of rather high frequency. Such currents may produce I^2R-losses of remarkable magnitude. Additional losses, as compared with rated frequency losses, occur in iron and conducting materials. Losses are to be considered in both stator and rotor because of skin effects at higher frequencies of current harmonics. Such skin effect is quite predominant in the rotor of squirrel-cage motors.

(ii) For constant torque operation, the harmonic losses increase with decreasing frequency, and in certain cases, such losses may be so high at lower frequencies that the motor may be overheated even at no load.

(iii) One advantageous point is that these variable speed motors need not have high starting torque as they can be started at a lower speed with their pull-out torque.

(iv) Increase in rotor leakage reactance reduces the amplitude of harmonic currents without any detriment to starting behaviour. Reduction of slot-depth and increase in slot-width for a particular slot-pitch (keeping an eye on tooth saturation due to reduction in tooth-section) has often produced good results.

(l)   *Cycloconverter-controlled motors*

(i) The cycloconverter introduces harmonics into the motor supply voltage and a small degree of discontinuity in the resulting motor current waveform. This is because of the necessity to switch, at zero current, from the converter bridge producing positive current to the bridge producing negative current, and vice versa.

The voltage harmonics are the harmonics of the cycloconverter supply frequency and their main contribution is to increase the copper loss of the cage winding. To minimise this loss, the effective rotor reactance is kept as high as possible, and the effective rotor resistance as low as possible. That is, the motor is designed with a high reactance rotor slot consistent with meeting the motor pull-out torque requirements. However, inspite of above steps, the motor may require derating specially with a 6-pulse cycloconverter in which the harmonic voltage amplitudes are larger.

(ii) Efficiency and thermal rating of the motor make it desirable to reduce to a minimum the rotor copper loss due to currents at the fundamental cycloconverter-output frequency. However, the speed control loop design may be made simpler if the slip at rated torque is made larger than that obtained through optimum design.

Example 7.1

Design of a 3-phase, 415 V, 5.5 kW, 4 pole, 25 s.r.p.s, 50 Hz squirrel-cage induction motor, foot-mounted, protected type. Materials to be used : stator and rotor conductors and end-rings-aluminium ; magnetic-42 quality stalloy. Mode of starting-star-delta.

Solution The design for standard frame and stampings is done using omputer program FRAME with the following designer's choice :-

B_{av} = 0.5 T; ac = 30,000 A.cond.m^{-1}; Efficiency = 0.85;

Full-load p.f. = 0.90; k_w = 0.955; T_{rmax} = 40°C.

From Table 7.1, we take the computed data for aluminium at 60°C :

$$D = 0.159 \text{ m.}; L_a = 0.127 \text{ ni.}; I_g = 0.34 \text{ mm.}; B_{av} = 0.512 \text{ T };$$
$$\phi_m = 7.93 \text{ mWh.}; \text{full-load current } I_1 = 5.775 \text{ A}; \text{ Pole-pitch } \tau = 0.125 \text{ m.}$$

Stator: slots S_1 = 36, turns/ph. N_1 = 240; Rotor: slots S_2 = 34, bar area = 39.2 mm^2; I^{2}R – loss = 243 W.

Standard stampings are used having stator outside diameter = 260 mm., inside diameter = 44.5 mm., and stator and rotor slots of dimensions as shown in Figure 7.4

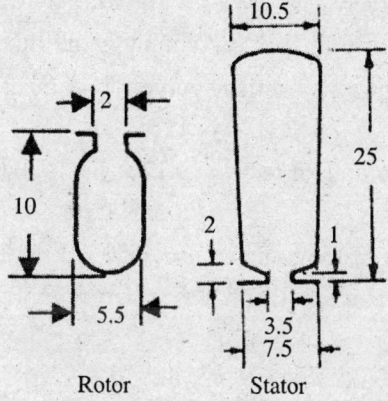

Figure 7.4 Rotor and stator slot dimensions.

I. For stator winding-subroutine

With aluminium as conducting material, we choose a current-density of 3.25 A. mm^2, giving a stator conductor section = 1.777 mm^2. From standard table, we choose :-

stator conductor section :   1.7674 mm^2 bare,

2.031 mm^2 with medium enamel insulation (IS 4800 II).

The modified value of current-density = 3.25 × 1.777/1.7674

$$= 3.27 \text{ A mm}^{-2}.$$

Check : Stator conductor/slot = 240 × 3 × 2/36 = 40.

Stator slot-area = 0.5 × (10.5 + 7.5) × 23 = 207 mm^2.

Space-factor = 1.7674 × 40/207 = 0.34 which is satisfactory.

We choose : Mush winding with a coil-span = 36/4 = 9 teeth. The pitch-factor is thus 1.0

Slots per pole per phase $g'_1 = 36/(3 \times 4) = 3$

$$\text{Distribution factor} = \frac{\sin 30°}{3 \sin 10°} = 0.96$$

That is, stator winding factor = $0.96 \times 1.0 = 96$

Approximate length of mean trun = $2.5 L_a + 1.5 L_a + 2\ \tau = 0.758$m.

Length of stator winding/ph. = $0.758 \times 240 = 182$ m.

Stator resistance/ph. = $0.0351 \times 182/1.7674 = 3.61$ Q

Stator I^2R –loss = $3 \times 5.775^2 \times 3.61 = 361$ W.

Iron loss in stator

(a) Tooth : Flux/slot-pitch = $7.93 \times 10^{-3} \times 4/36 = 8.81 \times 10^{-4}$ Wb.

Width = 0.0076 m.

Sectional area = $0.0076 \times 0.127 \times 0.95$

$$= 9.17 \times 10^{-4}\ m^2.$$

Density = $1.36 \times 8.81 \times 10^{-4}/9.17 \times 10^{-4}$

$$= 1.31\ T., \text{assuming whole flux passing through teeth.}$$

Specific loss, from the characteristic curve for 42 quality stalloy

at 1.31 T = 12 W. kg^{-1}

Total volume of teeth = $9.17 \times 10^{-4} \times 0.025 \times 36$

$$= 8.255 \times 10^{-4}\ m^3.$$

With specific weight of tooth-material = 7550 kg. m^{-3},

loss = $12 \times 8.255 \times 10^{-4} \times 7550$

$$= 75\ W.$$

(b) Core : Flux per path = $7.93 \times 10^{-3}/2 = 3.965 \times 10^{-3}$ Wb.

Depth = $0.26 - (0.159 + 0.025 \times 2) = 0.0255$ m.

Sectional area = $0.0255 \times 0.127 \times 0.95 = 3.076 \times 10^{-3}\ m^2.$

Density = $3.963 \times 10^{-3}/3.076 \times 10^{-3} = 1.29$ T.

Specific loss = 11.7 W. kg^{-1}.

Volume = $3.076 \times 10^{-3} \times \pi\ (0.159 + 2 \times 0.025 + 0.0255)$

$$= 2.25 \times 10^{-3}\ m^3.$$

Loss = $11.7 \times 2.25 \times 10^{-3} \times 7550 = 199$ W.

That is, total iron loss = $75 + 199 = 274$ W.

and total loss in stator = $274 + 361 = 635$ W.

Cooling surface

Outside surface = $\pi \times 0.26 \times 0.127 = 0.104$ m^2.

Inside surface = $\pi \times 0.159 \times 0.127 = 0.063$ m^2.

$$\text{Two sides} = (\pi/4) \, (0.262\text{-}0.1592) = 0.066 \text{ m.}$$
$$\text{Total cooling surface} = 0.104 + 0.063 + 0.066 = 0.233 \text{ m}^2.$$

Temperature-rise

$$\text{Rotor diameter} = 0.159 - 2 \times 0.0034 = 0.152 \text{ m}$$
$$\text{Peripheral velocity} = \pi \times 0.152 \times 25 = 11.94 \text{ m.sec}^{-1}$$
$$\text{The factor } h^* = (1 + 0.1 \times 11.94)/0.03 = 73.13$$

$$\text{Temperature-rise in stator} = \frac{635}{0.2135 \times 73.13} = 40.6°\text{C}$$

II. *For subroutine for magnetic circuit and no-load current.*

(a) *Gap :*

$$\text{Stator slot-pitch } y_{s1} = \pi \times 0.159/36 = 0.0139 \text{ m.}$$
$$\text{Slot opening } w_{o1} = 0.0035 \text{ m.}$$
$$w_{01}/l_g = 10.29$$

From Figure 4.7, Carter's coefficient kc_1 = 0.83 for semi-closed slot.

$$\text{From equation (4.9), } kg_1 = \frac{0.0139}{0.0139 - 0.83 \times 0.0035} = 1.26$$

$$\text{Again, rotor slot-pitch } y_{g2} = \pi \times 0.152/34 = 0.0147 \text{ m.}$$
$$\text{Slot opening } w_{02} = 0.002 \text{ m.}$$
$$w_{02}/lg = 5.88$$
$$\text{Carter's coefficient } k_{c2} = 0.65$$

$$\text{Hence, } k_{g2} = \frac{0.0147}{0.0147 - 0.65 \times 0.002}$$
$$= 1.1$$

$$\text{Effective gap length } l'_g = 0.00034 \times 1.26 \times 1.1 = 0.00047 \text{ m.}$$
$$\text{Gap m.m.f.} = 0.00047 \times 1.36 \times 0.512 \times 800,000$$
$$= 264 \text{ amp-turn/pole.}$$

(b) *Stator todth*
From the magnetising characteristic of 42 quality stalloy,

$$\text{for B} = 1.31 \text{ T, H} = 480 \text{ Amp.turn. m}^{-1}.$$
$$\text{Length of flux path} = \text{height of a tooth} = 0.025 \text{ m.}$$
$$\text{Hence, tooth m.m.f.} = 480 \times 0.025$$
$$= 12 \text{ amp-turn/pole.}$$

(c) *Stator core*

$$\text{For B} = 1.29 \text{ T, H} = 450 \text{ amp-turn. m}^{-1}.$$

---

* Sanraugasundaram A, Gangadharan G, Palani R : Electrical Machine Design Data Book, p. 190.

Length of flux-path per pole $= \frac{1}{4} \times 0.5 \times \pi \, (0.159 + 0.025 \times 2 + 0.0255)$

$$= 0.092 \text{ m.}$$

Hence, stator core mmf $= 450 \times 0.092$

$$= 41 \text{ amp-turn/pole.}$$

(d) *Rotor tooth*

Tooth-width at l/3rd from the bottom of the tooth

$$= \frac{\pi}{34} \, (0.152 - 2x \frac{2}{3} \times 0.01) - 0.0055$$

Corresponding tooth-section $= 0.0078 \times 0.127 \times 0.95$

$$= 9.41 \times 10^{-4} \text{ m}^2.$$

Flux per tooth-pitch $= 0.00793 \times 4/34 = 9.33 \times 10^{-4}$ Wb.

Density at that section $= 1.36 \times 9.33 \times 10^{-4}/9.41 \times 10 = 1.35$ T.

For $B = 1.35$ T, $H = 600$ amp-turn. m^{-1}.

Rotor tooth m.m.f. $= 600 \times$ height of rotor tooth

$$= 600 \times 0.01 = 6 \text{ amp.turn/pole.}$$

(e) *Rotor core*

Depth $= 0.5(0.152 - 0.01 \times 2 - 0.0445)$

$$= 0.044 \text{ m.}$$

Section $= 0.044 \times 0.127 \times 0.95 = 5.3 \times 10^{-3} \text{m}^2.$

Length of flux-path per pole $= 0.5 \times \pi \, [(0.152 - 0.12 \times 2) - 0.044]/4$
$$- 0.5 \times 0.044$$

$$= 0.0566$$

Flux per path $= 0.5 \times 7.93 \times 10^{-3} = 3.965 \times 10^{-3}$ Wb.

Density $= 3.965 \times 10^{-3}/5.3 \times 10^{-3} = 0.74$ T

For $B = 0.74$ T, $H = 110$ amp-tun m^{-1}.

Rotor core m.m.f. $= 110 \times 0.0566 = 6$ amp-turn/pole.

Hence, total m.m.f $= 264 + 12 + 41 + 6 + 6$

$$= 329 \text{ amp-turn/pole.}$$

Magnetising current, $I_{om} = \dfrac{329 \times 4}{2.34 \times 240 \times 0.955} = 2.45$ amp.

Core loss component of no-load current I_{oc} = Total iron loss/(3 × 415)

$$= 274/(3 \times 415)$$

$$= 0.22 \text{ amp.}$$

Hence, no load current $= [2.45^2 + 0.22^2]^{1/2}$

$$= 2.46 \text{ amp.}$$

Magnetising reactance/phase $= 415/2.45 = 169.4 \; \Omega$

III. For the subroutine for leakage reactances and short-circuit current

(a) Slot permeance coefficients (equn. 4.54) :

$$\text{Stator} : \lambda_{s1} = \frac{19}{3 \times 10.5} + \frac{4}{10.5} + \frac{2 \times 1}{10.5 + 3.5} + \frac{1}{3.5}$$

$$= 1.484$$

$$\text{Rotor} : \lambda_{s2} = \frac{10}{5.5} + \frac{1}{2} = 2.31 \text{ in rotor terms.}$$

$$= 2.31 \times \frac{0.955^2 \times 36}{1^2 \times 34} = 2.23 \text{ in siator terms.}$$

The factor $\dfrac{4\pi f s \mu_0 N_1^2}{(p/2) g_1} = \dfrac{4\pi \times 50 \times 4 \times \pi \times 10^{-7} \times 240^2}{2 \times 3}$

$$= 7.58$$

Hence, slot leakage reactance : stator $x_{gl1} = 7.58 \, \lambda_{s1}$

$$= 7.58 \times 0.127 \times 1.484$$

$$= 1.43 \ \Omega/\text{phase.}$$

rotor $xsi_2 = 7.58 \times 0.127 \times 2.23$

in stator terms $= 2.15 \ \Omega/\text{phase.}$

(b) Overhang (equn. 4.58) :

Coil span/pole-pitch = 1.0

Hence, $k_s = 0.75$

Thus, overhang leakage reactance $x_0 = 0.00475 \times 0.75 \times 0.159 \times (240/4)^2$

$$= 2.04 \ \Omega \ \text{phase.}$$

(c) Zigzag (equn. 4.57) :

$$\text{Zigzag leakage reactance } x_z = \frac{5}{6} \times 4^2 \times 169.2 \left(\frac{1}{36^2} + \frac{1}{34^2} \right)$$

$$= 3.69 \ \Omega/\text{phase}$$

With cage rotor, the belt leakage is negligibly small. Hence, total leakage reactance

$$= x_1 + x_{2s}$$

$$= x_{s1} + x_{s2} + x_0 + x_z$$

$$= 1.43 + 2.15 + 2.04 + 3.69 = 9.31 \ \Omega/\text{phase.}$$

For class A motor, $x_1 = x_2 = 4.66 \ \Omega/\text{phase.}$

Short-circuit current

Rotor equivalent phase current on full-load $= 5.775 \times 0.9$

$$= 5.2 \text{ Amp.}$$

$$\text{Rotor resistance/phase (in stator terms)} = \frac{\text{rotor } i^2r - \text{loss}}{3 \times 5.2^2}$$

$$= 3\Omega$$

$$\text{Blocked-rotor impedance} = (3.61 + 3) + j\,9.31$$

$$= 11.4\,\Omega \angle 54.6°$$

$$\text{Short-circuit current} = \frac{415}{11.4} = 36.4 \text{ amp.}$$

$$\text{Short-circuit p.f.} = 0.58 \text{ lag.}$$

IV. For subroutine for performance calculation

From the current locus diagram : on full-load,

$$I_1 = 5.77 \text{ amps.}$$

$$I_2 = 5.14 \text{ amps.}$$

$$P.f. = 0.89$$

Losses : stator I^2R-loss $= 3 \times 5.77^2 \times 3.61 = 361$ W

rotor I^2R-loss $= 3 \times 5.14^2 \times 3.0 = 238$ W

iron loss $= 274$ W

mechanical loss (assumed) $= 70$ W

Total loss $= 943$ W

$$\text{Full-load efficiency} = \frac{5500}{5500 + 943} \times 100 = 85.3\%$$

$$\text{Full-load slip} = \frac{238}{5500 + 238 + 70} = 0.041$$

At start, with stator star-connected

$$\text{starting current} = \frac{415/\sqrt{3}}{11.4 \angle 54.6°}$$

$$= 21.0 \text{ amp. at } 0.58 \text{ lag.}$$

Rotor current (in stator terms) $= 21.0 \times 0.58 = 12.2$ ampt,

Starting torque ∞ rotor i^2r-loss $= 3 \times 12.2^2 \times 3$

$$= 1340 \text{ syn. watts.}$$

Full-load torque ∞ (output power + rotor i^2r-loss)

$$= 5500 + 238 = 5738 \text{ syn. watts.}$$

Hence, starting torque/full-load torque $= 1340/5738$

$$= 0.23$$

Example 7.2

With the following Designer's choice, compare Copper and Aluminium when used in a 3-ph. 415V, 4-pole, 50-Hz squirrel-cage induction motor in the following items :

Stator : no. of slots, turns per phase;

Rotor : no. of slots, slot-area, bar-area, bar current-density, I^2r-loss, and Mean Temperature-rise.

Choose for maxm. temperature-rise TRMAX :

$B_{av} = 0.5$ T, $\overline{ac} = 30,000$ A-cond. m^{-1}, efficiency = 0.85.

P.f. = 0.90, full-load slip = 0.04, $k_w = 0.955$.

4 cases have been considered with the bar and end-ring material as copper/ aluminium, and the maximum value of the mean temperature-rise of rotor considering heating under starting condition at 35/40°C.

Table 7.1 Computed data for a 3PH, 415 V, 4 POLE 50 HZ Sq. cage induction motor.

Bar & end-ring	Copper		Aluminium	
ρ	0.0214 Amp. mm^{-2}		0.0351 Amp. mm^{-2}	
TRMAX	35°C	40°C	35°C	40°C
D(m)	0.159	0.159	0.159	0.159
L_a (m)	0.116	0.107	0.141	0.127
l_a (mm)	0.34	0.34	0.34	0.34
B_{av} (T)	0.504	0.509	0.51	0.512
ϕ_m (mWb)	7.3	6.8	9.0	7.93
$\overline{ac}$ (A.cond. m^{-1})	18668	20231	15369	17057
Stator :				
No. of slot	36	36	36	36
Turns/phase	264	288	216	240
Rotor :				
No. of slot	34	34	34	34
Slot area (mm2)	54.6	54.6	54.6	54.6
Bar area (mm8)	39.2	39.2	39.2	39.2
Bar current-density (A, mm^{-2})	5.9	6.4	4.83	5.36
$I^2 r$–loss (W)	187	208	216	243
Mean temp, rise (°C)	33.1	39.9	31.5	39.3

7.2.2 Design of Stator frame

Figure 9.9a gives a computer program developed with the constraints-use of Standard Frame size as per IS 1231-1967/IS 8223-1976 and class A insulation.

As before, to facilitate easy transfer of large number of data, use of switches and the header file "Sploss 3.h" friend-function has been invoked.

Governing Equations

<u>Step 1:</u> The stator inside diameter D and gross core length L are computed using Standard equations. For the separation of D and L, it is assumed that the L is equal to pole-pitch. For any other criterion, the program is to be modified.

<u>Step 2:</u> Choose Standard Frame and obtain values of modified D and L, and the outside diameter of stator, Do. (Table 7.1) Modifiy the maximum flux per pole, phim, and compute the no. of stator turns 'sturn' Choose number of slots.

<u>Step 3:</u> To determine depth of stator core 'dSc' and depth of stator slot 'dSs' Choose the maximum tooth-density BSt and stator core density BSc and obtain average tooth-width 'wtav' from the equation:

BSt = 1.36 phim x no. of poles 'p' x 10^6 / (L x Ss x wtav x fst), where, fst is the stacking factor (about 0.95).

Similarly, from the equation,

BSc, the stator core density = phim x 10^6 / (2 x L x dSc x fst), dSc is to be computed, and hence, dSs = 1/2 (Do - D) - dSc.

<u>Step 4:</u> The program is now poised to compute detail-dimensions of stator slot for which we'll have to select the type of slot. The given program is based on choice between two types (Figure 7.6(a)) but the program can include any number of types.

From the slot configuration, approximate values of w1, w2, wos, h2, h3 and h4 are computed. Considering all computed values and also those of slot depth and depth of core below slots, the Designer chooses the final values of all these slot-parameters.

<u>Step 5:</u> Stator turns, Conductors per slot, Conductor-section and Space factor.

The design calculations continues with the modification of number of stator turns per phase and number of conductors per slot both of which must have integer values and (no. of slots per phase x no. of conductors per slot) equals twice the no. of stator turns per phase. (*Note:* two conductors make one turn).

Again, knowing the stator current per phase and choosing the value of current density (art. 7.2.1(d)) letting a lower current density, if aluminium is chosen as conductor material, by a factor 1.6) the bare section of conductor is obtained. From the Standard table for conductors, bare and insulated diameters and sections together with the SWG no. are entered. In the computation, SCC - insulation of conductor is used.

The actual space in slots accupied by insulated conductors divided by the area of slot gives the slot space factor which should be 0.6 or below.

<u>Step 6:</u> Stator tooth and core losses at rated voltage and frequency.

For each of the two types of slots, the total tooth-area, volume of stator teeth, weight, average tooth-width, and maximum tooth density were computed with the designed parameters, and thus loss in stator teeth.

Similarly for the core below the slots.

```
// Design of Three-phase Cage-rotor Induction Motor - Stator Frame
// To obtain main dimensions and Stator slot details and Statr iron losses
// constraint: Use standard Frame as per IS 1231, 1967 for class A
// insulation.
#include <iostream.h>
#include <iomanip.h>
#include <math.h>
#include <conio.h>
#include "sploss3.h"



class spec
{
 public:
    int a, b, c ;
    float d, e, f ;
  void getdata(int a, int b, int c, float d, float e, float f) ;
  void display() ;
  void stator(int a, int b, int c, float d, float e, float f) ;
} ;

void spec :: getdata(int phase,int synsp,int pole,float powut,float volt,
          float slip )
  { a =phase ; b=synsp ; c=pole ; d=powut ; e=volt ; f=slip ; }
void spec :: display()
  { cout<<" No.of phase=  "<< a <<", synchro.speed= "<< b <<"r.p.s."
     <<", No. of pples= "<< c <<",  power output =  "<< d
     <<", voltage/ph=  "<< e <<",  slip=  "<<f; }

void spec :: stator(int a, int b, int c, float d, float e, float f)
{
    long g ;              // electric loading, amp-cond / m.sq.
    float h ;             // magnetic loading, Tesla.
    float pf, eff, akw, fst ;  //power-factor,efficiency,winding factor,
//                              stacking factor

  cout << endl <<"Enter the values of sp.elec.loading, sp. magn.loading,"
         " pf, eff, akw, fst " ;
  cin >> g >> h >> pf >> eff >> akw >> fst ;

  cout << endl << "Sp.Electric loading= "<< g <<"amp.cond/m";
  cout << endl << setw(30) << "Sp.Magnetic loading= "<< h <<" Tesla" ;
  cout << endl << setw(40) <<"power factor=  "<< pf ;
  cout << endl << setw(50) <<"efficiency=  "<< eff ;
  cout << endl << setw(60) <<"winding factor=  "<< akw ;
  cout << endl << setw(70) <<"stacking factor=  "<< fst ;

  int sturn ;
  long DsqL, dd ;
```

```
    float pin, D, L, phim, strn, Frame, Do, x, y, Ss, dSc ;
    const float PI = 3.14159 ;
    float k1 = pow(10,6) ;
    float k2 = pow(10,-3) ;

//Compute Stator inside diameter (D mm.) and core length (L,mm.)
    pin = d / (pf * eff) ;
    DsqL = pin * k1 / (1.11 * k2 * PI * PI * akw * b * g * h) ;
    cout << endl <<"DSQL = "<< DsqL ;
//              Choose standard frame :
//              Set gross core length(L)equal to
//              pole-pitch,to start with,which gives,
      dd = (DsqL) * c / PI ;
      D = pow(dd,0.33333) ;
      L = DsqL / (D * D) ;              //computed gross core length,mm.
    cout << endl <<"computed diameter= "<< D << setw(30) <<"computed length"
          " = "<< L ;
    cout << endl <<"Choose standard frame amd enter frame size(Frame), stator"
          " diameter(D - mm.),Length(L -mm.), stator outside diameter(Do -mm.)" ;
    cin >> Frame >> D >> L >> Do ;
    cout << endl <<" Frame Size= "<< Frame;
    cout << endl << setw(15) <<"Stator inside diameter= "<< D <<" mm." ;
    cout << endl << setw(30) <<"Gross length= "<< L <<" mm." ;
    cout << endl << setw(45) <<" Stator outside diameter= "<< Do <<" mm." ;

    phim = PI * h * D * L /(c * k1);              //flux per pole
    strn = e / ( 2.22 * akw * phim * b * c) ;   //computed stator turns per ph
    cout << endl <<"Flux per pole= "<< phim ;
    cout << endl << setw(30) <<"Stator turns/ph= "<< strn ;
    getch() ;

//Choose Number of Stator slots

      cout << endl <<"Enter designer's choice of no. of stator slots, Ss" ;

      cin >> Ss ;

// DESIGN OF STATOR SLOT
      int Zs ;
    float BSt, BSc, cbd, cid, cbs, cis, swg, spf, wtav ;
    float w1, w2, wos, dSs, h2, h3, h4, h34, Sss ;
    cout << endl <<"Enter Designer's choice of BSt and BSc" ;
    cin >> BSt >> BSc ;

    cout << endl <<"Maxm.flux-density in stator tooth= "<< BSt <<" T"
          << setw(30) <<" Flux-density in stator core= "<< BSc <<" T";
    getch() ;

    wtav = phim * c * k1/( Ss * L * fst * BSt / 1.36 ) ; //Av. width of tooth
    dSc = phim * k1 / ( 2. * BSc * L * fst ) ;
    dSs = 0.5 * (Do - D) - dSc ;

    float k4 = PI / Ss ;
//  Choose the type of slot
    int o1,o2 ;
```

```
cout << endl <<"Enter the value: o1 = 1 for TYPE1 ;o1 = 2 for TYPE2 Slot";
cin >> o1 ;
switch(o1)
{
 case 1 : {          //TYPE1
         w1 = ( k4 * ( D + 2. * dSs ) - wtav) / (1. + k4) ;
          cout << endl <<"Enter Designer's choice of h34 = h3 + h4" ;
           cin >> h34 ;
         w2 = k4 *.( D + 2. * h34 ) - wtav ;
         h2 = dSs - h34 - 0.5 * w1 ;
         }
         break ;
 case 2 : {         //TYPE2
         w1 = k4 * ( D + 2. * dSs ) - wtav ;
         cout << endl <<"Enter Designer's choice of h34 = h3 + h4" ;
          cin >> h34 ;
         w2 = k4 * ( D + 2. * h34 ) - wtav ;
         h2 = dSs - h34 ;
         }
         break ;
}
         cout << endl <<"Slot depth=  "<< dSs << setw(10) <<"Depth of "
         "core below slot=  "<< dSc << setw(10) <<"w1=  "<< w1
         << setw(10) <<" w2=  "<< w2 << setw(10) <<" h2=  "<< h2 ;
         cout << endl <<"Enter Designer's choice of dSs,dSc,w1,w2,wos,"
                 " h2,h3,h4" ;
         cin >> dSs >> dSc >> w1 >> w2 >> wos >> h2 >> h3>> h4 ;
getch() ;
cout << endl <<"Enter the value :o2 = 1 for TYPE1 ;o2 = 2 for TYPE2 Slot";
 cin >> o2 ;
switch(o2)
{
 case 1 : {
         Sss = (0.5 * PI * w1 * w1 / 4.) + ( 0.5 * (w1 + w2) * h2) ;
//         available stator slot section,neglecting wedge-area
         }
         break ;
 case 2 : {
         Sss = 0.5 * ( w1 + w2 ) * h2 ;
//          available slot sectional area, mm.sq.neglecting wedge-area
         }
         break ;
}
cout << endl << setw(50) <<" STATOR SLOT DESIGN DATA" ;
cout << endl <<"Number of stator slots=  "<< Ss ;
cout << endl << setw(10) <<"Slot depth =  "<< dSs <<" mm." ;
cout << endl << setw(20) <<"Slot width at bottom of slot=  "<< w1<<" mm.";
cout << endl << setw(30) <<"Slot with at top near taper pt.="<<w2<< "mm.";
cout << endl << setw(40) <<" Slot opening=  "<< wos <<" mm." ;
cout << endl << setw(30) <<"height h2= "<< h2 <<" mm." ;
cout << endl << setw(20) <<"height h3= "<< h3 <<" mm." ;
cout << endl << setw(10) <<"height h4= "<< h4 <<" mm." ;
cout << endl <<" Depth of Stator core below slot = "<< dSc <<" mm." ;
cout << endl << "Available Stator slot sectional area=  "<< Sss <<"mm.sq.";
```

```
//   Design of Stator winding

     float Scur,cd,css ;
     cout << endl <<"computed stator turns per phase= "<< strn ;
     cout << endl <<"Enter Designer's choice of stator turns/phase &"
             " conductors/slot (Zs), so that both are integers" ;
     cin >> strn >> Zs ;
     Scur = pin / (a * e) ;  //  stator current per phase

     cout << endl <<"Enter the chosen value of current-density, amp/mmsq" ;
      cin >> cd;

     cbs = Scur / cd ;        //  conductor-bare section, mm.sq.
     cout << endl <<" computed bare conductor section=  "<< cbs <<" mm.sq" ;

     cout << endl <<"Enter from Standard Table the bare diameter (cbd-mm.), "
          "bare section (cbs-mm.sq), insulated diameter (cid- mm.) , "
          "insulated section (cis -mm.sq.) & swg number" ;
      cin >> cbd.>> cbs >> cid  >>.cis >> swg ;
     css = Zs * cis ; // computed area per slot forinsulated conductors
     cout << endl<<"computed slot-area occupied by insul. conduc.= "<<css
             <<" mm.sq." ;

     spf = css /Sss ;
     cout << endl << "Stator slot space factor=  "<< spf ;

//   The above slot space factor should preferably be between 0.70 and 0.75.
//   If higher, increase the current density to  reduce the conductor section
//   If lower, decrease the current density.

//   compute stator core- and stator tooth- loss at rated voltage.

     int o3 ;
     float mSss,Sts,vSt,Stwt,wtb,wtnt,Stlos,vSc,Scwt,Sclos,Silos,spwt ;

     cout << endl <<"Enter the value of specific weight of core material"
             "(kg / mm.cubed)" ;
      cin >> spwt ;

     cout << endl <<"Enter the value:o3 =1 for TYPE1 ;o3 = 2 for TYPE2 Slot" ;
      cin >> o3 ;
     switch(o3)
     {
      case 1 : {      // Weight of stator teeth
             mSss = (0.5 * PI * w1 * w1 / 4.) + ( 0.5 * (w1 + w2) * h2) +
                 (0.5 * (w2 + wos) * h3) + wos * h4  ;
//             modified stator slot section,including wedge-area
             Sts = (PI * ((D+2*dSs)*(D+2*dSs) - D * D)/4.-(mSss * Ss)) ;
//                         total tooth-area
             vSt = L * Sts * fst ;  // volume of stator teeth
             Stwt = spwt * vSt * k2;    // weight of teeth,kgms.
//                 Tooth density
             wtb = (PI * (D+2*(h2 + h3 + h4))/ Ss) - w1 ; //tooth width at
//                             near slot-bottom
```

```
         wtnt = (PI * (D+2*(h3 + h4)) / Ss) - w2 ; // -do- at near-top
         wtav = 0.5 * (wtb + wtnt) ;
         BSt = 1.36 * phim * c * k1 / ( Ss * L * wtav * fst) ;

         y = BSt ;
         float sploss3(float) ;
         x = sploss3(y) ;
         Stlos = x * Stwt ;      // loss in stator teeth
```

```
//                 Weight of stator core
         vSc = PI * (Do*Do - (D+2*dSs)*(D+2*dSs)) * L * fst / 4. ;
         Scwt = spwt * vSc * k2;    // weight of stator core in kgms.
//                 Core density
         BSc = phim * k1 / (2. * dSc * L * fst) ;

         y = BSc ;
         float sploss3(float) ;
         x = sploss3(y) ;
         Sclos = x * Scwt ;
         Silos = Stlos + Sclos ;
         }
         break ;
case 2 : {        // Weight of stator teeth
         mSss = ( 0.5 * (w1 + w2) * h2) + (0.5 * (w2 + wos) * h3)
              + wos * h4 ;    // modified stator slot section,
                              including wedge-area
//       Sts = (PI * ((D+2*dSs)*(D+2*dSs) - D * D)/4.-(mSss * Ss)) ;
//                 total tooth-area, mm.sq.
         vSt = L * Sts * fst ; // volume of stator teeth
         Stwt = spwt * vSt * k2 ;    // weight in kgms.

//                 Tooth density
         wtb = (PI * (D+2*dSs)/ Ss) - w1 ;//tooth width at slot bottom
                         slot-bottom
         wtnt = (PI * (D+2*(h3 + h4)) / Ss) - w2 ; // -do- at near-top
         wtav = 0.5 * (wtb + wtnt) ;
         BSt = phim * c * k1 / ( Ss * L * wtav * fst) ;
         y = BSt ;
         float sploss3(float) ;
         x = sploss3(y) ;
         Stlos = x * Stwt ;      // loss in stator teeth

//                 Weight of stator core
         vSc = PI * (Do*Do - (D+2*dSs)*(D+2*dSs)) * L * fst / 4. ;
         Scwt = spwt * vSc * k2 ;       // weight in kgms.
//                 Core density
         BSc = phim * k1 / (2. * dSc * L * fst) ;

         y = BSc ;
         float sploss3(float) ;
         x = sploss3(y) ;
         Sclos = x * Scwt ;
         Silos = Stlos + Sclos ;
         }
         break ;
```

```
        }
        cout << endl <<"Weight of Stator teeth=  "<< Stwt <<"  kg." ;
        cout << endl << setw(15) <<"Loss in Stator teeth=  "<< Stlos <<"  watts" ;
        cout << endl << setw(30) <<"Weight of Stator core=  "<< Scwt <<" kg";
        cout << endl << setw(45) <<"Loss in Stator core =  "<< Sclos <<"  watts" ;
        cout << endl << setw(55) <<"Stator iron losses=  "<< Silos <<" watts" ;
   }

   void main()
   {
    spec x ;
    clrscr ;
    int phase, synsp, pole ;
    float powut, volt, slip ;

    cout << endl <<"Enter : a, b, c, d, e, f" ;
    cin >> phase >> synsp >> pole >> powut >> volt >> slip ;

    x.getdata(phase, synsp, pole, powut, volt, slip) ;
    x.display() ;
    x.stator(phase, synsp, pole, powut, volt, slip) ;
   }
```

Figure 7.5(b) A computer program for the design of Stator frame of a Three-phase cage-rotor Induction motor.

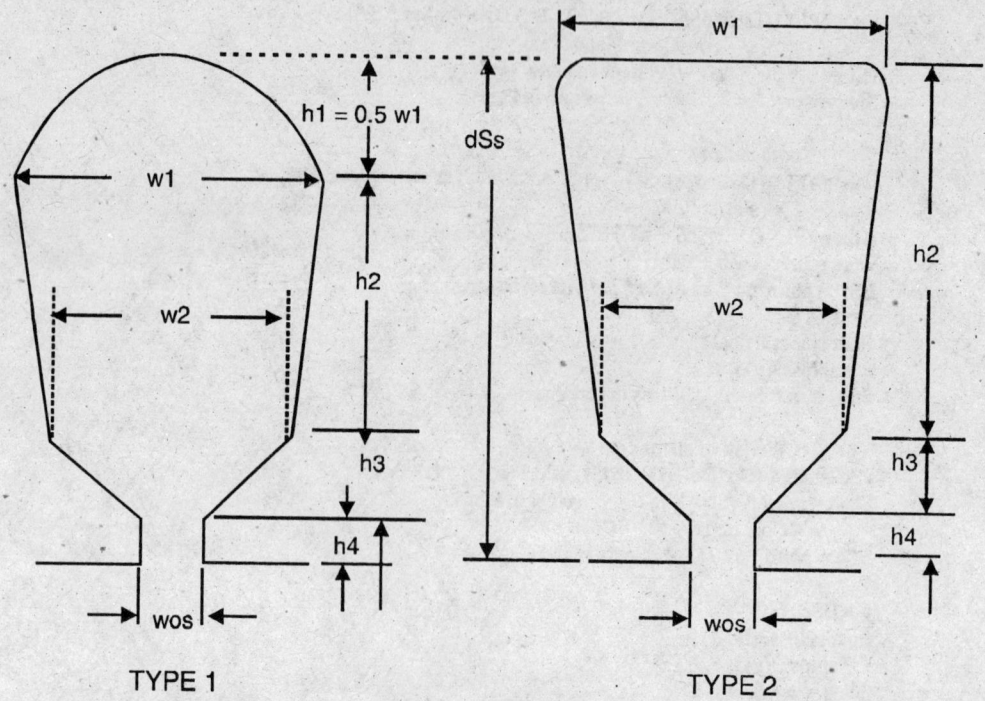

TYPE 1   TYPE 2

Figure 7.6 Stator slots considered in program (Figure 9.8).

RESULT

Enter (E) 3   25   4   7500   415   0.04
     (E) 25000   0.45   0.8   0.85   0.95   0.95
     (o) Dsq L = 3768028
         Computed diameter D = 168.65   Computed length L=132.4
         Choose Standard frame and enter frame size (Frame), stator diameter
         (D-mm), Length (L-mm.), stator outside diameter (Do)
     (E) 160   165   140   260
     (o) Frame Size = 160
             Stator inside diameter = 165 mm.
                 Gross length = 140 mm.
                     Stator outside diameter = 260 mm.
         Flux per pole = 0.008164
                 Stator turns / ph. = 241.02
         Enter Designer's choice of no. of stator slots, Ss
     (E) 36
     (o) Enter Designer's choice of BSt and BSc
     (E) 1.5   1.3
     (o) Maximum flux-density in stator tooth = 1.5 Flux-density in stator core = 1.3

Run 1 for TYPE 1 Slot

     Enter the value: 01 = 1 for TYPE 1 ; 01 = 2 for TYPE 2 Slot.
(E) 1
(o) Maxm. flux-density in stator tooth = 1.5 T
     Flux-density in stator core = 1.3 T
     Enter Designer's choice of of h34 = h3 + h4
(E) 3.5
(o) slot depth = 23.9 Depth of core below slot = 23.6 w1 = 11.4
                 w2 = 8.8   h2 = 14.7

     Enter Designer's choice of dSs, dSc, w1, w2, wos, h2, h3, h4
(E) 24   23.5   11   9   3   15   2.5   1
(o) Enter the value : o1= 1 for TYPE 1 ; o1 = 2 for TYPE 2 Slot.
(E) 1
(o)   STATOR SLOT DESIGN DATA

Number of stator slots = 36
    Slot depth = 24 mm.
        Slot width at bottom of slot = 11 mm.
            Slot width at top near taper pt. = 9 mm.
                Slot opening = 3 mm.
        Height h2 = 15 mm.

Height h3 = 2.5 mm.
Height h4 = 1 mm.
Depth of Stator core below slot = 23.5 mm.
Available Stator slot sectional area = 197.5 mm. sq.
Computed stator turns per phase = 241.02
Enter the chosen value of current density, amp/mm. sq
Enter Designer's choice of stator turns/phase and conductors/slot

(E)   240   40

(o)   Enter the chosen value of current-density, amp/mm. sq.

(E)   4.5

(o)   Computed bare conductor section = 1.9687 mm. sq.
Enter from STANDARD TABLE the bare diameter (cbd. mm.), bare section (cbs-mm.-sq.), insulated diameter (cid - mm.), insulated section (cis-mm.-sq.) and swg number

(E)   1.6256   2.0755   1.8056   2.5601   16

(o)   computed slot-area occupied by insul. Conductor = 102.4 mm. sq.
Stator slot space factor = 0.52
Enter the value of specific weight of conductor material (spwt - kg/mm. cubed)

(E)   0.00889
Enter the value : o1 = 1 for TYPE 1 ; o1 = 2 for TYPE 2 Slot.

(E)   1

(o)   Weight of Stator teeth = 7.68 kg.
Loss in Stator teeth = 84.6 watt
Weight of Stator core = 20.64 kg.
Loss in Stator core = 169.6 watt
Total Stator iron losses = 254.2 watt

Run 2 for TYPE 2 SLOT

Enter the value: o1 = 1 for TYPE 1 ; o1 = 2 for TYPE 2 Slot.

(E)   2

(o)   Maxm. flux-density in stator tooth = 1.5 T
Flux-density in stator core = 1.3 T
Enter Designer's choice of h34 = h3 + h4

(E)   3.5

(o)   slot depth = 23.9 Depth of core below slot = 23.6   w1 = 12.4
w2 = 8.8   h2 = 20.4
Enter Designer's choice of dSs, dSc, w1, w2, wos, h2, h3, h4

(E)   24   23.5   12   9   3   20   3   1

(o)   Enter the value : o1 = 1 for TYPE 1 ; o1 = 2 for TYPE 2 Slot.

(E)   2

(o)   STATOR SLOT DESIGN DATA

Number of stator slota = 36

Slot depth = 24 mm.

Slot width at bottom of slot = 12 mm.

Slot width at top near taper pt. = 9 mm.

Slot opening = 3 mm.

Height h2 = 20 mm.

Height h3 = 3 mm.

Height h4 = 1 mm.

Depth of Stator core below slot = 23.5 mm.

Available Stator slot sectional area = 210 mm. sq.

Computed stator turns per phase = 241.02

Enter Designer's choice of stator turns/phase and conductors/slot

(E)   240   40

(o)   Enter the chosen value of current-density, amp/mm. sq.

(E)   4.5

(o)   computed bare conductor section = 1.9687 mm. sq.

Enter from STANDARD TABLE the bare diameter (cbd. mm.), bare section (cbs - mm. - sq.), insulated diameter (cid. - mm.), insulated section (cis-mm. sq.) and swg number

(E)   1.6256   2.0755   1.8056   2.5601   16

(o)   computed slot-area occupied by insul. conductor = 102.4 mm. sq.

Stator slot space factor = 0.49

Enter the value of specific weight of conductor material (spwt - kg/mm. cubed)

(E)   0.00889

Enter the value: o1 = 1 for TYPE 1 ; o1 = 2 for TYPE 2 Slot.

(E)   2

(o)   Weight of Stator teeth = 7.6 kg.

Loss in Stator teeth = 40.2 watt

Weight of Stator core = 20.64 kg.

Loss in Stator core = 169.6 watt

Total Stator iron losses = 210 watt

7.2.3   Design of Rotor dimensions

In the computer program (Figure 7.8) for the design of 'rotor dimensions' of a three-phase general - purpose cage - rotor induction motor, input data from stator can be obtained from Figure 7.7. The program deals with two types of rotor slots (Figure 7.6) — parallel – sided and round, but any other slot configuration can be incorporated through minor modification in the program. In testing the program, aluminium has been chosen as the bar-conductor and end-ring materials. Through entry of data on specific resistance and specific weight, any other material can be incorporated.

Governing Equations

Step 1: Assuming a gap length, the rotor diameter, and from the shaft diameter Dsh (as Per IS 1231-1967) for the Frame 160M, rotor inside diameter are computed. DRcs is the sum of rotor slot depth and depth of core below slots. The depth of rotor below slot (dRc) is given by the equation,

Flux-density in rotor core, BRc = phim x 10^6/(2 x L x dRc x fst), which leads to the computation of depth dRs.

Step 2: For any type of rotor slot, the width of rotor tooth at 1/3rd from the bottom of slot (wRt) is then computed from the equation of maximum flux-density at 1/3rd tooth-height from the bottom of tooth,

BRt = 1.36 phim x no. of. Poles (p) x 10^6 / (L x wRt x no. of rotor slot x fst).

For a **parallel-sided slot**, the tooth-pitch at 1/3rd from the slot bottom (Yrt) is given by,

Yrt = Perimeter at 1/3rd from the slot-bottom/no. of rotor slots,

and, slot width = YRt - wRt.

For the **round slot**, Yrt is taken along the centers of the round slots (since along this line the tooth width is minimum),

$$= \pi \text{ [rotor dia. DR} - 2 (h4 + 0.5 \text{ w1)]/Rs}$$

$$= \pi \{DR - 2 [h4 + 0.5 (dRs - h4)]\}/Rs,$$

where w1 = slot diameter; h4 = rotor slot opening, and dRs = depth of rotor slot.

$$= \pi \text{ [DR - (dRs + h4)]/Rs}$$

Step 3: Area of conductor in the slot (abar) is assumed occupy 95% if slot area.

For **parallel-sided slot**, abar = 0.95 x slot width x (slot depth - height of tapered part and lip part (approximately).

For **round slot**, abar = 0.95 π (slot diameter)^2/4.

Also, bar corrent for any type of slot,

bcur = Rotor current x conductor/slot x winding factor x stator slot/ rotor slot, where,

Rotor current = Stator current x power-factor.

Bar current density = bcur/abar, which should be around 5 amps per mm. sq. for aluminium die-cast rotor.

```
// Design of Three-phase Cage-rotor Induction Motor- Rotor dimensions
// constraint: Given stator dimensions; Insulation:class A
#include <iostream.h>
#include <iomanip.h>
#include <math.h>
#include <conio.h>
#include "sploss3.h"


class spec
{
  public:
    . int a, b, c, d, e ;                  //given data
      float f, g, h, k, l, m, n, o, q ;      //given data
    void getdata(int a, int b, int c, int d, int e, float f, float g, float h,
          float k, float l, float m, float n, float o, float q) ;
    void display() ;
    void rotor(int a, int b, int c, int d, int e, float f, float g, float h,
          float k, float l, float m, float n, float o, float q) ;
} ;
```

```
void spec :: getdata(int p, int ph, int ns, int Ss, int Zs, float phim,
            float D, float L, float fst, float Dsh, float Scur,
            float pf, float et, float akw)
{
    a = p; b = ph ; c = ns ; d = Ss  ; e = Zs; f = phim; g = D  ;
    h = L; k = fst; l = Dsh; m = Scur; n = pf; o = et  ; q = akw;
}


void spec :: display()
{   cout <<"  No. of poles=  "<< a <<"  phase= "<< b <<" synchronous"
        " r.p.s.= "<< c <<"  Number of Stator slots=  "<< d
        <<"  Number of conductor/slot in Stator=  "<< e
        <<"  Maximum flux= "<<f<<",  Stator inside diameter=  "<< g
        <<", Gross armature length=  "<< h <<",  Stacking factor=  "
        << k <<",   Shaft extension diameter = "<< l
        <<",   Stator current= "<< m <<"  Power-factor = "<< n
        <<",   supply volt/phase= "<< o <<",  Winding factor=   "<< q ;
}


void spec :: rotor(int a, int b, int c, int d, int e, float f, float g, float h,
            float k, float l, float m, float n, float o, float q)
{
 int Rs ;
 float lg, DR, DRi, dRcs, BRc, BRt, dRs, dRc, wRt, YRt, wor, wr1, h2, h3, h4 ;
 float x, y, Rcur, bcur, cdbar, abar ;
 const float PI = 3.14159 ;
 float k1 = pow(10,6) ;

//Compute Rotor diameter (DR mm.)

 cout << endl <<"Enter Designer'c choice of Gap Length (lg-mm.)" ;
 cin >> lg ;

 DR = g - 2 * lg ;        // Rotor diameter
 cout << endl <<"Rotor diameter = "<< DR <<" mm.";

// Choice of Number of Rotor slots (See art. 7.2.1 g)
 cout << endl <<"Enter Designer's choice of number of rotor slots" ;
 cin >> Rs ;

// Choice of Rotor inside diameter, DRi ( mm.)
//      SPEC gives shaft diameter(Dsh) from IS 1231 - 1967
//      for standard frame of 3-phase induction motor
//      with class A insulation. From the rotor inside
//       diameter, the shaft is reduced in steps to
//       reach the given value of Dsh.
// Choose DRi on the basis of the valueof Dsh, say 1.15 times Dsh.

 DRi = 1.15 * l ;                      // inside dia. of rotor
 cout << endl <<"Inside diameter of rotor = "<< DRi <<" mm." ;
 cout << endl <<"Enter Designer's choice of DRi" ;
 cin >> DRi ;
//          dRcs = dRc + dRs ;
 dRcs = 0.5 * (DR - DRi) ;
```

```cpp
// DESIGN OF ROTOR SLOT

    float h34, YRs ;

    cout << endl <<"Enter Designer's choice of BRt and BRc" ;
    cin >> BRt >> BRc ;

    cout << endl <<"Maxm.flux-density at 1/3rd tooth-height from bottom= "
    << BRt <<" T"<< setw(30) <<" Flux-density in Rotor core= "<< BRc <<" T";
    getch() ;

    dRc = f * k1 / ( 2. * BRc * h * k ) ;      // computed depth of rotor core
    dRs = (0.5 * (DR - DRi)) - dRc ;           // computed rotorslot depth

    cout << endl <<"computed Depth of rotor core = "<< dRc <<" mm"<< setw(20)
            <<"computed deph of rotor slot= "<< dRs <<" mm";
    cout << endl <<"dRc + dRs = "<< dRcs ;

    YRs = PI * DR / Rs ;                       //slot-pitch at ROTOR SURFACE
    cout << endl <<"slot-pitch at ROTOR SURFACE= "<< YRs ;

    cout << endl <<"Enter Designer's choice of dRc and dRs" ;
    cin >> dRc >> dRs ;
    Rcur = m * n ;                             //rotor current
    bcur = Rcur * e * k * d / Rs ;             // Bar current
    cout << endl <<"Rotor current= "<< Rcur <<" amp"<< setw(20)
            <<"Bar current= "<< bcur <<" amp";

// Choose the type of slot
    int o1,o2 ;
    cout << endl <<"Enter the value : o1 = 1 for TYPE1-PARALLEL SIDED SLOT,"
            " ; o1 = 2 for TYPE2- ROUND SLOT " ;

    cin >> o1 ;
    switch(o1)
    {
    case 1 :          //For Parallel-sided slot
    {   cout << endl <<"Enter the values of h3, h4" ;
        cin >> h3 >> h4 ;
        h34 = h3 + h4 ;
        h2 = dRs - h34 ;
        wRt = 1.36 * f * k1 * a / (BRt * Rs * h * k) ; //width
//                          of tooth at 1/3rd from tooth bottom
        YRt = PI * (DR - 2 * (2/3) * dRs) / Rs ; // tooth-pitch at
//                          1/3rd from the tooth-bottom
        wr1 = YRt - wRt ;               // Slot width
        abar= 0.95 * wr1 * (dRs - h34) ; // bar area

        cout << endl <<"wRt= "<< wRt << setw(10) <<"wr1= "<< wr1 ;
        cout << endl <<"abar= "<< abar ;

        cdbar = bcur / abar ;
        cout << endl <<"Bar current density= "<< cdbar <<" amp/mm.sq";
//    current density should be around 5 amp/mm.sq. for aluminium
//                     die cast rotor.
```

```
    cout <<endl<<"Enter Designer's choice of rotor slot opening(wor)" ;
    cin >> wor ;

    cout << endl <<"Slot depth= "<< dRs << setw(10) <<"Depth of "
    "core below slot=  "<< dRc << setw(10) <<"wrl=  "<< wrl
    << setw(10) <<"wor=  "<< wor << setw(10) <<" h2= "<< h2 ;

    cout << endl <<"Enter Designer's Final choice of dRs,"
               "dRc, wrl, wor, h2, h3, h4" ;
    cin >> dRs >> dRc >> wrl >> wor >>  h2 >> h3 >> h4 ;
    BRc = f * k1 / (2. * dRc * h * k );
    wRt = YRt - wrl ;
    BRt = 1.36 * f * k1 * a /(wRt * Rs * h * k );
    cout << endl << setw(45) <<"ROTOR  DESIGN  DATA" ;
    cout << endl <<"Rotor diameter=  "<< DR <<" mm.";
    cout<<endl<<setw(15)<<"Inside diameter of rotor= "<<DRi<<"mm.";
    cout << endl << setw(30) <<"Slot depth=  "<< dRs <<" mm." ;
    cout<<endl<<setw(45)<<"Depth of core below slot=" <<dRc<<"mm." ;
    cout << endl << setw(60) <<"Slot width =  "<< wrl <<" mm." ;
    cout << endl << setw(70) <<"Slot opening= "<< wor <<" mm." ;
    cout<<endl<<setw(60)<<"Slot ht.-parll.side portion= "<<h2<<"mm";
    cout<<endl<<setw(45)<<"Bar curr. density= "<< cdbar<<"A/mm.sq";
    cout<<endl<<setw(30)<<"Slot ht.-tapered portion=  "<<h3<<" mm";
    cout<<endl<<setw(15)<<"Slot height- lip portion= "<< h4 <<" mm." ;
    cout<<endl<<"Flux-density at 1/3 tooth ht.from root= "<< BRt<<" T";
    cout<<endl<<"Rotor core flux-density=  "<< BRc <<" T" ;


    }
    break ;

    case 2 :          // For Round slot
    { cout << endl <<"Enter the value of h4" ;
      cin >> h4 ;
        float wRt1, wRt2 ;
      h2 = dRs - h4 ;                  // Diameter of slot

      wRt2 = 1.36 * f * a * k1 / ( BRt * Rs * h * k);
      YRt = PI * (DR -2.* h4 - h2) / Rs ;     // tooth pitch along
//                          the centre of round slots
      cout << endl<<"Tooth pitch along centre of slots= " << YRt <<" mm.";
      wRt1 = YRt - h2 ;  // tooth width from the value of h2
      cout << endl <<"wRt1=  "<< wRt1 << setw(10) <<"wRt2= ."<< wRt2 ;

      cout << endl <<"Enter choice of rotor slot dia.(h2),opening ( wor)"
               "height of lip (h4)" ;

      cin >> h2 >> wor >> h4 ;
      getch() ;
      abar = 0.95 * PI * h2 * h2 / 4.; // area of bar
      cdbar = bcur / abar ;
      cout << endl <<"Bar current density=  "<< cdbar ;
      getch() ;
//       current density should be around 5 amp/mm.sq. for aluminium
//                  die cast rotor.
```

```
                 YRt = PI * (DR -2,* h4 - h2) / Rs ;        // tooth pitch along
       //                                  the centre of round slots
                 cout << endl<<"Tooth pitch along centre of slots=  " << YRt <<" mm.";
                 wRt = YRt - h2 ;   //  tooth width from the value of h2
                 BRt = 1.36 * f * k1 * a /(wRt * Rs * h * k) ;
                 dRc = 0.5 * (DR - DRi) - h2 - h4 ;
                 BRc = f * k1 / (2. * dRc * h * k ) ;


                      cout << endl << setw(45) <<"ROTOR  DESIGN  DATA" ;
                 cout << endl <<"Rotor diameter=  "<< DR <<" mm.";
                 cout << endl << setw(15)<<"Inside diameter of rotor=  "<<DRi<<" mm";
                 cout << endl << setw(30)<<"Slot depth=  "<< dRs <<" mm." ;
                 cout << endl << setw(45)<<"Depth of core below slot=  "<<dRc<<" mm";
                 cout << endl << setw(60)<<"Slot diameter =   "<< h2 <<" mm.";
                 cout << endl << setw(45)<<"Bar curr.density=  "<<cdbar<<" A/mm.sq.";
                 cout << endl << setw(25)<<"Slot opening=  "<< wor <<" mm." ;
                 cout << endl <<"Slot height- lip portion=  "<< h4 <<" mm." ;
                 cout<<endl<<"Flux-density at 1/3 tooth ht.from root=  "<< BRt<<" T";
                 cout<<endl<<"Rotor core flux-density=  "<< BRc <<" T" ;
                 }
                 break ;
              }
        }

        void main()
        {
         spec x ;
         clrscr ;

         int p, ph, ns, Ss, Zs ;
         float phim, D, L, fst, Dsh, Scur, pf, et, akw ;
         cout << endl <<"Enter :a, b, c, d, e, f, g, h, k, l, m, n, o, q" ;
         cin >> p >> ph >> ns >> Ss >> Zs >> phim >> D >> L >> fst >> Dsh >> Scur
            >> pf >> et >> akw;

         x.getdata(p, ph, ns, Ss, Zs, phim, D, L, fst, Dsh, Scur, pf, et, akw) ;
         x.display() ;
         x.rotor(p, ph, ns, Ss, Zs, phim, D, L, fst, Dsh, Scur, pf, et, akw) ;
         }
```

Figure 7.7   A computer program for the design of rotor demensions of a three-phase cage-rotor Induction motor.

RESULT

Enter(E)   4   3   25   36   40   0.00816
                      165   140   0.955   45   7.87   0.87   415   0.955

(o)   Enter Designer's choice of Gap Length (lg - mm.)

(E)   0.4

(o).   Rotor diameter = 164.2 mm.

       Enter Designer's choice of number of rotor slots

(E)   44

[*Note:* Choice of rotor inside diameter, Dri is computed with the assumption that it is 1.15 times the specified shaft diameter, Dsh, (IS 1231 - 1967 for 3-phase induction motor with class A insulation), since the shaft is reduced in steps to reach the value of Dsh.]

(o) Inside diameter of rotor = 51.75 mm.
   Enter Designer's choice of Dri.

(E) 51.2

(o) Enter Designer's choice of BRt and BRc.

(E) 1.5   0.7

(o) Maxm. flux-density at 1/3rd tooth-height from bottom = 1.5 T Flux- density in Rotor core = 0.7 T.
   computed depth of rotor core = 43.82 mm. computed depth of rotor slot = 12.68 mm.
   Enter Designer's choice of depth of rotor core (dRc) and depth of rotor slot (dRs)

(E) 43.5   13

(o) Rotor current = 6.85 amp.   Bar current = 212.9 amp.
   Enter the value of : 01 = 1 for TYPE1-PARALLEL-SIDED SLOT; 01 = 2 for TYPE2 - ROUND SLOT.

Run 1: PARALLEL-SIDED slot

(E) 1

(o) Enter the values of h3 and h4

(E) 1.5   0.5

(o) wRt = 5.06   wrl = 6.67
   abar = 65.9 mm. sq.
   Bar current density = 3.06 amp / mm. sq.
   Enter Designer's choice of rotor slot opening (wor)

(E) 2

(o) Slot depth = 13   Depth of core below slot = 43.5 wrl = 6.67 wor = 2
   h2 = 11
   Enter Designer's choice of dRs, dRc, wrl, wor, h2, h3, h4

(o)   ROTOR DESIGN DATA

Rotor diameter = 164.2 mm.
   Inside diameter of rotor = 51.2 mm.
      Slot depth = 13 mm.
         Depth of core below slot = 43.6 mm.
            Slot width = 6.5 mm.
         Slot opening = 2 mm.
      Slot height-parll.-sided portion = 11 mm.
   Bar current-density = 3.06 amp/mm. sq.
Slot ht, - tapered portion = 1.5 mm.
Slot height-lip portion = 0.5 mm.

Run 2 - case 2: ROUND slot.

[*Note:* With 44 rotor slots as above, slot-pitch at the rotor surface is 11.7 mm. which is inadequate to accommodate round slots with bar current-density around 5 anps/mm. sq.

The slot combination (Appendix V) of 36/28 is thus selected. The result is as below:

(o)   Rotor current = 6.86   Bar current = 334.5
      Enter the value of : 01= 1 for TYPE1-PARALLEL-SIDED SLOT; 01 = 2
      for TYPE2 - ROUND SLOT.

(E)   2

(o)   Tooth-pitch along centres of slots = 17.08 mm.
      Bar current-density = 5.38 amp/mm. sq.
      Enter Designer's choice of rotor slot-opening (wor)

(E)   2

(o)   Slot depth = 13   Depth of core below slot = 43.5   wrl = 6.67
                          wor = 2   h2 = 11
      Enter Designer's choice of dRs, dRc, wrl, wor, h2, h3, h4

(o)   ROTOR DESIGN DATA

Rotor diameter = 164.2 mm.
    Inside diameter of rotor = 51.2 mm.
        Slot depth = 11.5 mm.
            Depth of core below slot = 45 mm.
                              Slot diameter = 14 mm.
            Bar current-density = 5.3 amp/mm. sq.
        Slot opening = 2 mm.
Slot height-lip portion = 0.5 mm.

7.2.4   Determination of M.M.F., magnetising current and magnetising reactance–Computer Program

Predetermination of performance after the design of physical parameters form an important component of the total design process so that a comparison can be made with the customer's requirement before the apparatus is manufactured. Figure 7.9 presents a computer program for determination of component magnetomotive forces in various parts of the magnetic circuit, the total, and hence, computation of the magnetising reactance.

The total magneto-motive force is due to 5 component parts of the machine magnetic circuit viz., stator core, stator teeth, gap, rotor teeth, and rotor core (Table 4.1).

Following Figure 7.7, the program is developed on the basis of data input to 'class data'. There are altogether 18 parameters, as below, entered as input the said class, from which whatever data required for computation of each component of magnetic circuit are accessed

1. No. of poles, p;	2. Gap flux /pole, phim;	3. Armature diameter D;
4. Gross armature length L;	5. Gap length lg;	6. No. of stator slots Ss:
7. No. of rotor slots Rs:	8. Stator slot opening wo1;	9. Rotor slot-opening wo2;
10. No. of radial vent duct, nd;	11. Width of duct wd;	12. Stator slot depth dSs;
13. Rotor slot depth dRs;	14. Stator tooth-width wSt;	15. Rotor slot-width wRs:

(Both 14 and 15 correspond to width at 1/3rd tooth- / slot-height from the bottom of the slot).

16. Stator core outside dia. Do;	17. Rotor inside dia. Dri;	18. Stacking factor fst.

The program uses 3 header files apart from usual library files:—"mcur 42st.h" which gives m.m.f. for flux-densities for the lamination material—42 quality stalloy. Obviously if any other material is used as lamination material, the corresponding file is to be used;

"carter_s.h" for the computation of Carter's coefficient for semi-closed slots; and "carter_o.h" for computation of Carter's coefficients for stator and rotor vent ducts (ref; art. 4.2.2). For simplicity, it has been assumed that both stator and rotor have the same number of vant ducts of equal width.

Again, in the program, it is assumed that the stator and rotor teeth are parallel-sided. In Figure 7.9, the part on 'class data' is omitted. A student can easily formulated the program.

Governing Equations

For computation of mmf. of any component of magnetic circuit, it is required to know the average flux-density in that component using which the amp-turn per metre is available from the material characteristics. Side by side, it is required to find the length of flux-path. The concept and equations have been developed in Chapter 4 and used in Example 7.1. The result obtained on running the program on a computer with input values of the parameters are given below.

```
// Computation of Magetising Force, Magnetising current & Magnetising
// reactance for Three-phase General-purpose Induction Motor.

#include <iostream.h>
#include <iomanip.h>
#include <math.h>
#include <conio.h>
#include "mcur42st.h"
#include "carter_s.h"
#include "carter_o.h"


class Statorcore
{
    public:
     int p ;
     float phim, D, L, Do, dSs, fst, dSc, ASc, BSc, lSc, y, x, FSc ;

    void getstatorcoredata()
    {
     cout << endl <<"Enter No. of poles (p), Maximun flux per pole (phim-Wb),"
        " Stator inside diameter (D- m), Stator outside diameter (Do-m), Gross"
        " armature length (L-m.),Depth of stator slot (dSs-m.), Stacking"
        " factor (fst)" ;
     cin >> p >> phim >> D >> Do >> L >> dSs >> fst ;
    }
    void seestatorcoredata()
    {
     cout<<endl<<"No. of poles= "<< p << setw(10) <<" Maximun flux per pole= "
        << phim << setw(10) <<" Stator inside diameter=  "<< D << setw(10)
        <<" Stator outside diameter=  "<< Do << setw(10) <<" Gross armature "
```

```
                 <<"length= "<< L << setw(10) <<" Depth of stator slot=   "<< dSs
                 << setw(10) <<" Stacking factor=   "<< fst ;
           }
      void computestatorcore()
      {
       const float PI = 3.1416 ;
        dSc = 0.5 * (Do - (D + 2. * dSs)) ;        // depth of stator core
        ASc = fst * dSc * L ;            // sectioal area of stator core
        cout<<endl<<"dSc=  "<<dSc<<setw(10)<<"ASc=  "<<ASc ;
        lSc = 0.5 *  PI * (D + 2. * dSs + dSc) / p ; //length of flux-path in
      //                                stator core/pole

       BSc = 0.5 * phim / ASc ;          //Flux-density, stator core in Tesla
       cout<<endl<<"lSc=  "<<lSc<<setw(10)<<"BSc=  "<<BSc ;
       getch() ;
       y = BSc ;
       float mcur42st(float) ;      //Function to find amp-turn / metre from
       x = mcur42st(y) ;          // magnetisation curve: 42 quality Stalloy
       cout<<endl<<"x=  "<< x ;
       getch() ;
       FSc = x * lSc ;
      }
      float statorcoremmf() { return FSc ; }
      } ;
      class statorteeth
      {
        public:
         int p ;
         float phim, L, fst, Ss, dSs, wSt, ASt, BSt, y, x, FSt ;
         void getstatorteethdata()
         {
         cout << endl <<"Enter No. of poles (p), Maximun flux per pole (phim-Wb),"
            " Gross armature length (L-m.),No. of Stator slot (Ss),Depth of stator"
            " slot (dSs-m.),Width of stator slot (wSt), Stacking factor (fst)" ;
         cin >> p >> phim >> L >> Ss >> dSs >> wSt >> fst ;
         }
         void seestatorteethdata()
         {
         cout<<endl<<"No. of poles= "<< p << setw(10) <<" Maximun flux per pole=  "
            << phim << setw(10) <<"Gross armature length=  "<< L << setw(10)
            <<"No. of Stator slot=  "<< Ss << setw(10) <<"Depth of stator slot=   "
            << dSs << setw(10) <<"Width of stator slot=  "<< wSt << setw(10)
            <<" Stacking factor=   "<< fst ;           ;
         }
      //              Stator teeth are parallel-sided,
      //              Rotor slot-section circular
        void computestatorteeth()
        {
        ASt = fst *  L * wSt ;    //sectional area of stator tooth at 1/3rd
      //                      tooth height from bottom
        BSt = 1.36 * phim * p / ( Ss * ASt) ;
        cout<<endl<<"ASt=  "<<ASt<<setw(10)<<"BSt=  "<<BSt ;

        y = BSt ;
        float mcur42st(float) ;       //Function to find amp-turn / metre
        x = mcur42st(y) ;      //from magnetisation curve:42 quality Stalloy
```

```
       cout<<endl<<"x=  "<< x ;
       getch() ;
       FSt = x * dSs ;                //length of flux-path = height of tooth
       }
       float statorteethmmf() { return FSt ; }
} ;


class gap
{
 public:
    int p, Ss, Rs, nd ;
    float phim, lg, D, L, wo1, wo2, wd ;
    float Leff, kd, DR, ys1, ys2, Bav, kcs, lgeff, kcr, kgd, kg1, kg2 ;
    float ratiod, ratios, ratior, Fg, r ;
    void getgapdata()
     {
      cout << endl <<"Enter No. of poles (p), Maximun flux per pole (phim-Wb),"
         " Armature diameter(D- m),Gross armature length (L-m.),No. of Stator"
         " slot (Ss),Gap length(lg-m.),Stator slot-opening (wo1),No. of Rotor"
         " slot(Rs), Rotor slot- opening (wo2), No.of ducts in stator( nd),"
         " Duct width (wd)" ;

      cin >> p >> phim >> D >> L >> Ss >> lg >> wo1 >> Rs >> wo2 >> nd >> wd ;
      }
     void seegapdata()
      {
      cout<<endl<<"No. of poles= "<< p << setw(10) <<" Maximun flux per pole=  "
         << phim << setw(10) <<"Armature diameter=  "<< D <<"Gross armature"
         " length=  "<< L << setw(10) <<"No. of Stator slot=  "<< Ss << setw(10)
         <<" Gap length=  "<< lg << setw(10) <<" Stator slot-opening=  "<<
         wo1 << setw(10) <<" No. of Rotor slot=  "<< Rs << setw(10) <<"Rotor "
         "slot-opening=  "<< wo2 << setw(10) <<"No. of ducts in stator=  "
         << nd << setw(10) <<" Duct  width=  "<< wd ;
      }

     void computegap()
      {
      const float PI = 3.1416 ;

//Compute effective core length, Leff:

      ratiod = 0.5 * wd / lg ;        // Ratio : duct width/ gap length
      cout << endl <<"ratiod=  "<< ratiod ;

      r = ratiod ;
      float carter_o (float) ;    //Function to find Carter's Coeff.:open slot
      kd = carter_o(r) ;          // Carter's Coefficient for duct opening
      cout << endl <<"kd=  "<< kd ;

      Leff = L - kd * nd * wd ;        // effective core length
      cout << endl <<"Leff=  "<< Leff ;
      getch() ;
      kgd = L / Leff ;              //gap coefficient due ducts in stator
//                as well as due to ducts in rotor since nd & wd are the
//                         same in the stator and rotor
```

```
//Compute Carter`s Coefficients for gap :  due to stator slot opening (kcs) &
//                                due to rotor slot opening (kcr)

    DR = D - 2. * lg ;                  // Rotor outside diameter
    ys1 = PI * D / Ss ;                 // Stator slot-pitch
    ys2 = PI * DR / Rs ;                // Rotor slot-pitch
    cout<<endl<<"DR= "<<DR<<setw(10)<<"ys1= "<<ys1<<setw(10)<<"ys2= "<<ys2 ;
    ratios = wo1 / lg ;          // Ratio: stator slot opening/gap length

    r  = ratios ;
    float carter_s (float) ;            //Function to find Carter's Coeff.
    kcs = carter_s(r) ;        // Carter's Coefficient for stator slot opening
    cout<<endl<<"ratios= "<<ratios<<setw(10)<<"kcs= "<<kcs ;
    getch() ;
    ratior = wo2 / lg ;             // Ratio: rotor slot opening/gap length

    r  = ratior ;
    float carter_s (float) ;            //Function to find Carter's Coeff.
    kcr = carter_s(r) ;          // Carter's Coefficient for rotor slot opening
    cout<<endl<<"ratior= "<<ratior<<setw(10)<<"kcr= "<<kcr ;
    kg1 = ys1 /( ys1 - kcs * wo1 ) ;    // gap coeff. due stator slots
    kg2 = ys2 /( ys2 - kcr * wo2 ) ;    // gap coeff. due rotor slots
    cout<<endl<<"kg1= "<<kg1<<setw(10)<<"kg2= "<<kg2 ;
    lgeff = lg * kg1 * kg2 * kgd * kgd ;        // effective gap length

    Bav = p * phim / (PI * D * Leff) ;   // Average gap-density at 30 degree
//                        (elec) from direct axis
    cout<<endl<<"lgeff= "<<lgeff<<setw(10)<<"lgeff= "<<lgeff ;
    getch() ;
    Fg = 800000 * Bav * lgeff ;
    }
  float gapmmf() { return Fg ; }
} ;

class rotorteeth
{
   public:
   int p ;
   float phim, D, L, fst, lg, Rs, dRs, wRs, nd, wd ;
   float DR, wRt, phiRt, ARt, BRt, y, x, FRt ;
   void getrotorteethdata()
   {
   cout << endl <<"Enter No. of poles (p), Maximun flux per pole (phim-Wb),"
      "Armature diameter(D- m),Gross armature length(L-m.),Gap length(lg-m.),"
      " No. of Rotor slot(Rs -m),Depth of Rotor slot(dRs-m.), Width of"
      " rotor slot(wRs-m.), No.of ducts in rotor( nd),Duct width (wd)" ;
   cin >> p >> phim >> D >> L >> lg >> Rs >> dRs >> wRs >> nd >> wd ;
   }
   void seerotorteethdata()
   {
   cout<<endl<<"No. of poles= "<< p << setw(10) <<" Maximun flux per pole=  "
      << phim << setw(10) <<"Armature diameter=  "<< D << setw(10) <<"Gross"
      " armature length= "<< L << setw(10) <<" Gap length=   "<< lg <<
      setw(10) <<" No. of Rotor slot=  "<< Rs << setw(10) <<"Depth of rotor"
      " slot=   "<< dRs << setw(10) <<"Width of rotor slot=   "<< wRs
```

```
       << setw(10)<<"No. of ducts in rotor=  "<< nd << setw(10)
       <<" Duct  width=  "<< wd ;
    }
    void computerotorteeth()
    {
    const float PI = 3.1416 ;
    DR = D - 2. * lg ;                // rotor diameter
    wRt = (PI * ( DR - 2. * 0.667 * dRs)/ Rs) - wRs ; //tooth-width at 1/3rd
//                    slot-depth from the bottom of rotor slot
    ARt = fst * wRt * (L - nd*wd) ;       // corresponding tooth-section
    cout<<endl<<"wRt=  "<<wRt<<setw(10)<<"ARt=  "<<ARt ;

    phiRt  = phim * p / Rs ;         //flux per rotor tooth-pitch
    BRt = 1.36 * phiRt / ARt ;        //flux-density at that tooth-section
    cout<<endl<<"phiRt=  "<<phiRt<<setw(10)<<"BRt=  "<<BRt ;
    getch() ;

    y = BRt ;
    float mcur42st(float) ;       //Function to find amp-turn / metre
    x = mcur42st(y) ;          //from magnetisation curve:42 quality stalloy
    cout<<endl<<"x=  "<<x ;
    getch ;
    FRt = x * dRs ;
    }
    float rotorteethmmf() { return FRt ; }
};


class rotorcore
{
 public:
   int p ;
  float phim, D, L, lg, dRs, DRi, fst ;
  float DR, dRc, DRcm, ARc, lRc, y, x, BRc, FRc ;
  void getrotorteethdata()
   {
   cout << endl <<"Enter No. of poles (p), Maximun flux per pole (phim-Wb),"
     "Armature diameter(D-m),Gross armature length(L-m.),Gap length(lg-m.),"
     " Depth of Rotor slot(dRs-m.),Inside diameter of rotor(DRi-m.),"
     " Stacking factor(fst)" ;
   cin >> p >> phim >> D >> L >> lg >> dRs >> DRi >> fst ;
   }
  void seerotorteethdata()
   {
   cout<<endl<<"No. of poles= "<< p << setw(10) <<" Maximun flux per pole=  "
     << phim << setw(10) <<"Armature diameter=  "<< D << setw(10) <<"Gross"
     " armature length=  "<< L << setw(10) <<" Gap length=  "<< lg <<
     setw(10) <<" Depth of Rotor slot=  "<< dRs << setw(10) <<"Inside"
     " diameter of rotor=  "<< DRi << setw(10) <<"Stacking factor=  "
     << fst ;
   }
  void computerotorteeth()
   {
   const float PI = 3.1416 ;
```

```
    DR = D - 2. * lg ;              // rotor diameter
    dRc = 0.5 * (DR - 2. * dRs - DRi) ;   // depth of rotor core
    DRcm = DRi + dRc ;             // mean diameter of rotor core
    ARc = fst * dRc * L ;          // rotor core section
    cout<<endl<<"dRc=  "<<dRc<<setw(10)<<"ARc=  "<<ARc<<"DRcm=  "<<DRc
    float kfl = 1.05 ;      // Factor to allow for curved nature of flux-path in
//                          rotor core at the back of rotor teeth
    lRc = 0.5 * kfl * PI * DRcm / p ;  // Length of flux-path in rotor core/pole
//                    = 0.5 pole-pitch on mean dia. of armature core
    BRc = 0.5 * phim / ARc ;        // Flux in armature core
    cout<<endl<<"lRc=  "<<lRc<<setw(10)<<"BRc=  "<<BRc ;
    getch() ;

    y = BRc ;
    float mcur42st(float) ;      //Function to find amp-turn / metre
    x = mcur42st(y) ;            //from magnetisation curve: 42 quality Stalloy
    cout<<endl<<" x=  "<< x ;
    FRc = x * lRc ;
    }
  float rotorcoremmf() { return FRc ; }
};
class total: public statorcore, public statorteeth, public gap,
        public rotorteeth, public rotorcore
{
  public:

    float summmf()
    {
    return statorcoremmf() + statorteethmmf() + gapmmf() + rotorteethmmf()
    + rotorcoremmf() ;
    }
};


void main()
{

  clrscr() ;
  total t ;

  t.statorcore :: getstatorcoredata() ;
  t.statorcore :: seestatorcoredata() ;
  t.statorcore :: computestatorcore() ;

  t.statorteeth :: getstatorteethdata() ;
  t.statorteeth :: seestatorteethdata() ;
  t.statorteeth :: computestatorteeth() ;

  t.gap :: getgapdata() ;
  t.gap :: seegapdata() ;
  t.gap :: computegap() ;

  t.rotorteeth  :: getrotorteethdata() ;
  t.rotorteeth  :: seerotorteethdata() ;
  t.rotorteeth  :: computerotorteeth() ;
```

```
t.rotorcore :: getrotorcoredata() ;
t.rotorcore :: seerotorcoredata() ;
t.rotorcore :: computerotorcore() ;

cout << endl <<"Total m. m. f= "<< t.summmf() ;

cout << endl <<"Enter values of winding factor, kw, and stator turns/ph,Ns"
        <<" No. of poles, p, and applied voltage/phase, et" ;
cin >> kw >> Ns >> p >> et;
Im = p * t.summmf() / ( 2.34 * kw * Ns ) ;
cout << endl <<"magnetising current per phase= "<< Im ;

Xm = et / Im ;
cout << endl <<"magneting reactance per phase= "<< Xm ;
}
```

Figure 7.8   A computer program for computation of Magnetising force, Magnetising current, and Magnetising reactance of a three-phase cage rotor induction motor.

RESULT

Enter(E)   4   0.00793   0.159   0.26   0.127   0.025   0.95
(o)   Magnetising Force due stator core = 41.25 amp. turns
(E)   4   0.00793   0.127   36   0.025   0.0076   0.95
(o)   478.1
(o)   Magnetising Force due stator teeth = 11.95 amp. turns
(E)   4   0.00793   0.159   0.127   36   0.0034 etc   etc.
               34   0.002   3   0.005
(o)   Magnetising Force due gap = 236.69 amp. turns
(E)   4   0.00793   0.159   0.127   0.0034   34   0.01
               0.0055   3   0.005   0.95
(o)   Magnetising Force due rotor teeth = 27.64 amp. turns
(E)   4   0.00793   0.159   0.127   36   0.0034   0.01   and so on.
(o)   Magnetising Force due rotor core = 3.7 amp. turns
               Total Magnetising Force   = 321

Enter values of winding factor (kw), stator turns/ph. (Ns), no. of poles (p), and applied voltage / ph. (et)

(E)   0.96   240   4   230
(o)   Magnetising current per phase = 2.38 amp.
       Magnetising reactance per phase = 96.6 ohms

7.2.5   Leakage reactances and performance at various slips Computer Programm

The computer program (Figure 7.10) is based on the concept discussed in art. 4.3 and is developed in the same way as the previous art. 7.2.4. The only difference is that in this program, the students have been initiated into reading and writing data to disk files in the form of calculating performance of the machine when one or more basic parameters are changed. No doubt, this could have been demonstrated separately.

For the computation of stator and rotor leakage reactances the following functions have been used:

1. For stator slot- 'stslot'; 2. For rotor slot- 'rtslot'; 3. For stator zigzag- 'stzigzag'; 4. For rotor zigzag- 'rtzigzag'; 5. For rotor skew- 'rtskew'; 6. overhang- 'overhang' (with the stipulation that the total overhang reactance will be equally shared by stator and rotor). The skew is only on rotor, but can b incorporated in the same way for stator if required.

In the program the following parameters are entered through 'class data' (not shown in Figure 9.10) —

1. No. of poles, p;
2. No. of phase, ph;
3. Frequency, fs;
4. Armature diameter D
5. Gross armature length L;
6. No. of stator slots Ss;
7. No. of rotor slots Rs;
8. Stator slot opening wo1;
9. Rotor slot-opening wo2;
10. Magnetising reactance Xm;
11. No of stator turns Ns;

As per requirement of a function, the data are accessed through invocation of 'friend function'. As mentional earlier, such invocation is necessary in view of accession of large number of data, use of functions through own created header files-"mcur 42 st.h", "carter_o.h", and "carter_s.h".

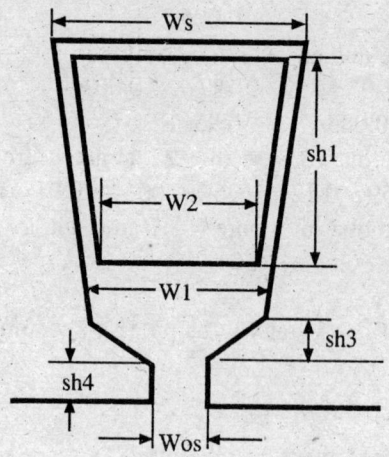

Figure 7.9 Configuration of chosen slot.

Governing Equations

The program is developed on the basis of equations in art. 4.3. The belt leakage reactance is assumed to be negligible.

Further, skewing has been considered on the rotor and the skew leakage reactance is determined as below:

Indicating Nsk = no. of slots skewed, the equation for skew leakage reactance is,

$$Xsk = 0.95 \, Xm \cdot \theta^2 / 12. \tag{7.1}$$

where, $\theta = \pi \cdot Nsk. / Rs$, Rs being the no. of rotor slots.

A factor 0.95 is used to take into account the effect of stator leakage flux.

The second part of the program is to store and read the performance data for various values of motor slip 's'. Output is recorded for slip varying from 0.02 to 0.1 in step of 0.02, and from 0.1 to 0.9 in step of 0.1.

```
// Determination of leakage reactance of a three-phase Induction Motor
// and Performance calculation at various slips.

#include <iostream.h>
#include <iomanip.h>
#include <math.h>
#include <conio.h>
#include "mcur42st.h"
#include "carter_s.h"
#include "carter_o.h"
#include <fstream.h>

class data
{
    public :


    data()
       { ....... }
    void display()
       { ....... }
    float get...() { return .. ; }
} ;

float stslot ( data d )
{
    int Ss, Ns, ph, fs.;
    float L, wo1, sh1, sh2, sh3, sh4, ws, w2, w1, rh4 ;
    float Lam1, Lam2, Lam3, Lam4, LamS1, LamR1, LamR1s, sxsl ;
    float k1, k2, kw1 ;

      ph = d. getph() ;
      fs = d. getfs() ;
      L  = d. getL() ;
      Ss = d. getSs() ;
      wo1 = d. getwo1();
      Ns = d. getNs() ;

    const float PI = 3.1416 ;
    k1 = pow (10.,-7.) ;
```

```
    cout << endl <<"Enter the dimensions of Stator slot : sh1, sh2, sh3, sh4,"
         <<" ws, w1- all in m." ;      // Stator slot as per Fig.4.29
    cin >> sh1 >> sh2 >> sh3 >> sh4 >> ws >> w1 ;


// Compute Stator slot permeance co-efficient
    w2 = ws - sh1 * (ws - w1)/( sh1 + sh2) ;
   Lam1 = 2. * sh1 / ( 3. * (ws + w2) ) ;
   Lam2 = 2. * sh2 / ( w2 + w1 ) ;
   Lam3 = 2. * sh3 / ( w1 + wo1 ) ;
   Lam4 = sh4 / wo1 ;
  LamS1 = Lam1 + Lam2 + Lam3 + Lam4 ;

   k2 = 4. * PI * fs * 4. * PI * k1 * Ns * Ns * 2.* ph / Ss ;
   sxsl = k2 * LamS1 * L ;    // Stator slot leakage reactance per phase
  return (sxsl) ;
};

float rtslot ( data d )
{
    int Ss, Rs, Ns, ph, fs ;
    float L, wo2, rh4,LamR1,LamR1s,rxsl ;
    float k1, k2, kw1 ;

     ph = d. getph() ;
     fs = d. getfs() ;
     L  = d. getL() ;
     Ss = d. getSs() ;
     Rs = d. getRs() ;

     wo2 = d. getwo2();
     Ns = d. getNs() ;

    const float PI = 3.1416 ;
    k1 = pow (10.,-7.) ;

    cout << endl <<"Enter the dimension of height of rotor lip( rh4- m.) " ;
    cin >> rh4 ;
// Compute Rotor slot permeance co-efficient
    LamR1 = 0.66 + rh4 / wo2 ;        // Circular rotor slot

    cout << endl <<"Enter the value of Stator winding factor, kw1" ;
    cin >> kw1 ;
    LamR1s = kw1 * kw1 * Ss * LamR1 / Rs ;  // Rotor slot permeance coeff.
//                        referred to the stator,knowing the
//                        winding constant for sq.cage winding = 1.
    k2 = 4. * PI * fs * 4. * PI * k1 * Ns * Ns * 2.* ph / Ss ;

    rxsl = k2 * LamR1s * L ;    // Rotor slot leakage reactance per phase
//                         in stator-terms
   return (rxsl) ;
 } ;
```

```
float stzigzag ( data  d )
{
      int p, Ss ;
      float Xm, k3s, sxz ;

      p  = d. getp() ;
      Ss = d. getSs() ;
      Xm = d. getXm() ;

      k3s = 1./(Ss*Ss) ;
      sxz = 5. * k3s * p * p * Xm  / 6. ;
      return (sxz) ;
} ;

float rtzigzag ( data  d )
{
      int p, Rs ;
      float Xm, k3r, rxz ;

      p  = d. getp() ;
      Rs = d. getRs() ;
      Xm = d. getXm() ;

      k3r = 1./(Rs*Rs) ;
      rxz = 5. * k3r * p * p * Xm  / 6. ;
      return (rxz) ;
} ;

float rtskew (data d)
{
      int p, Rs ;
      float Xm, Nsk, theta, rxsk ;     //Nsk = No. of slots skewed.

      p  = d. getp() ;
      Rs = d. getRs() ;
      Xm = d. getXm() ;

      const float PI = 3.1416 ;
   cout << "Enter the value of No. of slots skewed (Nsk)" ;
   cin >> Nsk ;
  theta = PI * Nsk / Rs ;
  rxsk = 59.4 * theta * theta * 0.95 / 12 ; // 0.95 is the assumed value
//                          stator leakage flux factor.
   return (rxsk) ;
};


float overhang (data d)
{
       int p, Ns, m ;
      float D, ks, yw, alpha, xo ;

      p  = d. getp() ;
      D  = d. getD() ;
      Ns = d. getNs() ;
```

```
        const float PI = 3.1416 ;

        cout << endl <<"Enter the value of Coil span factor (yw)" ;
         cin >> yw ;

        alpha = yw * p / ( PI * D ) ;

        cout << endl <<"Enter m = 0 for wide spread connection, m = 1 for"
                <<" narrow-spread connection" ;
         cin >> m ;

      if (m == 0)
        {
        if ( alpha < 0.67 )
        { ks = 1.12 * alpha ;}
        else
         if ( alpha < 1.0 )
         { ks = 0.75 ;}

        }
        else
        if (m == 1)
         {
        if ( alpha < 0.67 )
        { ks = (1.4 * alpha - 0.19) ;}
        else
         if ( alpha < 1.0 )
         { ks = (0.24 + 0.76 * alpha) ;}
         }
      xo = 0.00475 * ks * D * Ns * Ns / ( p * p ) ;
      return (xo) ;
};
void main()
{
  float sxsl, rxsl, sxz, rxz, xo, x1, x2s ,rxsk ;
  clrscr() ;
  data d ;
  d.getvalue() ;
  cout <<"\n"; d.display() ;

  sxsl = stslot(d) ;
  cout << endl <<" Stator Slot leakage reactance=      "<< sxsl <<" ohm";

  rxsl = rtslot(d) ;
  cout << endl <<" Rotor Slot leakage reactance=      "<< rxsl <<" ohm";

  sxz = stzigzag(d) ;
  cout << endl <<" Stator Zigzag leakage reactance=      "<< sxz <<" ohm";

  rxz = rtzigzag(d) ;
  cout << endl <<" Rotor Zigzag leakage reactance=      "<< rxz <<" ohm";

  rxsk = rtskew(d) ;
  cout << endl <<" Rotor skew leakage reactance=   "<< rxsk <<" ohm";
```

```
xo = overhang(d) ;
cout << endl <<" Overhang leakage reactance=  "<< xo <<" ohm";


x1 = sxsl + sxz + 0.5 * xo ;
cout << endl <<" Stator leakage reactance=  "<< x1 <<" ohm";

x2s = rxsl + rxz + rxsk + 0.5 * xo ;
cout << endl <<" Rotor leakage reactance=  "<< x2s <<" ohm";
//        PERFORMANCE CALCULATION AT VARIOUS SLIP
float et, r1, r2, Xm, x22, ns, s, Rf, Xf, Z, I1, spf ;
float Pi, Qi,rpf, Pg, Te, omegas, omegar, Pm, P2, Po, To ;
float c1, c2, c3, c4, c5, c6, c7, c8 ;

cout << endl <<" Enter  Stator resistance / ph. (r1), Rotor resistance / ph."
        " in stator terms (r2) & Magnetising reactance / phase (Xm)" ;
 cin >> r1>> r2>> Xm ;

cout << endl <<"Enter the values of Supply voltage / ph. and syn.r.p.s " ;
   cin >> et >> ns ;
   cout << endl <<"Enter the value of No-load rotational loss (Po)" ;
   cin >> Po ;
   cout << endl << setw(60) << "PERFORMANCE SHEET AT RATED VOLT & FREQ.";

   ofstream outf("RESULT") ;
  for( s=0.02; s<=0.06; s+=0.02)
  {
  x22 = Xm + x2s ;

  c1 = (r2/s) * (r2/s) + x22 * x22 ;
  c2 = (r2/s) * (r2/s) + x2s * x22 ;
  Rf = Xm * Xm * (r2/s) / c1 ;
  Xf = Xm * c2 / c1 ;
  c3 = (Rf + r1) * (Rf + r1) + (Xf + x1) * (Xf + x1) ;
  Z  = sqrt(c3) ;        //Effective impedance across input
  I1 = et / Z ;          //Input current per phase
  spf= (r1 + Rf) / Z ;        // stator power-factor,cos phi-1
  c4 = spf * spf ;
  c5 = 1. - c4 ;
  c6 = sqrt(c5) ;          // sin phi-1

  Pi = et * I1 * spf ;      // Input real power per phase
  Qi = et * I1 * c6 ;        // Input reactive power per phase

  c7 = r2 * r2 + (s * x2s) * (s * x2s) ;
  c8 = sqrt(c7) ;
  rpf = r2 / c8 ;          // rotor power-factor
  Pg = I1 * I1 * Rf ;      // per phase real power across the gap
  const float PI = 3.1416 ;

  omegas = 2. * PI * ns ;    // synch.speed in radians per second
  Te = 3. * Pg / omegas ;   // electromagnatic torque, N.m.for 3 phase.
  Pm = ( 1. - s ) * Pg ;     // Per phase internal mech. power
  P2 = Pm - Po ;          // Per phase output power
  omegar = (1. - s) * omegas ; // Rotor speed in radians per second
  To = 3. * P2 / omegar ;   // Output torque, N.m.
```

```
            outf<<s<<endl ;
            outf<<I1<<endl ;
            outf<<spf<<endl ;
            outf<<Pi<<endl ;
            outf<<Qi<<endl ;
            outf<<Te<<endl ;
            outf<<To<<endl ;
        }
        outf.close() ;
        ifstream inf("RESULT") ;
        while(!inf.eof())
        {
        inf >> s ;
        inf >> I1 ;
        inf >> spf;
        inf >> Pi ;
        inf >> Qi ;
        inf >> Te ;
        inf >> To ;

        cout << endl << setw(20) << " Slip= "<< s ;
        cout << endl << "Input current per phase=  "<< I1 ;
        cout << endl << setw(30) <<"Input power factor=  "<< spf ;
        cout << endl << setw(35) <<"Input real power per phase=  "<< Pi ;
        cout << endl << setw(45) <<"Input reactive power / phase=  "<< Qi ;
        cout << endl << setw(55) <<"Electromagnetic torque= "<< Te <<"N.m." ;
        cout << endl << setw(60) <<"Output torque=  "<< To <<"N.m." ;
        }
        inf.close() ;
    };
```

Figure 7.10   A computer program for the determination of leakage reactance of a Three-phase
cage-rotor induction motor and its performance calculation.

RESULT

(Enter)	4	3	50	0.159	0.127
	36	34	0.0035	0.002	96.6
	240				

(output) Enter the dimensions of Stator slot: sh1, sh2, sh3, sh4, ws, w1 - all in m.

(E)	0.022	0.001	0.001	0.001	0.0105
	0.0075				

(o)   Stator slot leakage reactance = 1.356

Enter the dimension of height of rotor lip (rh4 - m.)

(E)   0.0015

(o)   Enter the value of Stator winding factor, kw1

(E)   0.955

(o)   Rotor slot leakage reactance = 1.311

Stator zigzag leakage reactance = 0.99

Rotor zigzag leakage reactance = 1.11

Enter the value of No. of slots skewed (Nsk).

(E)   1.5
(o)   Rotor skew leakage reactance = 0.09
        Enter the value of coil-spanfactor (yw)
(E)   0.123
        Enter m = 0 for wide-spread connection, m = 1 for narrow-spread connection.
The remaining result is given below in tabular from to facilitate comparison.

	Run 1 (m = 0) (wide-spread)	Run 2 (m = 1) (narrow-spread)
Overhang leakage reactance =	2.04	2.69
Stator leakage reactance =	4.27	3.69
Rotor leakage reactance in stator terms =	4.41	3.86

(o)   Enter Stator resistance per phase (r1), Rotor resistance per phase in stator
        terms (r2) and Magnetising reactance/phase (Xm)
(E)   1.5   0.4   96.6
(o)   Enter the value of Supply voltage/phase and Syn. r.p.s.
(E)   230   25
(o)   Enter the value of No-load rotational loss
(E)   85

The RESULT is given in a tabular form below;

Slip	Current		power-fac.		Real power		reactive power		El.torque		Out.torque	
	W	N	W	N	W	N	W	N	W	N	W	N
.02	3.4	3.37	0.70	0.7	546	542	553	545	10.1	10	8.4	8.4
.04	5.4	5.4	0.85	0.84	1054	1044	651	662	19.3	19.1	17.6	17.4
.06	7.5	7.5	0.88	0.87	1519	1498	810	833	27.4	27	25.7	25.3
.08	9.5	9.4	0.89	0.87	1935	1897	1016	1054	34.4	34	32.6	32
.10	11.4	11.3	0.88	0.86	2297	2238	1256	1307	40.2	39	38.4	37.3
.30	23	22	0.70	0.67	3721	3405	3744	3746	56	51	54	49
.50	27.3	25.7	0.57	0.54	3609	3197	5135	4983	47.6	42	44.3	39
.70	29.2	27.4	0.49	0.46	3326	2900	5844	5583	39	34	33.6	28.5
.90	30.3	28.2	0.44	0.42	3081	2662	6245	5913	32.6	28.1	16.3	12

W : WIDE-SPREAD   N : NARROW-SPREAD

Salient-pole Machines

Four illustrations relating direct current and synchronous machines have been discussed in this article.

Resistance of wound armature

[as in the armature of a direct-current machine, synchronous machine with rotating poles and armature on the stator, small synchronous machines with armature on the rotor, stator winding of wound-rotor and squirrel-cage induction machines, and rotor winding of a wound-rotor induction motor.]

Required data: number of poles (p), armature voltage for d.c./voltage per phase for a.c. machines (et), armature diameter (D) and core length (L), number of armature slots (Rs on rotor and Ss when on stator), number of turns / turns per phase for a.c. (t), slot depth (dSs-on stator, dRs-on rotor), armature conductor-bare section (cbs), number of strands per armature coil (str), % coil pitch (pcp), and width of insulated coil (ciw).

Equations: The winding resistance is given by the classical formula:

$$R = \rho \cdot Lmt \cdot t / (cbs \cdot str) \qquad (9.2)$$

where,
ρ is the specific resistance of the conductor material at the operating temperature;
Lmt = mean leangth of one turn

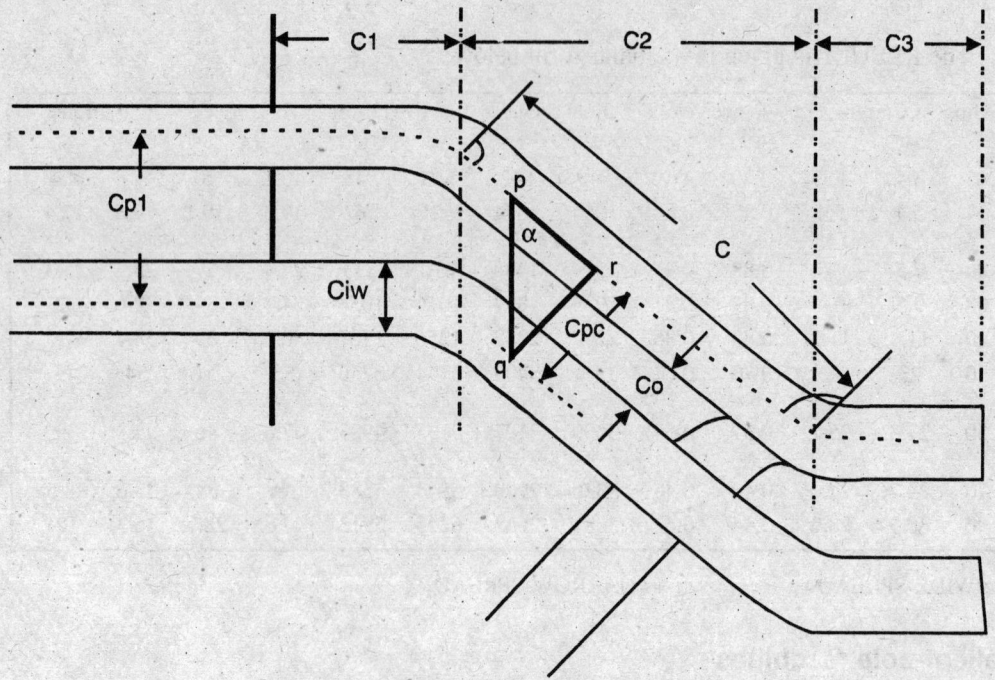

Figure 7.11 Configuration of overhang of a wound coil.

$$= 2 L + 4 C + 4 (C1 + C3) \qquad ... \text{ see Figure 7.11.}$$

C is determined as below:

Slot-pitch at the bottom of slot, Cpl

$$= \pi (D + 2 dSs) / Ss \quad ... \text{ when the winding is on stator;}$$
$$= \pi (Dr - 2 dRs) / Rs \quad ... \text{ when the winding is on rotor.}$$

Cpc = coil-pitch in the overhang area (Figure 7.11)
   = width of insulated coil + K,
   where, K is given by the empirical equation, K = 0.002 et + 4.3 mm.
   Defining sin α = Cpc/Cpl, from the triangle pqr in Figure 7.11,

C = ½ π Cpl. [(Ss or Rs as the case may be)/no. of turns per pole]. (pcp/100). Sec α.
(C1 + C3) in Figure 13 is given by an empirical equation,

$$C1 + C3 = 0.02293 \text{ et} + 55 \text{ mm.}$$

[*Note:* R as obtained above is the d.c. resistance. For large alternating current machines, eddy-currents in the conductors introduce a loss-factor by which the resistance as computed above should be multiplied to give a.c. resistance. The loss factor can be reduced by transposition as discussed in art. 7.3.1 (c)].

7.3   THREE PHASE SYNCHRONOUS MACHINE

7.3.1   Important Design Considerations are as below

(a) *Radial gap length.*

As discussed earlier, the length of radial gap is an important design parameter in a rotating machine as its contribution to the no-load mmf is large. For a synchronous machine, the gap length bears an additional importance as it affects stability. Modern turboalternators have long radial gap length for reason of operational stability.

The gap length can be computed using equations (4.18) and (4.19) :-

Gap mmf F_g = 800,000 $B_{av} k_g k_{gd} l_g$ amp-turn/pole, for non- salient pole machine;
$$= 800,000 \ (B_{av}/f_d) \ k_g \ k_{gd} \ l_g \text{ amp-turn/pole for salient-pole machine.}$$

From Table 4.1,
   B_{av} = 0.55 – 0.65 Tesla, for non-salient role machine ;
   B_{av} = 0.50 – 0.65 Tesla, for salient-pole machine.

The field distribution factor for a salient-pole machine can be assumed equal to the ratio of pole-arc and pole pitch. A value of 0.67 can be assumed to start with.

The gap mmf F_g is related to the armature-reaction mmf F_{ar} by the following equation,*

$$F_g = \text{Far}/x_{md} \tag{7.13}$$

---

*   Walker, J. H. : Large A. C. Machines, Bharat Heavy Electricals Ltd. pp. 79.

where,
$$F_{ar} = \frac{2.12\, I_1 N_s k_a}{p k_{0_1}\, k_{o_2}} \quad \text{amp-turn/pole†} \tag{7.14}$$

and x_{md} is the direct-axis magnetising reactance; I_1 and N_s are respectively the armature current/phase and turns/phase; k_a, an amplitude factor 1.05; K_{01} and K_{02} are factors dependent on the number of slots in positive and negative phase-bamls of the stator winding and each can be taken as 1.05 for preliminary design.

The ratio, no-load ampere-turn per pole F_g/armature-reaction ampere-turn per pole, F_{ar} are normally as below :-

Turbo-alternator, with AER   ...   ...   0.4 to 1.0
(smaller value for fast-acting type);
Water-wheel generator   ...   ...   1.0 to 1.5
Salient-pole generator, medium speed   ...   0.75 to 1.1
Synchronous motor   ...   ...   0.8 to 2.0

Example 7.2

Compute the minimum radial gap length at the pole-centre for a 175 MVA, 3 phase, 11 kV, 20 pole, 50 Hz., 0.9 lag water-wheel generator.

Solution   Designer's choice : $B_{av} = 0.05$ T.
$$\overline{ac} = 65,000 \text{ amp.cond m}^{-1}.$$
$$k_d = 0.955$$
$$k_p = 0.833$$

Using equn. 3.23 (a), noting that, $n_s = 2 \times 50/20 = 5\text{r.p.s.},$
$$175000 = 1.11\, \pi^2 \times 0.955\, D^2 L_a \times 5.0 \times 0.65 \times 65000 \times 10^{-3}$$
yielding $D^2 L_a = 79.18$ m^3.

For a maximum peripheral speed of 100 m. sec^{-1}.
$$D = \frac{100}{\pi \times 5.0} = 6.36 \text{ m.}$$
or $L_a = 1.95$ m.

Pole-pitch $= \pi\, 6.36/20 = 1.0$m.

Assuming pole arc $= 0.067\tau = 0.67$ m.,

Flux per pole $= B_{av} \cdot \pi\, D L_a/p$
$$= 0.65\ 7\pi \times 6.36 \times 1.95/20 = 1.95/20$$
$$= 1.27\text{Wb}.$$

Voltage per phase $= \dfrac{11000}{\sqrt{3}} = 6351$
$$= 4.44\, \phi m f_s\, N_s\, k_d\, k_p$$

---

† Alger, P : "The calculation of armature reactance of Synchronous Machines" Transactions American I EE, 1928, pp. 47.

or,
$$N_s = \frac{6351}{4.44 \times 1.27 \times 50 \times 0.955 \times 0.833}$$

$$= 28.3$$

We choose, $N_s = 28$

Modified flux per pole = $1.27 \times 28.3/28 = 1.284$ Wb.

Assuming armature winding consist of 4parallel circuits and single-turnbar winding, number of stator slots/ph. = $28 \times 4 = 112$;

Total number of stator slots = $112 \times 3 = 336$

$$\text{Slot-pitch} = \frac{\pi \times 0.36}{336} = 0.0595 \text{ in.}$$

Flux per slot-pitch, for a pole-arc = $0.67 \times$ pole-pitch,

$$= \frac{1.27 \times 20}{336 \times 0.67} = 0.113 \text{ Wb}$$

For a maximum tooth-density of 1.8 T, and assuming all flux passing through teeth,

$$\text{tooth-width (average)} = \frac{0.113}{1.8 \times (1.95 - 49 \times 0.01)} \text{ with 40 radial}$$

ducts each 0.01 m. wide

$$= 0.04 \text{ m.}$$

That is, slot-width (average) = $0.0595 - 0.04 = 0.02$ m.

With open slot, slot opening $w_0 = 0.02$ m.

$$\text{Stator current per phase} = \frac{175000 \times 10^3}{\sqrt{3} \times 11000} = 9185 \text{ amps.}$$

Armature reaction m.m.f (from equation 7.14),

$$\text{Far} = \frac{2.12 \times 9185 \times 28 \times 1.05}{20 \times 6.36 \times 1.95} = 25963 \text{ amp-turn/pole.}$$

Mean gap density $B_{av} = \dfrac{1.284 \times 20}{\pi \times 6.36 \times 1.95} = 0.659$ T

or, Peak gap m.m.f. on no-load (from equn. 4.19),

$$F_g = 800,000 \times \frac{0.659}{0.67} \times 1.0 \, l_g \text{ taking } k_g = 1.0$$

and $k_{ga} = 1.0$ to start with.

$$= 786865.7 \, l_g$$

With AER, no-load amp-turn per pole = $1.1 \times$ armature amp-turn per pole

$$= 25963 \times 1.1 = 28559$$

yielding, $l_g = 0.036$ m.

Now, $(l_g/w_0) = 1.8$ giving k_c (from Fig. 4.7) = 0.26
and for $(l_g/w_0)) = 3.6$, $k_{ca} = 0.42$.

That is,
$$k_g = \frac{0.0595}{0.0595 - 0.26 \times 0.02} = 1.095$$

and
$$k_{gd} = \frac{1.95}{1.95 - 40 \times 0.01 \times 0.42} = 1.094$$

Thus, modified value of the minimum radial gap length,

$$= \frac{0.036}{1.095 \times 1.094} = 0.03 \text{ m.}$$

(b) *Stator slot*

   (i) The primary requirement on the selection of number of Slots in stator is that a balanced three-phase winding is obtained.

   (ii) The choice of stator slots per pole and phase can be based on the following :-

   Salient-pole and small
   non-salient pole machine   ... 2. to 6;
   Turbo-alternator, 2 pole   ... 8 or 9.

   (iii) In practically all large water-wheel generators, the number of stator slots psr pole and phase is fractional.

   (iv) Stator slot-pitch on large water-wheel generator normally lies between 0.5 and 0.9 m. and that on turbo-alternator, 0.75 and 0.9 m. Larger the slot-pitch, the fewer are the number of slots, and thus wider and deeper the slots and coilsides to accomodate the required ampere-turns.

(c) *Stator coil*

   (i) The aim of arriving at the best arrangement of conductors is to obtain maximum pitch factor.

   (ii) The normal practice in the design of stator coils is to have the total eddy-current loss less than about 20% of the total conductor (d.c.) I^2R-loss.

   (iii) In deep coilsides, transposition of individual strands in each effective turn is essential to avoid excessive eddy-current effect. In multi-turn coils wth turn per coil larger than 3, transposition in 180°—turn effectively restricts eddy-current loss. With 2 or 3 turns, a semi-Roebel transposition is often satisfactory, whereas with single turn bar winding, full Roebel transposition is essential (Figure 7.12).

   (iv) In Roebel transposition, each strand is arranged to move continuously through all positions in the depth of the coilside.

The transposition can be best understood by tracing the path of a strand (say 1 in Figure 7.8) at one and of the straight position of the bar and in one of the two stacks. From its position in one stack, it is bent at a small angle and made to proceed downwards to the bottom of the adjoining stack. In this way, it traverses all positions in the cailside thereby eliminating eddy-current losses.

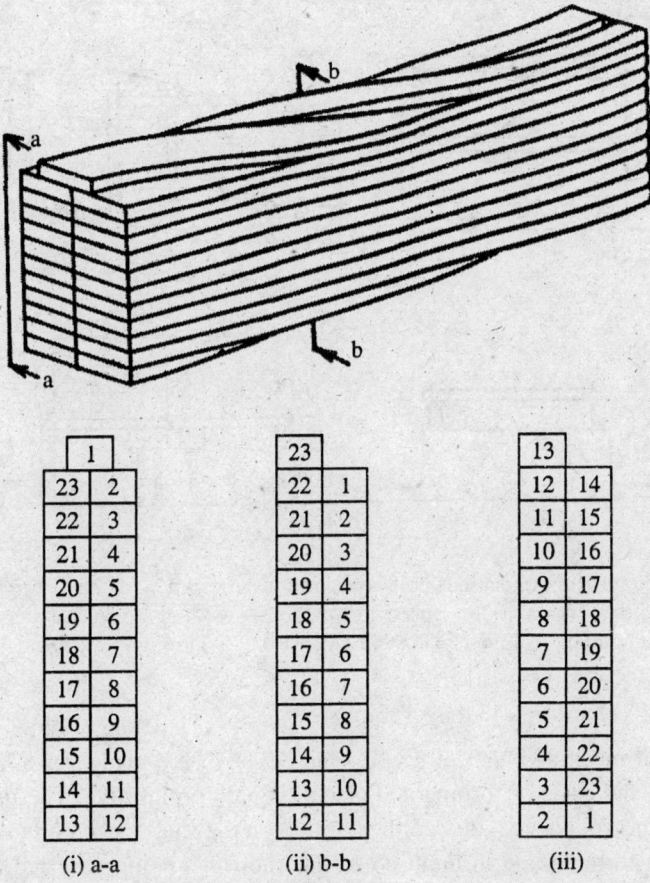

	1
23	2
22	3
21	4
20	5
19	6
18	7
17	8
16	9
15	10
14	11
13	12

(i) a-a

	23
22	1
21	2
20	3
19	4
18	5
17	6
16	7
15	8
14	9
13	10
12	11

(ii) b-b

	13
12	14
11	15
10	16
9	17
8	18
7	19
6	20
5	21
4	22
3	23
2	1

(iii)

Figure 7.12   Full Roebel transposition.

(d) *Rotor construction. Salient pole* :   The rotor spider construction depends on the peripheral velocity. Two broad divisions can be made ns below :-

(i) Peripheral velocity less than 130 m. sec^{-1}.

For small and medium size, punchings are shrunk on the shaft. (Figure 7.13(a)). For large size, punched rotor body is fixed on fabricated steel spider the arms of which are attached to a hub fitted on the shaft (Figure 7.13 (b)).

Note : (1) For peripheral speed less than 40 m. sec^{-1}., the spider consists of circular fabricated steel ring to which pole core and shoe are bolted. The shoes may be laminated below 30 m. sec^{-1}. Above this speed, dovetail or T-head construction is used.

(2) For large internal combustion engine-driven generators, stub-shafts are bolted direct to the engine flywheel in order to avoid fatigue failure due to forced torsional oscillations because of the nature of torque characteristic of the drive.*

---

*   Serl, S. K. : Electrical Machinery, 3rd Edition, p. 541.

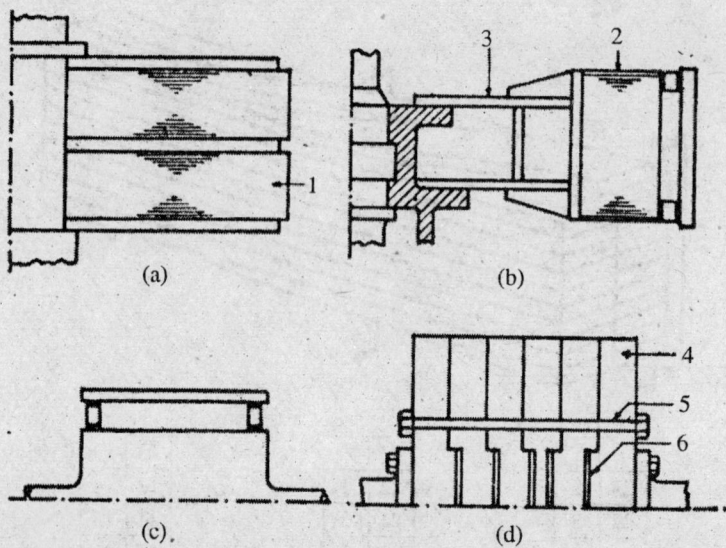

Figure 7.13 Rotor construction-salient-pole synchronous machine. (a) small and medium size-low speed, (b) large size-low speed, (c) medium size-high speed, (d) large size-high speed. 1,2-Punchings, 3-Spider, 4-Rolled steel disc, 5-Through and through bolt, 6-Spigot.

(ii) Peripheral velocity more than $130 m. sec^{-1}$.

For low output machines, the rotor body and the shaft are made out of single forging.

For large size, the costs of forging and machining are prohibitive. The forging is replaced by thick (12–18 cm.) rolled steel discs either shrunk on the shaft or spigotted to each other and held together with through and through holts, the end-rings being attached to the stub shafts (Figure 7.13c).

[*Note* : Water-wheel generators generally have vertical shaft and thus require thrust bearings along with guide bearings. The thrust bearings are placed above the rotors in top-bracket type, and below the rotor in the bottom-bracket or umbrella type. The former is used for high speed machines and the latter, for low speed generators.]

Turbo-generator :   Turbo-alternators are generally of two-pole type since at high speed, the efficiency of steam turbine is greatly increased, and the overall size of alternator and turbine is reduced. The peripheral speed is generally between 150 and 200 $m.sec^{-1}$. and large centrifugal force develops high mechanical stresses in some parts of the rotor. Rotor is thus made of massive steel forging with slots milled. In larger machines, special chrome-nickel-molybdemim steel is used because of its high tensile strength. Also, high speed necessitates the use of metal wedges, and non-magnetic steel wedges are normally used.

[*Note* : High peripheral speed results in length of a turbo-generator much larger than its diameter. On the other hand, in low speed water-wheel generators, axial length may be 1/5th to 1 /7th of its diameter.]

7.3.2   Computerised Design

The procedure for the design of synchronous machine using digital computer can be based on similar lines as the induction motor (Figure 7.6). Figure 7.14 shows a general flow-chart which consists of six subroutines, such as, flux wave analysis, armature winding, magnetic circuit, reactances, field winding, and performance.

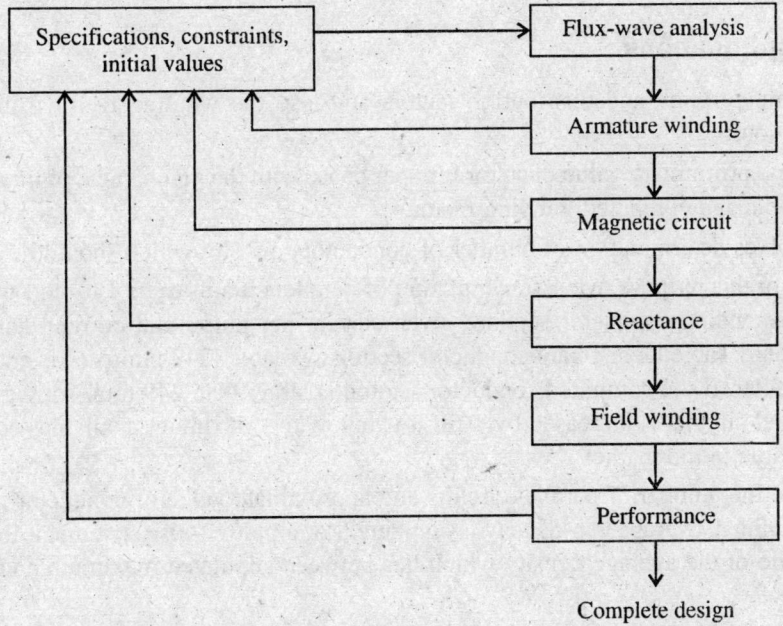

Figure. 7.14   General flow-chart for synchronous machine design.

The customer's specifications may consist of kVa, line voltage, number of phase, frequency, speed, field voltage, etc. The other constraints imposed on the designer may be efficiency, temperature-rise, per cent regulation, synchronising power, and standard stator and rotor stampings and conductor sections.

The design procedure starts with the choice of maximum and minimum gap density, specific electric loading, etc., and in each subroutine (such as, armature winding, magnetic circuit) there is a check so that suitable basic assumptions are modified to obtain the desired results.

7.3.3   Design of stator coil of a Water-wheel Generator

The problem becomes interesting when there is a constraint such as the average gap density must lie within a band given by a maximum and a minimum value. A computer program is given in Figure 7.14.

The nature and characteristics of the prime-mover is a strong determinant in the design of synchronous machines. In the case of water-wheel generator, inertia of the rotating mass is an important parameter so that the necessary 'flywheel effect' is ensured for satisfactory speed regulation of the hydro-turbine. This necessitates that a water-wheel generator should have large number of poles to match with the low speed condition, thus larger number of poles for a given bus frequency, and as such should be of salient-pole type and also large rotor diameter. Economic considerations demand that the poles should be on the rotor. The criterion of overspeeding when the full-load is thrown off with governor inoperative is very important, but this is outside the perview of this book.

Governing Equations

Step 1: Compute pitch and distribution factors and also the winding factor using standard equations.

Step 2: Assess priliminary value of stator turn per phase with the given value of maximum flux per pole and computed winding factor.

Step 3: Involves determination of number of conductors per slot which should be an integer.

For this purpose, we assume number of parallel circuits to be 1 to start with and the corresponding conductor section, given current per phase and current-density. From Standard Table largest bare conductor section available (249 mm.-sq.) is entered in the computer. If the computed conductor section is larger than 249 mm.-sq. the number of parallel circuits is increased by 1 till it is below the maximum available section (or as available in the market).

Step 4: Once the number of parallel circuits and approximate value of conductor per slot are determined, average gap-density is computed. A simple 'if-else statement' then ensures a value of the average density which lies between stipulated maximum and minimum value.

Step 5: On the basis of values of number of parallel circuits, number of conductors per slot and average gapdensity, coil throw, pitch factor and flux per pole are modified and preliminary value of slot width is determined.

Step 6: Choosing the shape of slots (in the given program, parallel-sided slots are assumed), the slot dimensions are designed and finally checked with the tooth-density.

```
// Design of Stator Coil of a Water-wheel generator.
// Constraint: Average gap density has a maximum and a minimum value.

# include <iostream.h>
# include <iomanip.h>
# include <math.h>
# include <conio.h>

class spec
  {
  public:
    int ph, p, fs, Ss ;
    float et, ap, D, L, phim ;
    void getdata( float et, int ph, int p, int fs, float ap, float D,
          float L, int Ss, float phim ) ;
```

```
        void display() ;
        void coil( float et, int ph, int p, int fs, float ap, float D,
                float L, int Ss, float phim)  ;
} ;

void spec :: getdata (float voltage, int phase, int pole, int freq,
                float current, float dia, float length, int slot,
                float maxflux )
{     et = voltage; ph= phase; p = pole; fs = freq; ap = current; D = dia;
      L = length; Ss = slot; phim = maxflux;
}

void  spec :: display()
{     cout <<"voltage/ph= "<< et <<", phase= "<< ph <<", pole= "
        << p <<", freq= "<< fs <<" current/ph= "<< ap
        <<" Arm.diameter= "<< D <<" Arm. length= "<< L
        <<" Stator slot= "<< Ss <<"  Maxm.flux /pole=  "<< phim ;
}

void spec :: coil (float et, int ph, int p, int fs, float ap, float D,
                float L, int Ss, float phim)
{
    int nd ;
    float cd,Bmax,Bmin,fdf,fst,win,din,wd ;

    cout <<"Enter the assumed values of current-density (cd),maximum"
        " flux-density(Bmax), minimum flux-density(Bmin),field distr.factor"

        "(fdf), stackingfactor(fst), width of slot insulation (win),"
        "depth of slot insulation(din), no.ofvent.duct(nd), width of vent"
        " duct (wd)" ;

    cin>>cd>>Bmax>>Bmin>>fdf>>fst>>win>>din>>nd>>wd ;

    cout<<"Current density= "<<cd<<" ,   Maxm.flux-density= "<<Bmax
        <<",  Minm.flux density=  "<<Bmin<<" ,  Field distribution factor=  "
        <<fdf<<" ,  Stacking factor= "<<fst<<" ,  Width of slot insulation= "
        <<win<<" , Depth of slot insulation= "<<din
        <<" ,  No.of vent duct= "<<nd<<" ,  Vent duct width= "<<wd;

        const float PI = 3.14159 ;

// Compute Pitch factor and Distribution factor
        int S, Sp, A, Z, Z2 ;
        float Cth, akp, gama, X, Y, akd, cwf ;

    Sp = Ss / p ;              //No. of stator slot per pole
     S = Ss / (p * ph) ;       //No. of stator slot per pole per ph.
    Cth = Sp;                  //Coil throw
    akp = sin( PI * Cth/ (Sp * 2.0)) ; //Pitch factor
    gama = PI / Sp ;
     X = sin( S * gama/ 2.0) ;
     Y = sin( gama/2.0) ;
    akd = X / ( S * Y ) ;       // Distribution factor
    cwf = akd * akp * fdf ;     // Winding factor
```

```cpp
// Compute Stator turns per phase
   float sturn, Ac, Acmax, Z1, Bav ;

   sturn = et /( 4.44 * cwf * fs * phim ); //Prelim.value of Stator turns/ph
     A = 1.0;                    // Assumed no. of parallel circuit
     Ac  = ap / ( A * cd );        // Bare Conductor section, mm.sq.
   cout << endl <<"conductor section=  "<< Ac ;

// On the basis of conductor current for 1 parallel circuit,largest conductor
// section is chosen from Standard Table; rectangular 25 mm x 10 mm, rounded
//              of bare section 249 mm.sq

   cout << endl <<"Enter the largest bare cond.section (Acmax)" ;
   getch() ;

   cin >> Acmax ;
   if (Ac>Acmax)
   {
     A = A + 1.;
     Ac = ap / ( A *  cd );
   }
   else
   {
     Z = A * ph * sturn * 2.0 / Ss;          // Conductor / slot
   }
//       Z should be an integer, and equal to or greater than 2.0
   if (Z<2.0)
   {
     Z = 2.0;
   }
   Bav = (et * p * ph) / (2.22 * PI * D * L * fs * cwf * Ss * Z);

//     Check whether  Bav is within the stipulated values of Bmax & Bmin
   if (Bav > Bmax)
   {
     Z = Z + 2.0;
     Bav = ( et * p * ph ) / (2.22 * PI * D * L * fs * cwf * Ss * Z );
   }
   else
   {
   if (Bav < Bmin)
   {
     Z = Z - 1.0;
     Bav = ( et * p * ph ) / (2.22 * PI * D * L * fs * cwf * Ss * Z );
   }
   }
   cout << endl <<"Av.gap density=  "<< Bav ;
   getch() ;
   int cthc, Zph;
   float phim1, Ratio, XX, cthcp,akpM, phimM, Ws;

   phim1 = Bav * PI * D * L / p;          //Modified flux per pole
   Ratio = phim1 / phim;
     XX = asin ( Ratio );
```

```
//   cthcp = Sp * 2.0 * XX / PI;          //Modified Coil-throw,should be an
//                                        integer
     cthc = cthcp;

     akpM = sin ( PI * cthc / (Sp * 2.0));    // Modified Pitch factor

     cout << endl << setw(60) <<"Distribution factor=  "<< akd ;
     cout << endl << setw(50) <<"Winding factor=  "<< cwf ;
     cout << endl << setw(40) <<"Coil throw=  "<< cthc ;
     cout << endl << setw(30) <<"Pitch factor=  "<< akpM ;
     getch() ;
     phimM = phim1 * akp / akpM;            //Modified flux per pole
       Bav = phimM * p /( PI * D * L );
       Zph = Z * A * Ss / ph;
     sturn = Zph / 2.0;
       Ws = 0.5 * PI * D / Ss;              // Slot width

//               Choosing Parallel-sided slots
     float cbw,cbh, AcM, cdM, WsM, dsM;

     cbw = Ws - win ;                       //Conductor width
     cbh = Ac / ( A * cbw * 1000. * 1000.);      //Conductor height
     cout<<endl<<"Ac= "<<Ac ;

     cout << endl <<"Enter the value of modified Ac(AcM)";
       cin >> AcM ;
     cout << endl <<"Chosen bare conduc.section=   "<< AcM <<" mm.sq" ;

       cdM = ap / ( A *  AcM );
       WsM = cbw + win;
       dsM = A * Z * cbh + din;
//             Check Tooth-density
     float wt3, Bt3;
       wt3 = ( PI * ( D + 0.67 * dsM)/Ss ) - WsM;
       Bt3 = phim * p / (wt3 * ( L - nd * wd ) * fst * Ss );

     cout << endl << setw(70) <<"Gap density =  "<< Bav ;
     cout << endl << setw(65) <<"Maxm. flux per pole= "<< phim ;
     cout << endl << setw(55) <<"Turns per phase= "<< sturn ;
     cout << endl << setw(50) <<"No. of conductor/slot=  "<< Z ;
     cout << endl << setw(40) <<"Number of slot= "<< Ss ;
     cout << endl << setw(35) <<"Current density= "<< cdM ;
     cout << endl << setw(30) <<"Slot depth=  "<< dsM ;
     cout << endl << setw(20) <<"Slot width=  "<< WsM ;
     cout << endl <<"Tooth density at 1/3rd from root= "<< Bt3 ;
}

void main()
{
 clrscr() ;
 spec m ;
 int phase, pole, freq,  slot ;
 float voltage, current, dia, length , maxflux ;
```

```
cout << endl <<"Enter the data: et, ph, p, fs, ap, D, L, Ss, phim" ;
cin>>voltage>>phase>>pole>>freq>>current>>dia>>length>>slot>>maxflux ;

m.getdata(voltage,phase,pole,freq,current,dia,length,slot,maxflux) ;
m.display() ;
m.coil(voltage,phase,pole,freq,current,dia,length,slot,maxflux) ;
};
```

Figure 7.15 A computer program for the design of stator coil of a Water-wheel Generator.

RESULT

(E)   3820   3   32   50   262   3.2
      0.44   288   0.083
(E)   4   0.67   0.57   0.67   0.95   0.007
      0.025   5   0.01
(o)   conductor section = 65.5
      Enter the largest bare cond. section (Acmax)
(E)   249
(o)   Av. gap density = 0.496

Distribution factor = 0.96

Winding factor = 0.643

Coil throw = 5

Pitch factor = 0.766

Enter the value of modified Ac(AcM).

(E)   65.5
(o)

Gap density = 0.658

Maxm. flux per pole = 0.083

Turns per phase = 384

No. of conductors/slot = 8

Number of slot = 288

Current density = 3.97

Slot depth = 0.075

Slot widh = 0.0175

Tooth density at 1/3rd from root = 1.383

7.4 DIRECT CURRENT MACHINE

7.4.1 Important Design Considerations are as Below

(a) *Number of poles :*

The choice of number of poles has important bearing on the design of a direct current machine.

(i) As in the cases of induction and synchronous machines, the number of poles of a d.c. machine does not bear a relation with speed and frequency. On the other hand, the frequency of flux reversal in the armature is given by the relation,

$$f = \rho . n_r$$

where, n_r is the armature speed in r.p.s. That is, for any constant speed fy, $n_r f$ is directly proportional to the number of poles. Thus, with large number of poles, a large frequency of flux reversal may result in large iron losses in armature. For the armature teeth, such losses may be excessive, since teeth are generally worked at a density near saturation. Generally, $f = 25$ Hz. is satisfactory, though for smaller machines, f can be as high as 50 Hz.

[*Note* : The choice of larger number of poles has the following advantages

(1) As indicated in equation (3.16), the total flux ($\phi . p$) in a machine for a given armature is approximately related to the rated voltage, of the machine and can be assumed constant, since armature resistance-drop on full-load is very small in comparison with the induced armature voltage. Thus increase in the number of poles reduces the flux per pole ϕ in direct proportion. This has an important implication in the fact that larger number of poles and thus smaller flux per pole in a machine has smaller flux through the yoke-section. Since the yoke-section is designed on the basis of chosen density in the yoke, the yoke-section required with larger number of poles is small.

(2) The eddy-current loss in armature is proportional to $B_a^2 . f^2$ i.e. $B_a^2 . p^2$, where, B_a is the armature flux-density at a particular speed. That is, eddy-loss is proportional to the total flux ($\phi . p$) and is unaltered with the change in the number of poles.

  The hysteresis loss, on the other hand, is proportional to $B_a^{1.6} . f$ i.e. $B_a^{1.6} . p$ and is thus reduced with increase in the number of poles.

(3) The height of field poles should be such that the field winding ampere-turns could be suitably accomodated. With the increase in number of poles, field ampere-turns per pole is reduced and generally it amounts to a reduction in the pole height. Thus, weight of field iron required is less.

(4) Smaller flux through the poles (as discussed in (1) above) leads to smaller pole-width, since pole-section is designed with a chosen pole-density and the length of a pole is approximately equal-to the armature core-length. The mean length of field turn is less and there is thus a saving in field copper.

(5) Larger number of poles leads to smaller overhang in armature coils leading to saving in armature copper].

(ii) Current per brush-arm : High value of current per brush-arm results in excessive sparking. Under normal condition, the current per brush-arm should not exceed 400 amperes with graphite brushes.

(iii) Pole-pitch $\tau = \pi D/p$ determines the diameter D and armature length L_a for a computed $D^2 L_a$ (equations 3.19 a & b) from customer's specifications. For a particular $D_2 L_a$, larger value of τ leads to larger diameter and smaller length. The peripheral speed is large which may be detrimental from mechanical considerations.

Again, a small (L_a/τ) leads to large armature ampere-turn per pole which in turn, may lead to inferior commutation and heavy sparking. On the other hand, a large (L_a/τ) means smaller diameter and larger length which is generally not economical.

The following data may be used in the design process :-

No. of pole	2	4	6	8 & above
Range of	upto 0.2	0.2–0.35	0.35–0.45	0.45–0.55
(in metre)				

(iv) Armature ampere-turn : A large value leads to increased distortion of the no-load field-form and thus inferior commutation. For normal design purpose, the following data may be used for machines upto 1500 kW output* :-

Output (kW)	upto 100	100–500	500–1500
Armature			
amp-turn.pole	upto 5000	5000-7500	7500-10000

(b) *Air Gap* : Choice of airgap length in a direct current machine is governed by :-

(i) The amount of distortion of no-load field-form by the armature reaction, the smaller the gap length, the larger is the distortion effect;

(ii) Cooling;

(iii) Bearings, and shaft deflection ;

(iv) Unbalanced magnetic pull.

The gap-length L_g (in mitre) may be chosen as below :-

$$\text{For 2 to 8 pole machine} : l_g = (0.04\text{-}0.047)\ (D/p)$$
$$\text{For 10 to 24 pole machine} : l_g = (0.048\text{-}0.053)\ (D/p) \qquad (7.16)$$

In equation 7.16, smaller value for the coefficient of (D/p) is to be chosen for smaller number of poles and smaller armature diameter :-

(c) *Armature Slot* : The choice of the number of slots, of their dimensions, insulation thickness, and of the dimensions of the conductors result from many considerations, both practical and theoretical. A few important ones are cited below :-

(i) For lap-wound armature, (1) the number of slots per pole-pair = an integer, so as to facilitate connection of equilising ring ; (2) the number of slots per pole = an integer, so that coils could be full-pitch.

(ii) To facilitate cooling, the slot-loading-(i.e. ampere-conductor per slot) should not be large. A value of 1500 may be taken as the higher limit.

(iii) Cost of punching slots in electrical stampings increases with the number of slots to be punched.

(iv) From commutation point of view, larger number of slots and smaller number of conductors per slot are better.

---

* Sawhney, A. K. : Electrical Machines Design (Book), pp. 575.

(v) The slot dimensions are subjected to the following conditions :-
   (1) The slot depth is related to the diameter and increases with it.
   (2) The minimum slot width is related to the voltage. This is because, the width of conducting part in the slot is equal to the slot-width less twice the thicknests of insulation, and the insulation thickness increases with the voltage. For a 10 kV machine, the minimum slot-width is about 20 mm. and the section of slot insulation and of conducting material in a slot are of the same order.
   (3) The tooth-width at the narrowest part of a tooth is determined by the allowable maximum tooth-density. With a chosen gap density B_{av} and a maximum tooth-density Btm, the ratio (tooth-width at the narrowest part of the tooth-slot-pitch) is somewhat determined by (B_{av}/B_{tm}).
   (4) The minimum armature tooth-section should be sufficient to resist the punching process and also the centrifugal stresses.

(d) *Current Density* :   From equation (3.28), it can be seen that in a design with lower specific electric loading, the current-density can be larger for the same I^2R- loss in the active parts of the armature conductors. The following values of current-density* could be a starting point for a design :-

For small wire-wound armature with very good normal ventilation	...	5.0 amp. mm^{-2};
For large strap-wound armature with very good normal ventilation	...	4.5 amp. mm^{-2};
For fairly high speed fan-ventilated machine	...	6–7 amp. mm^{-2}.

The above values should be suitably reduced for low peripheral speed machines.

(e) *Field System* :   For large machines, solid poles made of forged steel and bolted to the frame are normally used. The pole-shoes are invariably laminated, the laminations being insulated from each other by thin coating of varnish or paper. The laminations are rivetted, and the shoe is fixed to the pole core by countersunk screws.

For smaller ratings, laminated pole is often preferred in which pole and pole-shoe form part of the same lamination. Laminations are generally 1.0 to 1.5 mm. thick, insulated from each other by insulation obtained through manufacturing process. Laminations are held between stiffening -plates of sheet steel at both ends for small machines and thick malleable iron plates for large machines.

In the design of field system, the field ampere-turns should be stronger than the armature full-load ampere-turns so that the distortion due to armature-reaction is less effective. For normal design, the ratio (no-load field ampere-turns/armature full-load ampere-turns) is as below :-

Non-interpole machine	...	1.0 to 1.3
Interpole machine	...	0.6 to 1.0

Larger the above ratio, the lesser is the distorting effect produced by armature reaction. However, larger ratio leads to increased field copper and thus increased cost.

---

* Clayton, A. E. : The Performance and Design of Direct Current Machines, (Book), pp. 359.

For motors, with wide range of speed control, the ratio can be as high as 2.0 for non-interpole machine.

To obtain spatial flux-distribution of airgap flux near to sinusoid, the pole-shoes are properly shaped. Whereas, the gap at the pole-centre is obtained from equation (7.16), the gap at the pole-tip is approximately 1.15 times the minimum gap-length. The nature of the field distribution can be determined by the methods as discussed in art. 4.2.3.

(f) *Commutator :* The commutator material is hard drawn copper segments separated by thin sheets of mica or micanite, which should be of soft variety so that copper and mica would wear down at the same rate. Mica and copper-segments (Fig 7.16) are clamped between V-shaped clamping rings of cast iron or steel with micanite insulation between them.

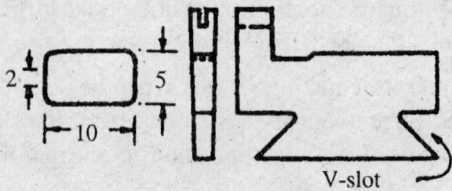

Figure 7.16

The segments are generally more than 3.5 mm. in thickness and the thickness of mica is between 0.6 mm. to 1.2 mm. depending on voltage between consecutive segments. The diameter of commutator is generally between 60 and 80 per cent of armature diameter with the peripheral speed generally about 15 meteres. $\sec^{-1}$.

(g) *Influence of thyristor supply :* With thyristor-controlled armature of a motor, the applied armature voltage can be considered to be made up of a d.c. component with superimposed ac. components. The magnitudes of a.c. components decrease with the increase in frequency of the components. Also, armature inductance reduces the effects of the higher frequency components considerably.

The important effects of the a.c. components are :-

(i) Increase in core losses. Thus, (1) motors for thyristor supplies are to be designed with completely laminated magnetic circuit to avoid the appearance of eddy-currents in the magnetic parts of the motor; (2) The specific magnetic loading to be chosen must be less than that given in Table 3.4.

(ii) The average torque increases by a small amount (1.0 to 7.4%)* which balances a part of the reduction in output caused by the increase of the rms. value of current.

(iii) Large a.c. components introduce torque pulsation, vibration and noise. At lower speeds, these may cause hunting in the drive system.

In such case, armature inductance is increased intentionally by having deeper slots so that currents due to a.c. components are reduced. Suitable compensating winding is used to neutralise the effects of increased armature inductance.

---

* Budig, P : Influence of Thyristor-supply to the design of d.c. and a.c. motors, International Conference on Electrical Machines–Design and Applications, IEE (London), 13-15 July, 1982, pp. 249.

(iv) Increase in I^2R-losses in armature and interpole windings due. to addi- tional skin effect. Smaller values of specific electric loading and current-densities of armature and interpole windings should be chosen. This results in larger arma- ture diameter to facilitate accomodation of larger slot-area and better cooling.

(v) Inferior commutation due to increase in reactance voltage.

Commutation can be improved upon by, (1) using larger number of commutator segments so that voltage between adjacent segments is low; (2) having reduced (pole arc/pole pitch) so that the effect of the main field poles on the coil undergoing commutation is less; (3) having larger air gap under the interpolts.

[*Note* : (1) The lamination of yoke and other solid parts not only improves the motor behaviour with thyristor supply but also its dynamic properties since the total effective time-constant is reduced.

(2) Improved performance can be obtained by using, (i) lesser number of poles so that the frequency of flux reversal in armature is less; (ii) properly shaped pole-shoes so that the spatial distribution of flux in the gap is as near to sinuosoid as possible; (iii) high class of insulation for armature, interpoles and compensating winding.

(3) Cooling by fan mounted on the shaft is not quite effective at lower speeds. Improved cooling technique such as, forced cooling, specially for large motors, should be adopted.]

(h) *For the design of large d.c. motors*

(i) Maximum continuous power output/armature diameter at any frequency is limited by commutator flash-over (i.e. average' voltage between consecutive segments) and temperature rise, the ratio, decreasing with the increase in frequency of flux reversal.

(ii) For a motor whose speed is controlled by variation in field excitation, the product (peak power at top speed x the ratio of top speed/base speed) for a particular armature diameter is limited by reactance voltage and commutation. However, the range increases with the increase in frequency.

(iii) The maximum continuous torque at rated speed per unit armature volume is limited by temperature-rise. The ratio decreases with the increase in frequency.

[*Note* : For (i) and (iii) above, better ratio is obtained with class F insulation instead of class A or B.]

I. Thus, in the design of d.c.motor, there is a minimum armature diameter (due to (i) above) and a corresponding minimum core length (due to (iii) above). The diameter may be increased to improve commutation or to reduce overall length, but top speed is governed by the centrifugal forces developed.

II. Restrictions imposed on the outside diameter can influence choice of armature diameter and the space available for main field. The modern trend in design is to have reduced radial pole height and hence, a larger armature diametercan be obtained for a given frame.

III. It is usual to have completely laminated magnetic circuit so that the motor can respond quickly to speed control, and thus commutate satisfactorily under high speed control and thyristor supply conditions.

IV. Glass F insulation is now commonly used for large motors

7.4.2 Computerised Design

Figure 7.17 shows a general flow-chart for the design calculation of a direct current machine which consists of 10 subroutines. As discussed earlier, the design process starts with the study of customer's specifications and corresponding manufacturer's constraints. For a standard design, the specifications could be : Type (motor or generator), full-load voltage, full-load armature current, base speed, connection of exciting winding (i.e. separately-excited, shunt, series, compound), shunt field exciting voltage, etc. The manufacturer's constraints may consist of available stampings and their characteristics, standard tables for conductors, available insulating materials, etc. Based on above and experience, the designer selects initial values, such as, specific loadings, efficiency, allowable field heating, desired per cent interpole, desired per cent series field ampere-turns, etc.

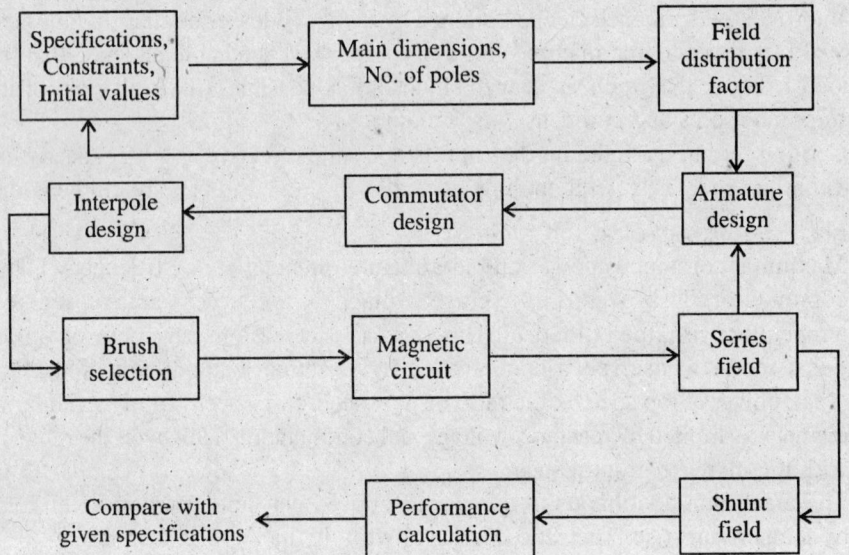

Figure 7.17 General flow chart for the design of d.c. machine.

M.M.F. of D.C./Salient-pole synchronous machine

Governing Equations

As discussed in chapter 4 the total magneto-motive or magnetising force is the sum of component forces due to (i) yoke, (ii) pole, (iii) gap, (iv) armature teeth, and (v) armature core.

```
# include <iostream.h>
# include <iomanip.h>
# include <math.h>
# include <conio.h>
# include "mcur42st.h"
# include "carter_o.h"
# include "carter_s.h"
```

```
//       mmf of a Salient pole machine - d.c. & synchronous
class mmf
{
    public:
      int a, b,  c;                // given data
      float d, e, f, g, h, k, l, m, n, o, q, r, t, u, v, w ;        // given data

      void getdata(int a,int b,int c,float d, float e, float f, float g,
              float h, float k, float l, float m, float n, float o,
              float q, float r, float t, float u, float v, float w ) ;
      void display() ;
      void yoke( int a, float d ,float u, float v ) ;
      void pole(float d, float r, float w ) ;
      void gap (int a,int b,int c,float d,float e,float f,float g,float h,
           float k) ;
      void armteeth(int a, int b,   int c, float d, float e, float f,
                  float h,float k, float o, float q, float t) ;
       void armcore(int a,float d, float l, float m) ;
} ;

void  mmf :: getdata(int p, int S, int nd, float phi, float D, float L,
                float lg, float wo, float wd,  float Ac,float Dc,
                float fst,float wt, float ds,  float Ap,float tou,
                float Ay, float Dy, float hp)
{
    a = p ; b = S ; c = nd ; d = phi ; e = D ; f = L ; g = lg ; h = wo ;
    k = wd; l = Ac; m = Dc ; n = fst ; o = wt; q = ds; r = Ap ; t = tou;
    u = Ay; v = Dy; w = hp ;
}


void mmf :: display()
{   cout << endl <<"No. of poles=  "<< a <<"No. of Armature slots=   "<< b
         <<" radial vent duct= "<< c <<"  Gap flux/pole=  " << d
         <<"  Armature diameter= "<< e <<"  Armature length= "<< f
         <<"  gap length= "<< g  <<"  Slot opening= "<< h
         <<"  Width of vent duct=  "<< k <<"  Armature effec.core area="
         " "<< l <<"  Mean dia.of armature core=  "<< m <<"  Stacking"
         " factor=  "<< n <<"  Tooth width at one-third from the bottom="
         "  "<< o <<"  Slot height= "<< q <<"  Effective Pole area=  "
         << r <<  "Ratio pole-arc to pole-pitch=  "<< t
         <<"  Effec.Yoke area=  "<< u <<"  Mean dia.of yoke= "<< v
         <<"  Effective pole height =  "<< w ;

}


void mmf :: yoke( int a, float d, float u, float v )


{
  float yleak, By, ly, y, x, Fy ;

  const float PI = 3.1416 ;
     float kfl = 1.05 ;  // Factor to allow for curved nature of
//                  flux-path in yoke at the back of the pole

  cout << endl <<"Enter the value of yoke leakage factor, yleak" ;
   cin >> yleak ;          // 1.05
```

```
        ly = 0.5 * kfl * yleak * PI * v / a ;
        By = 0.5 * d / u ;
        y = By ;
         float mcur42st(float) ;
         x = mcur42st(y) ;
        Fy = x * ly ;        //,1.05 is the Amplitude factor
        cout << endl <<"MMF due yoke=   "<< Fy ;
    }


    void mmf :: pole(float d, float r, float w )
    {
      float pleak, Bp, x, y, Fp ;

      cout << endl <<"Enter the value of Pole leakage factor (pleak)" ;
      cin >> pleak ;
      Bp = pleak * d / r ;

       y = Bp ;
       float mcur42st(float) ;

       x = mcur42st(y) ;
       cout <<endl <<" x =   "<< x ;
       getch() ;
      Fp = x * w ;
      cout << endl <<"MMF due Pole=  "<< Fp ;
    }

    void mmf :: gap(int a, int b, int c, float d, float e, float f, float g,
              float h, float k)
    {
      float ys, Bav, Bg, ratios, y, kc, kgs, ratiod, kd, kgd, Fg ;
      const float PI = 3.1416 ;

      ys = PI * e / b ;           //slot pitch
      Bav = a * d / ( f * b * ys ) ;      //average gap-density
      Bg = Bav / t ;              //gap-density at pole centre
      ratios = h / g ;           // slot opening /gap length

      y = ratios;
      float carter_s(float) ;
      kc = carter_s(y) ;

      kgs = ys / ( ys -(kc * h)) ;

      ratiod = h / k ;               //slot opening / duct width
      cout << endl <<"Ratio: slot opening/duct width=   "<< ratiod ;

      y = ratiod;
      float carter_o(float) ;
      kd = carter_o(y) ;

      kgd = f / ( f -( kd * c * k)) ;

      Fg = 800000 * Bg * kgs * kgd * g ;
      cout << endl <<" Gap m.m.f. (Fg) = "<< Fg ;
    }
```

```cpp
void mmf :: armteeth(int a,int b,int c,float d,float e,float f,float k,
              float n, float o, float q, float t)
{
  float At, Btap, ys, x, y, x1, K, Ft ;
  const float PI = 3.1416 ;
  At = n * t * (f - (c * k)) * o * b / a ;   // apparent tooth-area
  Btap = d / At ;
  cout << endl <<" Btap = "<< Btap ;
  getch() ;

  if (Btap < 2.0)
  {
   y = Btap ;
   float mcur42st(float) ;
   x1 = mcur42st(y) ;

   getch() ;
   Ft = x1 * q ;
   cout << endl <<" Armature tooth m.m.f.(Ft) =  "<< Ft ;
  }
  else
  if(
  Btap >= 2.)
  {
  ys = PI * e / b ;

  y = Btap ;
  float mcur42st(float) ;
  x1 = mcur42st(y) ;

  Ft = x1 * q ;
  cout << endl <<" Armature tooth m.m.f.(Ft) =  "<< Ft ;
  }
}
void mmf :: armcore(int a,float d, float l, float m)
{
  float x, Bc, lc, y,  Fc ;

  const float PI = 3.1416 ;
     float kfl = 1.05 ;  // Factor to allow for curved nature of
//                 flux-path in yoke at the back of the pole
  lc = 0.5 * kfl * PI * m / a ;   //length of flux-path in armature core

//                 = 0.5 pole-pitch on mean dia.of arm.core
  Bc = 0.5 * d / l ;         // flux in armature core

  y = Bc ;
  float mcur42st(float) ;
  x = mcur42st(y) ;
  getch() ;
  Fc = x * lc ;
  cout << endl <<" Armature core m.m.f.(Fc) =  "<< Fc ;
}

void main(void)
```

```
{
    mmf x ;
    int p, S, nd ;
    float phi,D,L,lg,wo,wd,Ac,Dc,fst,wt,ds,Ap,tou,Ay,Dy,hp ;

    cout << "Enter the data: a,b,c,d,e,f,g,h,k,l,m,n,o,q,r,t,u,v,w"<< endl ;
    cin >>p>>S>>nd>>phi>>D>>L>>lg>>wo>>wd>>Ac>>Dc>>fst>>wt>>ds>>Ap>>tou>>Ay
        >>Dy>>hp ;

    x.getdata(p,S,nd,phi,D,L,lg,wo,wd,Ac,Dc,fst,wt,ds,Ap,tou,Ay,Dy,hp) ;
    x.display() ;

    x.yoke(p,phi,Ay,Dy) ;
    x.pole(phi, Ap, hp ) ;
    x.gap(p,S,nd,phi,D,L,lg,wo,wd) ;
    x.armteeth(p,S,nd,phi,D,L,wd,fst,wt,ds,tou) ;
    x.armcore(p,phi,Ac,Dc) ;
}
```

Figure 7.18 A computer program for the determination of m.m.f.s of a Salient-pole Machine-D.C. and Synchronous.

RESULT

(E) 20   324   41   1.14   6.5   1.73
     0.036   0.001   0.0065   1.086   1.99   0.8
     0.0424   0.083   0.92   0.777   1.28   2.44
     1.67

(o) Enter the value of yoke leakage factor, yleak
(E) 1.1
(o) MMF due yoke = 10
     Enter the value of Pole leakage factor (pleak)
(E) 1.15
(o) MMF due pole = 541
     Gap m.m.f. (Fg) = 24102
     Btap = 1.82   Armature tooth m.m.f. (Ft) = 1402
     Bc = 0.525   Armature core m.m.f. (Fc) = 11.5

7.5 DESIGN CONSIDERATIONS FOR HIGH VOLTAGE ROTATING MACHINES

The most important consideration for a designer with respect to high voltage rotating machines is the insulation of armature conductors, and slot-lining. Whereas, with high voltage, the armature current for the same power rating is small and definite advantage is accured in terms of saving in copper, care must be taken to choose proper insulating materials.

Present trend is to use 'thermoset'–type insulating material which has resulted in improved processing, reduced insulation levels and smaller and more efficient machines operating at higher levels of electrical, mechanical and thermal stresses.

The outcome of continual process of development of improved insulation during the last decade has lead to the firm establishment of two basic processes :
'Resin Rich' and 'vacuum-Pressure Impregnation' (VPI)

The Resin Rich process involves the application of material which contain a high resin content and needs no further addition of resin. VPI, on the other hand, involves application of dry materials which are then subjected to a vacuum-impregnation process using synthetic resin.

The choice of process is dependent on many factors, the prime of which being the initial capital cost. The Resin Rich process has lower plant capital cost, needs no vacuum treatment, and can produce all types of coil with flexible end-winding. Further, in this process, each coil can be tested and repaired, if necessary, during winding or test. The VPI, on the other hand, ensures impregnation of all coils together. In this process, machines can be dry-wound with little chance of damage during winding and with possible tighter coils in slots giving better heat transfer.

Stator coil Insulating Material

Important insulating materials are :

   (i) mica paper, preferably produced by the thermo-chemical process so that the particle size is smaller, and having highest fatigue strength;
  (ii) synthetic resins of epoxy and polyester type. These have good consistency, long life, and excellent safe handling characteristics;
 (iii) supporting materials, such as, glass fibre, which gives good protection to mica paper in VPI insulation system. Glass fabric improves impregnation. Glass paper has better thermal properties but is difficult to handle. Polyester paper, glass cloth, and polyester film are also used as good supporting materials.

Stator Coil Manufacture

I. 3.3 kW, 500–1000 kW Motor

It is usual for these motors to have full diamond stator coils consisting of both conductors and turns. With both Resin Rich and VPI,

   (i) it is essential that conductor/turn insulation to bare copper be applied by machine. Thus the insulation material needs to be thin, having high dielectric and mechanical strength. Voltage stresses of 2.3 to 2.6 kV. mm^{-1} can be employed with the application of correct number of layers;
  (ii) The loop is then formed and pressed into slots and then pulled into diamond;
 (iii) Main insulation is applied throughout the coil by machine.

II. 13.2 kV, 80–120 MW Water-wheel Generator

Water-wheel generators are of slow speeds and thus need large number of stator coils. The coils are considerably large since stator diameter is large, and of diamond type with 'Roebel' construction in slot portion to reduce losses. Corona shields are generally employed at the conductor/turn stack -in order to reduce electrical discharge at the conductor/turn main insulation surface. With both Resin Rich and VPI,

(i) Copper conductor should be covered with either synthetic resin-coated glass braid or glass lapped, or with resin rich glass-backed micapaper.

(ii) Consolidation of conductor/turn-stack of coil is essential so as to eliminate voids and to provide solid base on which the main insulation can be applied.

III. 22 kV, 500–1000 MW turbogenerator

Windings of these machines are generally internally water-cooled with the stator being pressurised in hydrogen. Stator coils are very large (approximately 12 m. in length) and are of diamond type. Because of size and weight, these coils are processed as half-coils. Roebel construction with hollow copper conductors are employed having a covering of synthetic resin-coated glass braid or glass lap in both Resin Rich and VPI processes. Insulated conductors are first formed into Roebel transposition and then laced together into straight conductor stack. The stack is then formed into half-coil and end water-boxes are fitted. The transposition spaces in the Roebel conductor stack are fitted with micaceous putty and are cured to give a solid base for the application of the main insulation.

For detailed study, the following reference may consulated :-

Neal, J. E., and Whitnan, A. G. : "Insulation of High Voltage Rotating Machines", IEE conf. Publication No. 213 on Electrical Machines-Design and Applications, 13-15 July, 1982. ,

7.6 DESIGN OF HIGH-SPEED MOTORS

The difficulties in the design of high speed motors are :-

(i) Excessive centrifugal forces. Special rotor designs and constructions are required to overcome the mechanical stresses.

(ii) High windage losses, specially in salient-pole rotors. The air is preferably be evacuated-type.

(iii) Pole-face losses (in synchronous motors, due to pulsations of the flux-density on the surface of the rotor) can be quite large if open or semi-closed slots are employed in stator.

(iv) An evacuated air gap or a liquid-cooled rotor necessitates proper seals.

(v) At high speeds, a low torqut results in a comparatively large power output. Thus, the rotor and overall motor dimensions will be small, resulting in small surface area for cooling. Cooling by fan mounted on the shaft is often discarded in view of increased windage loss, noise and mechanical complications at very high speeds. Direct rotor cooling, though expensive may be the answer to this problem.

(vi) Generally, high speed motors are now-a-days started by a variable-frequency inverter or amplifier. For good operation, of the frequency converter, commutating reactance should be small. The stator leakage reactance of the motor, which is the major component of the commutating reactance, should thus be as small as possible.

(vii) The field winding of high speed synchronous machines are particularly difficult to ventilate properly.

(viii) For a turbo-alternator at 3000-rpm., considerable mechanical problems relating critical speed and vibration arise from high peiipheral speed (of the order of 150 m. sec), large weight of the rotating mass, and the distance between the bearings.

7.7 NOISE AND VIBRATION IN ELECTRICAL MACHINES

The modern trend in electrical machines design, in view of highly competitive market, is towards the adoption of lighter and cheaper construction for a given output. This means full utilisation of active materials and increased specific magnetic and electric loadings. As a result, electrical machines have become noisier. The sources of noise can be classified as below :-

I. Magnetic excitation, due to the magnetic force waves generated in the air gap of a rotating machine. These force waves will only cause high noise or vibration levels if their frequencies and modes coincide with or closely approach the natural frequencies of the stator.

II. Mechanical excitation forms the noise and vibration sources whose amplitude and frequency are functions of the machine speed only. These sources include cooling fans, ball and roller bearings, and mechanical unbalance in the rotor. Forces produced by these sources may be small in amplitude to produce noise only. However, high noise and vibration may result if they coincide with the natural frequencies of structural components.

III. Aerodynamic excitation is due to the flow of cooling air through the machine. Broad-band noise is caused by the flow of turbulent air over the machine surfaces and between the stator and rotor. Discrete-frequency noise may develop due to the air impinging on the stator teeth and ducts or high velocity air passing stationary supports.

IV. Increase in noise level with load. Noise and vibration components which are electromagnetically excited and associated with the mmf. Slot harmonics may increase with load.

Reduction Measures
   (i) Magneticdly-excited noise and vibration can be reduced in amplitude by the reduction of the amplitude of magnetic forces or by ensuring that the forcing frequencies are not close to the natural frequencies of the structure.
  (ii) Vibration can be reduced by careful design of shaft/rotor/bearing systems, by using resilient mounts between cover work and the machine, and by careful dynamic balancing.
 (iii) Fan noise can often be reduced by changing the number of blades or staggering them to avoid interactions with stationary parts.
 (iv) Aerodynamically excited noise can be reduced by,
       (a) using the smallest acceptable fan,
       (b) ensuring that the air-circuit is as streamlined as possible, by minimising projections on the stator and rotor surfaces.

For detailed study, the reader may consult,
   Tulleth, A : "The Design Triangle", International Conf. on Electrical Machines– Design and Applications, IEE Publication no. 213, 13-15 July, 1982, pp. 79, and the references in the paper.

SHORT QUESTIONS

7.1 Describe important features of modular construction of induction motor-frame.

7.2 On what factors do the choice of current-density in rotating machines depend ?

7.3 Comment on the choice of flux-density and current-density for induction motors to be used in rural areas.

7.4 Mention 3 points to be considered for the choice of slot-combination in induction motors.

7.5 Why is a short gap length so important to the operation of an induction motor ?

7.6 Why is it possible to construct alternators of much larger capacity than d.c. generator ?

   (Hints : Commutation is a serious limiting factor.)

7.7 Why is it possible to design alternators to generate much higher voltages than d.c. generators ?

7.8 Why, for rural use, a continuous-rated induction motor should be de-rated ?

7.9 Why is it advantageous to use Alnico bars in the rotor of an inverter-controlled induction motor ?

7.10 What is the effect of rotor leakage reactance on the amplitude of harmonic currents in au inverter-controlled induction motor ?

7.11 What are the effects of cycloconverter on motor supply voltage and current during the speed control of induction motor ?

7.12 The field distribution factor or field form factor for a normal salient-pole synchronous machine is approximately,

   (a) unity,

   (b) 0.67,

   (c) 0.40.

7.13 What is Roebel transposition ?

7.14 Why is the power system turbo-alternator of generally 2 poles ?

7.15 State why a turbo-alternator has smaller diameter and larger length whereas a water-wheel generator has larger diameter and smaller length.

7.16 Enumerate 5 advantages in the choice of larger number of poles in a d.c. machine.

7.17 Enumerate 5 considerations to be taken into account in the choice of number and dimension of slots for a d.c. machine.

7.18 What are the important effects of the a.c. components in the armature voltage of a thyristorised supply for a d.c. motor ?

7.19 Suggest measures for the improvement in performance of a thyristor-supplied d.c. motor.

7.20 Mention 3 points which limit the maximum continuous power output/torque of a direct current machine.

7.21 What is the most important consideration for the design of high voltage rotating machine ?

7.22 State the advantages of using 'thermoset'-type insulating materials in high voltage rotating machines.

7.23 Enumerate important insulating materials for the stator coil of a high voltage a.c. machine.

7.24 Mention 8 points of difficulty in the design of high speed motors.

7.25 Classify the sources of noise in electrical machines.

7.26 How are noise and vibration in electrical machines reduced ?

7.27 Indicate 'True' or 'False' :-

   (1) The value of current-density for short-time rating is smaller than for continuous rating.

(2) In an inverter-controlled induction motor, the harmonic losses increase with increasing frequency for constant torque operation.

(3) For slip-ring induction motors, the rotor winding is normally wave-connected.

(4) Variable speed induction motors can be started at a lower speed with their pull-out torque.

(5) Loss in cage rotor due to harmonics in cycloconverter supply voltage can be minimised by reduced rotor reactance and increased rotor resistance.

(6) Modern turbo-alternators have short radial gap length for reason of operational stability.

(7) The gap m.m.f. in a synchronous machine is inversely proportional to its direct-axis magnetising reactance.

(8) In a direct current machine, for a particular D^2L_a, smaller value of pole-pitch leads to larger diameter and smaller length.

(9) In a direct current machine, a small (L_a/τ) means inferior commutation.

(10) Specific magnetic loading for a direct current motor with thyristorised supply is more than that with normal d.c. supply.

(11) In a direct current machine, the maximum continuous torque at rated speed per unit armature volume is limited by the temperature-rise.

(12) Modern trend in the design of direct current machines is to have small pole-height.

(13) High stator leakage reactance gives good commutation in an inverter-operated high-speed motor.

CHAPTER 8

Finite Element Method

8.1 The importance of numerical analysis in electrical machines design has arisen in the last few decades, because of the impressive growth of the rating of electrical systems and consequently of their dimensions and costs. In early sixties, the maximum rating of turbo-generators was in the range of 200 MW while machines in the range of 1000 and 1500 MW have recently been built. Similar increases have taken place in the rating of transformers and associated electrical devices.

As discussed in Chapter 2, it is impractical for economic reasons to increase the volume of machines in proportion to their rating. It is necessary to use increased stresses in materials and adapt new design criteria. Elaborate use of superconductors has also been proposed. In many cases, new devices are notably different in shapes from the older ones.

Such a situation has posed new problems to the designer. For instance, in the past, valuable information pertaining the design was forthcoming from experimental data. With the increase in ratings and related costs, such availability of data has greatly reduced. The situation thus demands greater accuracy in the prediction of theoretical performance, specially in view of complexity of geometry of the problems involved and the characteristics of the materials used. Thus, a better knowledge of the time- and space-distribution of electric current and magnetic flux in an electrical machine is necessary so that a reliable description of reactances, power losses and electromechanical stresses can be obtained, from which the magnetic, thermral, and mechanical design of the machine can be derived.

Analytical methods such as separation of variables, the conformal mapping, leading to solution in closed form, have the limitation that problems with complex geometry cannot be handled with desired accuracy. The deficiency in analytical methods has been largely eliminated in numerical methods such as the finite-difference method (Chapter 4), the finite element method,[*] which has come to the fore with the advent of large digital computer.

---

[*] Chari, M.V.K. and Silvester, P.P. (Editors) : Finite Elements in Electrical and Magnetic Field Problems, John Wiley.   Silvester P.P. and Ferrari R.L Finite Elements for Electrical Engineers.

Like the finite-difference method, the finite-element method is a numerical analysis technique for obtaining solutions to a wide variety of engineering problems. Whereas the finite difference model of a problem gives *pointwise* approximation to the governing equation (art. 4.2.3 & Appendix), a finite element model gives *piecewise* approximation to the governing equation. That is, in the finite difference model, a solution region is envisaged as an array of grid points; in the finite element model, the solution region can be analytically modelled or approximated by replacing it with an assemblage of discrete elements. The advantage is that these elements can be used to represent exceedingly complex shapes since they can be put together in a variety of ways. Thus, for problems with irregular solution region or unusual specification of boundary conditions, finite element method is particularly well-suited requiring fewer nodes.

8.2   THE METHOD

Consider a body of matter or a region of space (termed as solution region) at a point of which it is desired to find a field variable (such as, potential, temperature, etc.) defined by a given equation and boundary conditions. The solution region is divided into elements and the unknown field variable is expressed in terms of assumed aproximating functions within each element.

The approximating function (often termed, interpolation function) are defined in terms of the values of field variables at specified points called nodes. The nodal values of the field variable and the interpolation functions of the elements completely define the behaviour of the field variable within the elements. Once the nodal values are found, the interpolation function defines the field variables throughout the assemblage of elements. Thus, the following steps may be considered :–

Step 1 – *Defining the elements :*   The solution region is divided into elements. A variety of element shapes-triangle, rectangle, quadrilateral, etc. for two-dimensional cases, may be employed in the same solution region.

Step 2 – *Selection of Interpolation function* :   Each element is assigned with nodes, and the interpolation function is chosen to represent the variation of the field variable over the element. A polynomial is often selected as an interpolation function as it is easy to integrate and differentiate, the degree of the polynomial depending on the number of nodes assigned to the element.

Step 3 – *Matrix Equation for elements :*   The matrix equations expressing the properties of individual elements are determined. The following four approaches are available :– Direct approach, the variational approach, the weighted residual approach, and the energy balance approach. The choice of approach depends on the nature of the problem.

Step 4 – *Assembly of elements :*   The matrix equations are combined to obtain the behaviour of the entire solution region.

[*Note* : (1) The matrix equations for the system have the same form as the equations for an invidual element except that they contain many more terms because they include all nodes.

(2) The procedure for assembly of matrix equations is based on continuity of field variable. That is, at a node where the elements are interconnected, the value of the field variable is the same for each element sharing that node].

Step 5 – Solution the assembly process in step 4 leads to a set of simultaneous equations which can be solved to obtain the unknown nodal values of the field variable, after assigning the boundary conditions.

Example 8.1

One dimensional heat flow

The example concerns heat flow through a composite body e.g. heat flow through insulation, core and frame in the stator of an induction motor) in which AA is maintained at a temperature T_i °C and BB, at T_0°C, $T_0 < T_i$). Heat flow is thus from AA to BB in the x-direction only (Figure 8.1).

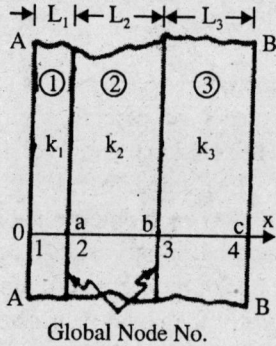

Global Node No.

Figure 8.1   One-dimensional heat-flow.

Solution

Step 1 – The solution region OG is divided into 3 elements oa, ab, and bc.

Step 2 – Element oa is assigned with global nodes 1 and 2, element ab with 2 and 3, and bc with 3 and 4.

Selection of interpolation function in this problem can be based on Fourier law (equation 2.1) instead of assuming an arbitrary polynomial.

Assuming constant thermal conductivity in a particular element, the quantity of heat flowing in the x-direction in each node,

$$qe = - k_e . S. \frac{\Delta T}{L_e}$$

(8.1)

where, ΔT is the drop in temperature across the element, and L_e, the length of the element along x–direction ; k_e, the thermal conductivity of the material of the element (in $W.m^{-1}.°C^{-1}$): and S m^2, the surface area of heat transfer per element.

Step 3 – In element 1 at the elemental node 1 (Figure 8.2),

$$q_{c_1} = - \frac{k_1 S}{L_1} (T_2 - T_1)$$

(8.2)

T_2 being the temperature at node 1.

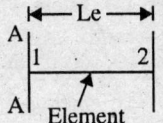

Figure 8.2 An element

With energy balance, heat flowing in at elemental node 1 = elemental node 2.

Thus, $\qquad\qquad\qquad\qquad q_{c2} = -q_{c1}$

$$= \frac{K_1 S}{L_1}\left(T_2 - T_1\right) \text{ watts} \qquad\qquad (8.3)$$

in matrix form, equation (8.3) reduces to,

$$\begin{bmatrix} qc_1 \\ qc_2 \end{bmatrix} = \frac{K_1 S}{L_1}\begin{bmatrix} 1 & -1 \\ -1 & 1 \end{bmatrix}\begin{bmatrix} T_1 \\ T_2 \end{bmatrix} \qquad\qquad (8.4)$$

or $\qquad\qquad\qquad\qquad \left[q_c^e\right] = \left[K^{(e)}\right]\left[T^e\right] \qquad\qquad (8.5)$

where, $\qquad [T^e] =$ Column matrix of nodal temperature.

$\qquad\left[q_c^e\right] =$ Column matrix of nodal heat flow.

$\qquad [K^{(e)}] =$ Matrix of thermal conductance coefficient, often termed as 'elemental stiffness matrix'.

$$= \left[K^{(1)}\right] = \begin{bmatrix} k_{11}^{(1)} & k_{12}^{(1)} \\ k_{21}^{(1)} & k_{22}^{(1)} \end{bmatrix} \text{ for element (1)} \qquad\qquad (8.6)$$

Thus, for element (2), the elemental stiffness matrix,

$$\left[K^{(2)}\right] = \begin{bmatrix} k_{11}^{(2)} & k_{12}^{(2)} \\ k_{21}^{(2)} & k_{22}^{(2)} \end{bmatrix} \qquad\qquad (8.7)$$

and for element (3),

$$\left[K^{(3)}\right] = \begin{bmatrix} k_{11}^{(3)} & k_{12}^{(3)} \\ k_{21}^{(3)} & k_{22}^{(3)} \end{bmatrix} \qquad\qquad (8.8)$$

Step 4 : (a) To assemble the elements, continuity of heat flow at the nodes is assumed. An arbitrary global numbering scheme is established from Figure 8.1 to identify the elements and nodes. A system topology, which establishes connections between the elements, can now be created as shown in Table 8.1. Such connections lead to the formation of global stiffness matrix, of which the elemental stiffness matrices are only sub-matrices.

(b) In this example, there are 4 global nodes so that the global stiffness matrix will be a square matrix of dimension 4 × 4. The global stiffness matrix is formed by addition of the 'expanded' elemental stiffness matrices which are also of dimension 4 × 4, as shown below :–

Table 8.1 System Topology

Element	Scheme	
	Elemental	Global
(1)	1	1
	2	2
(2)	1	2
	2	3
(3)	1	3
	2	4

For element (1), there is a direct correspondance between the elemental and global stiffness matrix coefficients, as shown in row 1–Table 8.1, and thus, the expanded elemental stiffness matrix,

$$\left[\bar{K}^{(1)} \right] = \begin{bmatrix} k_{11}^{(1)} & k_{12}^{(1)} & 0 & 0 \\ k_{21}^{(1)} & k_{22}^{(1)} & 0 & 0 \\ 0 & 0 & 0 & 0 \\ 0 & 0 & 0 & 0 \end{bmatrix} \tag{8.9}$$

For element (2), the correspondance between the elemental and global stiffness matrix coefficients are (row 2, Table 8.1) : -

elemental	*global*
$k_{11}^{(2)}$	$k_{22}^{(2)}$
$k_{12}^{(2)}$	$k_{23}^{(2)}$
$k_{21}^{(2)}$	$k_{32}^{(2)}$
$k_{22}^{(2)}$	$k_{33}^{(2)}$

$$\tag{8.10}$$

so that, from equation (8.7)

$$\left[\bar{K}^{(2)} \right] = \begin{bmatrix} 0 & 0 & 0 & 0 \\ 0 & k_{22}^{(2)} & k_{23}^{(2)} & 0 \\ 0 & k_{32}^{(2)} & k_{33}^{(2)} & 0 \\ 0 & 0 & 0 & 0 \end{bmatrix} \tag{8.11}$$

Similarly, for element (3),

$$\left[\bar{K}^{(2)} \right] = \begin{bmatrix} 0 & 0 & 0 & 0 \\ 0 & 0 & 0 & 0 \\ 0 & 0 & k_{33}^{(3)} & k_{34}^{(3)} \\ 0 & 0 & k_{43}^{(3)} & k_{44}^{(3)} \end{bmatrix} \tag{8.12}$$

Thus, the global stiffness matrix is,

$$[\bar{K}] = \sum_{e=1}^{3} [K^{(e)}] = \begin{bmatrix} k_{11}^{(1)} & k_{12}^{(1)} & 0 & 0 \\ k_{21}^{(1)} & \left(k_{22}^{(1)} + k_{22}^{(2)}\right) & k_{23}^{(2)} & 0 \\ 0 & k_{32}^{(2)} & \left(k_{33}^{(2)} + k_{33}^{(3)}\right) & k_{34}^{(3)} \\ 0 & 0 & k_{43}^{(3)} & k_{44}^{(3)} \end{bmatrix} \qquad (8.13)$$

(c) The expansion and summation principle applied to determine the global stiffness matrix can be applied to determine the column matrices of the global equation. For example, from equation (8.4), the expanded column matrices of nodal heat flow are :–

element (1)   element (2)   element (3)

$$\begin{bmatrix} q_{c1}^{(1)} \\ q_{c2}^{(1)} \\ 0 \\ 0 \end{bmatrix} \qquad \begin{bmatrix} 0 \\ q_{c2}^{(1)} \\ q_{c3}^{(2)} \\ 0 \end{bmatrix} \qquad \begin{bmatrix} 0 \\ 0 \\ q_{c3}^{(2)} \\ q_{c4}^{(3)} \end{bmatrix}$$

It is to be noted that,

    (i) heat flowing out at global node 2 of element (1) =

    = heat flowing in at global node 2 of element (2) =

    (ii) heat flowing out at global node 3 of element (2) =

    = heat flowing into node 3 of element (3)

    Thus, the global column matrix of heat flow is,

$$\begin{bmatrix} q_{c_1} \\ 0 \\ 0 \\ \vdots \\ q_{c4} \end{bmatrix} \qquad (8.15)$$

Similarly, the global column matrix of nodal temperature is,

$$\begin{bmatrix} T_1 \\ T_2 \\ T_3 \\ T_4 \end{bmatrix} \qquad (8.16)$$

Putting the boundary conditions, $T_1 = T_i$ and $T_4 = T_0$, the global matrix equation is,

$$\begin{bmatrix} q_{c1} \\ 0 \\ 0 \\ q_{c4} \end{bmatrix} = \begin{bmatrix} k_{11}^{(1)} & k_{12}^{(1)} & & \\ k_{21}^{(1)} & k_{22}^{(1)} + k_{22}^{(2)} & k_{23}^{(3)} & \\ & k_{32}^{(2)} & k_{33}^{(2)} + k_{33}^{(3)} & k_{34}^{(3)} \\ & & k_{43}^{(3)} & k_{44}^{(3)} \end{bmatrix} \begin{bmatrix} T_1 \\ T_2 \\ T_3 \\ T_4 \end{bmatrix} \qquad (8.17)$$

Example 8.2

A two-dimensional problem

The example concerns the steady-state potential distribution in a two-dimensional domain as defined in Figure 8.3 with the following boundary conditions :

$$\frac{dA}{dx} = 0 \text{ at } x = 0; \qquad\qquad \frac{dA}{dy} = 0 \text{ at } y = 0.$$

A $(L/2, y) = A_0$; A $(x, L) = A_0$,
A being the vector potential (field variable).

The governing equation for the potential distribution is given by the bi-harmonic equation,

$$\frac{\partial}{\partial x}\left(\sigma \frac{\partial A}{\partial x}\right) + \frac{\partial}{\partial y}\left(\sigma \frac{\partial A}{\partial y}\right) = J \qquad (8.18)$$

J being the current-density of the material of the domain, and σ, its reluctivity.

Step 1. As shown in Figure 8.3, the solution region is divided into 4 elements and the element shape is chosen to be triangular.

Step 2. The elements are assigned with global node numbers as shown.

[*Note* : The numbering of nodes is important as it determines the bandwidth of the global stiffness matrix, and hence computer time for the solution of the equations. While numbering, care must be taken to see that the maximum difference between the two node numbers of an element is minimum. For example, the numbering as in Figure 8.4 will give a maximum difference of 4, while that in Figure 8.3 gives 3.]

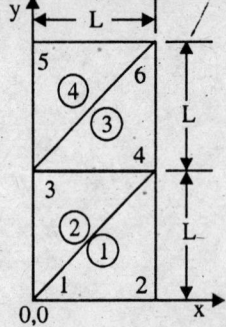

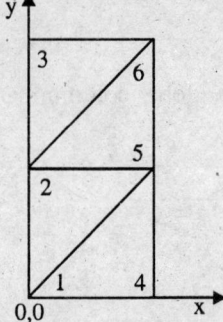

Figure 8.3 Pertaining Example 8.2   Figure 8.4 Pertaining numbering of nodes.

(ii) Interpolation function is chosen as below, assuming linear variation of potential A with respect to x and y within the element.

$$A^{(e)}(x,y) = C_1^{(e)} + C_2^{(e)} x + C_2^{(e)} y \qquad (8.19)$$

[*Note* : (1) The degree of polynomial is 3, the number of nodes of the chosen type of element.

(2) Rectangular element, rather than the triangular gives better accuracy.

(iii) The constants $C_1^{(e)}$, $C_2^{(e)}$ and $C_3^{(e)}$ for each elemental node can now be determined, as below :–

Writing (Figure 8.5).

for elemental node

$$i : A_i = C_1^{(e)} + C_2^{(e)} x_i + C_3^{(e)} y_i$$

$$j : A_j = C_1^{(e)} + C_2^{(e)} x_j + C_3^{(e)} y_j$$

$$m : A_m = C_1^{(e)} + C_2^{(e)} x_m + C_3^{(e)} y_m \qquad (8.20a)$$

we have,

$$C_1^{(e)} = \frac{1}{2\Delta^e}\left[A_i (x_j y_m - y_j x_m) + A_j (x_m y_i - y_m x_i) + A_m (x_i y_j - y_i x_j)\right]$$

$$C_2^{(e)} = \frac{1}{2\Delta^e}\left[A_i (y_j - y_m) + A_j (y_m - y_i) + A_m (y_i - y_j)\right]$$

$$C_3^{(e)} = \frac{1}{2\Delta^e}\left[A_i (x_m - x_j) + A_j (x_i - x_m) + A_m (x_j - x_i)\right]$$

$$2\Delta^e = \begin{vmatrix} 1 & x_i & y_i \\ 1 & x_j & y_j \\ 1 & x_m & y_m \end{vmatrix} = 2 \times [\text{Area of the elemental triangle with vertices } i, j \text{ and } m.]$$

$$(8.20b)$$

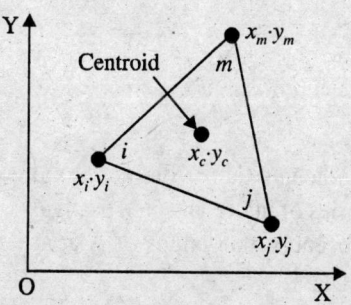

Figure 8.5 Triangular element.

Substituting $C_1^{(e)}$, $C_2^{(e)}$ and $C_3^{(e)}$ in equation (8.19a),

$$A^{(e)}(x, y) = \frac{1}{2\Delta^e}(a_i + b_i x + c_i y) A_i + (a_j + b_j x + c_j y) A_j + (a_m + b_m x + c_m y) A_m$$

where

$$a_i = x_j \cdot y_m - x_m \cdot y_j, \, b_i = y_j - y_m, \, c_i = x_m - x_j;$$
$$a_j = x_m \cdot y_i - x_i \cdot y_m, \, b_j = y_m - y_i, \, c_f = x_i - x_m;$$
$$a_m = x_i \cdot y_f - x_j \cdot y_i, \, b_m = y_i - y_j, \, c_m = x_j - x_i. \tag{8.22a}$$

In matrix form, equations (8.19a), (8.20a), (8.21a) and (8.22a) could be represented as,

(i) Equation (8.19a) : $[A^{(e)}(x, y) = [P^T][C^{(e)}]$

where,

$$[P^T] = [1 \; x \; y] \text{ and } [C^{(e)}] = \begin{bmatrix} C_1 \\ C_2 \\ C_3 \end{bmatrix} \tag{8.19b}$$

(ii) Equation (8.20a) : $[A^{(e)}] = [P^{(e)}][C^{(e)}]$

where,

$$\left[A^{(e)}\right] = \begin{bmatrix} A_i \\ A_j \\ A_m \end{bmatrix} \text{ and } \left[P^{(e)}\right] = \begin{bmatrix} 1 & x_i & y_i \\ 1 & x_j & y_j \\ 1 & x_m & y_m \end{bmatrix} \tag{8.20b}$$

(iii) Equation (8.21) : $[C^{(e)}] = [P^{(e)}]^{-1}[A^{(e)}]$

$$= [R^{(e)}][A^{(e)}]$$

where,

$$\left[R^{(e)}\right] = \frac{1}{2\Delta^e} \begin{bmatrix} a_i & a_j & a_m \\ b_i & b_j & b_m \\ c_i & c_j & b_m \end{bmatrix} \tag{8.21b}$$

(iv) Equation (8.22a) :

$$[A^{(e)}(x, y)] = [P^T][P^{(e)}]^{-1}[A^{(e)}] \tag{8.22b}$$

If the solution region contains M elements, the complete representation of the field variable over the whole region is given by,

$$A(x, y) = \sum_{e=1}^{M} A^{(e)}(x, y) \tag{8.23}$$

[*Note :* Equation (8.23) indicates that if the nodal values of A are known, the solution region can be represented by a series of interconnected triangular planes. There is no discontinuity at the interelement boundaries because the values of A at any two nodes determine the linear variation of A along that boundary.]

Step 3. In the solution region, considering energy balance,* we may define a functional,

$$I = dW_f - dW_e \tag{8.24}$$

---

* Sen, S.K. : Electrical Machinery, 3rd Edn., 1984, p. 19.

where, $d\,W_f$ is the change in the stored magnetic field energy, and dW_e, the change in electrical energy. For equilibrium, I should be minimum. That is, $(d\,I/d\,A)$ = is zero.

From equation (8.24), 1

$$I = \iint \frac{1}{2}\,B^2\,\sigma\,dx.dy - \iint J\,A\,dx.dy$$

$$= \frac{1}{2}\iint\left[\sigma\,By^2 + \sigma\,B_z^2\right]dx.dy - \iint J\,A\,dx.dy$$

Since, $\qquad\qquad B_y = \dfrac{\partial A}{\partial x},\ \text{ and } B_x = \dfrac{\partial A}{\partial y},$

$$I = \frac{1}{2}\iint_{\substack{\text{area}\\ \Delta}}\left[\sigma\left(\frac{\partial A}{\partial x}\right)^2 + \sigma\left(\frac{\partial A}{\partial y}\right)^2 - 2\,J\,A\right]dx.dy$$

[*Note* : It is evident from equation (8.25) that the functional,

$$I = \iint f\left(x,\,y,\,A,\,\frac{\partial A}{\partial x},\,\frac{\partial A}{\partial y}\right)dx.xy$$

has a minimum value when $(dI/dA) = 0$, that is, when $A\,(x,\,y)$ satisfies,

$$\frac{\partial}{\partial x}\left[\frac{\partial f}{\partial\left(\dfrac{\partial A}{\partial x}\right)}\right] + \frac{\partial}{\partial y}\left[\frac{\partial f}{\partial\left(\dfrac{\partial A}{\partial y}\right)}\right] - \frac{\partial f}{\partial A} = 0 \qquad\qquad (8.26)$$

Representing equation (8.25) as,

$$I = I_k - I_g,$$

where, $\qquad\qquad I_k = \dfrac{1}{2}\iint_{\Delta}\left[\sigma\left(\dfrac{\partial A}{\partial x}\right)^2 + \sigma\left(\dfrac{\partial A}{\partial y}\right)^2\right]dx.dy \qquad\qquad (8.27)$

and $\qquad\qquad I_g = \displaystyle\iint_{\Delta} J\,A\,dx.dy$

we have, for equilibrium, $\qquad\qquad \dfrac{dI}{dA} = \dfrac{dI_k}{dA} - \dfrac{dI_g}{dA} \qquad\qquad (8.28)$

But $\qquad\qquad \dfrac{dI_k}{dA} = \displaystyle\sum_{e=1}^{M}\dfrac{dI_k(e)}{dA^{(e)}} \qquad\qquad (8.29a)$

and $\qquad\qquad \dfrac{dI_g}{dA} = \displaystyle\sum_{e=1}^{M} = \dfrac{dI_g(e)}{dA^{(e)}} \qquad\qquad (8.29b)$

M being the number of elements in the solution region, and variations of $I_k^{(e)}$ and $I_g^{(e)}$ are taken only with respect to the nodal values associated with the element (e).

[*Note* :   For the example 8.2,

$$I_k = \frac{1}{2} \int\limits_{x=0}^{L} \int\limits_{y=0}^{2L} \left[\sigma \left(\frac{\partial A}{\partial x} \right)^2 + \sigma \left(\frac{\partial A}{\partial y} \right)^2 \right] dx.dy$$

$$I_g = \int\limits_{0}^{L} \int\limits_{0}^{2L} J \, A \, dx.dy$$

and $$\frac{dI_k}{dA} = \sum_{e=1}^{4} \frac{dI_k(e)}{dA^{(e)}} ; \quad \frac{dI_g}{dA} = \sum_{e=1}^{4} \frac{dI_g(e)}{dA^{(e)}} \tag{8.30}$$

(i) Assuming $\sigma = \sigma_e$ = constant within the element, we have from equation (8.22b) and (8.30),

$$I_k(e) = \frac{1}{2} \sigma_e \int\limits_{0}^{L} \int\limits_{0}^{2L} \left\{ [P_x^T \, R^{(e)} \, A^{(e)}]^2 + [P_y^T . R^{(e)} . A^{(e)}]^2 \right\} dx.dy$$

where, $$\left[P_x^T \right] = \frac{d}{dx} \left[P^T \right] = [0 \quad 1 \quad 0]$$

$$\left[P_y^T \right] = \frac{d}{dy} \left[P^T \right] = [0 \quad 0 \quad 1]$$

Thus,

$$\frac{dI_k^{(e)}}{dA^{(e)}} = \frac{1}{2} \sigma_e \int\limits_{0}^{L} \int\limits_{0}^{2L} \left\{ 2P_x^T \, R^{(e)} \, A^{(e)} \, [P_x^T \, R^{(e)}] + 2P_y^T \, R^{(e)} \, A^{(e)} \, [P_y^T \, R^{(e)}] \right\} dx.dy$$

$$= \sigma_e \int\limits_{0}^{L} \int\limits_{0}^{2L} \left\{ R^{(e)^T} \, [P_x . P_x^T + P_y . P_y^T] \, R^{(e)} \, A^{(e)} \right\} dx.dy$$

But $$[P_x] = \frac{d}{dx}[P] = \begin{bmatrix} 0 \\ 1 \\ 0 \end{bmatrix}, \text{ and } [P_y] = \frac{d}{dy}[P] = \begin{bmatrix} 0 \\ 0 \\ 1 \end{bmatrix}$$

Hence,

$$\int\limits_{0}^{L} \int\limits_{0}^{2L} [P_x \, P_x^T + P_y \, P_y^T] \, dx \cdot dy = \int\limits_{0}^{L} \int\limits_{0}^{2L} \begin{bmatrix} 1 & 0 & 0 \\ 0 & 1 & 0 \\ 0 & 0 & 1 \end{bmatrix} dx \cdot dy$$

$$= \Delta^{(e)} \begin{bmatrix} 1 & 0 & 0 \\ 0 & 1 & 0 \\ 0 & 0 & 1 \end{bmatrix}$$

since the above matrix is constant.

Thus,

$$\frac{dI_k^{(e)}}{dA^{(e)}} = K^{(e)} \cdot A^{(e)}$$

where,

$$K^{(e)} = \frac{1}{4} \frac{\sigma_e}{\Delta^{(e)}} \begin{bmatrix} b_i^2 + c_i^2 & b_i b_j + c_i c_j & b_i b_m + c_i c_m \\ & b_f^2 + c_j^2 & b_j b_m + c_j c_m \\ & & b_m^2 + c_m^2 \end{bmatrix} \qquad (8.31a)$$

and $2 \Delta^{(e)} = (a_i + a_j + a_m)$ from equations (8.21a) and (8.22a). $\qquad (8.31b)$

That is, the element of $K^{(e)}$ – matrix is given by,

$$K_{pq} = \frac{1}{4} \frac{\sigma_e}{\Delta^{(e)}} (b_p\, b_q + c_p\, c_q) \qquad (8.32)$$

   (ii) Assuming $J = J^{(e)} =$ constant within an element, we have from equations (8.27) and (8.22b).

$$I_g^{(e)} = J^{(e)} \int\limits_{\Delta^{(e)}} \int A^{(e)}\, (x, y)\, dx \cdot dy$$

$$= J^{(e)} \int\limits_{\Delta^{(e)}} \int P^T\, R^{(e)}\, A^{(e)}\, dx \cdot dy$$

$$\frac{dI_g^{(e)}}{dA^{(e)}} = J^{(e)} \int\limits_{\Delta^{(e)}} \int \left(P^T\, R^{(e))} \right)^T\, dx \cdot dy$$

$$= J^{(e)} \int\limits_{\Delta^{(e)}} \int \begin{bmatrix} 1 \\ x \\ y \end{bmatrix} dx \cdot dy \qquad (8.33)$$

At the centroid of the triangular element, $x = x_c$ and $y = y_c$, so that,

$$\frac{dI_g^{(e)}}{dA^{(a)}} = J^{(e)} \cdot R^{(e)T} \cdot \Delta^{(e)} \begin{bmatrix} 1 \\ x_c \\ y_c \end{bmatrix}$$

$$= J^{(e)} \cdot R^{(e)T} \cdot \Delta^{(e)} \cdot P_0^{(e)}$$

where,
$$[P_e]^{(e)} = \begin{bmatrix} 1 \\ x_e \\ y_e \end{bmatrix} \tag{8.34}$$

But,
$$x_{(c)} = \frac{1}{3}(x_i + x_j + x_m), \text{ and}$$

$$y_c = \frac{1}{3}(y_i + y_j + y_m)$$

That is,
$$\frac{dI_g^{(e)}}{dA^{(e)}} = g^{(e)}$$

where,
$$g^{(e)} = \frac{1}{3} J^{(e)} \Delta^{(e)} \begin{bmatrix} 1 \\ 1 \\ 1 \end{bmatrix} \tag{8.35}$$

Step 4. Assembly of elements.

   The system topology establishing the connections between the elements is shown in Table 8.2, based on which

Table 8.2 System Topology

		Nodal Scheme	
Element	Node No.	Elemental	Global
	i	1	1
(1)	j	2	2
	m	3	4
	i	1	1
(2)	j	2	4
	m	3	3
	i	1	3
(3)	j	2	4
	m	3	6
	i	1	3
(4)	j	2	6
	m	3	5

the equations for elemental stiffness matrices can be written down. That is, for element (1), from equations (8.32) and (8.35),

$$\begin{bmatrix} k_{11} & k_{12} & k_{14} \\ k_{21} & k_{22} & k_{24} \\ k_{41} & k_{42} & k_{44} \end{bmatrix} \begin{bmatrix} A_1 \\ A_2 \\ A_4 \end{bmatrix} = \frac{1}{3} J_1 \, \Delta_1 \begin{bmatrix} 1 \\ 1 \\ 1 \end{bmatrix}$$

For element (2)

$$\begin{bmatrix} k_{11} & k_{14} & k_{13} \\ k_{41} & k_{44} & k_{43} \\ k_{31} & k_{34} & k_{33} \end{bmatrix} \begin{bmatrix} A_1 \\ A_4 \\ A_3 \end{bmatrix} = \frac{1}{3} J_2 \, \Delta_2 \begin{bmatrix} 1 \\ 1 \\ 1 \end{bmatrix}$$

For element (3),

$$\begin{bmatrix} k_{33} & k_{34} & k_{36} \\ k_{43} & k_{44} & k_{46} \\ k_{63} & k_{64} & k_{66} \end{bmatrix} \begin{bmatrix} A_3 \\ A_4 \\ A_6 \end{bmatrix} = \frac{1}{3} J_3 \, \Delta_3 \begin{bmatrix} 1 \\ 1 \\ 1 \end{bmatrix}$$

For element (4),

$$\begin{bmatrix} k_{33} & k_{36} & k_{35} \\ k_{63} & k_{66} & k_{65} \\ k_{53} & k_{56} & k_{55} \end{bmatrix} \begin{bmatrix} A_3 \\ A_6 \\ A_5 \end{bmatrix} = \frac{1}{3} J_4 \, \Delta_4 \begin{bmatrix} 1 \\ 1 \\ 1 \end{bmatrix}$$

After assembly of elements, we have the global stiffness matrix,

$$\bar{K} = \begin{bmatrix} k_{11}^{(1)} + k_{11}^{(2)} & k_{12}^{(1)} & k_{13}^{(2)} & k_{14}^{(1)} + k_{14}^{(2)} & & \\ k_{21}^{(1)} & k_{22}^{(1)} & & k_{24}^{(1)} & & \\ k_{31}^{(2)} & & k_{33}^{(2)} + k_{33}^{(3)} + k_{33}^{(4)} & k_{34}^{(2)} + k_{34}^{(3)} & k_{35}^{(4)} & k_{36}^{(3)} + k_{36}^{(4)} \\ k_{41}^{(1)} + k_{41}^{(2)} & k_{42}^{(1)} & k_{43}^{(2)} + k_{43}^{(3)} & k_{44}^{(1)} + k_{44}^{(2)} + k_{44}^{(3)} & & k_{46}^{(3)} \\ & & k_{53}^{(4)} & & k_{55}^{(4)} & k_{56}^{(4)} \\ & & k_{63}^{(3)} + k_{63}^{(4)} & k_{64}^{(4)} & k_{65}^{(4)} & k_{66}^{(3)} + k_{66}^{(4)} \end{bmatrix}$$

$$(8.37)$$

For computation, inputs to the computer are (Figure 8.3),

Element	x_i	x_j	x_m	y_i	y_j	y_m
1	0	L	L	0	0	L
2	0	L	0	0	L	L
3	0	L	L	L	L	$2L$
4	0	L	0	L	$2L$	$2L$

The computer starts with element 1 and computes as below :

Element	1	2	3	4
$b_i = y_j - y_m$	$-L$	0	$-L$	0
$c_i = x_m - x_j$	0	$-L$	0	$-L$
$b_j = y_m - y_i$	L	L	L	L
$c_j = x_i - x_m$	$-L$	0	L	0
$b_m = y_i - y_j$	0	$-L$	0	$-L$
$c_m = x_j - x_i$	L	L	L	L
$b_i^2 + c_i^2$	L^2	L^2	L^2	L^2
$b_i b_j + c_i c_j$	$-L^2$	0	$-L^2$	0
$b_i b_m + c_i c_m$	0	L^2	0	$-L^2$
$b_j^2 + c_j^2$	$2L^2$	L^2	$2L^2$	L^2
$b_j b_m + c_j c_m$	$-L^2$	$-L^2$	$-L^2$	$-L^2$
$b_m^2 + c_m^2$	L^2	$2L^2$	L^2	$2L^2$

From equation (8.31), assuming $\sigma e = \sigma = $ constant, and knowing $\Delta^{(e)} = \Delta = $ constant area of the triangular element $= \dfrac{1}{2} L^2$, the elemental stiffness matrices are,

$$\text{For element (1), } K^{(1)} = \frac{1}{2}\sigma \begin{bmatrix} 1 & -1 & 0 \\ -1 & 2 & -1 \\ 0 & -1 & 1 \end{bmatrix}$$

$$\text{element (2), } K^{(2)} = \frac{1}{2}\sigma \begin{bmatrix} 1 & 0 & -1 \\ 0 & +1 & -1 \\ -1 & -1 & 2 \end{bmatrix}$$

$$\text{element (3), } K^{(3)} = \frac{1}{2}\sigma \begin{bmatrix} 1 & -1 & 0 \\ -1 & 2 & -1 \\ 0 & -1 & 1 \end{bmatrix}$$

$$\text{element (4), } K^{(4)} = \frac{1}{2}\sigma \begin{bmatrix} 1 & 0 & -1 \\ 0 & 1 & -1 \\ -1 & -1 & 2 \end{bmatrix}$$

so that, the global stiffness matrix is,

$$\bar{K} = \frac{1}{2}\sigma \begin{bmatrix} 2 & -1 & -1 & 0 & 0 & 0 \\ -1 & 2 & 0 & -1 & 0 & 0 \\ -1 & 0 & 4 & -2 & -1 & 0 \\ 0 & -1 & -2 & 4 & 0 & -1 \\ 0 & 0 & -1 & 0 & 2 & -1 \\ 0 & 0 & 0 & -1 & -1 & 2 \end{bmatrix}$$

The determination of $\bar{K}$ – matrix as shown above can be done analytically using connection matrix $D^{(e)}$ ($E \times 3$ matrix, E being the number of global nodes) for each element, defined as,

$$D^{(e)} = \begin{bmatrix} 0 & 0 & 0 \\ 0 & 0 & 0 \\ - & - & - \\ 1 & 0 & 0 \\ - & - & - \\ 0 & 1 & 0 \\ - & - & - \\ 0 & 0 & 1 \\ - & - & - \end{bmatrix} \begin{matrix} \\ \\ \\ \dots\dots i-\text{th row} \\ \\ \dots\dots j-\text{th row} \\ \\ \dots\dots m\text{-th row} \\ \end{matrix} \qquad (8.38)$$

so that,

$$\bar{K} = \sum_{e=1}^{M} D^{(e)} \cdot K^{(e)} \cdot D^{(e)T} \qquad (8.39)$$

For the example 8.2, using Table 8.2,

$$D^{(1)} = \begin{bmatrix} 1 & 0 & 0 \\ 0 & 1 & 0 \\ 0 & 0 & 0 \\ 0 & 0 & 1 \\ 0 & 0 & 0 \\ 0 & 0 & 0 \end{bmatrix}, D^{(2)} = \begin{bmatrix} 1 & 0 & 0 \\ 0 & 0 & 0 \\ 0 & 0 & 1 \\ 0 & 1 & 0 \\ 0 & 0 & 0 \\ 0 & 0 & 0 \end{bmatrix}, D^{(3)} = \begin{bmatrix} 0 & 0 & 0 \\ 0 & 0 & 0 \\ 1 & 0 & 0 \\ 0 & 1 & 0 \\ 0 & 0 & 0 \\ 0 & 0 & 1 \end{bmatrix}, D^{(4)} = \begin{bmatrix} 0 & 0 & 0 \\ 0 & 0 & 0 \\ 1 & 0 & 0 \\ 0 & 0 & 0 \\ 0 & 0 & 1 \\ 0 & 1 & 0 \end{bmatrix}$$

The corresponding $D^{(e)T}$ are,

$$D^{(1)^T} = \begin{bmatrix} 1 & 0 & 0 & 0 & 0 & 0 \\ 0 & 1 & 0 & 0 & 0 & 0 \\ 0 & 0 & 0 & 1 & 0 & 0 \end{bmatrix}, \quad D^{(2)^T} = \begin{bmatrix} 1 & 0 & 0 & 0 & 0 & 0 \\ 0 & 0 & 0 & 1 & 0 & 0 \\ 0 & 0 & 1 & 0 & 0 & 0 \end{bmatrix}$$

$$D^{(3)^T} = \begin{bmatrix} 0 & 0 & 0 & 1 & 0 & 0 \\ 0 & 0 & 0 & 1 & 0 & 0 \\ 0 & 0 & 0 & 0 & 0 & 1 \end{bmatrix}, \quad D^{(4)^T} = \begin{bmatrix} 0 & 0 & 1 & 0 & 0 & 0 \\ 0 & 0 & 0 & 0 & 0 & 1 \\ 0 & 0 & 0 & 0 & 1 & 0 \end{bmatrix} \quad (8.40)$$

Similarly, the global g – matrix is assembled from $g^{(e)}$ elemental matrices (equation 8.35) using.

$$\bar{g} = \sum_{e=1}^{M} D^{(e)} \cdot g^{(e)} \quad (8.41)$$

For the example 8.2, assuming $J_1 = J_2 = J_3 = J_4 = J$ in the right hand sides of equations 8.36, we have,

$$D^{(1)} \, g^{(1)} = \frac{1}{6} J \, L^2 \begin{bmatrix} 1 \\ 1 \\ 0 \\ 1 \\ 0 \\ 0 \end{bmatrix}, \qquad D^{(2)} \, g^{(2)} = \frac{1}{6} J \, L^2 \begin{bmatrix} 1 \\ 0 \\ 1 \\ 1 \\ 0 \\ 0 \end{bmatrix}$$

$$D^{(3)} \, g^{(3)} = \frac{1}{6} J \, L^2 \begin{bmatrix} 0 \\ 0 \\ 1 \\ 1 \\ 0 \\ 1 \end{bmatrix}, \qquad D^{(4)} \, g^{(4)} = \frac{1}{6} J \, L^2 \begin{bmatrix} 0 \\ 0 \\ 1 \\ 0 \\ 1 \\ 1 \end{bmatrix}$$

which after assembly yields,

$$g = \sum_{e=1}^{4} D^{(e)} \cdot g^{(e)} = \frac{1}{6} J \, L^2 \begin{bmatrix} 2 \\ 1 \\ 3 \\ 3 \\ 1 \\ 2 \end{bmatrix} \quad (8.42)$$

Figure 8.6 gives the computed K and g matrices for the example 8.2.

```
1.00    -0.50    -0.50     0.00     0.00     0.00
-0.50    1.00     0.00    -0.50     0.00     0.00
-0.50    0.00     2.00    -1.00    -0.50     0.00
0.00    -0.50    -1.00     2.00     0.00    -0.50
0.00     0.00    -0.50     0.00     1.00    -0.50
0.00     0.00     0.00    -0.50    -0.50     1.00
```

```
0.333             0.167              0.500         0.500     0.167
0.333
```

Figure 8.6 Computed output –K & g matrices, Example 8.2

Step 5 : *Introduction of Boundary conditions and Solution of Equations* From equations (8.28), (8.39) and (8.41), we have a set of system equations leading to:

$$\left[\overline{K}\right] \cdot \left[A\right] = \left[\overline{g}\right] \tag{8.43}$$

(*Note* : (1) In the formation of equation (8.43), the boundary conditions with respect to the 'forcing function at the nodes' or 'nodal actions' have not been taken into consideration.

For unique solution of the above set of equations, at least one nodal variable must be specified. Such specification can be for external nodes and/or internal nodes. Introduction of this will lead to a reduction of the number of variables. However, such introduction is generally achieved so that the original number of equations remain unchanged arid major re-structuring of. computer storage is avoided.

One method is to modify certain diagonal terms of the global stiffness matrix $\left[\overline{K}\right]$ and the corresponding terms in the $\left[\overline{g}\right]$ matrix. The procedure can be as below :–

(i) Multiply the diagonal term(s) in the matrix $\left[\overline{K}\right]$ corresponding to the specified nodal variable(s) by a large number, say 10^{12}.

(ii) Replace the corresponding term(s) in the $\left[\overline{g}\right]$ - matrix by the specified nodal variable(s) multiplied by 10^{12}.

For the example 8.2, with boundary coadition

$A = A_0$ at $x = 0$, and also at $y = 0$, i.e. $A_1 = A_0$, $A_2 = A_0$, $A_3 = A_0$, $A_5 = A_0$ equation (8.43) is modified as,

$$\begin{bmatrix} k_{11} \times 10^{12} & k_{12} & k_{13} & k_{14} & 0 & 0 \\ k_{21} & k_{22} \times 10^{12} & 0 & k_{24} & 0 & 0 \\ k_{31} & 0 & k_{33} \times 10^{12} & k_{34} & k_{35} & k_{36} \\ k_{41} & k_{42} & k_{43} & k_{44} & 0 & k_{46} \\ 0 & 0 & k_{53} & 0 & k_{55} \times 10^{12} & k_{56} \\ 0 & 0 & k_{63} & 0 & k_{65} & k_{66} \end{bmatrix} \begin{bmatrix} A_1 \\ A_2 \\ A_3 \\ A_4 \\ A_5 \\ A_6 \end{bmatrix} = \begin{bmatrix} A_0\, k_{11} \times 10^{12} \\ A_0\, k_{22} \times 10^{12} \\ A_0\, k_{33} \times 10^{12} \\ g_4 \\ A_0\, k_{55} \times 10^{12} \\ g_6 \end{bmatrix}$$

To test that this procedure gives, the desired result, consider the first 3 equations and the 5th, which gives, $A_1 = A_0$, $A_2 = A_0$, $A_3 = A_0$, and $A_5 = A_0$, since $k_{11} \times 10^{12} \gg k_{12}$, k_{13} etc., $k_{22} \times 10^{12} \gg k_{21}$, k_{24}, and so on.

Equation (8.44) can be rearranged as,

$$\begin{bmatrix} [\bar{K}_{11}] & [\bar{K}_{12}] \\ [\bar{K}_{21}] & [\bar{K}_{22}] \end{bmatrix} \begin{bmatrix} [A_1] \\ [\bar{A}_1] \end{bmatrix} = \begin{bmatrix} [\bar{g}_1] \\ [\bar{g}_2] \end{bmatrix} \qquad (8.45)$$

where,

$$[\bar{K}_{11}] = \begin{bmatrix} k_{11} \times 10^{12} & k_{12} & k_{13} & 0 \\ k_{21} & k_{22} \times 10^{12} & 0 & 0 \\ k_{31} & 0 & k_{33} \times 10^{12} & k_{35} \\ 0 & 0 & k_{53} & k_{55} \times 10^{12} \end{bmatrix} \qquad (8.46a)$$

$$[\bar{K}_{12}] = \begin{bmatrix} k_{14} & 0 \\ k_{24} & 0 \\ k_{34} & k_{36} \\ 0 & k_{56} \end{bmatrix} \qquad (8.46b)$$

$$[\bar{K}_{21}] = \begin{bmatrix} k_{41} & k_{42} & k_{43} & 0 \\ 0 & 0 & k_{63} & k_{65} \end{bmatrix} = [\bar{K}_{12}^T] \qquad (8.46c)$$

$$[\bar{K}_{22}] = \begin{bmatrix} k_{44} & k_{46} \\ k_{64} & k_{66} \end{bmatrix} \qquad (8.46d)$$

$$[\bar{g}_1] = \begin{bmatrix} A_0 & k_{11} \times 10^{12} \\ A_0 & k_{22} \times 10^{12} \\ A_0 & k_{33} \times 10^{12} \\ A_0 & k_{55} \times 10^{12} \end{bmatrix} \qquad (8.46e)$$

$$[\bar{g}_2] = \begin{bmatrix} g_4 \\ g_6 \end{bmatrix} \qquad (8.46f)$$

Solving the first set of equations in (8.45), $[\bar{A}_1]$ can be determined

Putting $[\bar{A}_1]$ in the secónd set in equation (8.45) and using equation (8.46c),

$$[\bar{K}_{22}], [\bar{A}_2] = [g_2] - [\bar{K}_{12}^T], [\bar{A}_1] \qquad (8.47)$$

## CHAPTER 9

Computer Programs in C++ Language in Linux Platform

9.1 C++– WINDOWS AND LINUX

9.1 Uptil Now we have discussed the programs in C++ language, which is in WINDOWS platform.

9.2 LINUX Open Source has since been popular and is being accepted by many as a platform alternative to WINDOWS. In this chapter we have developed similar programs with LINUX OS. The algorithms followed earlier are applicable here also.

The programs developed in C++ –LINUX more of less are the same as that in WINDOWS OS with the following exceptions*:

#include <iostream>

which is a standard file in Linux where "all elements of the standard C++ programs start their execution, independently of its location within the same code."

In C++– WINDOWS it is #include <iostream.h> .

using namespace std;

In Linux, "all the elements of the standard C++ library are declared within a 'namespace' with name 'std'. So in order to access its functionality the expression declares that we will be using these entities".

int main()

'This line corresponds to the begining of the defination of the main function. The main function is the point by where all C++ programs start their execution independently of its location within the source code."

---

*http:// www.cplusplus.com/doc/tutorial/
C++ –Linux readers may find this online tutorial useful.

"The word 'main' is followed in the code by a pair of parentheses (). This is because it is a function declaration. In C++, what differentiates a function declaration from other types of expressions are these parentheses that follow its name. Optionally, these parentheses may enclose within a list of parameters."

return (0);

"The return statement causes the main fuction to finish, **'return'** is followed by a return code (here it is (0). This is the most usual way to end a C++ console."

Apart from above, C++-Linux accepts the following 'constants' and 'instruction' :

C++ Linux	C++Windows
float PI = 3.14159L ;	constant float PI = 3.14159 ;
float k = 1e-6f ;	float k = pow(10,-6) ;
getchar () ;	getch() ;

9.2   APPROXIMATION OF A CURVE BY STRAIGHT LINES

The presentation under art. 4.2.2 is applicable to Linux.

It is to be noted that the header files are to be ended by .h,

(for example) :

<mcur42st.h>

The program (Figure 4.5b) and the flow-chart (Figure 4.5a) could be used.

9.3   SOME USEFUL COMPUTER PROGRAMS.

(a) **Optimisation program**

A computer program in Linux is given in Figure 9.1 [this matches with the program in Figure 5.3(b)] which starts with, as mentioned above,

```
#include<iostream>
using namespace std;
int main()
```

and ends with return (0) ;

[*Note* : Compare with Figure 5.3(b), Line 21 : 'else if' not acceptable in Lynax.]

Example 9.1(a)   Given, the equation : $c = ab - (a + b)$.

For a value of 'a', to find the value of 'b' for $c <=$ Cmax.

```
     #include<iostream>
     using namespace std;
     int main()
5    float a, b, c, Cmax;
     cout<<"¥n Enter the values of a, b, and Cmax :¥n¥n";
     cin >> a >> b >> Cmax ;
     c = a*b - (a + b );
```

```
10
        if (c >= Cmax);
        {
              do

        {
15      b = b - 0.001 ;
        c = a*b - (a + b );
        }
        while ( c > Cmax );
        cout<<"a: "<<a<<"b: "<<b<<"c: "<<c<<endl;
20      }
        if (c< Cmax);
        {
              do

        {
25      b = b + 0.001 ;
        c = a*b - (a + b );
        }
        while ( c < Cmax );
        b = b - 0.001 ;
30      c = a*b - (a + b );
        cout<<"a: "<<a<<"b: "<<b<<"c: "<<c<<endl;
        }
        return (0);
34      }
```

Figure 9.1 A computer program for optimisation.

(b) **Optimum value of a parameter.**
   [See Figure 5.3 to compare.]

Example 9.1(b). Given a fraction. To find the nearest integer.

```
#include<iostream>
using namespace std;

int main()
{
    int x, v ;
    float u, y ;
    cout<<endl<<" Enter the value of the fraction, u" ;
    cin >> u ;

    x = u ;
    y = u - x ;
        if (y >= 0.5 )
            {
            v = x + 1.0 ;
            } else
                {
                    v = x ;
                }
cout<<endl<<"The value of the nearest integer is = "<< v ;
    }
```

Figure 9.2 A computer program for obtaining nearest integer of a given fraction.

(c) Largest among four numbers.

Example 9.1(c)   Given 4 integers. To find the largest.

```
#include<iostream>
#include<math.h>
#include<stdio.h>
using namespace std ;

int main()
{
        int S5, S6, S7, S8, Smax, y ;
    cout<<endl<<"Enter the values : " ;
    cin >> S5 >> S6 >> S7 >> S8 ;
    cout<<endl<<" S5= "<<S5<<" S6= "<<S6<<"S7= "<<S7<<"S8= "<<S8 ;

    y = S6 – S5 ;

    if(y > 0)
        {
            Smax = S6 ;
        }
        else
        if(y < 0)
        {
            Smax = S5 ;
        }
        y = Smax - S7 ;
        if(y > 0)
        {
            Smax = Smax ;
        }
        else
        if(y < 0)
        {
            Smax = S7 ;
        }
        y = Smax - S8 ;
        if(y > 0)
        {
            Smax = Smax ;
        }
        else
        if(y < 0)
        {
            Smax = S8 ;
        }
}
```

Figure 9.3   Computer program for determining largest of 4 numbers.

Example 9.1(d)   Solution of Simultaneous Equations by Gauss Elimination Method.

```
#include<iostream>
#include<stdio.h>
#include<math.h>
```

```
using namespace std;

int main()
{
int i, j, k, n ;
float a[20][20], x[20] ;
double s, p ;

printf("Enter the number of equations : ") ;
scanf("%d", &n) ;
 printf("\nEnter the co-efficients of the equations :¥n¥n") ;
for(i = 0 ; i < n ; i++)
{
for(j = 0 ; j < n ; j++)
{
printf("a[%d][%d] = ", i + 1, j + 1) ;
scanf("%f", &a[i][j]) ;
}
printf("b[%d] = ", i + 1) ;
scanf("%f", &a[i][n]) ;
}
for(k = 0 ; k < n - 1 ; k++)
{
for(i = k + 1 ; i < n ; i++)
{
p = a[i][k] / a[k][k] ;
for(j = k ; j < n + 1 ; j++)
a[i][j] = a[i][j] - p * a[k][j] ;
}
}
x[n-1] = a[n-1][n] / a[n-1][n-1] ;
for(i = n - 2 ; i >= 0 ; i--)
{
s = 0 ;
for(j = i + 1 ; j < n ; j++)
{
s += (a[i][j] * x[j]) ;
x[i] = (a[i][n] - s) / a[i][i] ;
}
}
printf("¥nThe result is :\n") ;
for(i = 0 ; i < n ; i++)
printf("¥nx[%d] = %.2f", i + 1, x[i]) ;
}
```

Figure 9.4   A computer program for the Solution of Linear Simultaneous Equations by Gauss Elimination Method.

[Ref: http://cprogramming.language-tutorial.com/2012/01/simultaneous-equation-using-gauss.html]

RESULT
[data from equation A. 4.8]

```
Enter the number of equations : 7
Enter the co-efficients of the equations :
a[1][1] = -4
a[1][2] = 1
```

```
a[1][3] = 0
a[1][4] = 0
a[1][5] = 0
a[1][6] = 0
a[1][7] = 0
b[1] = −150

a[2][1] = 1
a[2][2] = −4
a[2][3] = 1
a[2][4] = 1
a[2][5] = 0
a[2][6] = 0
a[2][7] = 0
b[2] = −100

a[3][1] = 0
a[3][2] = 1
a[3][3] = −4
a[3][4] = 0
a[3][5] = 1
a[3][6] = 0
a[3][7] = 0
b[3] = −175

a[4][1] = 0
a[4][2] = 1
a[4][3] = 0
a[4][4] = −4
a[4][5] = 1
a[4][6] = 1
a[4][7] = 0
b[4] = 0

a[5][1] = 0
a[5][2] = 0
a[5][3] = 1
a[5][4] = 1
a[5][5] = −4
a[5][6] = 0
a[5][7] = 1
b[5] = −50

a[6][1] = 0
a[6][2] = 0
a[6][3] = 0
a[6][4] = 1
a[6][5] = 0
a[6][6] = −4
a[6][7] = 1
b[6] = 0

a[7][1] = 0
a[7][2] = 0
a[7][3] = 0
a[7][4] = 0
a[7][5] = 1
```

```
a[7][6] = 1
a[7][7] = -4
b[7] = -25
```

```
The result is :
x[1] = 53.31
x[2] = 63.24
x[3] = 70.16
x[4] = 29.50
x[5] = 42.40
x[6] = 12.36
x[7] = 19.94
```

9.4   SMALL SINGLE–PHASE TRANSFORMER

Example 9.2 :   It is required to develop a computer program for a small air-cooled single phase transformer having one primary winding and two secondary windings using standard stampings as in Figure 9.6.

```cpp
#include<iostream>
#include<stdio.h>
#include<math.h>
#include<stdio.h>
#include "loscrgo.h"
using namespace std ;

class spec
{
    private :
    float csf, wsf, effy ;
    public :
    int vp, fs, vs1, vs2 ;
    float as1, as2 ;
    float akeq ;

    void getdata(int vp,int fs,int vs1,int vs2,float as1,float as2) ;
    void display(int vp,int fs,int vs1,int vs2,float as1,float as2) ;
    void lamination(int vp,int fs,int vs1,int vs2,float as1,float as2) ;
};

void spec::getdata
    (int vprim, int freq, int vsec1, int vsec2, float asec1, float asec2)
{
    vp=vprim; fs=freq; vs1=vsec1; vs2=vsec2; as1=asec1; as2=asec2;
}

void spec::display(int vprim, int freq, int vsec1, int vsec2,
        float asec1, float asec2)
{
    cout<<endl<<"primary volts= "<<vp<<"frequency= "<<fs
        <<",secd.volts-winding1= "<<vs1<<",secd.volts-winding2= "<<vs2
        <<",secd.amps-winding1= "<<as1<<",secd.amps-winding2= "<<as2;
}
void spec::lamination(int vp,int fs,int vs1,int vs2,float as1,float as2)
{
```

```
//     Assumed quantities
       float csf, wsf, effy ;                    //csf:core stacking factor
//                                               wsf:window space factor
       float aKeq ;                              //output coefficient


//                     vz: primary volt-drop on full-load at u.p.f,
//     insulation thickness, metre: co1: between core & secondary 2.

    cout<<"Enter the values of core stacking factor (csf), window "
        "space factor (wsf) and efficiency (effy)" ;
        cin >> csf >> wsf >> effy ;
        cout<<"Enter the assumed value of output coefficient (aKeq)" ;
        cin >> aKeq ;
//                     Compute Primary Current at unity p.f.
       float va, wp, ap ;

       va = vs1 * as1 + vs2 * as2 ;
       wp = va / effy ;
       ap = wp / vp ;
       cout<<endl<<"va = "<<va<<"wp= "<<wp<<"ap= "<<ap ;
//                     Compute Iron area and Window area
       float BM, J ;
       float eT, ai, ac, aw ;
       float k = 1e6f ;

       for (BM=1.2; BM<=1.50; BM+=0.1)
    {
       eT = aKeq * (sqrt( va/1000.)) ;            //voltage per turn
       ai = eT / (4.44 * BM * fs ) ;              // Apparent Iron area
       ac = ai / csf ; // Core area
       float k1 = 2.22 * fs * BM * ai * wsf * k ;

       for (J=2.0; J<=2.40; J+=0.1)
       {
       cout<<endl<<"BM = "<<BM <<"J = "<<J ;
       aw = va / (k1 * J) ;                       // Window area
       cout<<endl<<"Window area= "<< aw ;
       }
    }
//                     CHOOSE STANDARD STAMPING,
//     so that B x W [see Figure6.17 and also types of stampings
//                     is LESS THAN 'aw'
       float type, A, B, C, D, E, Aw ;
       cout<<endl<<"Enter Lamination data- type, A, B, C, D & E" ;
       cin >> type >> A >> B >> C >> D >> E ;
       cout<<endl<<"type= "<< type <<"A= "<< A <<"B= "<< B <<"C= "<< C
       <<"D= "<< D <<"E "<< E ;
       Aw = B * D ;
       cout<<endl<<"Aw= "<< Aw ;

//     Check whether Aw is < than aw. If yes, proceed.
//     Otherwise change type or increase J.
//     Note: Increasing J will increase temperature-rise.
       cout<<endl<<"Enter BM and J corresponding to chosen aw ";
       cin >> BM >> J ;
```

```
        float thc ;
        thc = ac / C ;
        cout<<endl<<"Thickness of core= "<< thc ;

//      WINDING DESIGN : Assume primary volt-drop at upf=7.5% of et

        float vx,PN,S1N,S2N,PCS,S1CS,S2CS ;
        float swgP,swg1,swg2,PBS,PIS,PID,S1BS,S1IS,S1ID, S2BS, S2IS, S2ID ;
        float PBD,S1BD,S2BD ;
            vx = (1 - 0.075) * vp ;
            PN = vx / eT ;                      // Primary turns
            S1N = PN * vs1 / vx ;               // Turns of sced. 1 windg.
            S2N = PN * vs2 / vx ;               // Turns of sced. 2 windg.

//              COMPUTE CONDUCTOR SECTION in mm.sq.
        PCS = ap / J ;                          //Bare Conductor section, Primary
        S1CS = as1 / J ;                        //Bare Conductor section, secd. 1
        S2CS = as2 / J ;                        //Bare Conductor section, secd. 2
        cout<<endl<<"Pri.bare cond.sec=           "<<PCS<<"Sec.1 bare cond.sec= "<<S1CS
                    <<"Sec.2 bare cond.sec=       "<<S2CS ;
        cout<<endl<<"from Standard Wire guage Table enter chosen values of :"
        "Pri.windg: swg no,Bare cond.section PBS,        Bare cond.dia PBD,"
                    "Insul.cond.dia. PID,            Insul.cond.sec. PIS"
        "Secd.1 windg:swg no,Bare cond.section S1BS,     Bare cond.dia S1BD."
                    "Insul.cond.dia. S1ID,           Insul.cond.sec. S1IS"
        "Pri.windg: swg no,Bare cond.section S2BS,       Bare cond.dia S2BD,"
                    "Insul.cond.dia. ,S2ID,          Insul.cond.sec. S2IS";

        cin >> swgP >> PBS >> PBD >> PID >> PIS ;
        cin >> swg1 >> S1BS >> S1BD >> S1ID >> S1IS ;
        cin >> swg2>> S2BS >> S2BD >> S2ID >> S2IS ;
        cout<<endl<<"Pri.Bare.sec= "<<PBS <<"Pri.Bare.dia.= "<<PBD
                <<"Pri. Insul. dia.= "<< PID <<"Pri. Insul.sec.= "<<PIS;
        cout<<endl<<"Sec1.Bare.sec= "<<S1BS<<"Sec.1 Bare.dia= "<<S1BD
                <<"Sec1 insul.dia.= "<<S1ID<<"Sec.1 Insul.sec= "<<S1IS ;
        cout<<endl<<"Sec2.Bare.sec= "<<S2BS<<"Sec.2 Bare dia.= "<<S2BD
                <<"Sec2. Insul. dia.= "<<S2ID<<"Sec.2 Insul.sec.= "<<S2IS ;
//              COMPUTE OVERALL COILWIDTH
        float S2L, S1L, PL, CS2L, CS1L, CPL, CS2N, CS1N, CPN ;

        S2L = (S2N * S2ID * 0.001) / B ;
        S1L = (S1N * S1ID * 0.001) / B ;
        PL = ( PN * PID * 0.001) / B ;
        cout<<endl<<"Sec2.: No.of Layers= "<<S2L<<"No. of Turns= "<<S2N ;
        cout<<endl<<"Sec1.: No of Layers= "<<S1L<<"No. of Turns= "<<S1N ;
        cout<<endl<<"Prim.: No of Layers= "<<PL<<"No. of Turns= "<<PN ;
        cin >> CS2L >> CS2N ;
        cin >> CS1L >> CS1N ;
        cin >> CPL >> CPN ;

//              COIL WIDTH
        float CW, co1, co2, co3, co4 ;
        cout<<endl<<"Enter: co1, co2, co3, co4 " ;
        cin >> co1 >> co2 >> co3 >> co4 ;
        CW = co1+(CS2L*S2ID*0.001)+co2+(CS1L*S1ID*0.001)+co3+(PL*PID*0.001)+co4 ;
        cout<<endl<<"coil width= "<<CW ;
```

```
//                          LENGTH OF MEAN TURN
      float S2LM, S1LM, PLM ;
      float PI = 3.14159L ;
      S2LM = 2.*(C+thc)+PI*(2.*co1+(CS2L*S2ID*0.001)) ;
      S1LM = S2LM +PI*((CS2L*S2ID*0.001)+2.*co2+(CS1L*S1ID*0.001)) ;
      PLM = S1LM+PI*((CS1L*S1ID*0.001)+2.*co3+(CPL*PID*0.001)) ;
      cout<<endl<<"Pri.bare cond.sec= "<<PCS<<"Sec.1 bare cond.sec= "<<S1CS
      <<"Sec.2 bare cond.sec= "<<S2CS ;
//                          RESISTANCE AT 75 DEG. CELSIUS
      float RP, RS1, RS2 ;
      RP = 0.0214 * PLM * CPN / PBS ; //with copper conductor
      RS1 = 0.0214 * S1LM * CS1N / S1BS ;
      RS2 = 0.0214 * S2LM * CS2N / S2BS ;
      cout<<endl<<"Primary resistance= "<<RP<<"Secondary1 resistance= "<<RS1
               <<"Secondary2 resistance= "<<RS2 ;

//                          CORE WEIGHT
      float cvol, cvcl, cwol, cwcl, crwt ;
      cvol = 2. * thc * E * csf * ( B + A ) ;        //core volume of outer limbs
      cvcl = thc * C * csf * B ;                      //core volume of central limb
      cwol = 0.0078 * cvol * k ;                      //core weight-outer limb
      cwcl = 0.0078 * cvcl * k ;                      //core weight-central limb
      crwt = cwol + cwcl ;                            //total core weight
      cout<<endl<<"Vol.of outer limbs= "<<cvol<<"Vol.of central limb= "<<cvcl
               <<"Core weight= "<<crwt<<" kg " ;

//                          CONDUCTOR WEIGHT
      float cvP, cvS1, cvS2, conwt ;
      cvP = PLM * PBS * CPN ;                         //conduc.volume,primary
      cvS1 = S1LM * S1BS * CS1N ;
      cvS2 = S2LM * S2BS * CS2N ;
      conwt = 0.00889 * ( cvP + cvS1 + cvS2 ) ;
      cout<<endl<<"Vol.of Prim.coil= "<<cvP<<"Vol.of Secnd.! coil= "<<cvS1
               <<"Vol.of Secnd.2 coil= "<<cvS2
               <<"Conductor weight= "<<conwt<<" kg " ;
      float Twt ;
         Twt = crwt + conwt ;
      cout<<endl<<"Total weight of transformer= "<<Twt<<" kg " ;

//                                  CORE LOSS
//   (a)  Central limb : flux-density = BM ; length of fluxpath = B
      float y, x, Bol, crlos1, crlosT ;
      y = BM ;
         float loscrgo(float) ;
             x = loscrgo(y1) ;
         cout<<endl<<"x= "<<x ;
         crlos1 = x * cwcl ;
                      cout<<endl<<"Core loss= "<< crlos1 <<"watt";
//   (b)  Outer limbs
      float phi, crlos2 ;
      phi = BM * C * thc ;
      Bol = 0.5 * phi / ( E * thc ) ;
```

```
            y2 = Bol ;
            cout<<endl<<"Bol= "<< Bol ;
                float loscrgo(float) ;
                    x = loscrgo(y) ;
                cout<<endl<<"x= "<<x ;
                crlos2 = x * cwol ;
                    cout<<endl<<"Core loss= "<< crlos2 <<"watt";
            crlosT = crlos1 + crlos2 ;
            cout<<endl<<"Total Core loss= "<< crlosT <<"watt";

//                                  CONDUCTOR LOSS
            float cnlosP, cnlos1, cnlos2, cnlosT ;
            cnlosP = ap * ap * RP ;
            cnlos1 = as1 * as1 * RS1 ;
            cnlos2 = as2 * as2 * RS2 ;
            cnlosT = cnlosP + cnlos1 + cnlos2 ;
            cout<<endl<<"Total conductor loss= "<< cnlosT <<" watt" ;

//                                  EFFICIENCY
            float wout, win ;
            wout = ( vp * ap) + (vs1 * as1) + (vs2 * as2) ;
            win = wout + crlosT + cnlosT ;
            effy = wout / win ;

//                          LOSS DISSIPATING SURFACE
                float Sol, Scc, ST ;
                Sol = (4 * E + 2 * thc ) * (A + B) ;            //Exposed surface- outer limbs
                Scc = (2. * (C + thc) + PI * CW) * B ;          //Exposed surface- coil on central limb
                ST = Sol + Scc ;                                //Total cooling surface
                cout<<endl<<"Total cooling surface= "<< ST <<" m.sq." ;
}

int main()
{
    spec S ;
        int vprim, freq, vsec1, vsec2 ;
    float asec1, asec2 ;
    cout<<endl<<"Enter vprim,freq,vsec1,vsec2,asec1,asec2" ;
    cin>>vprim>>freq>>vsec1>>vsec2>>asec1>>asec2 ;

    S.getdata(vprim,freq,vsec1,vsec2,asec1,asec2) ;
    S.display(vprim,freq,vsec1,vsec2,asec1,asec2) ;
    S.lamination(vprim,freq,vsec1,vsec2,asec1,asec2);
}
```

Figure 9.5 A Computer program for small transformer.

The program is run with the following specifications of a transformer : Input primary voltage = 230v.

Secondary1 voltage = 30v, current = 10amp at u.p.f.

Secondary2 voltage = 3v, current = 1amp at upf.

The transformer is air-cooled ; standard stampings (Figure 9.6) to be used;

for temperature-rise within specified limit, the loss dissipating surface should not be less than 0.001950 metre-sq. per watt-loss.

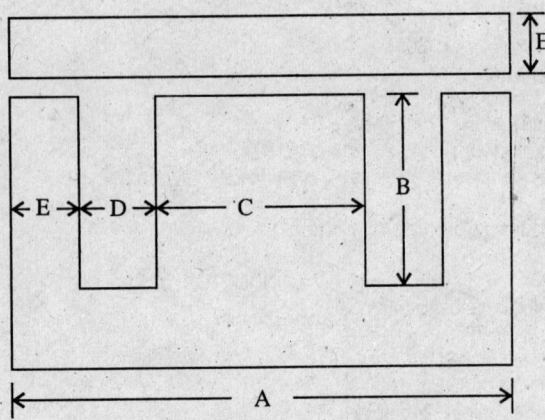

Figure 9.6 E. I. Lamination

(Dimensions in metre)

A	B	C	D	E
0.084	0.041	0.025	0.016	0.013
0.095	0.047	0.032	0.016	0.016
0.076	0.035	0.019	0.019	0.010
0.114	0.032	0.038	0.019	0.019

RESULT

Enter vprim, freq, vsec1, vsec2, asec1, asec2 0 : 230   50   30   10   3   1
primary volts= 230frequency= 50,secd.volts-winding1= 30,secd.volts-winding2= 10,secd.amps-winding1= 3,secd.amps-winding2= 1
Enter the values of core stacking factor (csf), window space factor (wsf) and efficiency (effy)
    0.95   0.5   0.85
Enter the assumed value of output coefficient (aKeq) 0.84

va = 100   wp= 117.647   ap= 0.511509
                 BM = 1.2   J = 2   Window area= 0.000752923
                 BM = 1.2   J = 2.1   Window area= 0.00071707
                 BM = 1.2   J = 2.2   Window area= 0.000684476
                 BM = 1.2   J = 2.3   Window area= 0.000654716
                 BM = 1.2   J = 2.4   Window area= 0.000627436
                 BM = 1.3   J = 2   Window area= 0.000752923
                 BM = 1.3   J = 2.1   Window area= 0.00071707
                 BM = 1.3   J = 2.2   Window area= 0.000684476
                 BM = 1.3   J = 2.3   Window area= 0.000654716
                 BM = 1.3   J = 2.4   Window area= 0.000627436
                 BM = 1.4   J = 2   Window area= 0.000752923
                 BM = 1.4   J = 2.1   Window area= 0.00071707
                 BM = 1.4   J = 2.2   Window area= 0.000684476
                 BM = 1.4   J = 2.3   Window area= 0.000654716
                 BM = 1.4   J = 2.4   Window area= 0.000627436

Enter Lamination data (from page 237)- type,   A,   B,   C,   D   &   E
                                          1   0.084   0.041   0.025   0.016   0.013

Aw= 0.000656

Enter BM and J corresponding to chosen aw   1.3   2.3

Thickness of core= 0.0359861

Pri.bare cond.sec= 0.222395   Sec.1 bare cond.sec= 1.30435   Sec.2 bare   cond.sec= 0.434783

From Standard Wire guage Table enter chosen values of :

Pri.windg: swg no,   Bare cond.sec. PBS,   Bare cond.dia PBD,   Insul.cond.dia.PID,   Insul.cond.sec. PIS

Secd.1 windg: swg no,   Bare cond. sec. S1BS,   Bare cond.dia S1BD.   Insul.cond.dia. S1ID,   Insul.cond.sec. S1IS

Secd.2 windg: swg no,   Bare cond. sec. S2BS,   Bare cond.dia S2BD,   Insul.cond.sec. S2IS   Insul. cond. dia. ,S2ID,

24.5	0.22346	0.5334	0.5714	0.2564
17.5	1.37	1.3208	1.3958	1.5302
21.5	0.456	0.762	0.824	0.5179

Sec2.: No.of Layers= 0.756596   No. of Turns= 37.6462   modified   value 1   38

Sec1.: No of Layers= 3.84487   No. of Turns= 112.938   4   113

Prim.: No of Layers= 11.1621   No. of Turns= 800.922   11   801

Enter: co1, co2, co3, co4   0.001   0.0001   0.0002   0.0002

coil width= 0.0142852

Pri. length of mean turn= 0.190144   Sec.1 length of mean turn = 0.151601

Sec. 2 length of mean turn 0.130844

Primary resistance= 14.5858   Sec.1 resistance= 0.267593

Sec.2 resistance= 0.23333

Core weight,   ol= 0.866635   cl= 0.273323   total= 1.13996 kg

Conductor weight= 0.531362 kg

Total weight of transformer= 1.67132 kg

Core loss, central limb= 0.273323watt   outer limb= 0.433316watt

Total Core loss= 0.70664watt

Total conductor loss= 6.45791 watt

Efficiency= 0.968131

Cooling surface /watt= 0.00311777 m.sq.

[*Note* :   The cooling surface per watt loss is more than the stipulated minimum 0.001950.]

9.5 SALIENT FIELD POLES

The computer program in Figure 9.6 is based on the Example 5.1.

```
#include<iostream>
#include<stdio.h>
using namespace std ;

class spec
{
        private :
        float wsf ;
        public :
        int n, Ff ;
        float d ;
        void getdata( int n, int Ff, float d, float wsf ) ;
        void display() ;

        void fieldcoil(int n, int Ff, float d, float wsf) ;
};
```

```cpp
void spec::getdata(int rpm, int amptur, float dia, float wsfac )
{
        n=rpm; Ff=amptur; d=dia; wsf=wsfac;
}

void spec::display()
{
        cout<<endl<<"speed= "<<n<<"rpm, .field ampere-turn/pole= "<<Ff
                <<", armature dia.= "<<d<<",winding space factor= "<<wsf ;
}

void spec::fieldcoil(int n, int Ff, float d, float wsf)
{
        float he, hp, k, k1, k2, k3 ;

        cout<<"Enter the values of pole height(hp) and width of"
        "coil-end insulation(hc)" ;
        cin>>hp>>he ;

// Assumed copper as winding material: sp.resis=0.0214x10pow-6
   k = (1e6f) ;
   k1 = (hp - 2.*he)*1000. ;
   k2 = Ff*0.0214*Ff/(wsf*k1*k1) ;

int o ;
float vr, cf;
float PI=3.14159L;

// Computation of Heat transfer Coefficient, cf
vr = PI*d*n/60. ;       //periferal speed
cout<<endl<<"Enter the value of o" ;
cin>>o ;
switch(o)               //o=1 for stationary poles, =2 for rotating poles
{
        case 1: cout<<endl<< "Stationary Poles" ;
             cf = 6.60*(1.+0.07*vr);
        cout<<endl<<"cf= "<<cf ;
             break ;
        case 2: cout<<endl<< "Rotating Poles" ;
             cf = 14.25*(1.+0.10*vr) ;
        cout<<endl<<"cf= "<<cf ;
        break ;
}

k3 = k2 / cf ;
cout<<endl<<"k3= "<<k3 ;

float lp,bp,de,dc,fs,k4,s,tmax, theta ;

cout<<endl<<"Enter the value of lp, bp, de" ;        //Figure 5.2
cin>>lp>>bp>>de ;

k4 = 2. * ((lp + bp) + PI*de) ;

cout<<endl<<"Enter the value of assumed coil-depth dc" ;
cin>>dc ;
```

```
s = k4 + PI*dc ;
cout<<endl<<"s= "<<s ;
fs = s/((s + de)*dc) ;
theta = k3*fs ;
cout<<endl<<"Computed temperature-rise= "<<theta ;
getchar() ;
cout<<endl<<"Enter maximum allowable temp.rise, tmax" ;
cin>>tmax ;
void optm(float,float&,float&,float&,float&,float&,float&) ;

float Dc ;
{
      float fs, s ;
      float PI = 3.14159L ;
      if (theta>=tmax)
  {
do
  {
      dc = dc + 0.0001 ;
      s = k4 + PI*dc ;
      fs = s/((s + de)*dc) ;
      Dc = dc ;
      theta = k3*fs ;
}
while(theta>tmax) ;
cout<<endl<<"s= "<<s<<"Dc= "<<Dc<<"temperature-rise= "<<theta ;
}
   else if(theta<tmax)
  {
  do
     {
      dc = dc - 0.0001 ;
      s = k4 + PI*dc ;
      fs = s/((s + de)*dc) ;
      theta = k3*fs ;
}
while(theta<tmax) ;
dc = dc + 0.0001 ;
      s = k4 + PI*dc ;
      fs = s/((s + de)*dc) ;
      Dc = dc ;
      theta = k3*fs ;
      cout<<endl<<"s= "<<s<<"Dc "<<Dc<<"temperature-rise= "<<theta ;
  }
}
   getchar() ;
// Compuite conductor section,af;field turns/pole Tf; field current If;
// field resistance per pole,rf; coppoer loss per pole,Wcf;
// volume of copper,per pole,vc; Weight of copper, per pole, wc
      int vf ;
      float swg, af, Tf, If, Wcf, mTf, maf, vc, wc ;
   cout<<endl<<"Enter field coil volt per pole";
      cin>> vf ;
//      conductor material chosen : Copper with enamelled insulation
//      sp.resis = 0.0214 ohm-mm.sq per metre
//           compute section of field conductor, af & swg
   af = 0.0214 * s * Ff / vf ;
```

```
        cout<<endl<<"computed conduc. section = "<< af ;
        getchar();
    cout<<endl<<"from standard table,choose nearest section & guage no.";
            cin>> maf >> swg ;
        cout<<endl<<"modified cond.section maf= "<<maf<<"guage no= "<<swg;

    Tf = wsf*Dc*(hp - 2.*he)* k/ maf ;
    cout<<endl<<"field turns per pole= "<< Tf ;
    cout<<endl<<"Enter modified fld.turns(whole no.) ";
            cin>> mTf ;

            If = Ff / mTf ;
            Wcf = vf * If ;
              vc = s * maf * Tf /k ; //m-cubed
            wc = 8890. * vc ; // kg.
            cout<<endl<< "Summary" ;
            cout<<endl<<"Temperature-rise = "<<theta ;
            cout<<endl<<"Field turns per pole = "<<mTf ;
            cout<<endl<<"Bare cond. section = "<<maf<<"mm-sq";
            cout<<endl<<"Conduc. guage no.= "<<swg ;
            cout<<endl<<"Field current= "<<If ;
            cout<<endl<<"Copper loss, per pole= "<<Wcf<<"watt" ;
            cout<<endl<<"volume of per pole copper- "<<vc<<"m.-cubed" ;
            cout<<endl<<"Weight of copper,per pole= "<<wc<<"kg" ;
    }

int main()
{
        spec S ;
    int rpm,amptur ;
    float dia,wsfac;
    cout<<endl<<"Enter rpm,armtur,dia,wsfac" ;
    cin>>rpm>>amptur>>dia>>wsfac ;

    S.getdata(rpm,amptur,dia,wsfac) ;
    S.display() ;
    S.fieldcoil(rpm,amptur,dia,wsfac) ;
}
```

Figure 9.6   Design of a Field coil for a given maxium temp. rise.

RESULT

[Note : The program is the same as under WINDOWS-C++ (Figure 5.4) with the measures incorporated as discussed above, in art. 9.2 and 9.3.

Enter the data as given in in art 5.2.1 (example 5.1) for Run 1: stationary poles, and those for Run 2 : rotating poles and find the results.]

9.6   TRANSFORMER WINDING DESIGN

9.6.1   Cross-Over Winding

[Matching with Figure 6.5(a), art. 6.5.1]

```cpp
//Design of Crossover Winding
#include <iostream>
using namespace std ;
class spec
{
public :
    int vo, to, ho, dico ;
    float ao ;
spec()
    { vo=0.0; ao=0.0; to=0.0; ho=0.0; dico=0.0; }
spec ( int v, float a, int t, int h, int dic)
    { vo=v; ao=a; to=t; ho=h; dico=dic; }
void get_value (void)
    { cout<<endl<<" Enter the values of vo,ao,to,ho,dico ";
      cin>>vo>>ao>>to>>ho>>dico;
    }
void display()
    {
        cout<<"volt per phase= "<<vo<<" amp per phase= "<<ao
        <<" turns per phase= "<<to<<" height of window= "<<ho
        <<" internal dia of coil= "<<dico;
    }
    int getv() {return vo;}
    float geta() {return ao;}
    int gett() {return to;}
    int geth() {return ho;}
    int getdic() {return dico;}
};

class winding:public spec
{
    public:
    float hyl;
    float dal;
    float cdo;
    float Hwdgo, nco, cso,hico,cbdo,cido,cbso,ciso,azho,azwo,Wwdgo;
    float doco,dmco,lmto,reso, wcpo,vcoo, cpwto,cwto,k1,roe,spwt;
    int anc, turns;

    winding()
    { cdo=0.0; Hwdgo=0.0; nco=0.0; cbso=0.0; hico=0.0; cbdo=0.0;
     cido=0.0; ciso=0.0; azho=0.0; azwo=0.0; doco=0.0; dmco=0.0;
     lmto=0.0; reso=0.0; wcpo=0.0; vcoo=0.0; cpwto=0.0;cwto=0.0;
     Wwdgo=0.0;}

    winding
    ( float cd, float Hwdg, float nc, float hic, float cbd,
        float cid, float cbs, float cis, float azh, float azw,
        float Wwdg, float doc, float dmc, float lmt, float res,
        float wcp, float vco, float cpwt, float cwt )

        { cdo=cd; Hwdgo=Hwdg; nco=nc; cbso=cbs; hico=hic; cbdo=cbd;
        cido=cid; ciso=cis; azho=azh; azwo=azw; doco=doc; dmco=dmc;
        lmto=lmt; reso=res; wcpo=wcp; vcoo=vco; cpwto=cpwt;cwto=cwt;
        Wwdgo=Wwdg;}
```

```
            winding (spec x)
            {
                  int anc, vpc;
                  float nco, k1, k2, cbdo,cido, cbso, ciso;

                  int v = x.getv();
                  float a = x.geta();
                  int t = x.gett();
                  int h = x.geth();
                  int dic = x.getdic();
//                float tap = x.gettap();


            cout<<endl<<"Enter the value of voltage per coil";
        cin>> vpc;
        cout<<endl<<"Enter the value of hyl and dal,in mm";
        cin>> hyl >> dal;
        cout<<endl<<"Enter assumed current density, amp/mm.sq.";
        cin>>cdo;

        nco = v/vpc;
        cout<<endl<<"nco= "<<nco;
        cout<<endl<<"Enter the integer value of number of coils";
        cin>>anc;
        k2 = 2. * hyl + (anc - 1.)*dal;
        Hwdgo = h - k2;
        hico = Hwdgo*1.03 / anc;
        cso = a / cdo;
        cout<<endl<<"bare conductor section= "<<cso;

        cout<<endl<<"Enter data for round conduc.from standard table:"
              "bare dia.(cbdo)-mm., insul.dia.(cido)-mm.,"
              "bare section(cbso)-mm.sq.,insulated sect.(ciso)-mm.sq";
        cin>>cbdo>>cido>>cbso>>ciso;

        cdo = cdo*cso/cbso;
        azho = hico/ciso;
        azwo = t/(anc*azho);
        cout<<endl<<"azho= "<<azho<<"azwo= "<<azwo;
        cout<<endl<<"Enter modified values of azho and azwo";
        cin>>azho>>azwo;
        turns = anc*azho*azwo;
        Hwdgo = cido*azho*anc/1'.03;
        Wwdgo = cido*azwo;

// Compute conductor loss and weight
        float roe, spwt;
        float PI= 3.14159L ;
        k1 = (1e-3f);
        cout<<endl<<"Enter values of sp.resis.-ohm.mm.sq/m &"
        <<"sp. weight,kg/mm-cubed";
        cin>>roe>>spwt;

        dmco = dic + Wwdgo;
        lmto = PI*dmco;
```

```
reso = roe*lmto*turns*k1/cbso;
wcpo = a*a*reso;
vcoo = turns*lmto*cbso;
cpwto = spwt*vcoo*k1;
}

void display()
{
cout<<endl<<"DESIGN SHEET";
cout<<endl<<"clearances";
cout<<endl<<"Axial between yoke and coil at each end"<<hyl<<"mm";
cout<<endl<<"Axial duct between two consecutive coils"<<dal<<"mm";
cout<<endl<<"modified current density= "<<cdo<<"amp per mm.sq,";
cout<<endl<<"Dia. of round bare conductor= "<<cbdo<<"mm";
cout<<endl<<"Dia. of round insulated conductor= "<<cido<<"mm";
cout<<endl<<"No. of axial coils = "<<anc;
cout<<endl<<"No. of radial strands in coil= "<<azho;
cout<<endl<<"No. of axial strands per coil= "<<azwo;
cout<<endl<<"No. of turns per phase= "<<turns;
cout<<endl<<"Winding height= "<<Hwdgo<<"mm.";
cout<<endl<<"Winding width= "<<Wwdgo<<"mm.";
cout<<endl<<"Conductor loss= "<<wcpo<<"watt";
cout<<endl<<"Conductor weight= "<<cpwto<<"g";
}
};

    int main() {
    spec m;
    m.get_value();
    cout<<"\n";m.display();

    winding n;
    n = m;
    cout<<"\n";n.display();

    return 0;
}
```

Figure 9.7(a) A computer program in Lynax for Cross-over winding.

RESULT

Enter the data from 'Result', Figure 6.5 and get the result.

9.6.2 Helix Winding

```
//Design of Helix Winding
#include<iostream>
#include<stdio.h>
#include<math.h>
using namespace std ;

class spec
{
    public :
    int to, ho, dico ;
```

```
    float ao, tapo ;
    spec()
    { ao=0.0 ; to=0.0 ; ho=0.0 ; dico=0.0 ; tapo=0.0 ; }
    spec( float a, int t, int h, int dic, float tap )
        { ao=a ; to=t ; ho=h ; dico=dic ; tapo=tap ; }
    void getvalue(void)
        { cout<<endl<<"Enter the values of : ao, to, ho, dico, tapo" ;
        cin>> ao >> to >> ho >> dico >> tapo ;
        }
    void display()
        { cout<<"amp/ph= "<< ao <<"turns/ph= "<< to <<"ht. of wondow= "
            << ho <<"internal dia. of coil= "<< dico <<"%tap= "<< tapo;
        }
    float geta() { return ao ; }
        int gett() { return to ; }
        int geth() { return ho ; }
        int getdic() { return dico ; }
        float gettap() { return tapo ; }
};

class winding : public spec
{
    public :
    float hyl ; // clearance between yoke and coil at each end, mm.
    float dal ; // axial duct width between two consecutive coils,mm.
    float cdo ; // assumed current density
    float Hwdgo, nco, cso, hico, cbho, ciho, cbwo, ciwo, cbso, ciso ;
    float Wwdgo, doco, dmco, lmto, reso, wcpo, vcoo, cpwto, cwto ;
    float k1, k2, roe, spwt ;

    winding()
    { cdo=0.0 ; Hwdgo=0.0 ; nco=0.0 ; hico=0.0 ; cbho=0.0 ; ciho=0.0 ;
    cbwo=0.0 ; ciwo=0.0 ; cbso=0.0 ; ciso=0.0 ; Wwdgo=0.0; doco=0.0 ;
    dmco=0.0 ; lmto=0.0 ; reso=0.0 ; wcpo=0.0 ; vcoo=0.0 ; cpwto=0.0;
    cwto=0.0 ; }

    winding ( float cd, float Hwdg, float nc, float hic, float cbh,
            float cih, float cbw, float ciw,float cbs, float cis,
            float Wwdg,float doc, float dmc,float lmt, float res,
            float wcp, float vco, float cpwt, float cwt )
    { cdo=cd ; Hwdgo=Hwdg ; nco=nc ; hico=hic ; cbho=cbh ; ciho=cih ;
    cbwo=cbw ; ciwo=ciw ; cbso=cbs ;ciso=cis ;Wwdgo=Wwdg; doco=doc ;
    dmco=dmc ; lmto=lmt ; reso=res ;wcpo=wcp ; vcoo=vco ;cpwto=cpwt;
    cwto=cwt ; }
    winding (spec x )
    {
    cout<<endl<<"Enter the number of axial coils" ;
    cin>> nco ;
    cout<<endl<<"Enter the values of hyl(mm.) & dal(mm.)" ;
    cin>> hyl >> dal ;
    cout<<endl<<"Enter the value of assumed current-density" ;
    cin>> cdo ;
    float a = x.geta() ;
      int t = x.gett() ;
      int h = x.geth() ;
    float dic= x.getdic() ;
```

```
//    float tap= x.gettap() ;

      float k1 = 1e-3f ;
      float k2 = 2.0*hyl + (nco-1.)*dal ;
      Hwdgo = h - k2 ;
      hico = Hwdgo* 1.03 / (( t + 1.0 )* nco ) ;//ht.of insul.conductor
      cso = a / cdo ; //conduc. section, mm-sq.
      cout<<endl<<"hico= "<<hico ;
      cout<<endl<<"cso= "<<cso ;
      getchar() ;
//                 since high current, rectangular strap conductor is used.

      cout<<endl<<"Enter from the standard table the values of :"
          <<" cbho (height-bare conduc.-mm.,)"
          <<" ciho (height-insul. conduc.- mm.)"
          <<" cbwo (width-bare conduc.-mm.)"
          <<" ciwo (width-insul. conduc.-mm.)"
          <<" cbso (section-bare conduc.- sq.mm.)"
          <<" ciso (section-insul. conduc.- sq.mm.)" ;
      cin >> cbho >> ciho >> cbwo >> ciwo >> cbso >> ciso ;
getchar() ;
      cdo = cdo * cso / cbso ;
Hwdgo = ciho * ( t + 1. ) / 1.03 ;
Wwdgo = ciwo ;

float PI = 3.14159L ;

cout<<endl<<"Enter the values of :"
        <<"roe (spec.resistance- ohm.mm-sq / metre )"
        <<"spwt (spec. weight- kg / mm.cubed )" ;
cin>>roe>>spwt ;
cout<<endl<< "roe= "<<roe <<"spwt= "<<spwt ;

doco = dic + 2. * Wwdgo ;                    //outside dia. of coil, mm.
dmco = 0.5 * (dic + doco) ;                  //mean dia. of coil, mm.
lmto = PI * dmco ;                           //length of mean turn, mm.
reso = roe * lmto * t * k1 / cbso ;          //coil resistance, ohm
wcpo = a * a * reso ;
vcoo = t * lmto * cbso ;                     //cond.volume- mm.cubed
cpwto = spwt * vcoo * k1 ;
}

void display()
    { cout<<endl<<" DESIGN DATA " ;

      cout<<endl<<"axial clearance between yoke & coil, at both ends= "<<hyl;
      cout<<endl<<"axial duct width between consecutive coils= "<<dal<<"mm" ;
      cout<<endl<<"Modified current-density= "<<cdo<<"amp/mm.sq" ;
      cout<<endl<<"No. of axial coils= "<<nco ;
      cout<<endl<<"height of rectangular bare conduc= "<<cbho<<"mm" ;
      cout<<endl<<"width of rectangular bare conduc= "<<cbwo<<"mm" ;
      cout<<endl<<"height of rectangular insul. conduc= "<<ciho<<"mm" ;
      cout<<endl<<"width of rectangular insul. conduc= "<<ciwo<<"mm" ;
      cout<<endl<<"radial width of coil= "<<Wwdgo<<"mm" ;
      cout<<endl<<"winding height= "<<Hwdgo<<"mm" ;
      cout<<endl<<"conductor loss= "<<wcpo<<"watt" ;
      cout<<endl<<"conductor volume= "<<vcoo<<"mm-cubed" ;
```

```
            cout<<endl<<"conductor weight= "<<cpwto<<" g ;
        }
};

int main()
{
        spec m ;
        winding n ;
        m. getvalue() ;

        n = m ;
        cout<<"\n"; n.display() ;
}
```

Figure 9.7(b) A computer program in Lynax for Helix winding.

(matching with the program in Windows in Figure 6.5(b)

RESULT

Enter the data from 'Result', Figure 6.5(b) and get the result.

9.6.3 Design of Multilayer Helix Type

(In match with Figure 6.5(c)

```
//Design of Multilayer helix winding
//Given specifications : current/phase (ap), %tap (tap), turns/ph. (tp),
//              height of window (ht- mm.), internal dia. of coil (dic-mm.)

#include<iostream>
#include<math.h>
#include<stdio.h>
using namespace std ;

class spec
{
        public:
            int c, d, e ;
        float a, b ;
        void getdata (float a, float b, int c, int d, int e) ;
        void display() ;
        void multilayerhelix (float a, float b, int c, int d, int e) ;
};

void spec::getdata(float ap, float tap, int tp, int ht, int dic)
                                { a=ap; b=tap; c=tp; d=ht; e=dic; }
void spec::display()
{
        cout<<endl<<"amp / ph= "<<a ;
        cout<<endl<<" % tap = "<<b ;
        cout<<endl<<"turns / ph= "<<c ;
        cout<<endl<<"winding ht= "<<d ;
        cout<<endl<<"internal dia. of coil= "<<e ;
}
```

```
void spec::multilayerhelix(float ap, float tap, int tp, int ht, int dic)
{
        float cd ;        //assumed current-density, amp per mm.sq.
        float hyl ;       //axial clearance bet. yoke & coil, mm.
        float drc ;       //radial width of ducts bet. layers, mm.
        float roe ;       //sp. resis. of conduc. material (Al),ohm/mm-sq.
        float spwt;       //sp. weight of conduc. material, kg/mm-cubed
        cout<<endl<<"Enter the values of : cd, hyl, drc, roe, spwt" ;
        cin>> cd >> hyl >> drc >> roe >> spwt ;
            int til ;
        float f, g, h, k, l, m, n, o, q, r, u ;
        float cbh, cih, cbw, ciw, cbs, cis ;
        int O, M ;

        float PI = 3.14159L ;
        float k1 = 1e-3f ;

        cout<<endl<<"Enter :        0 = 1 for TWO layer,"
                                   "0 = 2 for THREE later,"
                                   "0 = 3 for FOUR layer." ;

cin>> O ;

if (O==1)                                        // no. of layer = m
{
        cout<<endl<<"number of layer m = 2" ;
        void twolayer(int,int&)
            {
            twolayer(c,til) ;
            }
        } else
        if (O==2)
        {
            cout<<endl<<"number of layer m = 3" ;
            void threelayer(int,int&) ;
        {
            threelayer(c,til) ;
            }
        } else
        if (O==3)
        {
            cout<<endl<<"number of layer m = 4" ;
            void fourlayer(int,int&) ;
            {
            fourlayer(c,til) ;
            }
        }
        getchar() ;

        f = d + ( 2. * hyl ) ;
        h = f / (1.03 * ( til + 1.0 )) ;             //height of insulated coil
        cout<<endl<<"height of insulated coil= "<< h ;
        getchar() ;
        g = a / cd ;                                 //conduc. section, mm-sq.
        cout<<endl<<"conductor section=          "<< g ;

//    Rectangular strip conductors are chosen
```

```cpp
    cout<<endl<<"Enter the values of :          ht. of bare (cbh-mm.), "
                                        "ht. of insul.(cih-mm.),"
                                        "width of bare (cbw-mm),"
                                        "width of insul.(ciw-mm.)"
                                        " conductor, "
              " bare conductor section (cbs-mm.sq), &"
              " insulated conduc. section (cis-mm.sq)"
              " from Standard Table." ;
    cin>> cbh >> cih >> cbw >> ciw >> cbs >> cis ;
    getchar() ;

    cd = a / cbs ;                          //modified current-density
    f = cih * (til + 1.0) / 1.03 ;          //winding height,mm

    cout<<endl<<"Enter: Number of layer M" ;
    cin>>M ;
    k = M * ciw + (M - 1.) * drc ;          //radial width of coil,mm

//      Compute Copper loss, watt and copper weight, kg

    l = e + 2.0 * k ;                       //outside dia. of coil, mm.
    m = 0.5 * (e + l) ;
    n = PI * m ;
    o = roe * n * c * k1 / cbs ;
    q = a * a * o ;                         //conductor loss, watt
    r = c * n * cbs ;                       //conductor volume, mm-cubed
    u = spwt * r * k1 ;                     //conductor weight, kg

    cout<<endl<<"      DESIGN SHEET " ;
    cout<<endl<<"axial clearance between yoke & coil=        "<<hyl<<"mm" ;
    cout<<endl<<"radial width of ducts between layers=       "<<drc<<"mm" ;
    cout<<endl<<"modified current density=     "<<cd<<"amp/mm-sq" ;
    cout<<endl<<"height of rect. bare conduc.=  "<<cbh<<"mm" ;
    cout<<endl<<"width of rect. bare conduc.=   "<<cbw<<"mm" ;
    cout<<endl<<"height of rect.insul.conduc=   "<<cih<<"mm" ;
    cout<<endl<<"width of rect.insul.conduc=    "<<ciw<<"mm" ;
    cout<<endl<<"radial width of coil=          "<< k <<"mm" ;
    cout<<endl<<"winding height=                "<< f <<"mm" ;
    cout<<endl<<"conductor loss=                "<< q <<"ohm" ;
    cout<<endl<<"conductor volume=              "<< r <<"mm-cubed" ;
    cout<<endl<<"conductor weight=              "<< u <<"kg" ;
};
//                                          twolayer

void twolayer(int c,int&til)
{
    int tol, M ;
    M = 2.0 ;                               // number of layers
    float tpl ;                             // turns per phase per layer
    tpl = c / M ;
        cout<<endl<<"TWO LAYER" ;
cout<<endl<<"tpl = "<< tpl ;

cout<<endl<<"Enter turns/phase:inner layer(til)"<<"outer layer"
        "(tol)" ;

    cin>> til >> tol ;
```

```
        cout<<endl<<"turns/phase-inner layer= "<< til<<"outer layer= "
        <<tol ;
        }
//          threelayer
        void threelayer(int c,int&til)
        {
            int tol, tml, M ;
            M = 3.0 ;                       // number of layers
            float tpl ;                     // turns per phase per layer
            tpl = c / M ;
        cout<<endl<<"tpl = "<< tpl ;

        cout<<endl<<"Enter turns/phase-inner layer(til)"<<"middle layer"
            "(tml)"<< "outer layer(tol)" ;

            cin>> til >> tml >> tol ;
            cout<<endl<<"turns/phase-inner layer= "<< til ;
            cout<<endl<<"turns/phase-middle layer= "<< tml ;
            cout<<endl<<"turns/phase-outer layer= "<< tol ;
        }
//          fourlayer
        void fourlayer(int c,int&til)
        {
            int tol, t2l, t3l, M ;
            M = 4.0 ;                       // number of layers
            float tpl ;                     // turns per phase per layer
            tpl = c / M ;
            cout<<endl<<"tpl = "<< tpl ;
        cout<<endl<<"Enter turns/phase-inner layer(til)" ;
            cout<<endl<<"Enter turns/phase-second layer(t2l)" ;
        cout<<endl<<"Enter turns/phase-third layer(t3l)" ;
        cout<<endl<<"Enter turns/phase-outer layer(tol)" ;
        cin>> til >> t2l >> t3l >> tol.;
        cout<<endl<<"turns/phase-inner layer= "<< til ;
        cout<<endl<<"turns/phase-second layer= "<< t2l ;
        cout<<endl<<"turns/phase-third layer= "<< t3l ;
        cout<<endl<<"turns/phase-outer layer= "<< tol ;
        }

int main()
{
    spec S ;
    float ap, tap ;
    int tp, ht, dic ;

    cout<<endl<<"Enter the data: a, b, c, d, e" ;
    cin>> ap >> tap >> tp >> ht >> dic ;

    S.getdata(ap, tap, tp, ht, dic) ;
    S.display() ;
    S.multilayerhelix(ap, tap, tp, ht, dic) ;
}
```

Figure 9.7(c)   A computer program in Lynax for Multilayer Helix winding.

RESULT

Enter the data those under 'Result'– figure 6.5(c) and get the result.

9.7   DESIGN OF THREE PHASE CORE TYPE TRANSFORMER

In Figure 9.8, a computer program has been displayed for a 25-kVa three-phase core type Transformer (the same as in article 6.7),written in Linax, using 'class with friend function' as described in Figure 6.6(a).

```
DESIGN OF THREE PHASE CORE-TYPE TRANSFORMER
#include<iostream>
#include <stdio.h>
#include <math.h>
#include "loscrgo.h"
#include "MgVaC.h"
using namespace std ;

class SPEC
{
    private :
    int NLos, Llos ;
    public :
        int VAmp, phase, freq ;
        float PVph, SVph ;
    SPEC()
        { VAmp=0.0;PVph=0.0;SVph=0.0;phase=0.0;freq=0.0;NLos=0.0;Llos=0.0; }
    SPEC(int Q,float vh,float vl,int ph,int fr,int Po,float Pl )
        { VAmp=Q; PVph=vh; SVph=vl ; phase=ph ; freq=fr; NLos=Po; Llos=Pl; }
    void display()
        { cout<<"Rated VA= "<< VAmp<<"Primary voltage/ph= "<< PVph
            <<"Secondary voltage/phase= "<<SVph<<"No. of phase= "<<phase
            <<"frequency= "<<freq<<"No load loss= "<< NLos
            <<"Load loss at 75 deg. cel.= "<< Llos;
        }
                                int getQ() { return VAmp ; }
                               float getvh() { return PVph ; }
                              float getvl() { return SVph ; }
                               int getph() { return phase ; }
                               int getfr() { return freq ; }
                               int getPo() { return NLos ; }
                               int getPl() { return Llos ; }
            friend float TRAN(SPEC z) ;
};

float TRAN( SPEC z )
{
    int Q, ph, fr ;
    float vh, vl, q, Et, ac, ai, Bm, dia, cw, Aw, w, h, Ic, Im, Io ;
    float keq, cd, wsf, fst, fsp, opf ;

    float k1 = 1e-3f ;
    cout<<endl<<"Enter assumed values of:keq,Bmo,cd,wsf,fst,fsp";
    cin >> keq >> Bm >> cd >> wsf >> fst >> fsp ;

                        Q = z.getQ() ;
                        vh = z.getvh();
                        vl = z.getvl();
                        ph = z.getph();
                        fr = z.getfr();
```

```
//                          Po = z.getPo();
//                          P1 = z.getPl();
    q = Q * k1 /ph ;                            //Va per phase
    Et = keq * sqrt(q) ;                        //Voltage / turn
    cout<<endl<<"Voltage per turn= "<< Et ;


//          DESIGN OF CORE
//CRGO sheet steel, 0,35 mm. in thickness and mitred 45 deg. will be used

    float k2 = 1e6f ;
        ai = Et*k2 / (4.44*Bm*fr) ;             //Net core section area, mm.sq
    cout<<endl<<"Computed core section= "<< ai <<"mm.sq." ;

        ac = ai / fst ; //Gross core section, mm.sq,
    cout<<endl<<"Gross core section= "<< ac <<"mm.sq." ;

//          Compute no. of Steps and Width of Limbs
    float k3, k4, ratio, HA ;
    float PI = 3.14156L ;

    cout<<endl<<"Choose no. of steps in the core and enter Optimum fill" ;
    cin >> opf ;

    dia = sqrt(ac * 4 / (PI * opf)) ;           //for two-stepped core
    cw = 0.85 * dia ;
    cout<<endl<<"Enter: core width(cw)-mm.= "<<cw
            <<"diameter(dia)= "<<dia ;
    cin >> cw >> dia ;

        ai = PI * fst * fsp * dia * dia / 4. ; //modified ai
        Bm = Et * k2 /(4.44 * ai * fr) ; //modified Bm
    Aw = Q * k2 /(ph * 1.11 * ai * wsf * Bm * cd * fr); //window area
//  Aw is the product of window-width(w) and window height(b).
//  The ratio (h / w) is generally between 1.5 and 2.5 depending on
//  Customer's choice.
    cout<<endl<<"Enter ratio of window height and width";
    cin >> ratio ;
        k3 = Aw / ratio ;
        w = sqrt(k3) ;                          //Window width
        h = ratio * w ;                         //Window height
        HA = h + 2.* cw ;                       //Height of Limb


//                      CORE LOSSES & NO LOAD CURRENT
        k4 = 1e-6f ;
    float Crwt, Crlos, x, Mgva, y ;

        Crwt = 7.85 * k4 * opf * ai * (3.*HA+2.*w) ; //Total weight
//                      of core(kg) assuming yoke section=core section.
    cout<<endl<<"Core weight= "<< Crwt<<"kg" ;
    cout<<endl<<"Maxm. flux density= "<< Bm ;
        y = Bm ;
    float loscrgo(float) ;
        x = loscrgo(y) ;
        Crlos = x * Crwt ;
            cout<<endl<<"Core loss= "<< Crlos <<"watt";
        y = Bm ;
    float MgVaC(float) ;
```

```
        x = MgVaC(y) ;
Mgva = x * Crwt ;
                cout<<endl<<"Magnetising volt-ampere= "<< Mgva ;
                    Ic = Crlos / (ph*vh) ; //Core loss current
                    Im = Mgva / (ph*vh) ; //Magnetising current
                    Io = sqrt(Ic*Ic+Im*Im) ; //No load current
//                  k5 = Im*100./Io; //Magnetising current as
//                          per cent of no load current
        cout<<endl<<"DESIGN SHEET- CORE";
cout<<endl<<"gross Core section= "<< ac <<"mm.sq";
cout<<endl<<"Net iron section= "<< ai <<"mm.sq";
cout<<endl<<"Modified flux-density= "<< Bm <<"T";
cout<<endl<<"Window area= "<< Aw <<"mm.sq";
cout<<endl<<"Window width= "<< w <<"mm.";
cout<<endl<<"Window height= "<< h <<"mm.";
cout<<endl<<"Core weight= "<< Crwt <<"kg";
cout<<endl<<"Core loss= "<< Crlos <<"watt";
cout<<endl<<"Magnetising volt-ampere= "<< Mgva ;
cout<<endl<<"Core loss component of no-load current= "<<Ic<<"amp";
cout<<endl<<"Magnetising current= "<< Im <<"amp";
cout<<endl<<"No load current= "<< Io <<"amp";


        cout<<endl<<" DESIGN OF LOW VOLTAGE WINDING" ;
    int hyl, hy, dal, da, Ccl, C ;
float v, cdm, cdl, tl, t, ncl, nc, cbhl, cbh, cihl, cih, cbwl, cbw ;
float ciwl, ciw, cbsl, cbs, cisl, cis, zhl, zh, zwl, zw, dmcl, dmc ;
float Hwdgl, Hwdg, Wwdgl, Wwdg, resl, res, cpwtl, cpwt, wcp, wcpl ;
float vco, vcol, odlv ;


        v = vl ;

void winding(int, float, float&, float&, float&, float&, float&,
float&, float&, float&, float&, float&, float&, float&, float&,
float&, int&, int&, int&, float&, float&, float&, float&,float&,
float&, float&, float&, float&) ;

winding (ph, q, Et, v, cd, h, dia, cdm, t, nc, cbh, cih, cbw, ciw,
        cbs, cis, hy, da, C, zh, zw, dmc,Hwdg, Wwdg, res, wcp, vco,
        cpwt) ;
        cdl=cdm, tl=t, ncl=nc, cbhl=cbh, cihl=cih, cbwl=cbw,
        ciwl=ciw, cbsl=cbs, cisl=cis, zhl=zh, zwl=zw, dmcl=dmc,
        Hwdgl=Hwdg, Wwdgl=Wwdg, resl=res, cpwtl=cpwt, wcpl=wcp,
        vcol=vco, dal=da, Ccl=C, hyl=hy ;

        cout<<endl<<"DESIGN SHEET- L. V. WINDING";

cout<<endl<<"Current density= "<< cdl <<"amp/mm.sq";
cout<<endl<<"Number of turns= "<< tl ;
cout<<endl<<"Number of coils= "<< ncl ;
cout<<endl<<"Winding details: Rectangular strap:";
cout<<endl<<" Bare Insulated ";
cout<<endl<<"Height"       <<cbhl     << cihl     <<"mm.";
cout<<endl<<"Width"        <<cbwl     << ciwl     <<"mm.";
cout<<endl<<"Section"      <<cbsl     << cisl     <<"mm.sq";
cout<<endl<<"            clearences " ;
cout<<endl<<" Axial between yoke & winding= "<<hyl<<"mm.";
cout<<endl<<"Axial between 2 consecutive l.v.coils= "<<dal<<"mm.";
```

```
cout<<endl<<"Radial between limb & l.v. winding= "<< Ccl<<"mm.";
 cout<<endl<<"Number of axial strands= "<< zhl ;
 cout<<endl<<"Number of radial strands= "<< zwl ;
  cout<<endl<<"Winding height= "<< Hwdgl <<"mm." ;
  cout<<endl<<"Winding width= "<< Wwdgl <<"mm." ;
  cout<<endl<<"Resistance per phase= "<< resl <<"ohm" ;
  cout<<endl<<"Conductor loss= "<< wcpl <<"watt " ;
    cout<<endl<<"Conductor volume= "<< vcol <<"mm.cubed" ;
    cout<<endl<<"Conductor weight= "<< cpwt <<" kg " ;
odlv = dmcl + Wwdgl ;
cout<<endl<<"Outside diameter of lv coil= "<< odlv <<" mm. " ;

              cout<<endl<<"DESIGN OF HIGH VOLTAGE WINDING" ;

      int hyh, dah, Co ;
   float cdh, th, nch, cbdh, cidh, cbsh, cish, zhh ;
   float zwh, dmch, Hwdgh, Wwdgh, resh, cpwth, wcph, vcoh ;
      v = vh ;
   void winding(int, float, float&, float&, float&, float&, float&,
   float&, float&, float&, float&, float&, float&, float&, float&,
   float&, int&, int&, int&, float&, float&, float&, float&,float&,
   float&, float&, float&, float&) ;
winding (ph, q, Et, v, cd, h, dia, cdm, t, nc, cbh, cih, cbw, ciw,
      cbs, cis, hy, da, C, zh, zw, dmc,Hwdg, Wwdg, res, wcp, vco,
      cpwt) ;
      cdh=cdm, th=t, nch=nc, cbdh=cbh, cidh=cih,
      cbsh=cbs, cish=cis, hyh=hy, dah=da, Co=C, zhh=zh,
      zwh=zw, dmch=dmc, Hwdgh=Hwdg, Wwdgh=Wwdg, resh=res,
      cpwth=cpwt, wcph=wcp, vcoh=vco ;

   cout<<endl<<"DESIGN SHEET- H. V. WINDING";

   cout<<endl<<"Current density= "<< cdh <<"amp/mm.sq";
      cout<<endl<<"Number 0t turns= "<< th ;
      cout<<endl<<"Number of coils= "<< nch ;
      cout<<endl<<"Winding details: Round ";
      cout<<endl<<" Bare Insulated ";
      cout<<endl<<"Diameter"<<cbdh << cidh <<"mm.";
      cout<<endl<<"Section" <<cbsh << cish <<"mm.sq";
      cout<<endl<<" clearences " ;
      cout<<endl<<" Axial between yoke & winding= "<<hyh<<"mm.";
      cout<<endl<<"Axial between 2 consecutive h.v.coils= "<<dah<<"mm.";
      cout<<endl<<"Radial between h.v. & l.v. winding= "<< Co<<"mm.";
       cout<<endl<<"Number of axial strands= "<< zhh ;
        cout<<endl<<"Number of radial strands= "<< zwh ;
         cout<<endl<<"Winding height= "<< Hwdgh <<"mm." ;
          cout<<endl<<"Winding width= "<< Wwdgh <<"mm." ;
           cout<<endl<<"Resistance per phase= "<< resh <<"ohm" ;
            cout<<endl<<"Conductor loss= "<< wcph <<"watt " ;
              cout<<endl<<"Conductor volume= "<< vcoh <<"mm.cubed" ;
               cout<<endl<<"Conductor weight= "<< cpwth <<" kg " ;

//               COMPUTATION OF PERCENT IMPEDANCE
    float ah, k9, Dm, avHwdg, tWwdg, cx, KR, LL, X, req, r, R, Z ;

    k9 = Q /(ph * ph * Et * Et) ;
  float k6 = pow(10,3) ;
```

```
    ah = q * k6 / vh ; //current, h, v, winding
    Dm = 0.5 * (dmcl + dmch) ; //av. of mean dia.-hv & lv coils
avHwdg = 0.5 * (Hwdgl+Hwdgh) ; //av. height-hv & lv coils
tWwdg = Wwdgl + Wwdgh ; //total width of hv & lv coils
   cx = Co + tWwdg/3. ;
   KR = 1.0 - ((Co+tWwdg) / (PI*avHwdg)) ; //RAGAWOSKY CO-EFF.
   LL = avHwdg / KR ;
   X = 0.0395 * k9 * k1 * cx * PI * Dm / LL ; //%reactance
   r = vh / vl ; //ratio of primary to secondary ph.voltage
req = resl * r * r ; // lv resis./ph. in hv terms.
   R = (req + resh) * ah * 100. / vh ; //% resis. in hv terms
   Z = sqrt(R*R + X*X) ; //%impedance
```

```
//                          LOSSES IN WINDINGS

float tcpwt, cnlos, strlos, FLlos, Eff, Reg, theta ;
float k10 ;
    tcpwt = cpwtl + cpwth ; //total conductor weight
    cnlos = ph * vh * ah * R / 100. ; //total conduc.loss at upf
strlos = 0.07 * cnlos ;//considering stray loss=7%total cond loss
FLlos = Crlos + cnlos + strlos ; //full-load loss at u.p.f.
  k10 = (100.*R)*(100.*R) + (100.*X)*(100.*X) ;
  Reg = sqrt(k10) / 100. ; // % Regulation
  Eff = Q * 100. / (Q + FLlos) ; // % efficiency at fl & upf.

cout<<endl<<" PERFORMANCE SHEET ";

cout<<endl<<"Per cent resistance= "<< R ;
cout<<endl<<"Per cent reatance = "<< X ;
cout<<endl<<"Per cent impedance = "<< Z ;
cout<<endl<<"Per cent regulation on full-load & upf.= "<< Reg ;
cout<<endl<<"Total conductor weight at f.l.& upf== "<<tcpwt<<"kg";
cout<<endl<<"Total conductor loss at f.l.& upf.= "<<cnlos<<"watt";
cout<<endl<<"Total Load loss at upf. = "<<FLlos<<"watt" ;
cout<<endl<<"Per cent Efficiency on f.l.& upf.= "<<Eff ;
```

```
//                       CORE-WINDING ASSEMBLY

float Chh, k11, k12, Wc, LA, WA ;

cout<<endl<<"Enter Chh, mm.- clearance beteen to h.v.windings"
           " of two phases" ;
cin >> Chh ;
Wc = Ccl + tWwdg + Co, + 0.5 * Chh ; //Total width of windings
//                             plus radial clearances at any limb.
k11 = 0.5 * w - Wc ; //TO CHECK whether total window and
//                    clearance assembly is accomodated in half window-space
cout<<endl<<" k11= "<< k11 ;
//     [NOTE ; k11 should be ideally zero or a small positive value]
k12 = 2.* ( Ccl + tWwdg + Co ) ;
 LA = 2.* w + k12 + 3.* dia ; //Length of Assembly
 WA = k12 + dia ; //Width of Assembly
cout<<endl<<"Length of assembly= "<<LA<<"mm."
         <<"Width of assembly= "<<WA<<"mm."
         <<"height of assembly= "<<HA<<"mm." ;
```

```
//           DESIGN OF TANK
```

```
    int Ntube ;
float HT, WT, LT, KT, AT, AT1, k13, k14, qet, mCS ;
float Ltube, Atube, ct1, ct2, ct3, ct4, ct5 ;
    cout<<endl<<"Enter the clearences between the Core winding"
            "assembly and the tank walls- "
                        "lengthwise on each side (ct1- mm.),"
                        "widthwise on each side (ct2- mm. ),"
                        "at top - upto oil level (ct3- mm.),"
                        "between oil level and lid (ct4-mm.),"
                        "at base (ct5- mm.)." ;
cin >> ct1 >> ct2 >> ct3 >> ct4 >> ct5 ;

HT = HA + ct3 + ct4 + ct5 ;               // height of the tank, mm.
WT = WA + 2. * ct2 ;                       // width of the tank, mm.
LT = LA + 2. * ct1 ;                       // length of the tank, mm.
cout<<" Height= "<<HT<<"Width= "<<WT<<"Length= "<<LT<<"mm.";

AT = 2.* k4 * HT * (WT + LT) ; //cooling surface of the Tank m-sq.
qet = FLlos / AT ; //heat dissipation, watts/m.sq.
cout<<endl<<" AT= "<<AT<<"m.sq."<<"qet= "<<qet<<"m.sq.";
cout<<endl<<"Enter Empirical Coefficient :"
        "KT=5.14 for plain Tank,"
            "KT= 2.90 for Tank with cooling tubes,"
                                "KT= 1.56 for Tank with radiators." ;
        cin >> KT ;
k13 = qet / KT ;
cout<<endl<<"k13= "<< k13 ;

float temp(float);
    theta = temp(k13) ;
cout<<endl<<"Mean top-oil temp.rise= "<<theta<<"deg.celsius" ;
    if (theta <= 40.)//when maximum top-oil temp-rise is 40 deg. cel.
{
    cout<<endl<<"Mean top-oil temp.rise= "<<theta<<"deg.celsius" ;
}
else
{
        KT = 2.9 ;
    k14 = pow(40.,1.25);
    qet = KT * k14 ;
    AT1 = FLlos / qet ;
    mCS = AT1 - AT ; //reqd. minm.cooling surface for mean oil
//                              temp-rise of 40 deg, celsius
cout<<endl<<"Enter: Perimeter of Tube (Atube,m.),"
        "and Effective Length of each tube (Ltube,m)" ;
cin >> Atube >> Ltube ;
Ntube = mCS /( Atube * Ltube );
cout<<endl<<"Minimum number of tubes required= "<<Ntube ;
    }
return(Ntube) ;
}

void winding(int ph, float q, float Et, float& v, float& cd, float& h,
    float& dia, float& cdm, float& t, float& nc, float& cbh,
    float& cih, float& cbw, float& ciw, float& cbs, float& cis,
    int& hy, int& ds, int& C, float& zh, float& zw, float& dmc,
    float& Hwdg,float& Wwdg,float& res, float& wcp, float& vco,
```

```
        float& cpwt)
    {
        float k1, k6, k7, da ;
        float at, vpc, atc, tc, cols, hco, azh, azw, dic, lmt ;
        double a, ac, acs ;
        float PI = 3.14159L ;
            k1 = pow(10,-3) ;
            k6 = pow(10,3) ;

            a = q * k6 / v ;                    //winding current / phase
            at = v / Et ;                       //computed turns/phase
            acs = a /cd ;                       //conductor section, mm.sq.
        cout<<endl<<"conductor section (acs,mm.sq)= "<< acs
//          Conductor Section
//[Note: For the design, Rectangular Strap conductor is used for l.v.wndg
        cout<<endl<<"Enter conductor data fron STANDARD TABLE:"
                    "for RECTANGULAR STRAP conductor: "
        "bare height(cbh-mm.),bare width(cbw-mm.),bare section(cbs-mm.sq.),"
        "insulated height(cih-mm.),insulated width(ciw-mm.),insulated"
        "section(cis-mm.sq.)"
                    "for ROUND conductor: "
        "bare diameter(cbh-mm.),bare section(cbs-mm.sq.),"
        "insulated diameter(cih-mm.),insulated section(cis-mm.sq.)"
                    " PUT cbw = ciw = 0 " ;
        cin >> cbh >> cbw >> cbs >> cih >> ciw >> cis ;
            cdm = cd * acs / cbs ;
        cout<<endl<<"Modified winding current-density= "<<cdm<<"mm." ;

//                                        TURNS AND NUMBER OF COILS
        cout<<endl<<"Enter Voltage per coil (vpc)" ;
        cin >> vpc ;
            ac = v / vpc ;                      //number of coil
            atc = at / ac ;                     //number of turns per coil per phase
        cout<<endl<<"no. of coil= "<< ac
                <<"no. of turns per phase= "<< at
                    <<"no. of turns per coil per phase= "<< atc ;
        cout<<endl<<"Enter integer values of 'ac'-(nc), of 'at'-(t),"
                "and of 'atc'- (tc)" ;
            cin >> nc >> t >> tc ;
            cols = t * cis ; //sectional area of each coil

//                                        WINDING DIMENSIONS
//                                        Assign values of clearences:
//  hy = axial clearence between yoke and winding at top & at bottom
//  da = axial duct width between two consecutive coils
        cout<<endl<<"Enter the values of:"
        "axial clearence between yoke and winding(hy-mm.) and"
        "axial duct width between two consecutive coils(da-mm.)" ;
            cin >> hy >> da ;
            k7 = 2. * hy + (nc - 1.) * da ;
            hco = (h - k7) / nc ;               //height of each coil
            azh = hco / cih ;                   //axial strands,SHOULD BE AN INTEGER
            azw = tc / azh ;                    //radial strands,SHOULD BE AN INTEGER
        cout<<endl<<"No. of axial strands= "<< azh
                <<"No. of radial strands= "<<azw ;

        cout<<endl<<"Enter designer's choice of 'azh' and 'azw' " ;
```

```
    cin >> zh >> zw ;                                // SHOULD BE INTEGER

    Hwdg = (zh * cih) * ac + (nc - 1.) * da ; //Winding height
    Wwdg = zw * cih ; //Winding width

//                                    COMPUTED PARAMETERS
    float roe ;                             //Specific resistance, ohm-m.
    float spwt ;                            //Specific weight, kg/mm.cubed

    cout<<endl<<"Enter the values of 'roe' and 'spwt' ";
      cin >> roe >> spwt ;

    cout<<endl<<"Enter the value of:"
        "The radial clearance between Limb and Coil, for LV winding"
            "and also between HV coils (C-mm.)- Assummed the same " ;
        cin >> C ;
            dic = dia + 2.*C ;               //Internal dia. of coil, mm.
            dmc = dic + Wwdg ;               //Mean dia. of coil, mm.
            lmt = PI * dmc ;                 //Length of mean turn, mm.
            res = roe*lmt*t*k1/cbs ;         //Resistance per phase
            wcp = a * a * res ;              //Conductor loss, watt
            vco = t * lmt * cbs ;            //Vol.of conduc.material, mm-cubed
            cpwt = spwt * vco * k1 ;         //Weight,conduc.material,kg
    }

    float temp(float k12)
    {
        float z, k, theta ;
            z = log(k12) ;
            k = z / 1.25 ;
        theta = exp(k) ;
        return(theta) ;
    }

int main()
{
    SPEC s(25000., 11000., 250., 3.,50., 200., 1000.);

    cout<<"\n"; s.display() ;

    int Ntube ;
    Ntube = TRAN(s) ;
    cout<<endl<<"Number of Tubes required= "<< Ntube ;
    }
```

Figure 9.8   Design of Three-phase Core-type Transformer.

RESULT

Enter the data under the result from 'Result'– under Figure 6.7.

9.8   M.M.F. AND MAGNETISING REACTANCE OF A 3-PHASE INDUCTION MOTOR.

[This program matches with Figure 7.9]

```cpp
#include <iostream>
#include <string>
#include <stdio.h>
#include "mcur42st.h"
#include "carterOpen.h"
#include "cartersemi.h"
using namespace std ;

class statorcore
{
    public :

    int p ;
    float phim, D, L, Do, dSs, fst, dSc, ASc, BSc, ISc, y, x, FSc ;

    void getstatorcoredata()
    {
                                        cout<<endl<<"STATOR CORE" ;

    cout<<endl<<"Enter No. of poles(p),Maximum flux per pole (phim-Wb),"
      "Stator inside diameter(D-m),Stator outside diameter(Do-m),"
      "Gross Armature length(L-m),Depth of stator slot(dSs-m),"
      "Stacking factor(fst)" ;
    cin>>p>>phim>>D>>Do>>L>>dSs>>fst ;
    }
    void seestatorcoredata()
    {
    cout<<endl<<"No. of poles= "<<p        <<"maximum fluxper pole= "<<phim<<
    "Stator inside diameter= " <<D          <<"Stator outside diameter= "<<Do<<
    "Gross Armature length= "<<L            <<"Depth of stator slot= "<<dSs<<
    "Stacking factor= "<<fst ;

    }
    void computestatorcore()
    {
    dSc=0.5*(Do-(D+2.*dSs)) ;   // depth of stator core 0.5*(0.26-(0.159+2*0.025))
    cout<<endl<<"dSc= "<<dSc ;
    ASc=fst*dSc*L ;     // sectional area of stator core
    ISc = 0.5 * 3.1416 * (D+2.0* dSs+dSc) / p ; //length of flux-path in stator
    BSc=0.5*phim/ASc ;          // core flux-density in stator core
    getchar() ;

    y=BSc ;
    float mcur42st(float) ;
    x = mcur42st(y) ;
    getchar() ;
    string mystr ;
    FSc=x*ISc ;
    cout<<endl<<"FSc= "<<FSc ;
    }
    float statorcoremmf(){return FSc ;}
} ;

    class statorteeth
    {
    public :
```

```
    int p ;
    float phim, L, Ss, dSs, fst, wSt, ASt, BSt, lSt, y, x, FSt ;

    void getstatorteethdata()
    {
                                    cout<<endl<<"STATOR TEETH" ;
        cout<<endl<<"Enter No. of poles(p),Maximum flux per pole (phim-Wb),"
            "Gross armature length(L-m),No. of stator slots(Ss),Depth of stator"
            "Slot(dSs-m),width of Stator Slot(wSt-m),Stacking factor(fst)" ;
        cin>>p>>phim>>L>>Ss>>dSs>>wSt>>fst ;
    }
    void seestatorteethdata()
    {
    cout<<endl<<"No. of poles= "<<p<<
    "maximum fluxper pole= "<<phim<<
    "Gross Armature length= "<<L<<"No. of stator slots= "<<Ss<<
    "Depth of stator slot= "<<dSs<<"width of Stator Slot= "<<wSt<<
    "Stacking factor= "<<fst ;
    }
//  Stator teeth are parallel-sided, Rotor slot-section circular

    void computestatorteeth()
    {
    ASt =fst*L*wSt ;              // sectional area of stator tooth at
                                 // 1/3rd tooth height from bottom
    BSt = 1.36*phim*p/(Ss*ASt) ;

    y=BSt ;
    float mcur42st(float) ;
    x = mcur42st(y) ;
    FSt=x*dSs;                    // Height of fluxpath = height of tooth
    cout<<endl<<"FSt=            "<<FSt ;
    }
    float statorteethmmf(){return FSt ;}

};

class gap
{
    public :
    int p, Ss, Rs, nd ;
    float phim, lg, D, L, wo1, wo2, wd ;
    float Leff, kd, DR, ys1, ys2, Bav, kcs, lgeff, kcr, kgd, kg1, kg2 ;
    float ratiod, ratios, ratior, Fg, r ;

void getgapdata()
{
    string mystr ;
                                    cout<<endl<<"Gap" ;
cout<<endl<<"Enter No. of poles(p),Maximum flux per pole (phim-Wb),"
    "Armature diameter(D-m),Gross armature length(L-m),No. of stator "
    "slots(Ss),gap length(lg-m),stator slot-opening(wo1),No. of rotor"
    "Slot(Rs),ROtor slot-opening(wo2-m),No. of ducts on rotor(nd),"
    "Duct width(wd)";
cin>>p>>phim>>D>>L>>Ss>>lg>>wo1>>Rs>>wo2>>nd>>wd ;
}
void seegapdata()
{
```

```
        cout<<endl<<"No. of poles=      "<<p
        <<"maximum flux per pole=      "<<phim
        <<"Armature diameter=        "<<D<<"Gross Armature length=      "<<L
        <<"No. of stator slots=      "<<Ss<<"gap length=        "<<lg
        <<"stator slot opening=      "<<wo1<<"No. of Rotor Slot=      "<<Rs
        <<"Rotor Slot opening        "<<wo2<<"No. of ducts in stator=      "<<nd
        <<"Duct width=      "<<wd ;
        }
        void computegap()
        {
//     Compute Effective core length Leff
        ratiod = 0.5*wd/lg ; //ratio-duct width/gap length=0.735

        r=ratiod ;
        float carterOpen(float) ; //function to find carter's coeff-open slot
        kd= carterOpen(r) ; //carter's coeff. for duct opening
        cout<<endl<<"kd= "<<kd ;
        Leff= L- kd * nd * wd ; //Effective core length
        kgd = L/Leff ;       //gap cofficient due ducts in stator as well as
                             //in rotor since nd and wd are the same in stator
                             //and rotor

// Compute carter's coefficients for gap-due stator slot opening (kcs)
// and rotor slot opening (kcr)
        float PI = 3.14159L ;
        DR = D - 2. * lg ; // rotor outside diameter
        ys1 = PI * D / Ss ; // stator slot pitch
        ys2 = PI * DR / Rs ; // rotor slot pitch

        ratios = wo1 / lg ; //ratio : stator slot opening/gap length
        r= ratios ;
        float cartersemi(float) ;
        kcs = cartersemi(r) ; //carter's coefficient for stator slot opening

        ratior = wo2 / lg ; //ratio : rotor slot opening/gap length
        r = ratior ;
        float cartersemi(float) ;
        kcr = cartersemi(r) ; //carter's coefficient for rotor slot open opening
        kg1 = ys1 / ( ys1-kcs*wo1 ) ; //gap coeff. due stator slots
        kg2 = ys2 / ( ys2-kcr*wo2 ) ; //gap coeff. due rotor slots

        lgeff = lg * kg1 * kg2'* kgd * kgd ; //effective gap length
        Bav = 1.36 * p * phim / ( 3.1416 *.D * Leff ) ; //Av. density at 30 deg.
//                                    (elec) from direct axis
        Fg = 800000 * Bav * lgeff ;
        }
        float gapmmf(){return Fg ;}
        };

        class rotorteeth
        {
        public :

        int p ;
        float phim, D, L, fst, lg, Rs, dRs, wRs, nd, wd ;
        float DR, wRt, phiRt, ARt, BRt, y, x, FRt ;
```

```cpp
void getrotorteethdata()
{
    string mystr ;
                                cout<<endl<<"ROTOR TEETH" ;
    cout<<endl<<"Enter No. of poles(p),Maximum flux per pole (phim-Wb),"
        "Stator inside diameter(D-m),Gross Armature length(L-m),"
        "gap length(lg-m),No.of Rotor slot(Rs),Depth of rotor slot(dRs-m),"
        "width of rotor slot(wRs-m),no.of duct in rotor(nd),"
        "duct width(wd-m), stacking factor (fst))" ;
    cin>>p>>phim>>D>>L>>lg>>Rs>>dRs>>wRs>>nd>>wd>>fst ;
}
void seerotorteethdata()
{
    cout<<endl<<"No. of poles= "<<p <<"maximum fluxper pole= "<<phim<<
    "Stator inside diameter= " <<D <<"armature length= "<<L<<
    "Gaplength= "<<lg <<"No. of rotor slot= "<<Rs<<
    "depth of rotor slot= "<<dRs <<"width of rotor slot= "<<wRs<<
    "No.of duct in rotor= "<<nd <<"duct width= "<<wd <<
    "stacking factor= "<<fst ;
}
void computerotorteeth()
{
    float PI = 3.14159L ;
 DR = D - 2.*lg ; //rotor diameter
 float DR1 = DR - ( 2.*0.667*dRs ) ;
 wRt=(PI*DR1/Rs)-wRs ; //tooth width at 1/3
//                              slot depth from rotor slot bottom
 ARt= fst*wRt* (L -( nd * wd )) ;    //corresponding tooth section
 phiRt=phim*p/Rs ;                   //flux per rotor tooth pitch
 BRt=1.36*phiRt/ARt ;                // flux-density at the tooth section

 y=BRt ;
 float mcur42st(float) ; //function to find amp-turns/metre
 x = mcur42st(y) ;
 FRt=x * dRs ;
 cout<<endl<<"FRt= "<<FRt ;
 }
 float rotorteethmmf(){return FRt ;}


} ;

class rotorcore
{
    public :
    int p ;
    float phim, D, L, lg, dRs, DRi, fst ;
    float DR, dRc, DRcm, ARc, IRc, y, x, BRc, FRc ;

    void getrotorcoredata()
    {
        string mystr ;
                        cout<<endl<<"ROTOR CORE" ;
    cout<<endl<<"Enter No. of poles(p),Maximum flux per pole (phim-Wb),"
        "armature diameter(D-m),gross armature length(L-m),gaplength(lg-m),"
        "depth of RotorSlot(dRs-m),Inside dia, of rotor(DRi-m),"
        "stacking factor(fst)" ;
        cin>>p>>phim>>D>>L>>lg>>dRs>>DRi>>fst ;
    }
```

```cpp
void seerotorcoredata()
{
 cout<<endl<<"No. of poles= "<<p<<"maximum fluxper pole= "<<phim<<
 "Armature dia= "<<D<<"gross Armature length= "<<L<<
 "gap length= "<<lg<<"depth of rotor slots= "<<dRs<<
 "inside Dia.of rotor= "<<DRi<<"Stacking factor= "<<fst ;
}

void computerotorcore()
{
 float PI = 3.14159L ;
 DR=D-2. * lg ;                 //rotor diameter
 dRc=0.5*(DR-2.*dRs-DRi) ;      //depth of rotor core
 DRcm=DRi+dRc ;                 //mean dia. of rotor core
 ARc =fst*dRc*L ;              // sectional area of rotor core

 float kfl=1.05 ;       //factor to allow for curved nature of flux-path
//                             in rotor core at the back of rotor teeth
 IRc=0.5*kfl*PI*DRcm/p ; //length of fluxpath in rotor core/pole
//                             =0.5 pole-pitch on mean dia of armature core
 BRc = 0.5*phim/ARc ;        //flux in armature core

 y=BRc ;
 float mcur42st(float) ;   //function to find ampereturns/metre
 x = mcur42st(y) ;  //from magnetisation curve:42quality stalloy
 FRc=x*IRc;
 cout<<endl<<"FRc= "<<FRc ;

}
float rotorcoremmf(){return FRc ;}
//
};

class total: public statorcore, public statorteeth, public gap,
        public rotorteeth, public rotorcore
{
    public :
    float summmf()
    {
        return statorcoremmf()+statorteethmmf()+gapmmf()+
           rotorteethmmf()+rotorcoremmf();
    }
};

int main()
{
    total m ;

    m.statorcore :: getstatorcoredata() ;
    m.statorcore :: seestatorcoredata() ;
    m.statorcore :: computestatorcore() ;

    m.statorteeth :: getstatorteethdata() ;
    m.statorteeth :: seestatorteethdata() ;
    m.statorteeth :: computestatorteeth() ;

        m.gap :: getgapdata() ;
        m.gap :: seegapdata() ;
```

```
        m.gap :: computegap() ;

    m.rotorteeth :: getrotorteethdata() ;
    m.rotorteeth :: seerotorteethdata() ;
    m.rotorteeth :: computerotorteeth() ;

    m.rotorcore :: getrotorcoredata() ;
    m.rotorcore :: seerotorcoredata() ;
    m.rotorcore :: computerotorcore() ;

    cout<<endl<<"total mmf= "<< m.summmf() ;

    float kw ;
        int Ns, p, et, Im, Xm ;
    cout<<endl<<"Enter the values of winding factor (kw), "
            <<"Stator turns/ph. (Ns), No. of poles (p), "
            <<"Applied voltage/ph. (et)" ;
    cin >> kw >> Ns >> p >> et ;
    Im = p * m.summmf() / (2.34 * kw * Ns) ;
    cout<<endl<<"magnetising current per phase= "<< Im ;
    Xm = et / Im ;
    cout<<endl<<"magnetising reactance per phase= "<< Xm ;
}
```

Figure 9.9   A computer program for determining M.M.F.s and Magnetising reactance of a three-phase induction motor.

RESULT

STATOR CORE
Enter No. of poles(p),Maximum flux per pole (phim-Wb),Stator inside diameter(D-m),Stator outside diameter(Do-m),Gross Armature length(L-m),Depth of stator slot(dSs-m),Stacking factor(fst)
     4   0.00793 0.159   0.26   0.127   0.025   0.95

                                                            BSc= 1.28877

FSc= 41.251

STATOR TEETH
Enter No. of poles(p),Maximum flux per pole (phim-Wb),Gross armature length(L-m),No. of stator slots(Ss),Depth of statorSlot(dSs-m),width of Stator Slot(wSt-m),Stacking factor(fst)
     4   0.00793 0.127   36   0.025   0.0076   0.95

                                                      BSt= 1.78

FSt= 11.9524

GAP
Enter No. of poles(p),Maximum flux per pole (phim-Wb),Armature diameter(D-m),Gross armature length(L-m),No. of stator slots(Ss),gap length(lg-m),stator slot-opening(wo1),No. of rotorSlot(Rs),ROtor slot-opening(wo2-m),No. of ducts on rotor(nd),Duct width(wd)
   4   0.00793   0.159   .0.127 ·36   0.00034   0.0035   34   0.002 3 · 0.005
      Bav= 0.688969   lgeff= 0.000371502   Fg= 204.763

ROTOR TEETH
Enter No. of poles(p),Maximum flux per pole (phim-Wb),Stator inside diameter(D-m),Gross Armature length(L-m), gap length(lg-m), No.of Rotor slot(Rs),Depth of rotor slot(dRs-m),width of rotor slot(wRs-m),no.of duct in rotor(nd),duct width(wd-m), stacking factor (fst)
        4   0.00793 0.159   0.127   0.00034 34   0.01   0.0055 3   0.005
0.95
                                                       BRt= 1.51021

FRt= 27.6413

ROTOR CORE
Enter No. of poles(p),Maximum flux per pole (phim-Wb),armature diameter(D-m),gross armature length(L-m),gaplength(lg-m),depth of RotorSlot(dRs-m),Inside dia, of rotor(DRi-m),stacking factor(fst)
        4   0.00793 0.159   0.127   0.00034 0.01   0.0445   0.9

                                                      BRc= 0.700568
FRc= 3.69719
total mmf= 289.305
Enter the values of winding factor (kw), Stator turns/ph. (Ns), No. of poles (p), Applied voltage/ph. (et)
        0.96   240   4   230

magnetising current per phase = 2.144
magnetising reactance per phase= 107.26

9.9   LEAKAGE REACTANCES OF 3-PHASE INDUCTION MOTOR

Demonstration of 'Multiple Inheritance' in C++ was described under Windows was made in art. 7.2.6 for computing leakage reactances of a 3-ph. Induction motor. The same program is demonstrated here under Linux o.s.

```
Leakage reactances of 3-ph. Induction Motor
#include<iostream>
#include<math.h>
#include<stdio.h>
using namespace std ;

class data
{
      public :
          int pole, phase, freq, Sslot, Rslot, Strnph ;
       float Dia, Length, Ssopen, Rsopen, Mgreac ;

      data()
      {  pole=0.0; phase=0.0; freq=0.0; Sslot=0.0; Rslot=0.0; Strnph=0.0;
         Dia=0.0; Length=0.0; Ssopen=0.0; Rsopen=0.0; Mgreac=0.0;
      }
      data( int p, int ph, int fs, int Ss, int Rs, int Ns, float D,
          float L, float wo1, float wo2, float Xm, float yw )
          { pole=p; phase=ph; freq=fs; Sslot=Ss; Rslot=Rs; Strnph=Ns;
             Dia=D; Length=L; Ssopen=wo1; Rsopen=wo2; Mgreac=Xm; }
          void getvalue(void)
          { cout<<endl<<"Enter the values of:pole, phase, freq, Sslot, Rslot,"
             "Strnph, Dia, Length, Ssopen, Rsopen, Mgreac " ;
          cin>> pole >> phase >> freq >> Sslot >> Rslot >> Strnph >> Dia
             >> Length >> Ssopen >> Rsopen >> Mgreac ;
          }
      void display()
      { cout<<endl<<"No. of poles= "<<pole<<"No.of phase= "<<phase
          <<"frequency= "<<freq<<"Stator slot= "<< Sslot
          <<"Rotor slot= "<<Rslot<<"Stator turns/ph= "<<Strnph
          <<"Stator inside Dia= "<<Dia<<"Arm.length= "<<Length
          <<"Stator slot opening= "<<Ssopen
          <<"Rotor slot opening= "<<Rsopen
          <<"Magnetising reactance= "<<Mgreac ;
```

```
    }
            int getp()    {return pole; }
            int getph()   {return phase; }
            int getfs()   {return freq; }
            int getSs()   {return Sslot; }
            int getRs()   {return Rslot; }
            int getNs()   {return Strnph; }
        float getD()      {return Dia; }
        float getL()      {return Length; }
        float getwo1()    {return Ssopen; }
        float getwo2()    {return Rsopen; }
        float getXm()     {return Mgreac; }
    };

    float stslot( data d )
    {
        int Ss, Ns, ph, fs ;
    float L, sh1, sh2, sh3, sh4, wo1, ws, w1, w2 ;
    float Lam1, Lam2, Lam3, Lam4, LamS1,sxsl ;
    float PI, k1, k2 ;

            ph=d.getph() ;
            fs=d.getfs() ;
            L=d.getL() ;
            Ss=d.getSs() ;
            wo1=d.getwo1();
            Ns=d.getNs()
    PI = 3.14159L ;
    k1 = 1e-7f ;

    cout<<endl<<"Enter the values of: sh1, sh2, sh3, sh4, ws, w1 (mm.)";
    cin>> sh1 >> sh2 >> sh3 >> sh4 >> ws >> w1 ;

//          Compute Stator slot permeance coefficient

    w2 = ws - sh1*(ws - w1)/(sh1 + sh2) ;
    Lam1 = 2. * sh1 / (3* (ws + w2)) ;
    Lam2 = 2. * sh2 / (w2 + w1) ;
    Lam3 = 2. * sh3 / (w1 + wo1) ;
    Lam4 = sh4 / wo1 ;
    LamS1 = Lam1 + Lam2 + Lam3 + Lam4 ;
        k2 = 4. * PI * fs * 4. * PI * k1 * Ns * Ns * 2. * ph / Ss ;
        sxsl = k2 * LamS1 * L ; //stator slot leakage reactance/ph.
        return(sxsl) ;
    };

    float rtslot( data d )
    {
        int Ss, Rs, Ns, ph, fs ;
        float L, wo2, rh4 ;
        float LamR1, LamR1s, rxsl ;
        float PI, k1, k2, kw1 ;

                    ph=d.getph() ;
                    fs=d.getfs() ;
                    L=d.getL() ;
```

```cpp
                    Ss=d.getSs() ;
                    Rs=d.getRs() ;
                    wo2=d.getwo2();
                    Ns=d.getNs() ;

        PI = 3.14159L ;
        k1 = 1e-7f ;

        cout<<endl<<"Enter the values of: height of rotor lip, rh4 -m" ;
        cin>> rh4 ;

//        Compute rotor slot permeance coefficient

        LamR1 = 0.66 + rh4 / wo2 ; // for circular rotor slot
        cout<<endl<<"Enter the value of: Stator winding factor kw1" ;
        cin>> kw1 ;
        LamR1s = kw1 * kw1 * Ss * LamR1 / Rs ; //rotor slot permeance coeff.
//                              referred to stator, knowing winding const.
//                              for cage winding = 1
        k2 = 4. * PI * fs * 4.*PI * k1 * Ns * Ns * 2. * ph /Ss ;
        rxsl = k2 * LamR1s * L ; //Rotor slot leakage reactance / phase
//              in stator terms
        return(rxsl);
};

        float stzigzag( data d )
        {
            int p, Ss ;
            float Xm, k3s, sxz ;


                        p = d.getp() ;
                        Ss = d.getSs() ;
                        Xm = d.getXm() ;
            k3s = 1. / (Ss * Ss) ;
            sxz = 5.* k3s * p * p * Xm / 6 ;
            return(sxz) ;
        };

        float rtzigzag( data d )
        {
            int p, Rs ;
            float Xm, k3r, rxz ;


                        p = d.getp() ;
                        Rs = d.getRs() ;
                        Xm = d.getXm() ;
            k3r = 1. / (Rs * Rs) ;
            rxz = 5.* k3r * p * p * Xm / 6 ;
            return(rxz) ;
        };

        float rtskew( data d )
        {
            int Rs ;
            float Nsk, theta, rxsk ; //Nsk: No. of slots skewed

                        Rs = d.getRs() ;
```

```cpp
    float PI= 3.14159L ;
    cout<<endl<<"Enter: No. of slots skewed (Nsk)" ;
    cin>>Nsk ;
    theta = PI*Nsk / Rs ;
    rxsk = 59.4* theta*theta* 0.95/12.; //0.95 is the assumed value
//                                of stator leakage flux factor
    return(rxsk) ;
};

float overhang( data d )
{
    int p, Ns, m ;
    float D, ks, yw, alpha, xo ;

                    p = d.getp() ;
                    D = d.getD() ;
                    Ns = d.getNs() ;

    float PI= 3.14159L ;
    cout<<endl<<"Enter: Coilspan factor(yw)" ;
    cin>>yw ;
alpha = yw * p / ( PI*D ) ;

cout<<endl<<"Enter: m = 0 for wide-spread connection"
            "m=1 for narrow-spread connection" ;
cin>>m ;
if (m==0)
{
if (alpha<0.67)
    {ks = 1.12 * alpha;}
    else
    if (alpha<1.0)
    {ks = 0.75 ;}
}
else
if(m==1)
    {
    if (alpha<0.67)
    {ks = (1.40 * alpha - 0.19);}
    else
    if (alpha<1.0)
    {ks = (0.24 + 0.76*alpha);}
    }
xo = 0.00475 * ks * D * Ns * Ns / (p * p) ;
return(xo) ;
};

int main()
{
    float sxsl, rxsl, sxz, rxz, x1, xo, x2s, rxsk ;

    data d ;
    d.getvalue() ;
    cout<<"\n";d.display();

    sxsl = stslot(d) ;
    cout<<endl<<"Stator slot leakage reactance= "<<sxsl<<" ohm";
```

```
            rxsl = rtslot(d) ;
            cout<<endl<<"Rotor slot leakage reactance= "<<rxsl<<" ohm";

            sxz = stzigzag(d) ;
            cout<<endl<<"Stator zigzag leakage reactance= "<<sxz<<" ohm";

            rxz = rtzigzag(d) ;
            cout<<endl<<"Rotor zigzag leakage reactance= "<<rxz<<" ohm";

            rxsk = rtskew(d) ;
            cout<<endl<<"Rotor skew leakage reactance= "<<rxsk<<" ohm";

            xo = overhang(d) ;
            cout<<endl<<"Overhang leakage reactance= "<<xo<<" ohm";

            x1 = sxsl + sxz + 0.5 * xo ;
            cout<<endl<<"Stator leakage reactance= "<<x1<<" ohm";

            x2s = rxsl + rxz + rxsk + 0.5 * xo ;
            cout<<endl<<"Rotor leakage reactance= "<<x2s<<" ohm";
};
```

Figure 9.10   Computation of Leakage reactances of a 3-phase induction motor.

RESULT

Put the data from art. 7.2.6 and find the result.

9.10 PERFORMANCE OF A 3-PHASE INDUCTION MOTOR AT VARIOUS SLIP

Problem : It is required to write down a computer program in Linax for determination of performance of an induction motor at various values of slip given its parameters.

[Note : the program matches art. 7.2.6]

```
//Given main parameters of a 3-ph. induction motor, to find
//its performance at various slips.
#include <iostream>
#include <fstream>
#include<math.h>
#include<stdio.h>
using namespace std;

int main()
{
        float et, r1, r2, x1, Xm, x22, x2s, ns, s, Rf, Xf, Z, I1, spf ;
        float Pi, Qi, Pg, Te, omegas, omegar, Pm, P2, Po, PI, To ;
        float c1, c2, c3, c4, c5, c6 ;

        cout<<endl<<"Enter : Stator resis./ph(r1), Rotor resis./ph in Stator"
                "terms(r2), Magnetising reactance/ph(Xm)"
                ", Stator leakage reactance/ph(x1)"
                ", Rotor leakage reactance/ph in stator terms(x2s)" ;
                cin >> r1 >> r2 >> Xm >> x1 >> x2s ;
```

```
cout<<endl<<"Enter: Supply voltage / ph(et), syn. r.p.s.(ns)";
cin >> et >> ns ;

cout<<endl<<"Enter No load rotational loss (Po)" ;
cin >> Po ;
getchar();
cout<<endl<<" PERFORMANCE SHEET AT RATED VOLTAGE ANDFREQUENCY " ;
ofstream outf("RESULT");
for (s=0.01; s<=0.50;s+=0.01)
{
    x22 = Xm + x2s ;
    c1 = (r2/s) * (r2/s) + x22 * x22 ;
    c2 = (r2/s) * (r2/s) + x2s * x22 ;
    Rf = Xm * Xm * (r2 / s) / c1 ;
    Xf = Xm * c2 / c1 ;
    c3 = (Rf + r1) * (Rf + r1) + (Xf + x1) * (Xf + x1) ;
    Z = sqrt(c3) ; //Effective impedance across input
    I1 = et / Z ; //Input current per phase
    spf = (r1 + Rf) / Z ; //Stator p.f.
    c4 = spf * spf ;
    c5 = 1.0 - c4 ;
    c6 = sqrt(c5) ;

Pi = et * I1 * spf ; //Input real power per phase
    Qi = et * I1 * c6 ; //Input reactive power per phase
    Pg = I1 * I1 * Rf ; //Real power / phase, across the gap
    PI = 3.14159L ;
omegas = 2. * PI * ns ; //Syn. speed, radians / sec.
omegar = (1-s) * omegas ; //Rotor speed, radians / sec.

    Te = 3.* Pg / omegas ; //Electromagnetic torque, N.m.
    Pm = (1 - s) * Pg ; //Internal mechanical power / phase
    P2 = (Pm - Po) ; //Output power / phase
    To = 3. * P2 / omegar ;//Output torque of the motor, N.m.

outf<< s <<endl ;
outf<<I1 <<endl ;
outf<<spf<<endl ;
outf<<Pi <<endl ;
outf<<Qi <<endl ;
outf<<Te <<endl ;
outf<<To <<endl ;
}
outf.close() ;
ifstream inf("RESULT") ;
while (!inf.eof())
    { inf>> s ;
    inf>> I1 ;
    inf>> spf ;
    inf>> Pi ;
    inf>> Qi ;
    inf>> Te ;
    inf>> To ;

cout<<endl<<"slip= "<<s ;
cout<<endl<<"input current/ph= "<<I1 ;
```

```
    cout<<endl<<"input power factor= "<<spf ;
    cout<<endl<<"input real power/ph.= "<<Pi ;
    cout<<endl<<"input reactive power/ph= "<<Qi ;
    cout<<endl<<"Electromagnetic torque= "<<Te<<"N.m." ;
    cout<<endl<<"Output torque= "<<To<<" N.m." ;
}
inf.close() ;
};
```

Figure 9.11   Performance of 3-phase Induction Motor at various slips

RESULT

Get the inputs from art. 7.2.6 to find the result.

9.11   DESIGN OF STATOR COIL OF A WATER-WHEEL GENERATOR

[The program matches with that in art. 7.3.3]

```
//Design of Stator coil of a Water-wheel generator
// To maintain Gap density between Bmax and Bmin
#include<iostream>
#include<math.h>
#include<stdio.h>
using namespace std ;

class spec
{
    public :
        int ph, p, fs, Ss ;
      float et, ap, D, L, phim ;

    void getdata (float et, int ph, int p, int fs, float ap, float D,
                float L, int Ss, float phim) ;

    void display();

    void coil (float et, int ph, int p, int fs, float ap, float D,
            float L, int Ss, float phim) ;
};

    void spec::getdata (float voltage, int phase, int pole, int frequency,
                    float current, float Diameter, float Length,
                    int slot, float maxmflux)
    {
        et=voltage, ph=phase, p=pole, fs=frequency, ap=current, D=Diameter,
    L=Length, Ss=slot, phim=maxmflux ;
    }

    void spec::display()
      {
    cout<<"voltage per ph. "<<et<<",phase= "<<ph<<",pole= "<<p<<
            ",frequency= "<<fs<<",current/ph= "<<ap<<",arm.diameter= "
```

```cpp
            <<D<<",armature length= "<<L<<",stator slot= "<<Ss<<
            ",maxm.flux/pole= "<<phim;
    }

    void spec::coil(float et, int ph, int p, int fs, float ap, float D,
                    float L, int Ss, float phim)
    {
        int nd ;
    float cd, Bmax, Bmin, fdf, fst, win, din, wd ;

    cout <<"Enter the assumed value of current-density (cd), maximum"
            "flux-density (Bmax), minimum flux-density (Bmin),"
            "field distribution factor (fdf), stacking factor (fst),"
            "width of slot insulation (win),depth of slot insul.(din),"
            "no. of vent duct(nd),width of ventilating duct (wd)" ;

    cin >>cd>>Bmax>>Bmin>>fdf>>fst>>win>>din>>nd>>wd ;

    cout <<"current-density= "<<cd
        <<", maximum flux-density= "<<Bmax
        <<", minimum flux-density= "<<Bmin
        <<", field distribution factor= "<<fdf
        <<", stacking factor= "<<fst
        <<",width of slot insulation= "<<win
        <<",depth of slot insul.= "<<din
        <<",no. of ventilating duct= "<<nd
        <<",width of ventilating duct= "<<wd ;

        float PI = 3.14159L ;

// Compute Pitch Factor and Distribution Factor
        int S, Sp, A, Z ;
        float Cth, k2, akp, gama, X, Y, akd, cwf ;

        Sp = Ss / p ;                          // number of stator slots per pole
        S = Ss / (p * ph) ;                    // no. of stator slots per pole per phase
        Cth = Sp ;                             // coil throw=
        k2 = (PI * Cth) / (Sp * 2.0) ;
        akp = sin(k2) ;                        // pitch factor
        cout<<endl<<"akp= "<<akp ;
        getchar() ;
        gama = PI / Sp ;
            X = sin(S*gama/2.0) ;
        getchar() ;
        Y = sin(gama / 2.0) ;

    akd = X / (S*Y) ;                          // distribution factor
    cwf = akd * akp * fdf ;                    // winding factor

// Compute Stator turns per phase
    float sturn, Ac, Acmax, Bav ;

    sturn = et / (4.44*cwf*fs*phim) ;          //prelim. value of stator turns
                                               // per phase
        A = 1.0 ;                              // assumed no. of parallel circuit
        Ac = ap / (A*cd) ;                     // bare conductor section, mm.sq.
```

```
        cout<<endl<<"bare conductor section= "<<Ac ;
    getchar() ;

//On the basis of conductor current for 1 parallel circuit, largest
//conductor section is chosen from Standard Table :
// Rectangular 25mm. x 10 mm.rounded, of bare section 249 sq. mm.

    cout<<endl<<"enter the largest bare conductor section (Acmax)" ;
    cin>> Acmax ;
    if(Ac> Acmax)
    {
        A = A+1.;
        Ac = ap / (A * cd) ;
    }
    else
    {
     Z = A * ph * sturn * 2.0/ Ss ; //conductor /slot
    }
// Z should be an integer, and equal to or greater than 2.0
    if(Z < 2.0)
    {
     Z = 2.0 ;
    }
    Bav = (et*p*ph)/(2.22*PI*D*L*fs*cwf*Ss*Z) ;
// Check whether Bav is within the stipulated values of Bmax & Bmin
    if(Bav>Bmax)
    {
    Z = Z + 2.0 ;
    Bav = (et*p*ph)/(2.22*PI*D*L*fs*cwf*Ss*Z) ;
    }
    else
    {
    if(Bav<Bmin)
    {
    Z = Z - 1. ;
    Bav = (et*p*ph)/(2.22*PI*D*L*fs*cwf*Ss*Z) ;
    }
    }
    cout<<endl<<"Average gap density= "<<Bav ;

    int cthc, Zph ;
    float phim1, Ratio, XX, cthcp, akpM, phimM, Ws ;

    phim1 = Bav * PI * D * L /p ;
    Ratio = phim1/phim ;
        XX = asin( Ratio ) ;
    getchar() ;
    cthcp = Sp * 2.0 * XX / PI ; //modified coil-throw
//                                      should be an integer
    cthc = cthcp ;
    akpM = sin( PI*cthc/(Sp * 2.0)) ; //modified pitch factor

    cout<<endl<<"Distribution Factor= "<<akd ;
    cout<<endl<<"Winding Factor= "<<cwf ;
    cout<<endl<<"coil throw= "<<cthc ;
    cout<<endl<<"Pitch Factor= "<<akpM ;
    getchar() ;
```

```
      phimM = phim1 * akp / akpM ; //modified flux per pole
      Bav = phimM * p /(PI * D * L) ;
      Zph = Z * A * Ss / ph ;
      sturn = Zph / 2.0 ;
      Ws = 0.5 * PI *D/ Ss ; // slot width

//                                    Choosing Parallal-sided Slots

      float cbw, cbh, AcM, cdM, WsM, dsM ;
      long k1 = 1e6f ;
      cbw = Ws - win ; //conductor width
      cbh = Ac / (k1*A*cbw) ; //conductor height
      cout<<endl<<"cbh= "<<cbh ;
      cout<<endl<<"Ac= "<<Ac ;

      cout<<endl<<"Enter the value of modified Ac (AcM)" ;
      cin>>AcM ;
      cout<<endl<<"Chosen bare conductor section= "<<AcM<<"mm-sq" ;

          cdM = ap / (A * AcM) ;
          WsM = cbw + win ;
          dsM = (A * Z * cbh) + din ;
//               Check Tooth-density
      float wt3, Bt3 ;
      wt3 = (PI * (D + 0.667 * dsM) / Ss) - WsM ;
      Bt3 = phim * p / (wt3 *(L - nd*wd)* fst * Ss) ;

      cout<<endl<<"gap density=                        "<<Bav ;
      cout<<endl<<"maxm. flux per pole=                "<<phim ;
      cout<<endl<<"turns per phase=                    "<<sturn ;
      cout<<endl<<"no. of conductors/slot=             "<<Z ;
      cout<<endl<<"number of slot=                     "<<Ss ;
      cout<<endl<<"current density=                    "<<cdM ;
      cout<<endl<<"slot depth=                         "<<dsM ;
      cout<<endl<<"slot width=                         "<<WsM ;
      cout<<endl<<"tooth density at 1/3rd from root= "<<Bt3 ;
      }

int main()
{
      spec m ;
      int phase, pole, freq, slot ;
      float voltage, current, dia, length, maxflux ;
      cout<<endl<<"Enter the data : et, ph, p, fs, ap, D, L, Ss, phim" ;
      cin>>voltage>>phase>>pole>>freq>>current>>dia>>length>>slot>>maxflux ;

      m.getdata(voltage,phase,pole,freq,current,dia,length,slot,maxflux) ;
      m.display() ;
      m.coil(voltage,phase,pole,freq,current,dia,length,slot,maxflux) ;
};
```

Figure 9.12   Design of Stator coil of a Water-wheel generator.

RESULT

Enter the data : et, ph, p, fs, ap, D, L, Ss, phim								
3820	3	32	50	262	3.2	0.44	288	0.083

Enter the assumed value of current-density (cd), maximumflux-density (Bmax), minimum flux-density (Bmin),field distribution factor (fdf), stacking factor (fst),width of slot insulation (win),depth of slot insul.(din),no. of vent duct(nd),width of ventilating duct (wd)

    4   0.67   0.57   0.67   0.95   0.007   0.025   5   0.01

bare conductor section= 65.5

Enter the largest bare conductor section (Acmax) 249

Distribution Factor= 0.959795

Winding Factor= 0.643063

coil throw= 5

Pitch Factor= 0.766044

Ac= 65.5

Enter the value of modified Ac (AcM)   66

Chosen bare conductor section= 66mm-sq

gap density=	0.658067
maxm. flux per pole=	0.083
turns per phase=	384
no. of conductors/slot=	8
number of slot=	288
current density=	3.9697
slot depth=	0.0751278
slot width=	0.0174533
tooth density at 1/3rd from root=	1.38286

YOU MAY TRY

[In this chapter, computer programs have been developed using Lynax o.s.s. for the Examples as in chapter except : (a) Disc Winding -art., (b) Disc-Helix winding-art., for a three-phase transformer; (c) Design of a Three-phase Cage-rotor Induction motor- Stator frame - art., (d) Design of a Three-phase Cage-rotor Induction motor- Rotor dimensions- art.,

(e) M.M.F. of D.C./ Salient-pole Synchronous machine- art.

It is stipulated that the students on their own, would develop these programs.]

## CHAPTER 10

Design of Single Phase Induction Motor

10.1 Single phase induction motors* are, in general, fractional horsepower size ranging from a few VA to 700 VA. The airgap field in the motor is not rotating as in a polyphase induction motor, but pulsating, i.e. fixed in space but varying in magnitude with time. As such, these motors are not self-starting.

Auto-start feature can be incorporated with the use of an additional winding, generally known as 'auxiliary winding' which enables the motor to start as an unbalanced two-phase induction motor, and in many appliactions, this winding is disconnected from the mains by means of centrifugal switch when the motor reaches a desired speed.

Depending on the nature of start, a single-phase induction motor is classified as,

   (i) resistance splitphase induction motor;
   (ii) capacitor splitphase induction motor;
   (iii) capacitor motor, when the auxiliary winding with the capacitor are not disconnected but kept in the circuit under running condition;
   (iv) shaded-pole induction motor, and so on.

In this chapter, design analysis of the first two types which are similar will be discussed. Only the general-purpose motors are considered.

For the performance analysis of single-phase induction motors, two equally important theories exist, viz.

The Double-revolving Field Theory;
The Cross-field Theory.

Here we shall use the former theory.

• S. K. Sen, Electrical Machinery, Khanna Publishers, Delhi.

10.2 IMPORTANT DESIGN CONSIDERATIONS

The design of single-phase induction motor can be done, in general, following the same methodology as for a polyphase induction motor, taking into consideration the following important considerations.

(a) The output equation (3.23a) for 1-phase is applicable to a single-phase induction motor with the modification that for a single phase IM,

$$q = 1.11 \, \eta \, \pi \cdot K \cdot (D \cdot La) \, ns \, Bav \, ac \, / \, 1.50 \text{ volt-amp.} \tag{11.1}$$

in which q is the input volt-amp.

This means that (D ∙ La) as obtained from equn. (11.1) for a single-phase induction motor is 1.50 times more than that obtained from equn. (3.23a) for one phase of a three- phase induction motor for the same output.

This is because in a polyphase motor, the conductors are uniformly distributed along the stator periphery, the field is rotating with a constant average flux-density where as, in a single-phase induction motor, the auxiliary winding distribution is different from that of the main winding (Figure 11.1), the field is pulsating i.e. stationary in space with magnitude varying with time

$$q(\text{Va}) = 0.74 \, \pi \cdot K_\omega \, (D \cdot La) \, ns \, Bav \, ac \text{ volt-amp.} \tag{11.2}$$

where, K_ω is the winding factor which depends on the type of winding used.

The nameplate power rating of a single-phase motor is = Po 'watts'. To get the input Va, we divide output power by (effy × power-factor), i.e.,

$$q = \text{Po} \, /(\text{effy} \times \text{p.f.}).$$

For a single phase induction motor, the design can be started with a value of K_ω equal to 0.955;

$$\text{That is, } q = 6.97 \, (D \cdot La) \, ns \, Bav \, ac \text{ volt-amp} \tag{11.3}$$

where, D and La are respectively the internal diameter of the stator and gross armature length, both in metres, ns is the synchronous speed of the rotor in r.p.s., Bav, Tesla, the average flux-density in the gap, and ac, the ampere-conductors per metre periphery of the armature.

As for polyphase induction motor, the design of single-phase induction motor starts with the assumptions of Bav and ac for the determination of D and La.

(b) Because of the nature of the field mentioned above,
  (a) a general-purpose standard split-phase induction motor has a starting torque and pull-out torque between 200 to 275 per cent of the full-load torque;
  (b) a special-purpose high torque split-phase induction motor has starting torque/full-load torque = 200 to 300 percent, and pull-out torque/full-load torque = 250 to 350 percent.
  (c) Capacitor-start motors have been designed with starting torque/ full-load torque = 300 to 450 percent; pull-out/full-load torque = 225 to 300 percent with low starting current.

(d) Capacitor-run or simply Capacitor motors have the capacitor left in the circuit and no centrifugal switch is used.* In such case, size of the capacitor is much less than that in (c) and starting torque is also much smaller.

(e) Preferred Rated Outputs, as per IS 996- 1964 are given in Table 10.1.

While, the Indian Standard mentions the methods of estimation of efficiency from Test, for the purpose of design we may consider the suggested values given in Table I, to start with [line 34 in the program].

Table 10.1

Output, watts	12	18	25	40	60	90	120	180	250	370	550	750	1100	1500
%Effy	0.16	0.25	0.3	0.36	0.38	0.44	0.48	0.55	0.57	0.63	0.75	0.78	0.80	0.80
p. f.	.625	.52	.55	0.57	0.61	0.62	0.63	0.65	0.70	0.74	0.76	0.77	0.78	0.80

10.3   A COMPUTER PROGRAM

Here a computer program written in Lynax is presented.

The whole program has been divided into 4 sections, as it is thought that it will be easier to handle smaller programs. The division into sections is purely arbitrary, and the whole program can be obtained by judicious assembly of the sections.

The sections are:

(a) computation of main and basic dimensions ;

(b) the squirrel-cage rotor ;

(c) the computation of performance ;

(d) the auxiliary or starting winding design.

Section (a)

Reference to the computer program in Figure 10.3,

(i) important assumption is the choice of relationship between armature diameter -D and gross armature length - L. Following the practice for larger machines, we have used the formula : L = pole-pitch, π .D / p. There is a check at line 98 in the program.

However, for general purpose single-phase induction motor L varies widely with respect to D, L = 0.60 to 1.40 D, depending on the nature of use, especially, torque requirements.

(ii) choice of number of slots in stator and rotor [lines 65 and 152 in computer program] is based on "slot combination- Appendix A1.6.

(iii) In order to evaluate the rotor-tooth density, the One-third density method is used (art 4.2.4.3 (C).

Reference Figure 10.1, point a indicates the level one-third tooth-height i.e. Rtht from the bottom of a tooth (or slot). With the radius of a slot = 0.5 × radius, dRs, of a slot (note : slots are circular with diameter dRs), oa = 0.5 × dRs − (Rtht / 3.)   line 211

With ob, the radius of the slot,

ba = sqrt ((0.5 × dRs × 0.5xdRs) − (oa × oa))

But, tooth-pitch at the point a is,

spa = PI * Da / Rs ; where Da is the 'rotor-diameter' at point a given

by,   Da = DR - (4.*(Rtht / 3.)).

Hence, the required tooth width ewt = spa − 2. × ba.

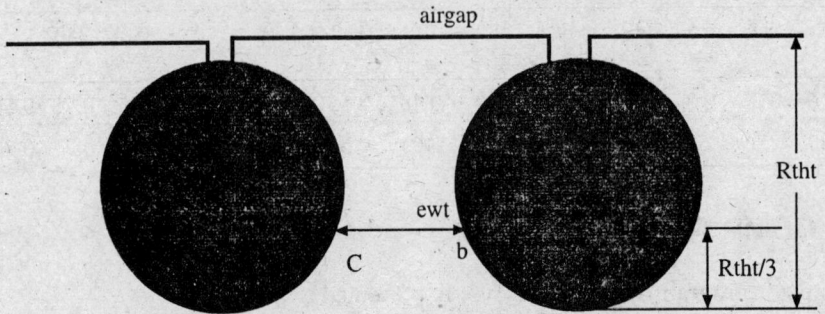

Figure 10.1 Computation of effective tooth-width.

(iv) Design of Armature Winding- The Skein Winding

In a single-phase induction motor, the armature winding is generally of concentric type. But, as in three-phase induction motor, if the conductors are evenly placed with the same number of coil-sides per slot, harmonics in the airgap flux will increase. A method of distribution of coil-sides per pole based on 'sine rule' has been evolved (known as 'skein wiinding') where slot-harmonics are visibly reduced . The process may better be explained taking an example.

A 9-slot per pole 'full-pitch' winding (say, for a 36-slot, 4-pole machine) is shown in Figure 10.2. There are four 'concentric' coils per pole with 'unequal coil-spread'; coil 1-9, coil 2-8, coil 3-7, coil 4-6, coil 5- empty [lines 265 to 280 in the computer program]; each group under one pole carries coil-sides different from others, the numbers determined by a 'sine rule' as discussed below :

The program starts with the computation of number of slots per pole Ssp and number of turns per pole Nmp.

Then coil-spans, in terms of number of slot are converted ino electric radians using the sine-law :

For coil 4 - 6 spanning 2 slots, ⋯   C46 = sinf((2.*PI)/(2.*Ssp)) elec.radian ;

Similarly, for coil 3-7 spanning 4 slots, ⋯   C37 = sinf((4.*PI)/(2.*Ssp)) ;

coil 2-8,spanning 6 slots, ⋯   C28 = sinf((6.*PI)/(2.*Ssp)) ;

and coil 1-9, spanning 8 slots, ⋯   C19 = sinf((8.*PI)/(2.*Ssp)) .

Knowing the total number of turns per pole Nmp, number of turns required in each of the 4 coils are computed, in proportion of the total as, For coil 4 − 6 : cN46 = C46 * Nmp / add; add = C46 + C37 + C28 + C19 being the total turns per pole.

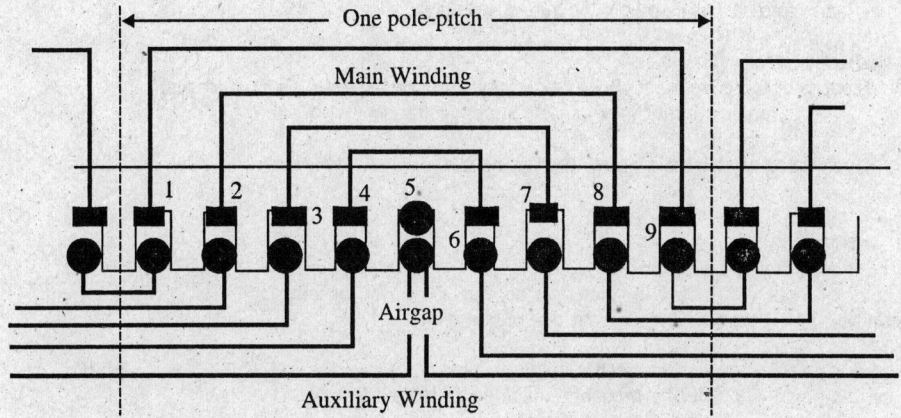

Figure 10.2 Single phase Induction Motor windings- 9 slots per pole.

Similarly, for coils C37, C28 and C19.

For the auxiliary winding, it is displaced through half pole-pitch as shown in the Figure.

The computed turns are then *suitably and judiciously* modified so that each is an integer number, and they in turn have a good 'highest common factor' (an integer also) to facilitate easy and economic winding (Appendix A.3).

(v) Computation of magnetising current and magnetising reactance [lines 401 to 406 in computer program] have been termed ' Preliminary values'. The reason will be evident when we compute their values in the next section (Figure 10.4).

Computer Program

```
//TO OBTAIN BASIC PARAMETERS, GIVEN THE KVA-RATING, SUPPLY VOLTAGE, FREQUENCY AND SPEED.

        #include<iostream>
        #include<math.h>
        #include<stdio.h>
        #include "mmf.h"
        #include "sploss3.h"
        #include "mcur42st.h"
        #include "cartersemi.h"
        using namespace std ;

        class spec
        {
            public:
                int a, b ;
            float c, d ;

        void getdata(int a, int b, float c, float d );
        void display();
        void stator(int a, int b, float c, float d );
        };
```

```
            void spec::getdata(int freq,int synsp,float powut,float volt )
              { a = freq; b = synsp; c = powut; d = volt; }

            void spec::display()
              {cout<<endl<<"frequency= "<<a<<", synchronous speed= "<<b<<"r.p.s."
                  <<", power outout= "<<c<<", Supply voltage= "<<d ; }

              void spec::stator(int a, int b, float c, float d )
            {
                long J ;
                float Bav ;
                float pf, effy, Cwm, fst ;
            cout << endl <<"Enter the values of: ac, Bav, Cwm, fst, pf, effy " ;
34          cin >> J >> Bav >> Cwm >> fst >> pf >> effy ;

            cout << endl <<"Specific electric loading= "<< J ;
            cout << endl <<"Specific magnetic loading= "<< Bav ;
            cout << endl <<"Winding factor- main wndg= "<< Cwm ;
            cout << endl <<"Core Stacking factor= "<< fst ;
            cout << endl <<"Power factor = "<< pf ;
            cout << endl <<"Per cent Efficiency= "<< effy ;


            cout<<"COMPUTE STATOR INSIDE DIA. (D,m.) and GROSS ARMATURE LENGTH (L,m.)" ;

                int p ;
            float pin, DsqL, dd, D, L, Do ;
            float PI = 3.14159L ;
                pin = c / (effy * pf) ;
50      .   DsqL = pin * 0.137 /( b * J * Bav );          //equn.(11.2) g= J h = Bav
            cout<<endl<<"DsqL= "<<DsqL ;

                p = 2. * a / b ;                    // number of poles

            //  Set Gross core length L equal to pole-pitch, i.e.L = PI * D / p ;

                dd = ( DsqL ) * p / PI ;          // dd = D-cubed.
                D = pow(dd,0.333) ;
59              L = PI * D / p ;

            cout<<endl<<"Diameter in m.= "<<D<<"Gross Length in m.= "<<L ;
            cout<<endl<<"MODIFY diameter & length, if necessary " ;
            cin >> D >> L ;

            cout <<"CHOOSE NUMBER OF STATOR SLOTS" ;
                int Ss ;
            float dSs, ASs, h4, wos, WSt, dSc, Dsh, phim, Strn ;
            cout<<endl<<"Enter Designer's choice of number of Stator slots";
69              cin >> Ss ;
                                            cout<<"Choose STANDARD LAMINATIONS" ;
//          based on the values of Diameter, Length and No. of stator
//       .  slots. choose standard laminations available in the market.For the
//          purpose of this design, take stator slots tapered with
//          parallel-sided tooth, and enter the following data from measurement:

                cout<<endl<<"Depth of slot (dSs, m.), Area of slot(ASs, m.sq.,"
                    "Height of lip (h4,m.),"                // Figure 9.12(a)
```

```
                "Slot opening (wos,m.),"
                "Width of tooth(WSt,m.),"
                "Depth of core below teeth(dSc,m.)"
                "Outside diameter of stator(Do,m.)"
                "Dia. of shaft at lamination(Dsh,m.)" ;
        cin >> dSs >> ASs >> h4 >> wos >> WSt >> dSc >> Do >> Dsh;


        phim = PI * D * L * Bav / p ; // in wb.
        cout<<endl<<"Maximum flux per pole (phim) = "<< phim ;
        Strn = d / (4.44 * phim * a * Cwm ) ;
        cout<<endl<<"Computed Main winding turns in series= "<<Strn ;


        int Nm, Nmp ;
        float mphim, mBav ;
        cout<<endl<<"MODIFY total main winding turns " ;
            cin >> Nm ;
        Nmp = Nm / p ;      //stator mainding turns in series per pole
        cout<<endl<<"Stator main wndg. turns in series/pole= "<<Nmp ;
        mphim = phim * Strn / Nm ; //modified flux/pole
        mBav = p * mphim /(PI * D * L) ; //modified av. gap density
        cout<<endl<<"Modified gap density= "<<mBav ;
//      IF mBav is high, increase L
        cout<<endl<<"Enter new L if necessary " ;
100         cin >> L ;
            cout <<"Main winding conductor parameters ";
        float I1, Ir, cdm, mcs, mcloss ;
            I1 = pin / d ; //Stator main winding current on full load
        Ir = I1m * pf ; //Correponding rotor current
        cout<<endl<<"Stator main winding current = "<< I1 ;
        cout<<endl<<"rotor current= "<< Ir ;


        cout<<endl<<"Enter: Current density, main winding= " ;
        cin >> cdm ;                          //in amp. per mm.sq.


        mcs = I1 / cdm ; //main winding bare cond. section, mm.sq.
        cout<<endl<<"main winding bare cond. section= "<<mcs<<" mm.sq.";


        float swg,cbdm,cbsm,cidm,cism,r1m ;
        cout<<endl<<"Assuming copper conductor, Enter:"
                " Swg No. for Enamelled copper wire (swg)"
                "main winding bare cond.section,(cbsm- mm.sq.)"
                "bare cond.dia,(cbdm-mm.)"
                "insulated cond. dia,(cidm- mm.), with enamel"
                "insulated cond. section,(cism mm,sq.)"
                " from Standard Table ";
125         cin >> swg >> cbsm >> cbdm >> cidm >> cism ;
            cout<<endl<<"Swg no. of enamelled copper wire= "<<swg
                <<"main winding bare cond.dia= "<<cbdm<<" mm."
                <<"bare cond.section= "<<cbsm<<" mm.sq."
                <<"insulated cond. dia= "<<cidm<<" mm.with enamel"
                <<"insulated cond. section= "<<cism<<" mm,sq." ;


        cdm = I1 / cbsm ; //modified curr. density, main wndg.
        cout<<endl<<"modified curr.density, main wndg= "<<cdm <<"mm. sq.";


                        cout <<"CHOICE OF ROTOR LAMINATION" ;
//      Number of rotor slots and slot area should be best selected on the
```

```
//      basis of stator-rotor slot combination and estimated bar area, aRs.
//      One may assume total rotor conductor section = around 80% with
//      copper and 110% with aluminium, of total main winding conductor
//      section in stator.


                              cout<<" Rotor Conductor Section" ;
        float TRcs, Rs, ARs, dRs, Rtht, wor, hlip ;
            TRcs = 1.1 * (2. * Nm) * cbsm ; //when single circuit
//                            winding with Aluminium conductor is assumed.
//        [For 2 parallel paths, conductor current will be halved and
//                        TRcs doubled.]
        cout<<endl<<"Estimated total rotor conduc. section= "<<TRcs<<" mm.sq " ;
        cout<<endl<<"Choose No. of Rotor slots (Rs) and slot-area (aRs)"
150           "so that TRcs = Rs * aRs (approx.),"
                    "& Enter :"
                  "No. of Rotor slot (Rs)" ;
                      cin >> Rs ;
        cout<<"Rotor slot area (aRs,in sq.mm.) " ;
                      cin >> ARs ;
        cout<<"Diameter of rotor slot (dRs, mm)" ; //Assumed circular slot
                      cin >> dRs ;
        cout<<"Rotor tooth height, incl. (Rtht, in m.)";
                      cin >> Rtht ;
        cout<<"Rotor-slot opening (wor, in m.)" ;
                      cin >> wor ;
        cout<<"height of rotor slot lip (hlip, m)" ;
                      cin >> hlip ;


                cout <<" GAP LENGTH ";
        float lg, DR ;
        cout<<endl<<"Enter Designer's choice of gap length (lg,m.)" ;
            cin >> lg ;
170           DR = D – (2. * lg ) ;               //Rotor diameter, m
                cout<<endl<<"Rotor diameter,m.= "<< DR ;


        cout<<"GAP DENSITY & STATOR & ROTOR TOOTH & CORE DENSITIES" ;
                        cout<<endl<<" (a) STATOR CORE " ;
        float BSc ;
        BSc = phim /( 2. * dSc * L * fst ) ;
        cout<<endl<<"Stator Core density= "<<BSc<<" Tesla" ;


                        cout<<" (b) STATOR TEETH " ;
        float aSt, BSt ;
        aSt = fst * L * WSt ; //sectional area of one stator tooth
        cout<<endl<<"Sect. area of tooth, aSt = "<< aSt ;
        BSt = phim * p / (Ss * aSt * fst) ; //maxm. Stator tooth density
        cout<<endl<<"Stator tooth density, BSt = "<< BSt <<"Tesla";

//      CHECK FOR STATOR TOOTH DENSITY-should not be more than 1.6T.
        cout<<endl<<"If BSt higher than 1.6 T, lower value of BSt " ;
        float y ;
            y = BSt- 1.60 ;
            if(y >= 0)
            {
               BSt = 1.55 ;
            }
            else
```

```
            if(y < 0)
            {
               BSt = BSt ;
            }
            cout<<endl<<"Average Stator tooth density= "<<BSt<<" Tesla" ;

200                            cout<<"(c) GAP DENSITY " ;
        float Bg ;
        Bav = p * phim / ( PI * D * L * fst) ; //Av. gap density
        Bg = PI * Bav / 2 ; //gap density at pole-axis
        cout<<endl<<"Gap Density at pole-axis= "<<Bg ;

                            cout<<"(d) ROTOR TOOTH DENSITY" ;

        float oa, ba, Da, spa, ewt, BRt, BRtm ;
209        //To find 'estimated tooth width ( ewt)' by One third density method
        //
        – art.4.2.4.3(c)
            oa = 0.5 * dRs - (Rtht / 3.) ; // ref. Figure 11.1
        //            point a is at one-third slot-height Rtht from slot-bottom.
                     ba = sqrt((0.5*dRs * 0.5*dRs) - (oa*oa)) ;
                     Da = DR - (4.*(Rtht / 3.)) ;        //diameter at point a
                     spa = PI * Da / Rs ;                //slot-pitch at a
                     ewt = spa - 2.*ba ;                 //tooth-width at 1/3 from tooth-bottom
                     cout<<endl<<"ewt= "<< ewt ; //note; ewt is more than the tooth-width
        //            at slot-centre
            BRt = phim*p /( fst * L * ewt * Rs) ;
            cout<<endl<<"av.flux density in rotor teeth,(BRt) = "<<BRt<<" Tesla ";
        BRtm = PI * BRt / 2.;
            cout<<endl<<"max. density in rotor teeth(BRtm)= "<<BRtm<<" Tesla " ;

                            cout<<"(e) ROTOR CORE DENSITY" ;
        float BRc, ARc, dRc ;

        dRc = 0.5 * (DR - Dsh - 2.*(dRs + hlip)) ; //depth of rotor core
        ARc = fst * dRc * L ;
        BRc = 0.5 * phim / ARc ; //flux-density in armature core
        cout<<endl<<"flux-density in armature core(BRc)= "<<BRc ;

            cout<<"COMPUTATION OF IRON WEIGHTS & IRON LOSSES" ;
                    cout<<"(a) STATOR CORE ";

        float spwt, Scloss, Scwt, vsc, x ;

            vsc = PI * ((Do * Do) - ((D+2.*dSs)*(D+2.*dSs))) * L * fst/4 ;
            cout<<endl<<"Enter: Specific weight of core material" ;
            cin >> spwt ;
            Scwt = spwt * vsc ;

        y = BSc ;
        float sploss3(float) ;
        x = sploss3(y) ;
        Scloss = x * Scwt ;
                cout<<endl<<"Loss in Stator core= "<<Scloss<<"W" ;

                    cout<<"(b) STATOR TEETH" ;
        float tASt, vSt, Stwt, Stloss ;
```

```cpp
// Total Teeth area in Stator
tASt = aSt * Ss ;
    cout<<endl<<"Total tooth-area in Stator = "<< tASt <<"m.-sq";
// Volume of Stator teeth
vSt = L * tASt ;
cout<<endl<<"Volume of Stator teeth, vSt = "<< vSt <<"m.cubed";
// Weight of Stator teeth
Stwt = spwt * vSt ;
cout<<endl<<"Weight of Stator teeth, Stwt = "<< Stwt <<"kg";

y = BSt ;
float sploss3(float) ;
x = sploss3(y) ;
            cout<<endl<<"Loss in Stator teeth= "<<Stloss<<"W" ;

                    cout<<"STATOR WINDING - LENGTH OF MEAN TURN" ;
float Ssp ;
Ssp = Ss / p ;                               // stator slots per pole

//              SKEIN WINDING ARRANGEMENT FOR STATOR MAIN WINDING

float C46, C37, C28, C19;
float cN46, cN37, cN28, cN19, add ;
float N46 ,N28, N37, N19, add2 ;
C46 = sinf((2.*PI)/(2.*Ssp)) ;           // 1/2Coil span :slots 1-9; elec.radian
C37 = sinf((4.*PI)/(2.*Ssp)) ;           //      –do–      3–7,     –do–
C28 = sinf((6.*PI)/(2.*Ssp)) ;           //      –do–      2–8,     –do–
C19 = sinf((8.*PI)/(2.*Ssp)) ;           //      –do–      4–6,     –do–
add = C46 + C37 + C28 + C19 ;
getchar();
cN46 = C46 * Nmp / add ;                  //computed turns in coil 4-6
cN37 = C37 * Nmp / add ;                  //      – do –   3–7
cN28 = C28 * Nmp / add ;                  //      – do –   2–8
cN19 = C19 * Nmp / add ;                  //      – do –   1–9
cout<<endl<<"cN46= "<<cN46<<"cN37= "<<cN37<<"cN28= "<<cN28
            <<"cN19= "<<cN19;
cout<<endl<<"modify computed turns suitably so that total is approx Nmp" ;
                cin >> N46 >> N37 >> N28 >> N19;
                    add2 = N46 + N37 + N28 + N19 ;
cout<<endl<<"Turns per pole= "<<add2 ; //add2 gives modified turns/pole
Nmp = add2 ;
    Nm = Nmp * p ; //total turns
    Cwm = ( N46 * C46 + N37 * C37 + N28 * C28 + N19 * C19 )/Nmp ;
cout<<endl<<"Cwm = "<< Cwm ;

//                      MEAN LENGTH OF TURN OF MAIN WINDING

float MLhTC46, MLhTC37, MLhTC28, MLhTC19, add3, MLhT ;
    int S46, S37, S28, S19 ; // coil span in terms of slot
cout<<endl<<"Enter coil-span in slot-terms" ;
    cin >> S46 >> S37 >> S28 >> S19 ;

float k1 = 4.2 * (D + dSs) / Ss ;
MLhTC46 = ( k1 * S46 + L ) * N46 ;        //Turns in coil 4–6
MLhTC37 = ( k1 * S37 + L ) * N37 ;        //Turns in coil 3–7
MLhTC28 = ( k1 * S28 + L ) * N28 ;        //Turns in coil 2–8
MLhTC19 = ( k1 * S19 + L ) * N19 ;        //Turns in coil 1–9
```

```
add3 = MLhTC46 + MLhTC37 + MLhTC28 + MLhTC19;
MLhT = add3 / Nmp ;
cout<<endl<<"Mean Length of 1/2 Turn= "<< MLhT ;

phim = d / (4.44 * Nm * Cwm * a) ; // Modified phim
Bav = phim * p / (PI * D * L) ; // Modified Bav.
cout<<endl<<"Modified Av. Gap Density= "<< Bav<<" Tesla" ;
cout<<endl<<"Flux per pole = "<< phim <<" wb ";

//                         COMPUTATION OF COMPONENT & TOTAL MMF
        float FSt,Fg,FRt,FRc;
//                              (a) STATOR CORE
     float lSc, FSc ;

     lSc = 0.5 * PI * (D+2.0* dSs+dSc) / p ; //length of flux-path in stator core
     y=BSc;
     float mcur42st(float) ;
     x = mcur42st(y) ;
     FSc= x * lSc ;
     cout<<endl<<"FSc= "<<FSc ;
//                              (b) STATOR TEETH
     y=BSt ;
     float mcur42st(float) ;
     x = mcur42st(y) ;
     FSt= x * dSs;                        // Height of fluxpath = height of tooth
     cout<<endl<<"FSt=                    "<<FSt ;

//                                  (c) GAP

// Compute carter's coefficients for gap-due stator slot opening (kcs)
// and rotor slot opening (kcr)
     float lgeff, ratios, ratior, kcs, kcr, kg1, kg2 ;
     float ysS, ysR ;

     ysS = PI * D / Ss ; // stator slot pitch
     ysR = PI * DR / Rs ; // rotor slot pitch
//   cout<<endl<<"DR=      "<<DR<<"ys1=      "<<ys1<<"ys2=      "<<ys2 ;

     ratios = wos / lg ; //ratio : stator slot opening/gap length
     cout<<endl<<"ratios= "<<ratios ;

     float r= ratios ;
     float cartersemi(float) ;
     kcs = cartersemi(r) ; //carter's coefficient for stator slot opening
     cout<<endl<<"kcs= "<<kcs ;

     ratior = wor / lg ; //ratio : rotor slot opening/gap length
     cout<<endl<<"ratior= "<<ratior ;

     r = ratior ;
     float cartersemi(float) ;
     kcr = cartersemi(r) ; //carter's coefficient for rotor slot opening
     cout<<endl<<"kcr= "<<kcr ;

     kg1 = ysS / ( ysS-kcs*wos ) ; //gap coeff. due stator slots
     kg2 = ysR / ( ysR-kcr*wor ) ; //gap coeff. due rotor slots
     cout<<endl<<"kg1=         "<<kg1         <<"kg2=         "<<kg2 ;
```

350

```cpp
        lgeff = lg * kg1 * kg2 ; //effective gap length

        cout<<endl<<"Bav=          "<<Bav        <<"lgeff=        "<<lgeff ;
        Fg = 800000 * Bav * lgeff ;
        cout<<endl<<"Fg= "<<Fg ;

//                                          (d) ROTOR TEETH
          int F ;
        float kfl ;

        y=BRt ;
        float mcur42st(float) ; //function to find amp-turns/metre
        x = mcur42st(y) ;
        FRt= x * dRs ;
        cout<<endl<<"FRt= "<<FRt ;

//                                          (e) ROTOR CORE
        float lRc, DRcm ;
        kfl=1.05 ; //factor to allow for curved nature of flux-path
//              in rotor core at the back of rotor teeth
        DRcm = Dsh + dRc ; //mean dia. of armature core
        1Rc=kfl*PI*DRcm/p ; //length of fluxpath in rotor core/pole
//                                = pole-pitch on mean dia of armature core
        cout<<endl<<"DRcm= "<<DRcm ;
        cout<<endl<<"lRc= "<<lRc ;
        y=BRc ;
        float mcur42st(float) ;        //function to find ampereturns/metre
        x = mcur42st(y) ;              //from magnetisation curve:42quality stalloy
        FRc= x * lRc;
        cout<<endl<<"FRc= "<<FRc ;

            cout<<endl<<"FSc= "<<FSc ;
            cout<<endl<<"FSt= "<<FSt ;
            cout<<endl<<"Fg= "<<Fg ;
            cout<<endl<<"FRt= "<<FRt ;
            cout<<endl<<"FRc= "<<FRc ;

        F = FSc + FSt + Fg + FRt + FRc ; //Total mmf.
        cout << endl << "Total Magnetomotive Force= "<< F ;
400
//                      MAGNETISING REACTANCE (PRELIMINARY VALUE)
        float Img, Xmpr ;

        Img = p * F /( 2.34 * Cwm * Nm ) ;
        cout<<endl<<"Magnetising current= "<< Img ;

        Xmpr = d / Img ;
        cout<<endl<<"Magnetising Reactance(preliminary value)= "<< Xmpr ;

//                      RESISTANCE OF MAIN WINDING & CONDUCTOR LOSS

        cout<<"For 75 deg. cel., use multiplying factor 1.22" ;

        cout<<endl<<"MLht = "<<MLhT<<" m. "<<" Nm= "<<Nm ;
        r1m = 1.22 * 0.0271 * 2.* MLhT * Nm ;
        //Stator main wndg. resistivity // 0.0271 ohm/m./mm.sq (IS) at 20 deg. cel.
```

```
        cout<<endl<<"Stator main wndg. resistance= "<<r1m<<"ohm" ; //6.39

        mcloss = I1 * I1 * r1m ; //Main winding conductor loss
        cout<<endl<<"Stator main wndg. conductor loss= "<<mcloss<<"Watt" ;
}

int main ()
{
        spec m ;
         int freq, synsp ;
         float powut, volt ;

        cout<<endl<<"Enter: a, b, c, d ";
         cin >> freq >> synsp >> powut >> volt ;

        m.getdata(freq, synsp, powut, volt );

        m.display () ;

        m.stator(freq, synsp, powut, volt );
}
```

Figure 10.3 A computer program for main parameters of a single-phase induction motor.

The program stated in Figure 10.3 has been applied to a 230-volt, 180-watt, 25 syn. r.p.s., 50 Hz. General-purpose single-phase induction motor. The results are :

```
Enter: a,  b,  c,  d      50      25      180       230
frequency= 50, synchronous speed= 25r.p.s., power outout= 180, Supply voltage= 230
Enter the values of: ac, Bav, Cwm, fst, pf, effy
15000    0.4     0.795    0.95      0.65     0.53
Specific electric loading= 15000
Specific magnetic loading= 0.4
Winding factor- main wndg= 0.795
Core Stacking factor= 0.95
Power factor = 0.65
Per cent Efficiency= 0.53
COMPUTE STATOR INSIDE DIA. (D,m.) and GROSS ARMATURE LENGTH (L,m.)
DsqL= 0.000477213
Diameter in m.= 0.0849076 Gross Length in m.= 0.0666862
MODIFY diameter & length, if necessary 0.085 0.07
CHOOSE NUMBER OF STATOR SLOTS
Enter Designer's choice of number of Stator slots 36
Choose STANDARD LAMINATIONS
Depth of slot (dSs, m.), Area of slot(ASs, m.sq.,Height of lip (h4,m.),Slot opening (wos,m.),Width of tooth(WSt,m.),Depth
of core below teeth(dSc,m.)Outside diameter of stator(Do,m.)Dia. of shaft at lamination(Dsh,m.)
0.0225    137     0.0005     0.003      0.0032    0.0125     0.15      0.02
Depth of slot, incl. lip (dSs, m.)= 0.0225Area of ONE slot(ASs,mm.sq.)= 137 mm.sq Height of lip (h4,m.) =
0.0005 m. Slot opening (wos,m.) = 0.003 m. Av.Width of tooth(WSt,m.) = 0.0032 m. Depth of St core below
teeth= 0.0125 m. Outside diameter of stator= 0.15 m. Dia. of shaft at lamination(Do,m.)= 0.02 m
Maximum flux per pole (phim) = 0.00186925
Computed Main winding turns in series= 697.174
MODIFY total main winding turns 696
```

Stator main wndg. turns in series/pole= 174

Modified gap density= 0.400675

Enter new L if necessary 0.07

Main winding conductor parameters

Stator main winding current = 2.27172

rotor current= 1.47662

Enter: Current density, main winding= 4

main winding bare cond. section= 0.567931 mm.sq.

Assuming copper conductor, Enter: Swg No. for Enamelled copper wire (swg)main winding bare cond.section,(cbsm-mm.sq.)bare cond.dia,(cbdm-mm.)insulated cond. dia,(cidm- mm.), with enamelinsulated cond. section,(cism mm,sq.) from Standard Table 20.75 0.5518 0.8282 0.8882 0.6196

Swg no. of enamelled copper wire= 20.75main winding bare cond.dia= 0.8282 mm.bare cond.section= 0.5518 mm.sq.insulated cond. dia= 0.8882 mm.with enamelinsulated cond. section= 0.6196 mm,sq.

modified curr.density, main wndg= 4.11693mm. sq.

CHOICE OF ROTOR LAMINATION Rotor Conductor Section

Estimated total rotor conductor section= 844.916 mm.sq

Choose No. of Rotor slots (Rs) and slot-area (aRs)so that TRcs = Rs * aRs (approx.),& Enter :No. of Rotar slot (Rs) 28

Rotor slot area (aRs,in sq.mm.) 31

Diameter of rotor slot (dRs, mm)0.0063

Rotor tooth height, incl. (Rtht, in m.)0.0065

Rotor-slot opening (wor, in m.) 0.0002

height of rotor slot lip (hlip, m)0.0002

GAP LENGTH

Enter Designer's choice of gap length (lg,m.) 0.0003

Rotor diameter,m.= 0.0844

GAP DENSITY & STATOR & ROTOR TOOTH & CORE DENSITIES

(a) STATOR CORE

Stator Core density= 1.12436 Tesla

(b) STATOR TEETH

Sect. area of tooth, aSt = 0.0002128

Stator tooth density, BSt = 1.02737Tesla

If BSt higher than 1.6 T, lower value of BSt

Average Stator tooth density= 1.02737 Tesla

(c) GAP DENSITY

Gap Density at pol-axis= 0.661387

(d) ROTOR TOOTH DENSITY

ewt= 0.00251209

av.flux density in rotor teeth,(BRt) = 1.5985 Tesla

max. density in rotor teeth(BRtm)= 2.51091 Tesla

(e) ROTOR CORE DENSITY

flux-density in armature core(BRc)= 0.546867

COMPUTATION OF IRON WEIGHTS & IRON LOSSES

(a) STATOR CORE

Enter: Specific weight of core material 8890

vsc= 0.000292482x = 6.36538

Loss in Stator core= 16.551W

(b) STATOR TEETH

Total tooth-area in Stator = 0.0076608m.-sq

Volume of Stator teeth, vSt = 0.000536256m.cubed

Weight of Stator teeth, Stwt = 4.76732kg

Loss in Stator teeth= 24.7954W
STATOR WINDING - LENGTH OF MEAN TURN
cN46= 20.987cN37= 39.4426cN28= 53.1409cN19= 60.4296
modify computed turns suitably so that total is apprcx Nmp
20   40   50   60
Turns per pole= 170
Cwm = 0.793774
Enter coil-span in slot-terms 2   4   6   8
Mean Length of 1/2 Turn= 0.142299
Modified Av. Gap Density= 0.410736 Tesla
Flux per pole = 0.00191942 wb
FSc= 13.1927

FSt=   4.01062
ratios= 10
kcs= 0.845
ratior=   0.666667
kcr=   0.126667
kg1=   1.51919   kg2=   1.00268
Bav=   0.410736   Igeff= 0.000456978
Fg=   150.158

FRt= 28.1671
DRcm= 0.0457
IRc= 0.0376873

FRc= 2.76997

FSc= 13.1927
FSt= 4.01062
Fg= 150.158
FRt= 28.1671
FRc= 2.76997
Total Magnetomotive Force= 198
Magnetising current= 0.627052
Magnetising Reactance(preliminary value)= 366.796 For 75 deg. cel., use multiplying factor 1.22

MLht = 0.142299 m. Nm= 680
Stator main wndg. resistance= 6.39838ohm
Stator main wndg. conductor loss= 33.0203Watt

Section (b) the squirrel-cage

Except for a few types, single-phase induction motors have invariably a squirrel-cage rotor.

A computer program, as developed is shown below. On line 70 for determination of stator reactances, the dimensions of the stator slots has been called for. We have chosen a tapered slot with parallel-sided tooth, of dimensions as shown in Figure 7.9.

A Computer Program

```
//TO FIND RESISTANCES AND LOSSES OF A CAGE ROTOR
#include <iostream>
#include <math.h>
#include <stdio.h>
using namespace std ;

class spec
{
        public:
                int d, p, Rs, Nm, Ss ;
        float DR, sk, D, L, fst, I1, pf, Ars, Cwm, r1 ;

void getdata(int d, int p,float DR,int Rs,int Nm,float sk,float D,float L,
        float fst,float I1,float pf,float Ars,float Cwm,int Ss,
        float r1);
void display();
void sqcage(int d,int p,float DR,int Rs,int Nm,float sk,float D,float L,
        float fst,float I1,float pf,float Ars,float Cwm,int Ss,float r1 );
};

void spec::getdata(int volt,int pole,float DiaR,int slotR,int Sturn,
        float skew,float diaS,float Length,float stfac,float Scur,
        float pfac,float ARsl,float wfacm,int SlotS,float ResisS)
        { d=volt;p=pole;DR=DiaR;Rs=slotR;Nm=Sturn;sk=skew;D=diaS;L=Length;fst=stfac;
I1=Scur;pf=pfac;Ars=ARsl;Cwm=wfacm;Ss=SlotS;r1=ResisS; }

void spec::display()
        {cout<<endl<<"Supply voltage= "<<d
                <<"No. of poles= "<<p<<" ,Rotor Dia= "<<DR<<" m."
                <<" ,No.of Rotor slots= "<<Rs
                <<" ,Total Stator turns= ."<<Nm
                <<" ,Amount of skew in number of Rotor slot= "<<sk
                <<" ,Stator internal diameter= "<<D<<" m."
                <<" ,Gross armature length= "<<L<<" ,stacking factor= "<<fst
                <<" ,Stator current= "<<I1<<" ,power factor= "<<pf
                <<" ,Area. of rotor slot= "<<Ars<<" sq.mm. "
                <<" ,Winding factor, main wndg.= "<<Cwm
                <<" ,No. of Stator slots= "<<Ss
                <<" ,Stator main winding resistance= "<<r1;
}

void spec::sqcage( int d,int p,float DR,int Rs,int Nm,float sk,float D,
        float L,float fst,float Is,float pf,float Ars,float Cwm,int Ss,float r1)
{

        float yr, ysk, Lb, roe, rb, De, Ae, re, r2 ;

            float PI = 3.14159L ;

        yr = PI * DR / Rs ;                      //one rotor slot-pitch
        ysk = sk * yr ;                          //pitch for sk no. of skewed-slot
50      Lb = sqrt(L * L + ysk * ysk) ;           //Length of rotor bar
        cout<<endl<<"Length of bar= "<<Lb<<" m. " ;

        cout<<endl<<"Enter: Resistivity of material of bars and rings"
```

```
                          "(ohm.mm.sq per m.)" ;
        cin >> roe ;

    rb = roe * Lb / Ars ;                        //resistance of one rotor bar
    cout<<endl<<"resistance of 1 bar= "<<rb ;

    cout<<endl<<"Enter : mean Diameter (in m.) and "
            "cross-section of End ring (in mm.sq)" ;
     cin >> De >> Ae ;

    re = roe * PI * De / Ae ;                    //resistance of 1 end-ring
    cout<<endl<<"resistance of 1 endring= "<<re ;

    r2 = 4. * Cwm * Cwm * Nm * Nm *2.* (rb + 0.2 * re * Rs/(p*p)) / Rs ;
    cout<<endl<<"rotor resistance in stator terms= "<<r2 ;

70  //                        DETERMINATION OF REACTANCES
    float ws, w1, w2, wos, sh1, sh2, sh3, sh4 ;
    float Lam1, Lam2, Lam3, Lam4, LamS1, k3, sxsl ;

        float k2 = 1e-6f ;
//    Stator slot Premeance
    cout<<endl<<"Enter slot dimensions: ws, w1, w2, wo2, sh1, sh2, sh3 sh4 " ;
    cin >> ws >> w1>> w2 >> wos >> sh1 >> sh2 >> sh3 >> sh4 ; //Figure 9.12(a)

//    STATOR SLOT LEAKAGE
        w2 = (ws * sh2 + w1 * sh1)/(sh1 + sh2) ;
    Lam1 = 2. * sh1 / (3* (ws + w2)) ;
    Lam2 = 2. * sh2 / (w2 + w1) ;
    Lam3 = 2. * sh3 / (w1 + wos) ;
    Lam4 = sh4 / wos ;
    LamS1 = Lam1 + Lam2 + Lam3 + Lam4 ;

        k3 = 4. * PI * p * 4. * PI * k2 * Nm * Nm * Cwm * Cwm * 2. / Ss ;
    sxsl = k3 * LamS1 * L ; //stator slot leakage reactance
     cout<<endl<<" k3= "<< k3 <<" LamS1= "<<LamS1 ;
     cout<<endl<<"stator slot leakage reactance= "<< sxsl ;

//    OVERHANG LEAKAGE
    float xo, ks ;

//    The factor ks varies between 0.75 and 1.12 for Wide-spread connection
    cout<<endl<<"Enter the values of the factor ks ";
        cin >> ks ;
    xo = 0.00475 * ks * D * Nm * Nm / (p * p) ;
    cout<<endl<<"Overhang leakage reactance= "<< xo ;

//    STATOR ZIGZAG REACTANCE (preliminary value)
    float Xmpr, sxzpr ;
    cout<<endl<<"Enter the value of Magn.reactance(prelim.value)";
    cin >> Xmpr ;
      sxzpr = 5.* p * p * Xmpr / ( 6. * Ss * Ss ) ;
    cout<<endl<<"Stator zigzag reactance(prelim.value)= "<< sxzpr ;

//                        MAGNETISING REACTANCE (FINAL VALUE)
        float k4, E, x1mpr, Xm, Img ;
    cout<<endl<<"Enter the value of Img ";
```

```cpp
        cin >> Img ;
        k4 = sqrt( 1 - pf*pf) ;
        x1mpr = sxsl + sxzpr + 0.5 * xo ;          // Stator winding leakage reactance(preliminary value)
          E = I1m * ( r1m * pf + x1mpr * k4 ) ;

            Xm = (d - E) / Img ; // Modified Magnetising Reactance
        cout<<endl<<"Modified Magnetising Reactance= "<< Xm ;

//      STATOR ZIGZAG REACTANCE (Modified value)
        float sxz ;

            sxz = 5.* p * p * Xm / ( 6. * Ss * Ss) ;
        cout<<endl<<"Stator zigzag reactance(modified value)= "<< sxz ;

//      ROTOR SLOT LEAKAGE
        float hlip, wor, LamR1, rxsl ;
        cout<<endl<<"Enter : wor and hlip ";
            cin >> wor >> hlip;

        LamR1 = 0.66 + hlip / wor ;
        cout<<endl<<" k3= "<< k3 <<" LamR1= "<<LamR1 ;
        rxsl = (Ss / Rs) * k3 * LamR1 * L ;
            cout<<endl<<"Rotor slot leakage reactance= "<< rxsl ;

//      ROTOR ZIGZAG REACTANCE
        float rxz ;
        rxz = 5.* p * p * Xm / ( 6. * Rs * Rs) ;
            cout<<endl<<"Rotor zigzag reactance= "<< rxz ;

//      ROTOR SKEW LEAKAGE REACTANCE
        float rxsk, gama ;

        gama = PI * sk / Rs ;
            rxsk = 59.4* gama*gama* 0.95/12.; //0.95 is the assumed value
//
        of stator leakage flux factor
            cout<<endl<<"Rotor Skew leakage reactance= "<< rxsk ;

            float x1m, x2s ;

            x1m = sxsl + sxz + 0.5 * xo ;
            cout<<endl<<"Stator leakage reactance= "<<x1m<<" ohm";

            x2s = rxsl + rxz + rxsk + 0.5 * xo ;
            cout<<endl<<"Rotor leakage reactance in stator terms= "<<x2s<<" ohm";

        cout<<endl<<"Magnetising Reactance= "<< Xm ;
}

int main ()
{
        spec m ;

            int volt, pole, slotR, Sturn, SlotS ;
            float DiaR,skew,diaS,Length,stfac,Scur,pfac,ARsl,wfacm,ResisS ;

        cout<<endl<<"Enter: d, p, DR, Rs, Nm, sk, D, L, fst, Is, pf, Ars, Cwm, Ss, r1 ";
```

```
        cin >> volt >> pole >> DiaR >> slotR >> Sturn >> skew >> diaS >> Length >> stfac >> Scur >>
pfac >> ARsl >> wfacm >> SlotS >> ResisS;

        m.getdata(volt,pole,DiaR,slotR,Sturn,skew,diaS,Length,stfac,Scur,
                            pfac,ARsl,wfacm,SlotS,ResisS) ;

        m.display () ;

        m.sqcage(volt,pole,DiaR,slotR,Sturn,skew,diaS,Length,stfac,Scur,
                            pfac,ARsl,wfacm,SlotS,ResisS) ;
}
```

Figure 10.4 A computer program for the squirrel-cage induction motor-split-phase / capacitor-start types.

Result when applied to a 180W, 230-volt, 25 syn.r.p.s, 50Hz motor :

```
Enter: d, p, DR, Rs, Nm, sk, D, L, fst, Is, pf, Ars, Cwm, Ss, r1
230   4   0.0844  28  680  1  0.085  0.07  0.95  2.2717  0.65  31  0.7938  36  6.4
Supply voltage= 230No. of poles= 4, Rotor Dia= 0.0844 m., No.of Rotor slots= 28, Total Stator turns= .680,
Amount of skew in number of Rotor slot= 1, Stator internal diameter= 0.085 m., Gross armature length= 0.07,
stacking factor= 0.95, Stator current= 2.2717, power factor= 0.65, Area. of rotor slot= 31 sq.mm., Winding
factor, main wndg.= 0.7938, No. of Stator slots= 36, Stator main winding resistance= 6.4

Length of bar= 0.0706376 m.

Enter: Resistivity of material of bars and rings(ohm.mm.sq per m.) 0.0341
resistance of 1 bar= 7.77014e-05
Enter : mean Diameter (in m.) and cross-section of End ring (in mm.sq)
0.0769     75
resistance of 1 endring= 0.000109842
rotor resistance in stator terms= 9.66889

Enter slot dimensions:    ws,    w1,    w2,    wo2,    sh1,    sh2,    sh3    sh4
        8.6       5.1     3      1     19      1       2      0.5

stator slot leakage reactance= 1.61855
Enter the values of the factor ks 0.75
Overhang leakage reactance= 8.75128
Enter the value of Magn.reactance(prelim.value) 366.8
Stator zigzag reactance(prelim.value)= 3.77366
Enter the value of Img 0.627

Modified Magnetising Reactance= 324.86
Stator zigzag reactance(modified value)= 3.34218
Enter : wor and hlip 0.001 0.0005

Rotor slot leakage reactance= 0.830238
Rotor zigzag reactance= 5.52483
Rotor Skew leakage reactance= 0.0591987
Stator leakage reactance= 9.33637 ohm
Rotor leakage reactance in stator terms= 10.7899 ohm
Magnetising Reactance= 324.86
------------------
(program exited with code: 0)
```

Section (c) Performance calculation.

For the performance at any value of slip, the Double-revolving field theory will be used. The input to the program are the parameters computed in earlier two sections.

Attention is drawn to the methodology of computing Losses and Efficiency at any slip s - line 66 and onwards in the program below.

With reference to the equivalent circuit of a single-phaqse induction motor in Figure 10.5 (based on the double-revolving field theory), the input current I1 can be calculated. We may term this current as *apparent* as the iron and F & W losses in the motor are not considered.

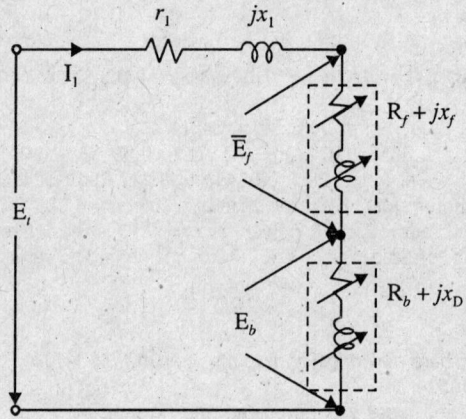

Figure 10.5 Equivalent circuit of single-phase induction motor by double-revolving field theory.

A simple method to account for these losses is to write down,

Effective stator main winding current, IM =

Apparent stator main windg. current I1 + loss current due to Stator Iron loss, Pim.

I1 = et / Z, where, Z = sqrt[(r1 + Rf + Rb)^2 + (x1 + Xf + Xb)^2]

and loss current due to stator core and teeth losses (Pim) = (Pim / et) * ((r1 + Rf + Rb) / Z) .... here we neglect rotor iron losses as negligibly small.

Thus, writing,   stator i^2 r loss = IM^2 r1 ;

forward-field i^2 r loss = I1^2 s Rf ;

backward-field i^2 r loss = I1^2 (2-s) Rb ;

we have, total loss power   Pl = IM^2 r1 + I1^2 [s * Rf + (2-s)* Rb] + Pim ;

Also, Power across the gap Pg = I1^2 (Rf – Rb) ;

Mechanical power output Pm = (1-s) Pg ;

and   Input power Pi = Pm + Pl ;

For computing output power Po, a factor 1.2 is used as below because of additional loss due to pulsating field. That is,

Po = Pm – (1.2 Pim + FWloss);

Computer Program

```
//Given main parameters of a single-ph. induction motor, to find
//its performance at various slips.
#include <iostream>
#include <fstream>
#include<math.h>
#include<stdio.h>
using namespace std;

int main()
{
        float r1, r2, x1, Xm, x22, x2s, s, Rf, Xf, Rb, Xb, Zsq, Z, I1, spf ;
        float Scloss, Stloss, Pim, FWloss, Pg, omegas, omegar, Pm, Po, PI, To ;
        float effy, pf, Tout ;
        float c1f, c2f, c3f, c1b, c2b, c3b, b, c, d ;

        cout<<endl<<"Enter : Stator resis.(r1), Rotor resistance in Stator"
                "terms(r2), Magnetising reactance(Xm)"
                ", Stator leakage reactance(x1)"
                ", Rotor leakage reactance in stator terms(x2s)" ;
        cin >> r1 >> r2 >> Xm >> x1 >> x2s ;
        cout<<endl<<"Enter: Supply voltage (et), syn. r.p.s.(ns)"
                "specified power output(powut)";
        cin >> d >> b >> c ;

        cout<<endl<<"Enter Loss in stator core(Scloss)"
                        <<"Loss in stator teeth(Stloss)";
        cin >> Scloss >> Stloss ;
            Pim = Scloss + Stloss ;                     //iron loss in Stator
//              ASSUME F & W losses 5 to 7.5% of output power
        cout<<endl<<"F & W losses(FWloss)" ;
        cin >> FWloss ;
        getchar();
        cout<<endl<<" PERFORMANCE SHEET AT RATED VOLTAGE AND FREQUENCY " ;
        ofstream outf("RESULT");
        for (s=0.01; s<=0.10; s+=0.01)
        {
            x22 = Xm + x2s ;
            c1f = r2 / s ;
            c2f = c1f * c1f + x22 * x22 ;
            c3f = c1f * c1f + x2s * x22 ;
            Rf = 0.5 * Xm * Xm * c1f / c2f ;
            Xf = 0.5 * Xm * c3f / c2f ;

            c1b = r2 / (2.-s) ;
            c2b = c1b * c1b + x22 * x22 ;
            c3b = c1b * c1b + x2s * x22 ;
                Rb = 0.5 * Xm * Xm * c1b / c2b ;
            Xb = 0.5 * Xm * c3b / c2b ;

                Zsq = (Rf+Rb+r1) * (Rf+Rb+r1) + (Xf+Xb+x1) * (Xf+Xb+x1) ;
            Z = sqrt(Zsq) ;                             //Effective impedance across input
            I1m = d / Z ;                               //Stator main windg. current
            spf = (r1 + Rf + Rb) / Z ; //Stator p.f.
            cout<<endl<<"Slip = "<<s<<" ,Stator main wndg. current= "<<I1 ;
            cout<<endl<<"stator power factor= "<<spf ;
```

```
      Pg = I1 * I1 * ( Rf - Rb) ;              //Real power / phase, across the gap

      PI = 3.14159L ;

            omegas = 2. * PI * b ; //Syn. speed, radians / sec.
            omegar = (1-s) * omegas ; //Rotor speed, radians / sec.
            cout<<endl<<"Rotor speed= "<<omegar<<"radians/sec" ;
//                           INTERNAL MECHANICAL POWER
            Pm = (1 - s) * Pg ; //Internal mechanical power

66          //                            LOSSES & EFFICIENCY
            float IM, Pl, Pi, Px, effy, pf, Tfl, Tout ;
//                        STATOR INPUT CURRENT IN MAIN WINDING IM
//                = Stator main windg.current I1m + loss current due to Stator Iron
//                                      loss, Pim.
//                      Loss current due to Stator Iron losses =
//                               (Pim / et ) * ((r1 + Rf + Rb) / Z)
      IM = I1 + (Pim/d) * ((r1+Rf+Rb)/Z) ; // Loss current due to Stator Iron
      losses + Stator main wndg current
            Pl = IM * IM * r1 + I1 * I1 * (s * Rf + (2-s)* Rb) + Pim ;
            Pi = Pm + Pl ; // Input power
            Po = Pm - ( 1.2 * Pim + FWloss ) ; //Output power ; the factor 1.2 is because of
      additional loss due to pulsating field
            Px = Po / c ;                 //output power as % of full load
            effy = Po / Pi ;
            pf = Pi / (IM * d) ;
            Tfl = c / 150.8 ; //full-load torque at slip 0.04, when omegar=150.8
            To = Po / omegar ; //Output torque of the motor, N.m.
      Tout = To / Tfl ;                   //Output torque of the motor,as of Tfl
      cout<<endl<<"Internal mechanical power="<<Pm<<"watt"
            <<"Input power= "<<Pi<<"watt"
            <<"Output power= "<<Po<<"watt"
            <<"Po as % of full load= "<<Px
            <<"Efficiency= "<<effy
            <<"Power factor= "<<pf
            <<"Output Torque= "<<To<<" N.M. "
            <<"Output Torque= "<<Tout<<" % 0f full-load torque " ;

      outf<< s <<endl ;
      outf<<I1 <<endl ;
      outf<<Pm <<endl ;
      outf<<Pi <<endl ;
      outf<<Po <<endl ;
      outf<<Px <<endl ;
      outf<<effy <<endl ;
      outf<<pf <<endl ;
      outf<<To <<endl ;
      outf<<Tout <<endl ;
   }
};
```

Figure 10.6 A computer program for determination of performance of a single-phase split-phase / capacitor-start induction motor.

Result as applied to the 180 W, 230 volt, 50Hz, 25 syn.r.p.s. Motor is :

Enter : Stator resis.(r1m), Rotor resistance in Statorterms(r2), Magnetising reactance(Xm), Stator leakage reactance(x1m), Rotor leakage reactance in stator terms(x2s) 6.4 9.67 325 9.34 10.79

Enter: Supply voltage (et), syn. r.p.s.(ns)specified power output(powut)

230   25   180

Enter Loss in stator core(Scloss)Loss in stator teeth(Stloss)

16.55   24.8

F & W losses(FWloss) 12

PERFORMANCE SHEET AT RATED VOLTAGE AND FREQUENCY

Slip = 0.01 ,Stator main wndg. current= 1.35175

stator power factor= 0.337429

Internal mechanical power= 84.0486wattInput power= 147.331 wattOutput power= 22.4286wattPo as % of full load= 0.124604Efficiency= 0.152233Power factor= 0.453527Output Torque= 0.144228 N.M. Output Torque= 0.121472 % 0f full-load torque.

Slip = 0.02 ,Stator main wndg. current= 1.52855

stator power factor= 0.547451

Internal mechanical power= 163.488wattInput power= 235.803wattOutput power= 101.868wattPo as % of full load= 0.565934Efficiency= 0.432006Power factor= 0.630147Output Torque= 0.661748 N.M. Output Torque= 0.557339 % 0f full-load torque.

Slip = 0.03 ,Stator main wndg. current= 1.78047

stator power factor= 0.675587

Internal mechanical power= 234.542wattInput power= 320.87wattOutput power= 172.922wattPo as % of full load= 0.960679Efficiency= 0.538916Power factor= 0.733514Output Torque= 1.1349 N.M. Output Torque= 0.955842 % 0f full-load torque.

Slip = 0.04 ,Stator main wndg. current= 2.0704

stator power factor= 0.74975

Internal mechanical power= 297.399wattInput power= 402.073wattOutput power= 235.779wattPo as % of full load= 1.30989Efficiency= 0.58641Power factor= 0.79272Output Torque= 1.56356 N.M. Output Torque= 1.31687 % 0f full-load torque.

Slip = 0.05 ,Stator main wndg. current= 2.37608

stator power factor= 0.792817

Internal mechanical power= 352.373wattInput power= 479.087wattOutput power= 290.753wattPo as % of full load= 1.61529Efficiency= 0.606889Power factor= 0.827039Output Torque= 1.94841 N.M. Output Torque= 1.641 % 0f full-load torque.

Slip = 0.06 ,Stator main wndg. current= 2.6847

stator power factor= 0.818112

Internal mechanical power= 399.87wattInput power= 551.712wattOutput power= 338.25wattPo as % of full load= 1.87916Efficiency= 0.613091Power factor= 0.84708Output Torque= 2.29081 N.M. Output Torque= 1.92938 % 0f full-load torque.

Slip = 0.07 ,Stator main wndg. current= 2.98923

stator power factor= 0.832881

Internal mechanical power= 440.363wattInput power= 619.848wattOutput power= 378.743wattPo as % of full load= 2.10413Efficiency= 0.611025Power factor= 0.858559Output Torque= 2.59264 N.M. Output Torque= 2.18358 % 0f full-load torque.

Slip = 0.08 ,Stator main wndg. current= 3.28563

stator power factor= 0.841115

Internal mechanical power= 474.366wattInput power= 683.483wattOutput power= 412.746wattPo as % of full load= 2.29304Efficiency= 0.603887Power factor= 0.864648Output Torque= 2.85612 N.M. Output Torque= 2.40548 % 0f full-load torque.

Slip = 0.09 ,Stator main wndg. current= 3.57162

stator power factor= 0.845107

Internal mechanical power= 502.415wattInput power= 742.675wattOutput power= 440.795wattPo as % of full load= 2.44886Efficiency= 0.593523Power factor= 0.867189Output Torque= 3.08372 N.M. Output Torque= 2.59718 % 0f full-load torque.

Slip = 0.1 ,Stator main wndg. current= 3.84595

stator power factor= 0.846234

Internal mechanical power= 525.045wattInput power= 797.54wattOutput power= 463.425wattPo as % of full load= 2.57459Efficiency= 0.581069Power factor= 0.867305Output Torque= 3.27807 N.M. Output Torque= 2.76086 % 0f full-load torque.

Section (d) Computer program for Auxiliary winding and Starting Torque.

```
//STARTING TORQUE OF SINGLE PHASE INDUCTION MOTOR

#include<iostream>
#include<math.h>
#include<stdio.h>
//#include "mmf.h"
//#include "sploss3.h"
//#include "mcur42st.h"
//#include "cartersemi.h"
using namespace std ;

class spec
{
        public:
            int p, Nmp, Ss ;
        float Cwm, D, dSs, L ;

        void getdata(int p,int Nmp,int Ss,float Cwm,float D,float dSs,float L);
        void display();
        void stator(int p,int Nmp,int Ss,float Cwm,float D,float dSs,float L);
};

        void spec::getdata(int pole,int sturnp,int Stslot,float wfacm,float dia,
                                        float htStslot,float length )
            {p=pole;Nmp=sturnp;Ss=Stslot;Cwm=wfacm;D=dia;dSs=htStslot;L=length;}

        void spec::display()
            {cout<<endl<<"No. of poles= "<<p
                <<", Stator main wndg. turns per pole= "<<Nmp
                <<"No. of Stator slots = "<<Ss
                <<"Winding factor- main wndg(Cwm)= "<< Cwm
                <<"Stator Internal dia= "<<D
                <<"Height of Stator slot= "<<dSs
                <<" Arm.Length= "<<L;}

        void spec::stator(int p,int Nmp,int Ss,float Cwm,float D,float dSs,
                    float L )
{

        float Cwa, K ;

        cout << endl <<"Enter the values of: Cwa,K " ;
        cin >> Cwa >> K ;

        cout << endl <<"Winding factor- auxl. wndg(Cwa)= "<< Cwa ;
        cout << endl <<"Effective auxl./main. windg turns(K)= "<< K ;
```

```
      float Nap ;
           int Ssp ;
 Nap = Nmp * Cwm * K / Cwa ;
 Ssp = Ss / p ; // stator slots per pole
 float PI = 3.14159L ;
```

//   SKEIN WINDING ARRANGEMENT FOR STATOR AUXILIARY WINDING

```
 float C811, C712, C613, C514,add, add2, cN811, cN712, cN613, cN514 ;
 float N811, N712, N613, N514 ;

 C811 = sinf((3.*PI)/(2.*Ssp)) ; // 1/2Coil span :slots 8-11, elec.radian
 C712 = sinf((5.*PI)/(2.*Ssp)) ; //      –do–        7–12,      –do–
 C613 = sinf((7.*PI)/(2.*Ssp)) ; //      –do–        6–13,      –do–
 C514 = sinf((9.*PI)/(2.*Ssp)) ; //      –do–        5–14,      –do–

      add = C811 + C712 + C613 + C514 ;
 getchar();
 cN811 = C811 * Nap / add ; //computed turns in coil 8-11
 cN712 = C712 * Nap / add ;
 cN613 = C613 * Nap / add ;
 cN514 = C514 * Nap / add ;
 cout<<endl<<"cN811= "<< cN811<<" , cN712= "<<cN712
                    <<" , cN613= "<<cN613<<" , cN514= "<<cN514 ;
```

//   cout<<endl<<"modify computed turns suitably so that total is approx Nap" ;
```
 cin >> N811 >> N712 >> N613 >> N514 ;

 add2 = N811 + N712 + N613 + N514 ;

 cout<<endl<<"Turns per pole= "<<add2 ; //add2 gives modified turns/pole
 Nap = add2 ;

 Cwa = ( N811 * C811+N712 * C712+N613 * C613+N514 * C514 )/Nap ;
 cout<<endl<<"modified Cwa= "<< Cwa ; //modified winding factor, //auxiliary

      K = Nap * Cwa / (Nmp * Cwm) ;
 cout<<endl<<"Modified K= "<< K ;
```

//   MEAN LENGTH OF TURN OF AXILIARY WINDING

```
 float MLhTC811, MLhTC712, MLhTC613, MLhTC514, add3, MLhT ;
 int S811, S712, S613, S514 ; // coil span in terms of slot

 cout<<endl<<"Enter coil-span in slot-terms" ;
 cin >> S811 >> S712 >> S613 >> S514 ;

 float k1 = 4.2 * (D + dSs) / Ss ;

 MLhTC811 = ( k1 * S811 + L ) * N811 ; //Turns in coil 8-11
 MLhTC712 = ( k1 * S712 + L ) * N712 ; //Turns in coil 7-12
 MLhTC613 = ( k1 * S613 + L ) * N613 ; //Turns in coil 6-13
 MLhTC514 = ( k1 * S514 + L ) * N514 ; //Turns in coil 5-14
      add3 = MLhTC811 + MLhTC712 + MLhTC613 + MLhTC514 ;
 MLhT = add3 / Nap ;
      cout<<endl<<"Mean Length of 1/2 Turn= "<< MLhT <<" m." ;
```

```
//                        CHOOSE AUXILIARY WINDING CONDUCTOR SIZE

        float r1a, swg, cbda, cbsa, cida, cisa ;
//           Generally auxiiary winding conductors have a size around 2 to
//              2 1/2 swg.-size larger than that of the main winding.
        cout<<"With main winding conductor of bare diameter 9 mm., choose "
        "conduc.BARE dia cbda, section cbsa,INSUL.dia cida & section cisa" ;
        cin >> swg >> cbda >> cbsa >> cida >> cisa ;
        cout<<endl<<"swg of chosen conductor= "<<swg
            <<"Auxl.bare cond.dia= "<<cbda<<" mm.sq. "
               <<"bare section= "<<cbsa<<" mm.sq. "
                   <<"insulated dia= "<<cida<<" mm.sq, "
                       <<"Insulated section= "<<cisa<<" mm.sq. " ;


//       cout<<"For 75 deg. cel.temp. rise., use multiplying factor 1.22" ;
        r1a = 1.22 * 0.04355 * 2.* MLhT * Nap * p ; //resistivity of Al = 0.04355 ohm/mm.sq. at 75 deg.cel.
            cout<<endl<<"Stator auxl. wndg. resistance= "<<r1a<<"ohm" ; //13.4

//    TO CHECK WHETHER MAIN & AUXL.CONDUCTORS CAN BE ACCOMODATED IN SLOTS
//    Reference FIG 11.2(b), find slot under one pole are accomodating
//    stator main and auxiliary coils
      float S5, S6, S7, S8, Smax, z, y ;
      float N46, N37, N28, N19, cism, ASs ;

      cout<<endl<<"Put the values of: N46, N37, N28, N19, cism, ASs" ;
      cin >> N46 >> N37 >> N28 >> N19 >> cism >> ASs ;

      S5 = 2. * N514 * cisa ;
      S6 = N46 * cism + N613 * cisa ;
      S7 = N37 * cism + N712 * cisa ;
      S8 = N28 * cism + N811 * cisa ;
      cout<<"S5= "<<S5<<" S6= "<<S6<<" S7= "<<S7<<" S8= "<<S8 ;
//    Select the largest from above, divide that by ASs. If the answer is
//    less than 0.5, you are happy.

      y = S6- S5 ;
      if(y > 0)
          {
              Smax = S6 ;
          }
          else
          if(y < 0)
          {
              Smax = S5 ;
          }
      y = Smax - S7 ;
          if(y > 0)
              {
                  Smax = Smax ;
              }
              else
              if(y < 0)
              {
                  Smax = S7 ;
              }
      y = Smax - S8 ;
```

```
            if(y > 0)
            {
                Smax = Smax ;
            }
            else
            if(y < 0)
            {
                Smax = S8 ;
            }
        cout<<"Smax= "<<Smax ;

        z = Smax * 100. / ASs ;
        cout<<"Per cent fillup of stator slots by main & auxl. conduc= "<<z
            <<" which is satisfactory " ;

//      PERFORMANCE COMPUTATION WITH MOTOR UNDER LOCKED ROTOR CONDITION

        float a, b, r1, r2, rr2, RM, rra, x1, x1a, RA, Zm, d, IM, Xc ;
        float rA, radd, Xm, C, Tmax ;

        float k5 = 1e6f ;
        cout<<endl<<"Enter the values of: frequency(a), syn. rps (b)"
            "input voltage (d),main wndg.resis(r1), "
            "main windg.leakage reactance(x1) ,"
            "rotor resis.in Stator terms(r2), "
            " magnetising reactance (Xm) ";
        cin >> a >> b >> d >> r1 >> x1 >> r2 >> Xm ;

//      EFFECTIVE ROTOR RESISTANCE UNDER LOCKED CONDITION
        rr2 = 1.1 * r2 ; // 10% increase under rated frequency condition

//      TOTAL MAIN WINDING RESISTANCE
        RM = r1 + rr2 ;
//      TOTAL LOCKED ROTOR RESISTANCE, in terms of Auxiliary winding
        rra = K * K * rr2 ;
//      TOTAL LEAKAGE REACTANCE in terms of Auxiliary winding
        x1a = K * K * x1 ;
//      TOTAL LEAKAGE RESISTANCE in terms of Auxiliary winding
        RA = r1a + rra ;

//      MAIN WINDING LOCKED ROTOR IMPEDANCE
        Zm = sqrt(RM * RM + x1 * x1) ; // Ref. Equivalent circuit, Fig, 11.5(d)SKSEN:Electrical
//
Machinery
//          MAIN WINDING LOCKED ROTOR CURRENT
        IM = d / Zm ;

//              MAXIMUM STARTING TORQUE : RESISTANCE SPLIT PHASE MOTOR
//      Resistance of the Auxiliary winding
        rA = x1a * Xm / ( Zm - r1 ) ; // Ref. Equation 11.51SKSEN:Electrical Machinery
        cout<<endl<<"Total resistance of Auxl. windg for maximum "
            " starting torque = "<< rA ;

//      Additional resistance to be added in auxl. winding
        radd = rA - r1a ;
        cout<<endl<<"Resistance to be added in Auxl. windg for maximum "
            " starting torque = "<< radd ;
```

```
//        MAXIMUM STARTING TORQUE : CAPACITANCE SPLIT PHASE MOTOR

//        CAPACITIVE REACTANCE FOR MAXIMUM STARTING TORQUE
          Xc = x1a + (RA / RM) * ( Zm - x1 ) ; //Ref: Equn. 11.55 SKSEN:Electrical Machinery
//        REQUIRED CAPACITANCE for maximum starting torque
          C = k5 / ( 2. * PI * a * Xc ) ; //in microfarad
          cout<<endl<<"Capacitance for maxm. starting torque= "<<C<<"microfarad" ;

//                               MAXIMUM STARTING TORQUE
          float Tmax , Tfl, c, omegar, k6 ;
          cout<<endl<<"Enter: output power, c , on full-load "
                << rotor-speed,omegar, at full-load slip 0.04" ;
          cin >> c >> omegar ;
              Tfl = c / omegar ; //Full-load power, Full load speed=150.8

          Tmax = ( r2 / (2. * PI * b))*(d * IM)*((Zm + RM)/(RA*Zm)) ;
//Ref: Equn. 11.56 SKSEN:Electrical Machinery
          cout<<endl<<"Maximum. starting torque= "<<Tmax<<" N. m." ;

              k6 = Tmax * 100. / Tfl ;
          cout<<endl<<"Starting torque as % of full-load torque" << k6 ;


}

int main ()
{
          spec m ;
              int pole, sturnp, Stslot ;
           float wfacm, dia, htStslot, length ;
          cout<<endl<<"Enter: a, Nmp, Ss, Cwm, D, dSs, L ";
              cin >> pole >> sturnp >> Stslot >> wfacm >> dia >> htStslot >> length ;

          m.getdata( pole, sturnp, Stslot, wfacm, dia, htStslot, length);

          m.display () ;

          m.stator( pole, sturnp, Stslot, wfacm, dia, htStslot, length );
}
```

Figure 10.7 A computer program for Auxiliary winding and Starting torque of a single-phase induction motor.

The program is applied to a 230-v, 180-W, 50 Hz. 25 s.r.p.s. Single-phase induction motor. Initial values fed to the computer are taken from sections (a), (b) and (c).

RESULT

Enter: a, Nmp, Ss, Cwm, D, dSs, L   4   170   36   0.7938   0.085   0.0225   0.07
No. of poles= 4, Stator main wndg. turns per pole= 170 No. of Stator slots = 36Winding factor- main wndg(Cwm)= 0.7938S tator Internal dia= 0.085 Height of Stator slot= 0.0225 Arm.Length= 0.07

Enter the values of: Cwa,K   0.85   1.3
Winding factor- auxl. wndg(Cwa)= 0.85
Effective auxl./main. windg turns(K)= 1.3

cN811= 32.1904 , cN712= 49.3186 , cN613= 60.4982 , cN514= 64.3808
32   48   60   60
Turns per pole= 200
modified Cwa= 0.845758
Modified K= 1.25348
Enter coil-span in slot-terms   2   4   6   8
Mean Length of 1/2 Turn= 0.138728 m.With main winding conductor of bare diameter 9 mm., choose conduc.BARE dia cbda, section cbsa,INSUL.dia cida & section cisa 23 0.6096 0.29186 0.6476 0.3294
swg of chosen conductor= 23Auxl.bare cond.dia= 0.6096 mm.sq. bare section= 0.29186 mm.sq. insulated dia= 0.6476 mm.sq, Insulated section= 0.3294 mm.sq.
Stator auxl. wndg. resistance= 11.7932ohm
Put the values of: N46, N37, N28, N19, cism, ASs
        20   40   50   60   0.6196   137
S5= 39.528 S6= 32.156 S7= 40.5952 S8= 41.5208Smax= 41.5208Per cent fillup of stator slots by main & auxl. conduc= 30.3072 which is satisfactory

Enter the values of: frequency(a), syn. rps (b)input voltage (d);main wndg.resis(r1m), main windg.leakage reactance(x1m) ,rotor resis.in Stator terms(r2), magnetising reactance (xm)
        50   25   230   6.4   9.34   9.67   325.
Total resistance of Auxl. windg for maximum starting torque = 366.053
Resistance to be added in Auxl. windg for maximum starting torque = 354.26

Capacitance for maxm. starting torque= 100.871microfarad
Enter the value of output power, c = 180
Maximum. starting torque= 11.0358 N. m.
Starting torque as % of full-load torque 5.91642

Appendix

(A-1) SOME USEFUL INFORMATIONS

A1.1 Properties of conducting materials-copper and aluminium at 20°C

	Copper		Aluminium	
	Hard-drawn	Annealed	Hard-drawn	Annealed
1. Resistivity-ohm. mm.-sq. per metre				
a. volume	0.0177	0.0172	0.0282	0.028
b. mass	0.157	0.153	0.076	
2. Density-g. per cm.-cubed.	8.89	8.89	2.1	2.7
3. Melting point- °C	1083	1083	660	660
4. Thermal conductivity - watts per metre-cubed/°C	350	350	200	200
5. Co-eff. of linear expn.	17×10^{-6}	17×10^{-6}	23×10^{-6}	23×10^{-6}
6. Specific heat, g. cal/°C	0.093	0.093	0.217	0.217

A1.2 Class of insulation and rating

Table 1.7, page 38 gives 'limits of temperature-rise' for rotating electrical machines, as adopted by the Indian Standard Institution. It can be seen that the allowable maximum temperature-rise in a.c. winding is minimum for class A insulation, and maximum for class H. In other words, for a particular frame size, use of improved grade of insulation will lead to larger kw-rating for the same voltage, speed, etc. For such a machine, the specific magnetic and electric loadings would be higher; number of stator turns will be lower, i.e. the conductor section will be larger (or there will be more number of parallel paths in armature winding or both) to

Table A.1

Class	Current-density	Temp, rise
A	4 to 5 amp. per mm2	60°C
E	6 to 7 amp. per mm2	75°C
B	7 to 8 amp. per mm2	80°C
F	8 to 10 amp. per mm2	95°C
H	8 to 10 amp. per mm2	130°C

[NOTE: In the above Table, with aluminium as the conducting material, the values of current-densities should be divided by 1.6.]

Slot insulation: For low voltage machines, the approximate thickness of insulation which can be used in the design of general purpose machines is given in Table A.2.

Table A.2

Class of Insulation	Insulating Material	Thickness in slot in mm. on eacli side
A	Pressphan, Leatheroid	0.7
E	Milinex with elephantkle nomex	0.5
B	Pressphan with polyster film, Puccro paper.	0.5
F	Polyster mat with polyster film	0.25
H	Flexible mica with vulcanised Glass Fibre/Flexible micanite	0.25

A1.3 Gap length in rotating machines

There is no hard and fast formulae for the choice of gap length in a rotating electrical machines. Factors which influence the choice are type of machine, generator/motor operation, specification, expected characteristics, bearing conditions, unbalanced magnetic pull, etc. From time to time efforts have been made to evolve equations for general purpose induction motors and one such effort has been described in art. 7.2.1(b) on the basis of a research work clone in 1959. With time, improved manufacturing techniques and materials, it has been possible to ensure lower gap length than obtained from equation (7.1). However, data for obtaining gap-length along the pole-axis for d.c machines, and gap length for general purpose induction motors with ball/rollar bearings have been presented in Figure A. 1 which could be fed into a computer program as discussed in art. The values obtained from these programs are approximate but very well serve as a starting point of design.

In Figure A.1, D is the armature diameter and lg, the gap length, both in mm.

1. D.C. Machine:

D (15):	100	200	250	400	500	750	900	1020	1150	1300
	1400	1500	1650	1700	1900					
lg(15):	1.25	2.0	2.5	3.5	3.8	4.6	4.9	5.0	5.5	5.7
	5.8	6.0	6.1		6.2	6.3				

2. 3-phase Induction motor:

D (14):	75	100	200	250	380	500	600	800	990	1010
	1150		1300		1400		1600			
Ig(14):	0.3	0.4	0.5	0.6	0.7	0.8	0.9	1.0	1.1	1.2
	1.25		1.3		1.4		1.5			

Figure A.1 Data for the choice of gap length for a d.c. machine and 3-phase induction motor.

A1.4 Transformer-design of tank

(i) As per Indian Standard, the temperature-rise of windings as measured by resistance method should no exceed:

For natural cooling (ON/OB/OW)	...	55°C
For forced oil cooling (OFN/OFB)	...	60°C
For forced water cooling (OFW)	...	65°C

Maximum top oil temperature-rise (by resistance method) 45°C

(ii) For determining the tank dimension for an ON-cooled transformer, sufficient clearances are to be maintained between,

(a) the outer surface of h.v. coil and tank wall along the length of the tank;

(b) the outer surface of h.v. coil and tank wall breadthwise, - should be more than that for (a) to accommodate tappings and for taking out leads);

(c) between the core and tank-wall at the base;

(d) top of core and oil level; and

(e) oil level and the top lid of the tank.

The amounts of clearances depend coolant properties, such as, thermal Conductivity, density, viscosity, specific heat, etc.

A1.5 Power rating of electric motors

One of the important principle of energy conservation-is the proper choice of electrical motors for the drive. This aspect has important bearings on the capital and running costs. For example, a motor of higher power rating used for a drive needing lower capacity will definitely serve the purpose but will beat the cost of efficiency of the motor and economic efficiency of the installation. With alternating current motors choice of excess capacity may lead to operation at inferior power-factor which will adversely affect the supply system and the generator. The enormity of this problem can be appreciated if we consider the very large number of electroc motors used as drives in industries, municipalities, commercial and domestic houses, agriculture, etc. etc.

For the design point of view, it is ultimately the losses and temperature-rise which form the important parameters, and as such, the standard practice is to specify the motors in terms of 'class of duty': *Continuous, short time and intermitent.* Continuous rating.

Continuous duty specifies that the motor during operation will finally reach its steady-state temperature-rise (Figure A.2a).

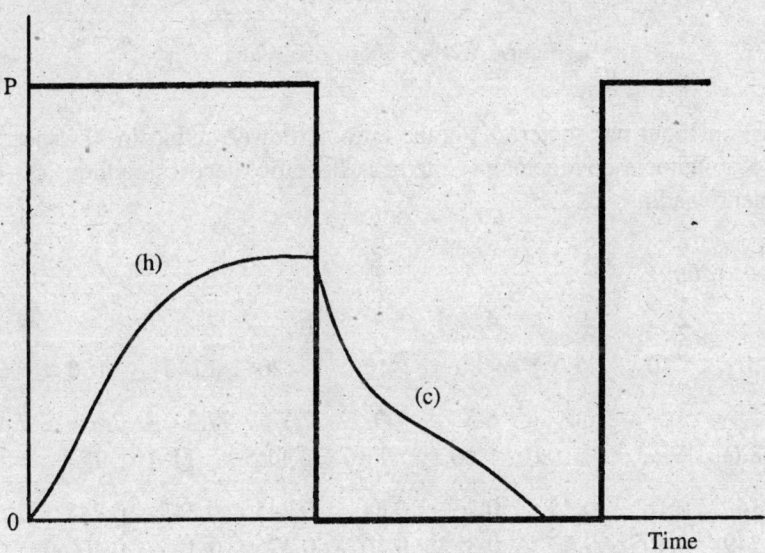

Figure A.2(a) Continuous rating.

Figure A.3 gives data for computer programs for selection of specific loadings for the design of continuous duty and short-term duly motors, from which computer programs could be prepared following technique stated in art. (J.2. The data for short-term duty corresponds to operating period of 5 minutes. For the intermediate duty condition, values in between continuous and short-term could be a good starting point. The choices should ultimately be checked by the computed temperature-rise at the end of each operating period. Again, the efficiency (n) for any power output for the short-term rating is less than that for continuous duty for which a computer program can also be set.

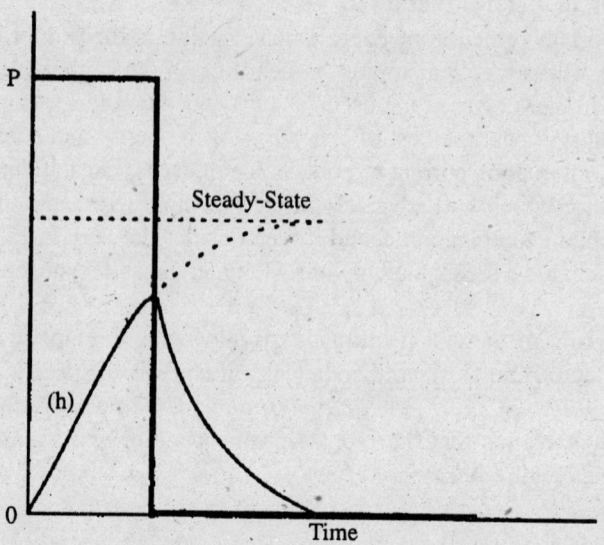

Figure A.2(b)   Short-time rating.

In Figure A.3 Input parameter (X) is the ratio of Power output in Kw/speed in rev. per minute, (ac) Kilo-ampere-conductor per metre is Specific electric loading, and B (Tesla) is specific magnetic loading.

1. *Continuous rating:*

X (18):

1	2	3	4	5	6	7	8	9	10
20	40	60	80	100	120	140	160		

(ac)(18):

4.5	5.5	6.2	6.5	7.0	7.1	7.2	7.3	7.4	7.5
8.6	9.3	10.0	10.5	11.0	11.5	11.8	12.0		

B(18):

0.14	0.18	0.22	0.23	0.24	0.245	0.247	0.248	0.248	0.248
0.249	0.305	0.33	0.355	0.367	0.375	0.385	0.40		

2. *Short-time (upto 5 minutes) rating:*

X (18):

1	2	3	4	5	6	7	8	9	10
20	40	60	80	100	120	140	160		

(ac)(18):

9.0	9.5	10.3	10.8	11.2	11.7	12.0	12.2	12.3	12.35
6.1	17.3	18.4	19.1	20.0	20.7	21.3	21.8		

B(18):

0.30	0.315	0.345	0.35	0.365	0.385	0.4	0.415	0.428	0.435
0.54	0.57	0.615	0.63	0.67	0.69	0.715	0.73		

3. Input: Power output, P watts; Output-Efficiency: Continuous rating (C), Short-term rating (S)

P watts (20):	5	10	20	30	40
	50	60	70	80	90
	100	200	300	400	500
	600	700	800	900	1000
η (C) (20)	27	36	42	46	49
	51	52	53.5	55	56
	56.5	64	66	68	70
	73	74	75	75.5	76.5
η (S) (20)	20	27	35	37.5	41
	43	45	45.5	46	46.5
	47	52	55	56	57.5
	58.5	60	61.5	62	62

Figure A.3 Data pertaining continuous and short-time (upto 5 mins.) ratings.

Short time duty is for a specified short period under which there is no chance for the motor, designed on the basis of continuous rating, to reach ils steady-state temperature-rise (Figure A.2b). Typical examples are motor operated turntable, lock gate, etc.

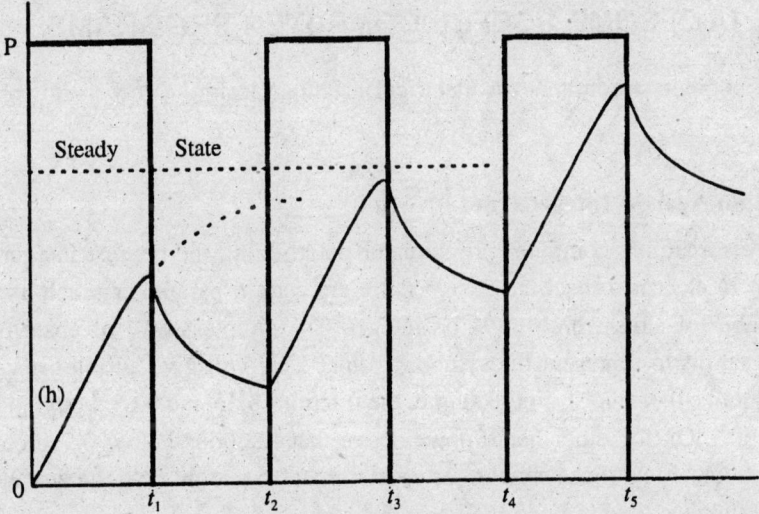

Figure A.2(c) Intermitent rating.

Intermittent or periodic duty demands repetitive operation of short duration too short to allow the motor o reach its steady-state temperature-rise followed by standstill period of short duration which, again, does not allow the motor to cool down to its ambient temperature (Figure A.2c). Typical examples are rolling mills, electric hammer, electric presses, etc.

The choice of specific electric and magnetic loadings are important points in the design of short-term- and intermittent-rated motors.

A1.6   Slot combination

This has been discussed very briefly in art. 7.2.1(g) for three-phase induction motor. Table A.3 gives various slot combinations as per I.S. It is recommended that the slots of squirrel-cage rotors be always skewed.

Table A.3   Stator Slot/Rotor Slot Combination

No. of Pole	Cage rotor I.M	Slip-ring I.M.
2	24/33; 36/27;	
4	24/34; 36/28; 36/44; 36/45; 48/35; 48/36; 48/38; 48/40; 48/57; 60/38; 60/48; 60/76; 72/60;	60/48; 72/60;
6	36/45; 36/48; 36/57; 54/39; 72/82; 72/93;	72/54;
8	72/44; 72/55; 96/72.	96/72

(A-2) SOME USEFUL COMPUTER PROGRAMS

In developing computer programs for electrical machine design, a few useful programs are often helpful.

A2.1   To obtain Nearest Integer from Fraction

Figure A.4 gives a simple computer program for determining the nearest integer value of a given fraction. In electrical machine design there are certain parameters such as, number of turns and number of slots which must be integer. The simplest way of ensuring this in a computer program is to represent the symbol as 'int'. This will give an integer value which, however, will not be the nearest. For example, the fractions 9.15 and 9.63 would give the same integer value of 9. On the other hand, if we desire that fractions below 0.5 should give the lower interger number and those equal to or greater than 0.5 should give the next higher value of integer, we may use the program in Figure A.4.

A2.2   A Program for Optimisation

Figure A.5(a) gives the flow-chart for a simple computer program, in which c is a given function of a and b. The parameter a is invariant and b is to be varied so that the optimum value of c equal to Cm is obtained.

A computer program written in C++ language is given in Figure A.5(b).

// General program for obtaining nearest integer value of a given fraction.

```
# include <iostream.h>

void main()
{
 int x, v ;
 float u, y ;
 cout << endl <<·"Enter the value of the fraction, u" ;
  cin >> u ;

x = u ;
y = u - x ;
  if ( y >= 0.5 )
  {
  v = x + 1.0 ;
  } else
    {
      v = x ;
    }
 cout << endl << " The value of the nearest integer is  =  " << v ;
}
```

Figure A.4 A computer program for determining nearest integer value of a given fraction.

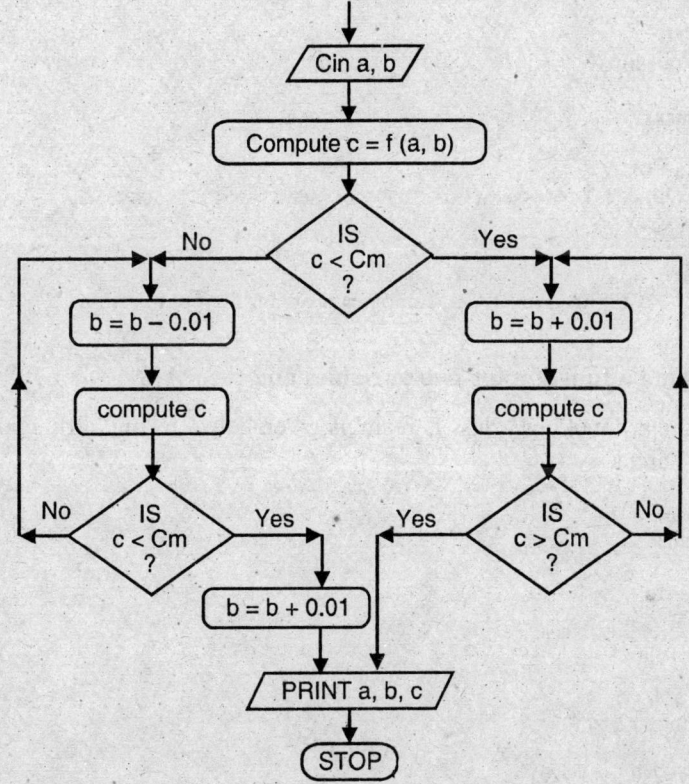

Figure A.5(a) Flow Chart for optimisation.

```
//Optimisation programme: To find optimum value of parameter b for a given cmax
#include <iostream.h>
#include <iomanip.h>


void main()
{
 float a,b,c,Cmax;
 cout<<"Enter the values of a , b, and Cmax : \n";
cin >> a >> b >> Cmax;
 c = a * b - (a + b);
 if ( c >= Cmax)             //
 {
 do
 {
 b = b - 0.001;
 c = a * b - (a + b );
 }
 while(c>Cmax);
 cout<<setw(20)<<"a: "<<a<<setw(19)<<"b: "<<b<<setw(18)<<"c: "<<c<<endl;
 }
 else if ( c < Cmax )
 {
 do
 {
 b = b + 0.001;
 c = a * b - (a + b);
 }
 while(c<Cmax);
 b = b - 0.001;
 c = a * b - (a + b);
 cout<<setw(20)<<"a: "<<a<<setw(19)<<"b: "<<b<<setw(18)<<"c: "<<c<<endl;
 }
}
```

Figure A.5(b) A computer program for optimisation.

A2.3 To Compute a function for two variables and record

A computer program, invoking ' class fstream' is given below to find a function I for varying two quantities Vt and s.

```
#include <iostream>
#include <fstream>
include <string>
using namespace std;

int main()
{
float xm, r2, Rf, s, I, vt;
cout<<endl<<"Enter r2, xm";
cin>>r2>>xm;
for (vt=200.; vt<=240.; vt+=20)
{
ofstream outf("RESULT");
```

```
for(s=0.02; s<=0.06;s+=0.02)
{
        Rf = xm * xm * r2 / s ;
        I = vt/Rf;
        outf"vt"endl;
        outf"s"endl;
        outf"Rf"endl;
        outf"I"endl;
}
outf.close();
ifstream inf("RESULT");
while (!inf.eof())
{
        inf>>vt;
        inf>> s;
        inf>>Rf;
inf>> I;
cout<<endl<<"voltage= "<<vt;
cout<<endl<<"slip= "<<s;
cout<<endl<<"Forward R= "<<Rf;
cout<<endl<<"amps- "<<I;
}
  inf.close();
}
};
```

RESULT

Enter r2, xm 2.5 -50.0

voltage = 200	slip= 0.02	Forward R= 312500	amps- 0.00064
	slip= 0.04	Forward R= 156250	amps- 0.00128
	slip= 0.06	Forward R= 104167	amps- 0.00192
voltage= 220	slip= 0.02	Forward R= 312500	amps- 0.000704
	slip= 0.04	Forward R= 156250	amps- 0.001400
	slip= 0.06	Forward R= 104167	amps- 0.002112
voltage= 240	slip= 0.02	Forward R= 312500	amps- 0.000768
	slip= 0.04	Forward R= 156250	amps- 0.001536
	slip= 0.06	Forward R= 104167	amps- 0.002304

(program exited with code: 0)

(A-3) SKEIN WINDING

Skein winding for a single-phase induction motor is better discussed through an example. Let us take the case of the 36-slot, 4-pole, 180-watt, 230-volt, 20-Hz. Motor (chapter 10) in which the conductor distribution of the main winding are: slot 1-9: 60 turns; slot 2–8 : 50 turns; slot 3–7 : 40 turns; slot 4–6 : 20 turns; total turns = 170, with mean length of half-turn = 0.143 m.

The highest common factor (h.c.f.) of 60, 50, 40 and 20 is 10.

For the winding, we take a coil of 10 turns (equal to h.c.f.) of suitable half length and wind through slots 4 and 6 twice (thus forming 20 turns- Figure A.3.(i), move to next pair of slots-3 and 7 and wind 4 times (forming 40 turns - Figure A.3.(ii) and then to slots 2 and 8, and 1 and 9 winding respectively 5 and 6 times.

Obvious advantages of skein winding are ease in winding and small and economic winding-time.

For a wide-spread winding, the coil-throws are : slot 1-10, 2-9, 3-8, 4-7 and 5-6 for the main winding.

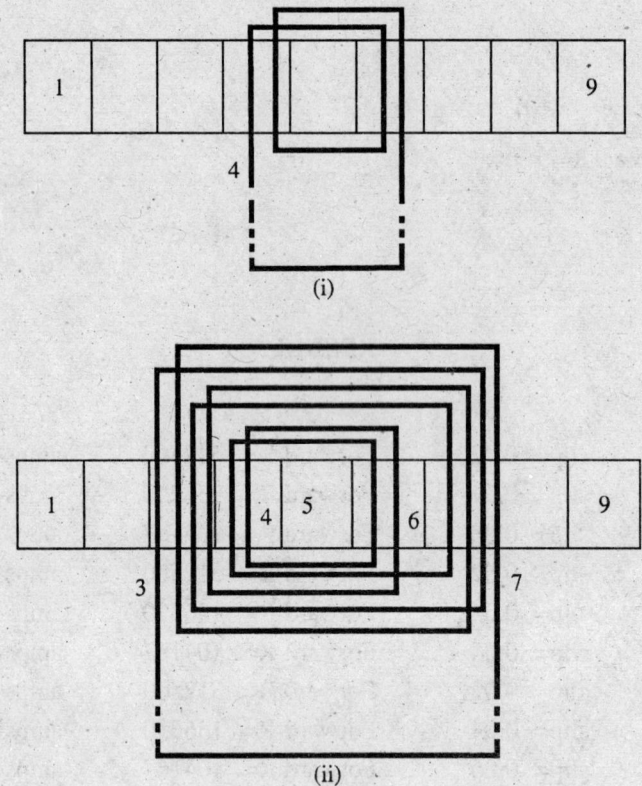

Figure A7   Skein winding for 9-slots per pole, full-pitch.

(A-4) FINITE - DIFFERENCE METHOD

In this method, the derivatives of the partial differential equation (4.22) is replaced by difference quotients, converting the equation to a difference equation. The difference equation corresponding to each point at the nodes of a grid network (Figure A7) that subdivides the solution region or domain at which the function values are unknown.A set of simultaneous equations are obtained, the solution of which gives values of the function at each node.

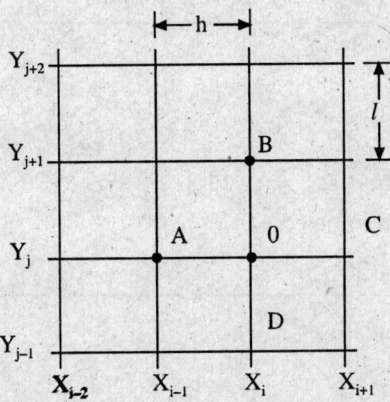

Figure A8 Grid of equal mesh-length

Let $h = \Delta x$ = equal spacing of the gridwork in the x-direction and $l = \Delta y$ = equal spacing in the y-direction. Assuming that the function $U(x)$ has a continuous fourth derivative, we have, by Taylor's series, (indicating $U'(x) = \partial U/\partial x$, $U''(x) = \partial^2 U/\partial x^2$, and so on),

$$U(x_n + h) = U(x_n) + U'(x_n)h + \frac{1}{2} U'(x_n)h^2 + \frac{1}{6} U'''(x_n) + \frac{1}{24} U^{iv}(x_n)h^4,$$

$$U_n < \zeta 1 < (Un + h) \tag{A.1}$$

and $\quad U(x_n - h) = U(x_n) - U'(x_n)\,h + \frac{1}{2} U''(x_n)\,h^2 - \frac{1}{6} U''(x_n)h^3 + \frac{1}{24}U^{iv}(xn)h^4$

$$(U_n - h) < \xi < U_n \tag{A.2}$$

Adding equations (A 4.1) and (A 4.2).

$$\frac{U(x_n + h) - 2U(x_n) + U(x_n - h)}{h^2} = U''(x_n) + \frac{1}{2}U^{iv}(x_n)h^2$$

$$(U_n - h) < \xi < (U_n + h)$$

For any node O defined by the coordinates (x_i, y_i), the above equation can be expressed as,

$$\frac{\partial^2 U}{\partial x^2} \equiv \frac{U_{i+1,j} - 2U_{i,j} + U_{i-1,j}}{h^2} + O(h^2)$$

The order relation $O(h^2)$ signifies that the error approaches proportionality to h^2 as $h \to O$. That is,

$$\frac{\partial^2 U}{\partial x^2} \equiv \frac{U_{i,i+1} - 2U_{i,j} + U_{i-1,j}}{h^2} \tag{A.3}$$

Again, subtracting equation (A 4.2) from equation (A 4.1),

$$\frac{\partial U}{\partial x} \equiv \frac{1}{2}\frac{U_{i,1+j} - U_{i-1,j}}{h} \tag{A.4}$$

Similarly,
$$\frac{\partial^2 U}{\partial y^2} \equiv \frac{U_{i,j+1} - 2U_{i-1,j} + U_{i,i-1}}{l^2}$$
(A.5)

and
$$\frac{\partial U}{\partial y} = \frac{1}{2}\frac{U_{i,j+1} - U_{i,i-1}}{l}$$
(A.6)

so that equation (4.23) is approximated as,
$$\frac{U_{i+1,j} - 2U_{u,j} + U_{i-1,j}}{h^2} + \frac{U_{i,j-1} - 2U_{i,j} + U_{i,j-1}}{l^2}$$
(A.7)

Taking $h = l$, we have,
$$U_{i+1,j} + U_{i-1,j} + U_{i,\,j+1} + U_{i,\,j-1} - 4\,U_{i,\,j} = 0$$
(A.8)

Example A

Datermine the potential distribution in a domain ABCDEF (Figure A9).

The solution region is divided into 10 mm. square so that, $h = l = 10$ mm. The nodes are numbered as shown in the figure. There are seven unknown potentials at nodes (2,2), (3,2), (4,2), (3,3), (4,3), (3,4), and (4,4).

Using equation (4.24), equations at the above nodes are,

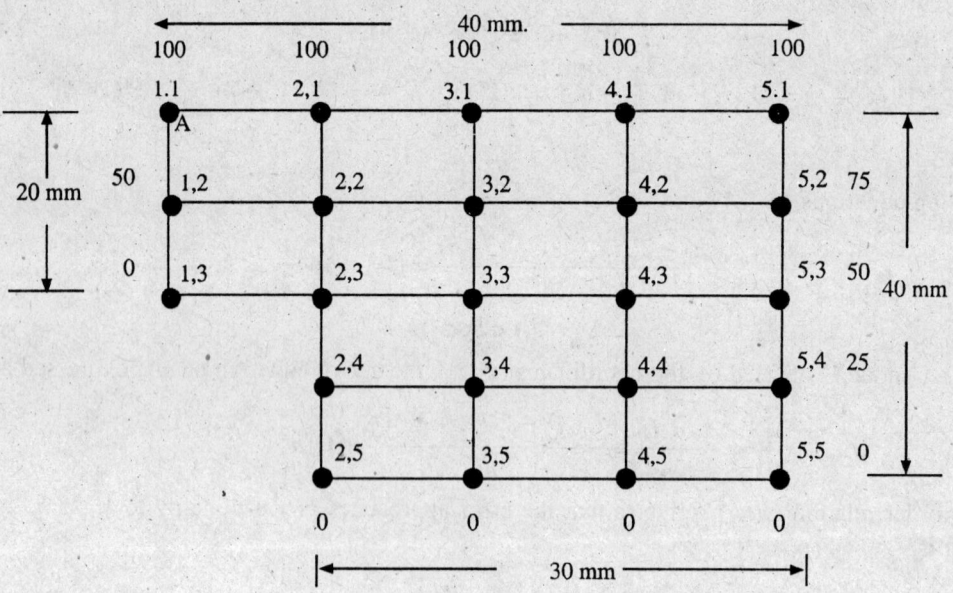

Figure A.9

node (2,2)	:	$150 - 4U_{22} + U_{32}$	$= 0$
node (3,2)	:	$100 + U_{22} - 4U_{32} + U_{42} + U_{33}$	$= 0$
node (4,2)	:	$175 + U_{32} + 4U_{42} + U_{43}$	$= 0$

node (3,3) :   0 $+ U_{32}$ $- 4U_{33} + U_{43} + U_{34}$ $= 0$
node (4,3) :   50 $+ U_{42} + U_{33} - 4U_{43}$ $+ U_{44} = 0$
node (3,4) :   0 $+ U_{33}$ $- 4U_{34} + U_{44} = 0$
node (4,4) :   25 $+ U_{43} + U_{34} + 4U_{44} = 0$

$$(A\ 4\ 8)$$

The problem reduces to solving the above set of linear simultaneous equations Using the elimination method, the augmented coefficient matrix is,

(2,2)	(3,2)	(4,2)	(3,3)	(4,3)	(3,4)	(4,4)	
−4	1						−150
1	−4	1	1				−100
	1	−4	0	1			−175
	1	0	−4	1	1		0
		1	1	−4	0	1	− 50
			1	0	−4	1	0
			1	1	−4		−25

$$(A.9)$$

Equation A.9 can be solved using the computer program of Gauss Elimination method, (Figure 9.4).

Lielvnann's Method

The relatively large number of zero coefficients in the matrix (equation A 4.9) suggests that iterative method is well-suiled. Well-known iterative methods are Liebmann's method and relaxation method, of which the former is well-suited for computer programming. In the Liebmann's method, equation (4.24) is written in the form,

$$U_{i,j}^{k+1} = \frac{1}{4}\left[U_{i+1,j}^k + U_{i-1,j}^k + U_{i,j-1}^k + U_{i,j-1}^k\right] \tag{A.10}$$

where, the superscripts show the order of iteration, i.e. the $(k + 1)$ iteration uliliijs the k-ih set of values around the node (i,j). The relaxation residue R is determined as,

$$R_{i,j}^{k+1} = \left[U_{i+1,j}^k + U_{i-1,j}^k + U_{i,j+1}^k + U_{i,j-1}^k\right] - 4U_{i,j}^k$$

or
$$\frac{1}{4} R_{i,j}^{k+1} = \frac{1}{4}\left[U_{i+1,j}^k + U_{i-1,j}^k + U_{i,j+1}^k + U_{i,j-1}^k\right] - U_{i,j}^k$$

That is, equation (A 4.10) yields,

$$U_{i,j}^{k+1} = U_{i,j}^k + \frac{1}{4} R_{i,j}^k \tag{A.11}$$

The Liebmann's process is thus a relaxation process wherein the residuals R are brought to zero by successive iteration.

However, for accelerating the rate of convergence a factor larger than 1/4 can be applied to the residual. For this, we use instead of equation .(A 4.11),

$$U_{i,j}^{k+1} = U_{i,j}^k + \alpha R_{i,j}^k$$

$$= U_{i,j}^k + \alpha\left[U_{i+1,j}^k + U_{i-1,j}^k + U_{i,j+1}^k + U_{i,j-1}^k - 4U_{i,j}^k \right]$$

$$= \alpha\left[U_{i+1,j}^k + U_{i-1,j}^k + U_{i,j+1}^k + U_{i,j-1}^k - \left(4 - \frac{1}{\alpha}\right)U_{i,j}^k \right] \qquad \text{(A.12)}$$

The value of α generally lies between 1/4 and 1/2.

Irregular boundary

When the boundary of the solution region is such that the grid cannot be drawn to have the boundary coincide with the nodes of the mesh, equation (A 4.7) is modified as below.

Consider the case when the spacing of nodes A, B, C, D from the node O (i,j) are unequal (Figure A 4.3). We may write,

$$\frac{\partial U}{\partial x}\bigg|\text{node A to O} = \frac{U_{i,j} - U_{i-1,j}}{h_1}$$

$$\frac{\partial U}{\partial x}\bigg|\text{node } O \text{ to } B = \frac{U_{i+1,j} - U_{i-1,}}{h_2}$$

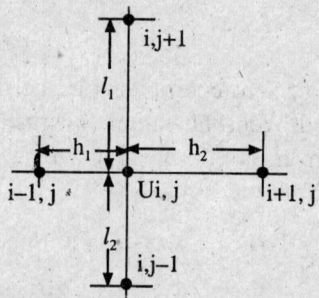

Figure A.10   Unequal mesh-length.

Since,

$$\frac{\partial^2 U}{\partial x^2} = \frac{\partial}{\partial x}\left(\frac{\partial U}{\partial x}\right)$$

we have,

$$\frac{\partial^2 U}{\partial x^2} \equiv \frac{1}{\frac{1}{2}(h_1 + h_2)}\left[\frac{\partial U}{\partial x}\bigg|\text{node } O \text{ to } B - \frac{\partial U}{\partial x}\bigg|\text{node A to O} \right]$$

$$\equiv 2\left[\frac{U_{i+1,j} - U_{i,j}}{h_2(h_1 + h_2)} - \frac{U_{i,j} - U_{i-1,j}}{h_1(h_1 + h_2)} \right]$$

Similarly,

$$\frac{\partial^2 U}{\partial y^2} \equiv 2\left[\frac{U_{i+1,j} - U_{i,j}}{l_2(l_1 + l_2)} - \frac{U_{i,j} - U_{i-1,j}}{l_1(l_1 + l_2)} \right]$$

Hence, equation (4.23) reduces to,

$$\frac{\partial^2 U}{\partial x^2} + \frac{\partial^2 U}{\partial y^2} \equiv 2\left[\frac{U_{i-1,j}}{h_1(h_1+h_2)} + \frac{U_{i+1,j}}{h_2(h_1+h_2)} + \frac{U_{i,j-1}}{l_1(l_1+l_2)} + \frac{U_{i,j+1}}{l_2(l_1+l_2)} - \left(\frac{1}{h_1 h_2} + \frac{1}{l_1 l_2}\right)U_{i,j}\right]$$

(A.13)

For application of Liebmann's method, the relaxation residue,

$$R_{i,j}^k = 2\left[\frac{U_{i-1,j}}{h_1(h_1+h_2)} + \frac{U_{i+1,j}}{h_2(h_1+h_2)} + \frac{U_{i,j-1}}{l_1(l_1+l_2)} + \frac{U_{i,j+1}}{l_2(l_1+l_2)}\right] - 2\left(\frac{1}{h_1 h_2} + \frac{1}{l_1 l_2}\right)U_{i,j}^k$$

and equation (4.24) is modified into,

$$U_{i,j}^{k+1} = \alpha\left\{2\left[\frac{U_{i+1,j}^k}{h_1(h_1+h_2)} + \frac{U_{i-1,j}^k}{h_2(h_1+h_2)} + \frac{U_{i,j+1}^k}{l_1(l_1+l_2)} + \frac{U_{i,j-1}^k}{l_2(l_1+l_2)}\right]\right.$$

$$\left. - \left[2\frac{(h_1 h_2 + l_1 l_2)}{h_1 h_2 l_1 l_2} - \frac{1}{\alpha}\right]U_{i,j}^k\right\}$$

(A.14)

QUESTIONS

9.1. Leaving the cost considerations aside, what benefits are accrued if an induction motor is designed with E-class of insulation instead of class A for the same voltage, frequency, slip, stator and rotor slot - numbers and dimensions and frame.

9.2. The computer programs in Figs. 9.5(a) to (e) and Figure 9.8 have been run with aluminium as the conductor material. Compare the results if copper conductors are used instead.

9.3 The program developed in Figure 9.8 considers the transformer as a core -type. How will you modify it if the transformer is of,
    (a) distribution type; ·
    (b) short-time rating?

9.4. The program developed in Figure 9.8 considers the transformer as a core -type. Mention important factors you will have to consider if the transformer if of Shell type.

9.5. The computer programs give ample opportunity to study the effects of various assumptions involved in the design process. You may takeup small projects to study the effects of following parameters in the design of transformer and rotating machines:

    (i) specific magnetic loading, (ii) specific electric loading, (iii) class of insulation, (iv) variation in supply frequency between 47.5 Hz. and 52 Hz. (v) duty: intermittent, and short-time, and how the final results compare with a continuously rated apparatus.

9.6. Slot combination and shape and size of slot form important parameters of Designer's choice. Study the effects of these choices on the design of an induction motor.

9.7. Customers often specify the type of enclosure. How are the basic assumptions and the design process are to be modified for a TEFC induction motor vis-a-vis the screen protected type?

9.8. The computer programs in Figure 9.9(a) and 9.10 has been run using class 'A' insulation which restricts the designer to a choice of current density between 4 and 5 amp. per mm. sq. and also to a maximum temperature-rise of 60 degree celcius.

The design has chosen a frame size 160 M which has:

Stator bore diameter 165 mm, outer diameter 260 mm, outer shaft diameter 45 mm.

In the design a slot-combination 36/44 has been adopted. With the same specified values for applied voltage, frequency, full-load slip and designed values of stator and rotor slot dimensions (assumed parallel-sided stator tooth and parallel-sided rotor slot), determine the maximum possible kw-rating which could be extracted from the same frame size if H-type insulation is used instead of A-type.

(Hint: Table A. 1 in Appendix gives the allowable temperature-rise for H type as 130 deg. Cel and current density between 8 and 10 amps, per sq. mm. We can safely assume that the temperature-rise is directly proportional to total losses. As such, the total losses with class H can be (130/00) = 2.167 limes that with class A insulation. Again, with choice of stator current-density twice the value for class A insulation, slots would accomodate larger number ampere-conductors.]